Mathematics for Machine Technology

Mathematics for Machine Technology

Fourth Edition

Robert D. Smith

Delmar Publishers

an International Thomson Publishing company I(T)P®

Albany · Bonn · Boston · Cincinnati · Detroit · London · Madrid · Melbourne
Mexico City · New York · Pacific Grove · Paris · San Francisco · Singapore
Tokyo · Toronto · Washington

NOTICE TO THE READER

Cover Design: Nicole Reamer

Delmar Staff:
Publisher: Michael McDermott
Acquisitions Editor: Gregory C. Clayton
Production Manager: Larry Main
Art Director: Nicole Reamer
Senior Project Editor: Christopher Chien
Assistant Editor: Julie Waite

COPYRIGHT © 1999
By Delmar Publishers
a division of International Thomson Publishing Inc.

The ITP logo is a trademark under license.

Printed in the United States of America

For more information, contact:

Delmar Publishers
3 Columbia Circle, Box 15015
Albany, New York 12212-5015

International Thomson Publishing Europe
Berkshire House
168-173 High Holborn
London WC1V 7AA
United Kingdom

Nelson ITP, Australia
102 Dodds Street
South Melbourne
Victoria, 3205 Australia

Nelson Canada
1120 Birchmont Road
Scarborough, Ontario
M1K 5G4, Canada

International Thomson Publishing France
Tour Maine-Montparnasse
33 Avenue du Maine
75755 Paris Cedex 15, France

International Thomson Editores
Seneca 53
Colonia Polanco
11560 Mexico D. F. Mexico

International Thomson Publishing GmbH
Königswinterer Strasse 418
53227 Bonn
Germany

International Thomson Publishing Asia
60 Albert Street
#15-01 Albert Complex
Singapore 189969

International Thomson Publishing Japan
Hirakawa-cho Kyowa Building, 3F
2-2-1 Hirakawa-cho, Chiyoda-ku,
Tokyo 102, Japan

ITE Spain/Paraninfo
Calle Magallanes, 25
28015-Madrid, Espana

1 2 3 4 5 6 7 8 9 10   XXX   05 04 03 02 01 00 99 98

Library of Congress Cataloging-in-Publication Data

Smith, Robert Donald, 1931–
     Mathematics for machine technology / Robert D. Smith. — 4th ed.
          p.   cm.
     Includes index.
     ISBN 0-8273-7942-0
     1. Shop mathematics.   I. Title.
TJ1165.S713   1998
621.9′02′0151313—dc21
                                                                    98-6660
                                                                        CIP

CONTENTS

Preface   ix

SECTION 1	**Common Fractions, Decimal Fractions, and Percentage**		**1**
	UNIT 1	Introduction to Common Fractions and Mixed Numbers.	1
	UNIT 2	Addition of Common Fractions and Mixed Numbers	6
	UNIT 3	Subtraction of Common Fractions and Mixed Numbers	11
	UNIT 4	Multiplication of Common Fractions and Mixed Numbers	15
	UNIT 5	Division of Common Fractions and Mixed Numbers	19
	UNIT 6	Combined Operations of Common Fractions and Mixed Numbers	22
	UNIT 7	Computing with a Calculator: Fractions and Mixed Numbers	27
	UNIT 8	Introduction to Decimal Fractions.	32
	UNIT 9	Rounding Decimal Fractions and Equivalent Decimal and Common Fractions.	36
	UNIT 10	Addition and Subtraction of Decimal Fractions	41
	UNIT 11	Multiplication of Decimal Fractions	45
	UNIT 12	Division of Decimal Fractions	47
	UNIT 13	Powers	51
	UNIT 14	Roots	57
	UNIT 15	Table of Decimal Equivalents and Combined Operations of Decimal Fractions	62
	UNIT 16	Computing with a Calculator: Decimals.	68
	UNIT 17	Introduction to Percents.	74
	UNIT 18	Basic Calculations of Percentages, Percents, and Rates.	77
	UNIT 19	Percent Practical Applications.	82
	UNIT 20	Achievement Review—Section One.	88
SECTION 2	**Linear Measurement: English and Metric**		**95**
	UNIT 21	English and Metric Units of Measure.	95
	UNIT 22	Degree of Precision, Greatest Possible Error, Absolute Error, and Relative Error	102
	UNIT 23	Tolerance, Clearance, and Interference.	107
	UNIT 24	English and Metric Steel Rules.	115
	UNIT 25	English Vernier Calipers and Height Gages	121
	UNIT 26	Metric Vernier Calipers and Height Gages	129
	UNIT 27	English Micrometers	134

UNIT 28   Metric Micrometers 141

UNIT 29   English and Metric Gage Blocks....................... 146

UNIT 30   Achievement Review— Section Two 149

SECTION 3   Fundamentals of Algebra   155

UNIT 31   Symbolism 155

UNIT 32   Signed Numbers..................................... 162

UNIT 33   Algebraic Operations of Addition, Subtraction,
and Multiplication 175

UNIT 34   Algebraic Operations of Division, Powers, and Roots 182

UNIT 35   Introduction to Equations.............................. 192

UNIT 36   Solution of Equations by the Subtraction, Addition,
and Division Principles of Equality....................... 199

UNIT 37   Solution of Equations by the Multiplication, Root,
and Power Principles of Equality........................ 208

UNIT 38   Solution of Equations Consisting of Combined Operations and
Rearrangement of Formulas............................ 216

UNIT 39   Ratio and Proportion 224

UNIT 40   Direct and Inverse Proportions......................... 231

UNIT 41   Applications of Formulas to Cutting Speed, Revolutions
Per Minute, and Cutting Time 237

UNIT 42   Applications of Formulas to Spur Gears 246

UNIT 43   Achievement Review—Section Three..................... 254

SECTION 4   Fundamentals of Plane Geometry   258

UNIT 44   Introduction to Geometric Figures....................... 258

UNIT 45   Protractors—Simple and Vernier 271

UNIT 46   Angles... 277

UNIT 47   Introduction to Triangles 284

UNIT 48   Geometric Principles for Triangles and Other Common
Polygons.. 290

UNIT 49   Introduction to Circles 301

UNIT 50   Arcs and Angles of Circles 309

UNIT 51   Fundamental Geometric Constructions 319

UNIT 52   Achievement Review—Section Four...................... 328

SECTION 5   Trigonometry   333

UNIT 53   Introduction to Trigonometric Functions.................. 333

UNIT 54   Analysis of Trigonometric Functions 342

UNIT 55   Basic Calculations of Angles and Sides of Right Triangles 347

UNIT 56   Simple Practical Machine Applications................... 354

UNIT 57   Complex Practical Machine Applications   361

UNIT 58   The Cartesian Coordinate System   371

UNIT 59   Oblique Triangles: Law of Sines and Law of Cosines   375

UNIT 60   Achievement Review—Section Five   385

SECTION 6   Compound Angles   390

UNIT 61   Introduction to Compound Angles   390

UNIT 62   Drilling and Boring Compound-Angular Holes: Computing
Angles of Rotation and Tilt Using Given Lengths   393

UNIT 63   Drilling and Boring Compound-Angular Holes: Computing
Angles of Rotation and Tilt Using Given Angles   397

UNIT 64   Machining Compound-Angular Surfaces: Computing Angles
of Rotation and Tilt   404

UNIT 65   Computing Angles Made by the Intersection of Two
Angular Surfaces   412

UNIT 66   Computing Compound Angles on Cutting and Forming Tools   418

UNIT 67   Achievement Review—Section Six   425

SECTION 7   Computer Numerical Control (CNC)   428

UNIT 68   Introduction to Computer Numerical Control (CNC)   428

UNIT 69   Control Systems, Absolute Positioning, Incremental
Positioning   432

UNIT 70   Binary Numeration System   439

UNIT 71   Achievement Review—Section Seven   443

Appendix   446

Answers to Odd-Numbered Applications   450

Index   477

PREFACE

Mathematics for Machine Technology is written to overcome the often mechanical "plug in" approach found in many trade-related mathematics textbooks. An understanding of mathematical concepts is stressed in all topics ranging from general arithmetic processes to oblique trigonometry, compound angles, and numerical control.

Both content and method are those used by the author in teaching applied machine technology mathematics classes for apprentices in the machine, tool-and-die, and tool design trades. Each unit is developed as a learning experience based on preceding units—making prerequisites unnecessary.

Presentation of basic concepts is accompanied by realistic industry-related examples and actual industrial applications. The applications progress from the simple to those with solutions which are relatively complex. Many problems require the student to work with illustrations such as are found in machine trade handbooks and engineering drawings.

An analytical approach to problem solving is emphasized in the geometry, trigonometry, compound angle, and numerical control sections. This approach is necessary in actual practice in translating engineering drawing dimensions to machine working dimensions. Integration of algebraic and geometric principles with trigonometry by careful sequence and treatment of material also helps the student in solving industrial applications. The Instructor's Guide provides answers and solutions for all problems.

Changes from the previous edition have been made to improve the presentation of topics and to update material.

A survey of instructors using the third edition was conducted. Based on their comments and suggestions, changes in the text revision were made. The result is an updated and improved fourth edition in both content and sequencing of content.

- The majority of instructors surveyed stated that their students are required to perform basic arithmetic operations on fractions and decimals prior to calculator usage. Thereafter, the students use the calculator almost exclusively in problem-solving computations. The structuring of calculator instructions and examples in this text reflect in the instructors' preferences. Calculator instruction and examples have been updated and greatly expanded in this edition. The scientific calculator is introduced in the Preface. Extensive calculator instruction and examples are given directly following the units on fractions and mixed numbers and the units on decimals. Further calculator instruction and examples are given throughout the text wherever calculator applications are appropriate to the material presented. A Calculator Applications Index is provided at the end of the Preface. It provides a convenient reference for all the material in the text for which calculator usage is presented. Often there are differences in the methods of computation among various makes and models of calculators. Where there are two basic ways of performing calculations, both ways are shown.

- Three new units on percent, units 17–19, have been added. The survey indicated that the topics of percent and percentages are relevant to manufacturing technology. Applications are presented, primarily in problems involving machining times, material shrinkage and expansion, alloy material composition, manufacturing costs, production rates, and percent defective product.

- In Section 5, Trigonometry, trigonometric function tables and interpolation have been eliminated. All computations involving trigonometric functions are done with a calculator. Calculator examples have been greatly increased.

- Section 7, Computer Numerical Control, has been updated. The revised material is centered on computer numerical control (CNC) rather than on numerical control (NC) as in previous editions.

Robert D. Smith has experience in both the manufacturing industry and in education. He held positions as tool designer, quality control engineer, and chief manufacturing engineer prior to teaching. Mr. Smith has taught applied mathematics, physics, and industrial materials and processes on the secondary school level and in Machine Trade Apprentice Programs. Mr. Smith is Associate Professor Emeritus of Industrial Technology at Central Connecticut State University, New Britain, Connecticut. He is the author of Delmar's *Technical Mathematics*.

Introduction to the Scientific Calculator

Scientific Calculator

A scientific calculator is to be used in conjunction with the material presented in this textbook. Complex mathematical calculations can be made quickly, accurately, and easily with a scientific calculator.

Although most functions are performed in the same way, there are some differences among different makes and models of scientific calculators. In this book, generally, where there are two basic ways of performing a function, both ways are shown. However, not all of the differences among the various makes and models of calculators can be shown. It is very important that you become familiar with the operation of your scientific calculator. An owner's manual or reference guide is included with the purchase of a scientific calculator. The manual explains the essential features and keys of the specific calculator and provides detailed information on the proper use of the calculator. *It is essential that the owner's manual be studied and referred to whenever there is a question regarding calculator usage.*

For use with this textbook, the most important feature of the scientific calculator is the Algebraic Operating System (AOS™). This system, which uses algebraic logic, permits you to enter numbers and combined operations into the calculator in the same order as the expressions are written. The calculator performs combined operations according to the rules of algebraic logic, which assigns priorities to the various mathematical operations. *It is essential that you know if your calculator uses algebraic logic.*

Most scientific calculators, in addition to the basic arithmetic functions, have algebraic, statistical, conversion, and program or memory functions. Some of the keys with their functions are shown. Scientific calculators have functions in addition to those shown.

General Information

Since there is some variation among different makes and models of scientific calculators, your calculator function keys may be different from the descriptions that follow. *To repeat, it is very important that you refer to the owner's manual whenever there is a question regarding calculator usage.*

- Solutions to combined operations shown in this text are performed on a calculator with algebraic logic (AOS™).

- Turning the Calculator On and Off

  The method of turning the calculator on with battery-powered calculators depends on the calculator make and model. When a calculator is turned on, 0 and/or other indicators are displayed. Basically, a calculator is turned on and off by one of the following ways.

  With calculators with an on/clear, $\boxed{\text{ON/C}}$, key, press $\boxed{\text{ON/C}}$ to turn on. Press the $\boxed{\text{OFF}}$ key to turn off.

SOME TYPICAL KEY SYMBOLS AND FUNCTIONS FOR A SCIENTIFIC CALCULATOR	
Key(s)	**Function(s)**
$+$, $-$, $\times$, $\div$, $=$, or EXE or ENTER	Basic Arithmetic
$+/-$ or $(-)$	Change Sign
π	Pi
$($, $)$	Parentheses
EE or EXP	Scientific Notation
Eng	Engineering Notation
STO, RCL, EXC	Memory or Memories
X^2, $\sqrt{x}$	Square and Square Root
$\sqrt[x]{y}$ or $\sqrt[x]{\ }$	Root
y^x or x^y	Power
$1/x$ or x^{-1}	Reciprocal
$\%$	Percent
a^b/c	Fractions and Mixed Numbers
log	Logarithm
DRG	Degrees, Radians, and Graduations
DMS or ° ′ ″	Degrees, Minutes, and Seconds
sin, cos, tan	Trigonometric Functions

With calculators with an all clear power on/power off, AC , key, press AC to turn on. Generally, the AC key is also pressed to turn off.

With calculators that have an on-off switch, move the switch either on or off. The switch is usually located on the left side of the calculator.

➤ **Note:** In order to conserve power, most calculators have an automatic power off feature that automatically switches off the power after approximately 5 minutes of nonuse.

- Clearing the Calculator Display and All Pending Operations

  To clear or erase *all* entries of previous calculations, depending on the calculator, either of the following procedures is used.

  With calculators with an on/clear, ON/C , key, press ON/C *twice.*

  With calculators with the all clear, AC , key, press AC .

- Erasing (Deleting) the Last Calculator Entry

  A last entry error can be removed and corrected without erasing previously entered data and calculations. Depending on the calculator, one of the following procedures is used.

  With calculators with the on/clear, ON/C , key, press ON/C .

  With calculators with a delete, DEL , key, press DEL .

  With calculators with a clear, C , key, press C .

- Alternate-Function Keys

  Most scientific calculator keys can perform more than one function.

  Depending on the calculator, the 2nd and 3rd keys or SHIFT key enable you to use alternate functions. The alternate functions are marked above the key and/or on the upper half of the key. Alternate functions are shown and explained in the book where their applications are appropriate to specific content.

➤ DECISIONS REGARDING CALCULATOR USE

The exercises and problems presented throughout the text are well suited for solutions by calculator. However, it is felt decisions regarding calculator usage should be left to the discretion of the course classroom or shop instructor. The instructor best knows the unique learning environment and objectives to be achieved by the students in a course. Judgments should be made by the instructor as to the degree of emphasis to be placed on calculator applications, when and where a calculator is to be used, and the selection of specific problems for solution by calculator. Therefore, exercises and problems in this text are *not* specifically identified as calculator applications.

Calculator instruction and examples of the basic operations of addition, subtraction, multiplication, and division of fractions are presented in Unit 7. They are presented for decimals in Unit 16. Further calculator instruction and examples of mathematics operations and functions are given throughout the text wherever calculator applications are appropriate to the material presented.

The index that follows lists the mathematics operations or functions and the pages on which the calculator instruction is first given for the operations or functions. It provides a convenient reference for all material in the text for which calculator usage is presented. The operations and functions are listed in the order in which material is presented in the text.

CALCULATOR APPLICATION INDEX	
Operations/Functions	**Page**
Addition, subtraction, multiplication, and division of fractions and mixed numbers	28–30
Combined arithmetic operations of fractions and mixed numbers	30–32
Addition, subtraction, multiplication, and division of decimals	68–69
Powers of positive numbers (square key, universal power key)	69
Roots of positive numbers (square root key, root key)	70
Combined operations of decimals	71–74
Pi key	157–158
Negative numbers (change sign key, negative key)	164–165
Powers of negative numbers and negative exponents	167–168
Roots of negative numbers	169–170
Fractional exponents (positive and negative)	170
Combined operations of signed numbers	170
Scientific notation (scientific notation key, exponent entry key)	187–188
Decimal-degrees and degrees, minutes, seconds conversion	262–263
Arithmetic operations with degrees, minutes, seconds	263–266
Sine, cosine, tangent functions	335–336
Cosecant, secant, cotangent functions	336–337
Angles of given functions (inverse functions)	337–338
Functions of angles greater than 90°	372–373

Section One
Common Fractions, Decimal Fractions, and Percentage

UNIT 1 Introduction to Common Fractions and Mixed Numbers

Objectives After studying this unit you should be able to

- **Express fractions in lowest terms.**
- **Express fractions as equivalent fractions.**
- **Express mixed numbers as improper fractions.**
- **Express improper fractions as mixed numbers.**

Most measurements and calculations made by a machinist are not limited to whole numbers. Blueprint dimensions are often given as fractions and certain measuring tools are graduated in fractional units. The machinist must be able to make calculations using fractions and to measure fractional values.

Fractional Parts

A *fraction* is a value which shows the number of equal parts taken of a whole quantity or unit. The symbols used to indicate a fraction are the bar (—) and the slash (/).

Line segment AB as shown is divided into 4 equal parts.

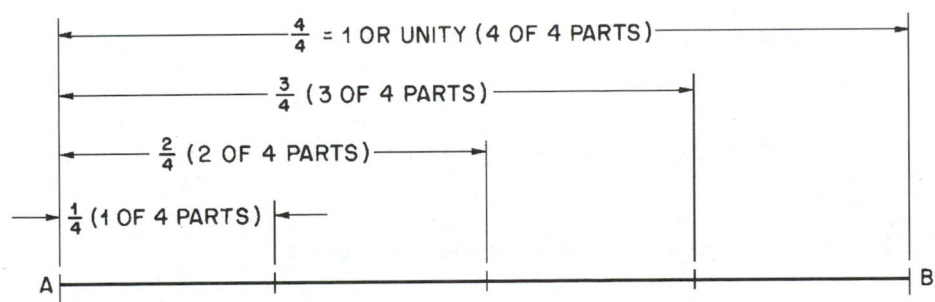

$$1 \text{ part } = \frac{1 \text{ part}}{\text{total parts}} = \frac{1 \text{ part}}{4 \text{ parts}} = \frac{1}{4} \text{ of the length of the line segment.}$$

$$2 \text{ parts} = \frac{2 \text{ parts}}{\text{total parts}} = \frac{2 \text{ parts}}{4 \text{ parts}} = \frac{2}{4} \text{ of the length of the line segment.}$$

$$3 \text{ parts} = \frac{3 \text{ parts}}{\text{total parts}} = \frac{3 \text{ parts}}{4 \text{ parts}} = \frac{3}{4} \text{ of the length of the line segment.}$$

$$4 \text{ parts} = \frac{4 \text{ parts}}{\text{total parts}} = \frac{4 \text{ parts}}{4 \text{ parts}} = \frac{4}{4} = 1, \text{ or unity (4 parts make up the whole).}$$

Each of the 4 equal parts of the line segment AB is divided into 8 equal parts. There is a total of 4×8 or 32 parts.

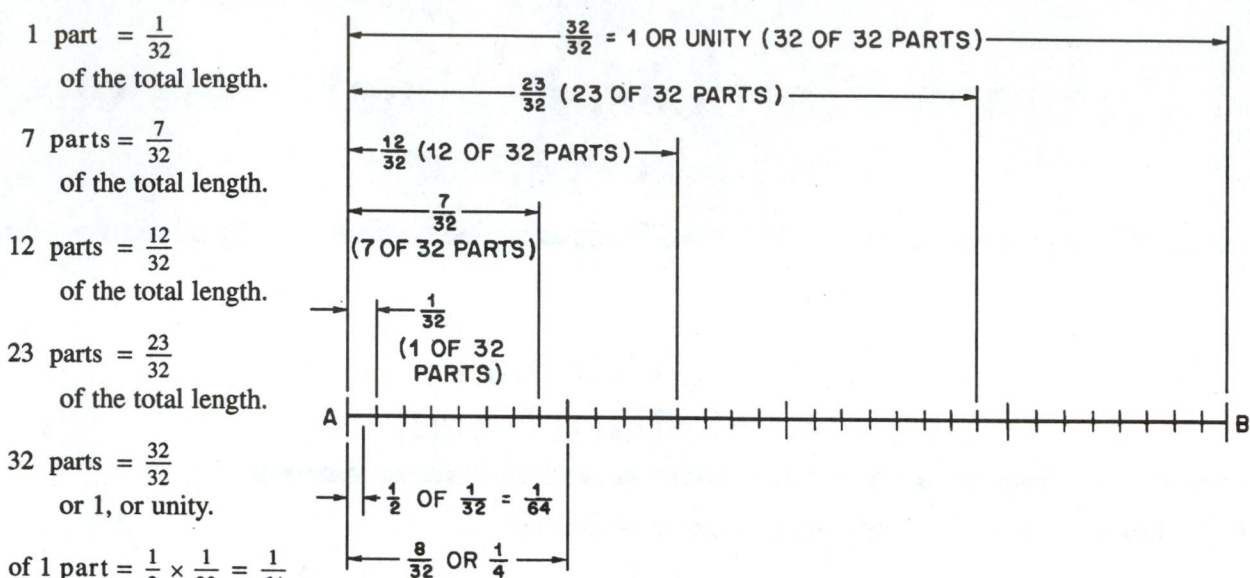

1 part $= \frac{1}{32}$ of the total length.

7 parts $= \frac{7}{32}$ of the total length.

12 parts $= \frac{12}{32}$ of the total length.

23 parts $= \frac{23}{32}$ of the total length.

32 parts $= \frac{32}{32}$ or 1, or unity.

$\frac{1}{2}$ of 1 part $= \frac{1}{2} \times \frac{1}{32} = \frac{1}{64}$ of the total length.

➤ **Note:** 8 parts $= \frac{8}{32}$ of the total length and also $\frac{1}{4}$ of the total length. Therefore, $\frac{8}{32} = \frac{1}{4}$.

Definitions of Fractions

A *fraction* is a value which shows the number of equal parts taken of a whole quantity or unit.

The *denominator* of a fraction is the number that shows how many equal parts are in the whole quantity. The denominator is written below the bar.

The *numerator* of a fraction is the number that shows how many equal parts of the whole are taken. The numerator is written above the bar.

The numerator and denominator are called the *terms* of the fraction.

$\underline{3}$ ← numerator
4 ← denominator

An *improper* fraction is a fraction in which the numerator is larger than or equal to the denominator, as $\frac{3}{2}, \frac{5}{4}, \frac{5}{8}, \frac{6}{6}, \frac{17}{17}$.

A *mixed number* is a number composed of a whole number and a fraction, as $3\frac{7}{8}, 7\frac{1}{2}$.

➤ **Note:** $3\frac{7}{8}$ means $3 + \frac{7}{8}$. It is read as three and seven-eighths. $7\frac{1}{2}$ means $7 + \frac{1}{2}$. It is read as seven and one-half.

A *complex fraction* is a fraction in which one or both of the terms are fractions or mixed numbers, as $\dfrac{\frac{3}{4}}{6}, \dfrac{32}{\frac{15}{4}}, \dfrac{8\frac{3}{4}}{3}, \dfrac{\frac{7}{16}}{2\frac{2}{5}}, \dfrac{4\frac{1}{4}}{7\frac{5}{8}}$.

Expressing Fractions as Equivalent Fractions

The numerator and denominator of a fraction can be multiplied or divided by the same number without changing the value. For example, $\frac{1}{2} = \frac{1 \times 4}{2 \times 4} = \frac{4}{8}$. Both the numerator and denominator are multiplied by 4. Because $\frac{1}{2}$ and $\frac{4}{8}$ have the same value, they are *equivalent*. Also, $\frac{8}{12} = \frac{8 \div 4}{12 \div 4} = \frac{2}{3}$. Both numerator and denominator are divided by 4. Since $\frac{8}{12}$ and $\frac{2}{3}$ have the same value, they are *equivalent*.

A fraction is in its *lowest terms* when the numerator and denominator do not contain a common factor, as $\frac{5}{9}, \frac{7}{8}, \frac{3}{4}, \frac{11}{12}, \frac{15}{32}, \frac{9}{11}$. *Factors* are the numbers used in multiplying. For example, 2 and 5 are each factors of 10; $2 \times 5 = 10$. Expressing a fraction in lowest terms is often called *reducing* a fraction to lowest terms.

Procedure   To reduce a fraction to lowest terms

- Divide both numerator and denominator by the greatest common factor (GCF).

Example   Reduce $\frac{12}{42}$ to lowest terms.

Both terms can be divided by 2. $\frac{12 \div 2}{42 \div 2} = \frac{6}{21}$

➤ **Note:**   The fraction is reduced, but not to lowest terms.

Further reduce $\frac{6}{21}$.
Both terms can be divided by 3. $\frac{6 \div 3}{21 \div 3} = \frac{2}{7}$ Ans

➤ **Note:**   The value $\frac{2}{7}$ may be obtained in one step if each term of $\frac{12}{42}$ is divided by 2×3 or 6. Six is the greatest common factor (GCF).

     $\frac{12 \div 6}{42 \div 6} = \frac{2}{7}$ Ans

Procedure   To express a fraction as an equivalent fraction with an indicated denominator which is larger than the denominator of the fraction

- Divide the indicated denominator by the denominator of the fraction.
- Multiply both the numerator and denominator of the fraction by the value obtained.

Example   Express $\frac{3}{4}$ as an equivalent fraction with 12 as the denominator.

Divide 12 by 4.
$12 \div 4 = 3$
Multiply both 3 and 4 by 3. $\frac{3 \times 3}{4 \times 3} = \frac{9}{12}$ Ans

Expressing Mixed Numbers as Improper Fractions

Procedure   To express a mixed number as an improper fraction

- Multiply the whole number by the denominator.
- Add the numerator to obtain the numerator of the improper fraction.
- The denominator is the same as that of the original fraction.

Example 1   Express $4\frac{1}{2}$ as an improper fraction.

Multiply the whole number by the denominator.

Add the numerator to obtain numerator for the improper fraction.

The denominator is the same as that of the original fraction.

$$\frac{4 \times 2 + 1}{2} = \frac{9}{2} \quad \text{Ans}$$

Example 2   Express $12\frac{3}{16}$ as an improper fraction.

$$\frac{12 \times 16 + 3}{16} = \frac{195}{16} \quad \text{Ans}$$

Expressing Improper Fractions as Mixed Numbers

Procedure   To express an improper fraction as a mixed number

- Divide the numerator by the denominator.

Examples   Express the following improper fractions as mixed numbers.

$$\frac{11}{4} = 11 \div 4 = 2\frac{3}{4} \quad \text{Ans}$$

$$\frac{43}{3} = 43 \div 3 = 14\frac{1}{3} \quad \text{Ans}$$

$$\frac{931}{8} = 931 \div 8 = 116\frac{3}{8} \quad \text{Ans}$$

APPLICATION

Fractional Parts

1. Write the fractional part which each length, A through F, represents of the total shown on the scale.

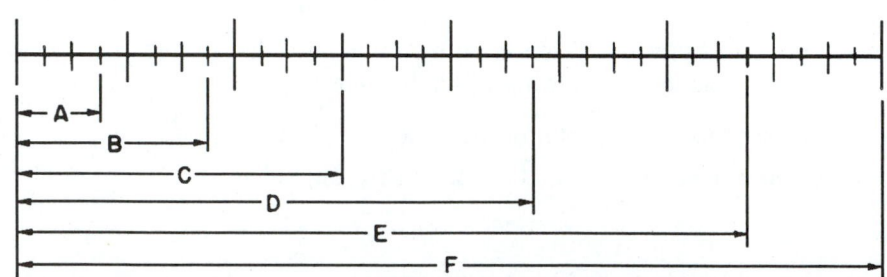

A = _____

B = _____

C = _____

D = _____

E = _____

F = _____

2. A welded support base is cut in four pieces. What fractional part of the total length does each of the four pieces represent? All dimensions are in inches.   _____

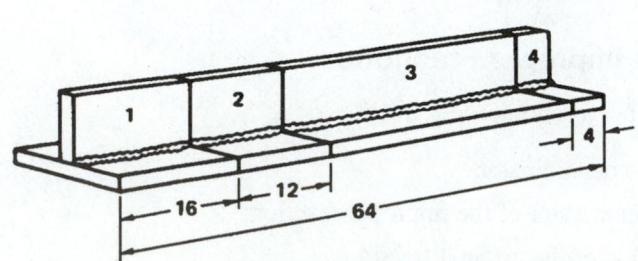

3. The circle is divided into equal parts. Write the fractional part each of the following represents.

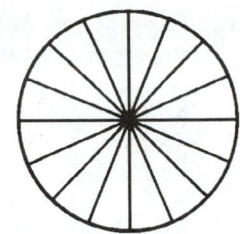

a. 1 part _____

b. 3 parts _____

c. 7 parts _____

d. 5 parts _____

e. 16 parts _____

f. $\frac{1}{2}$ of 1 part _____

g. $\frac{1}{3}$ of 1 part _____

h. $\frac{3}{4}$ of 1 part _____

i. $\frac{1}{10}$ of 1 part _____

j. $\frac{1}{16}$ of 1 part _____

Expressing Fractions as Equivalent Fractions

4. Reduce to halves.

a. $\frac{4}{8}$ _____

b. $\frac{9}{18}$ _____

c. $\frac{100}{200}$ _____

d. $\frac{121}{242}$ _____

e. $\frac{25}{10}$ _____

f. $\frac{18}{12}$ _____

g. $\frac{126}{36}$ _____

h. $\frac{225}{50}$ _____

5. Reduce to lowest terms.

a. $\frac{6}{8}$ _____

b. $\frac{12}{4}$ _____

c. $\frac{6}{10}$ _____

d. $\frac{30}{5}$ _____

e. $\frac{11}{44}$ _____

f. $\frac{14}{6}$ _____

g. $\frac{24}{8}$ _____

h. $\frac{65}{15}$ _____

i. $\frac{25}{150}$ _____

j. $\frac{14}{105}$ _____

6. Express as thirty-seconds.

a. $\frac{1}{4}$ _____

b. $\frac{3}{4}$ _____

c. $\frac{11}{8}$ _____

d. $\frac{7}{16}$ _____

e. $\frac{21}{16}$ _____

f. $\frac{19}{2}$ _____

g. $\frac{197}{16}$ _____

h. $\frac{21}{8}$ _____

7. Express as equivalent fractions as indicated.

a. $\frac{3}{4} = \frac{?}{8}$ _____

b. $\frac{7}{12} = \frac{?}{36}$ _____

c. $\frac{6}{15} = \frac{?}{60}$ _____

d. $\frac{17}{14} = \frac{?}{42}$ _____

e. $\frac{20}{9} = \frac{?}{45}$ _____

f. $\frac{14}{3} = \frac{?}{18}$ _____

g. $\frac{7}{16} = \frac{?}{128}$ _____

h. $\frac{13}{8} = \frac{?}{48}$ _____

i. $\frac{21}{16} = \frac{?}{160}$ _____

Mixed Numbers and Improper Fractions

8. Express the following mixed numbers as improper fractions.

a. $2\frac{2}{3}$ _____

b. $1\frac{7}{8}$ _____

c. $5\frac{2}{5}$ _____

d. $3\frac{3}{8}$ _____

e. $5\frac{9}{32}$ _____

f. $8\frac{3}{7}$ _____

g. $10\frac{1}{3}$ _____

h. $9\frac{4}{5}$ _____

i. $100\frac{1}{2}$ _____

j. $4\frac{63}{64}$ _____

k. $49\frac{3}{8}$ _____

l. $408\frac{13}{16}$ _____

9. Express the following improper fractions as mixed numbers.

a. $\frac{5}{3}$ _____

b. $\frac{21}{2}$ _____

c. $\frac{9}{8}$ _____

d. $\frac{87}{4}$ _____

e. $\frac{72}{9}$ _____

f. $\frac{127}{124}$ _____

g. $\frac{127}{32}$ _____

h. $\frac{57}{15}$ _____

i. $\frac{150}{9}$ _____

j. $\frac{235}{16}$ _____

k. $\frac{514}{4}$ _____

l. $\frac{401}{64}$ _____

10. Express the following mixed numbers as improper fractions. Then express the improper fractions as the equivalent fractions indicated.

a. $2\frac{1}{2} = \frac{?}{8}$ _____

b. $3\frac{3}{8} = \frac{?}{16}$ _____

c. $7\frac{4}{5} = \frac{?}{15}$ _____

d. $12\frac{2}{3} = \frac{?}{18}$ _____

e. $9\frac{7}{8} = \frac{?}{64}$ _____

f. $15\frac{1}{2} = \frac{?}{128}$ _____

11. Sketch and redimension this plate. Reduce all proper fractions to lowest terms. Reduce all improper fractions to lowest terms and express as mixed numbers. All dimensions are in inches.

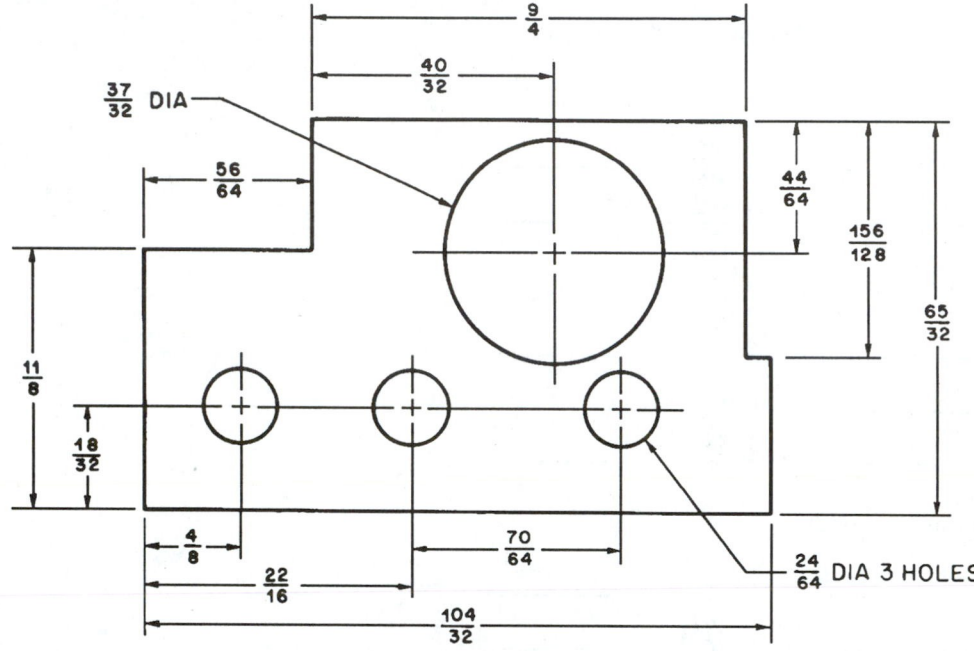

UNIT 2 Addition of Common Fractions and Mixed Numbers

Objectives After studying this unit you should be able to

- **Determine least common denominators.**
- **Express fractions as equivalent fractions having least common denominators.**
- **Add fractions and mixed numbers.**

A machinist must be able to add fractions and mixed numbers in order to determine the length of stock required for a job, the distances between various parts of a machined piece, and the depth of holes and cutouts in a workpiece.

Least Common Denominators

Fractions cannot be added unless they have a common denominator. *Common denominator* means that the denominators of each of the fractions are the same, as $\frac{5}{8}, \frac{7}{8}, \frac{15}{8}$.

In order to add fractions which do not have common denominators, such as $\frac{3}{8} + \frac{1}{4} + \frac{7}{16}$, it is necessary to determine the least common denominator.

The *least common denominator* is the smallest denominator which is evenly divisible by each of the denominators of the fractions being added. Or, stated in another way, the *least common denominator* is the smallest denominator into which each denominator can be divided without leaving a remainder.

Procedure   To find the least common denominator

- Determine the smallest number into which all denominators can be divided without leaving a remainder.

- Use this number as a common denominator.

Example 1   Find the least common denominator of $\frac{3}{8}$, $\frac{1}{4}$, and $\frac{7}{16}$.

The smallest number into which 8, 4, and 16 can be divided without leaving a remainder is 16.
Write 16 as the least common denominator.

Example 2   Find the least common denominator of $\frac{3}{4}$, $\frac{1}{3}$, $\frac{7}{8}$, and $\frac{5}{12}$.

The smallest number into which 4, 3, 8, and 12 can be divided is 24.
Write 24 as the least common denominator.

➤ **Note:**   In this example, denominators such as 48, 72, and 96 are common denominators because 4, 3, 8, and 12 divide evenly into these numbers, but they are not the least common denominators.

Although any common denominator can be used when adding fractions, it is generally easier and faster to use the least common denominator.

Expressing Fractions as Equivalent Fractions with the Least Common Denominator

Procedure   To change fractions into equivalent fractions having the least common denominator

- Divide the least common denominator by each denominator.

- Multiply both the numerator and denominator of each fraction by the value obtained.

Example 1   Express $\frac{2}{3}$, $\frac{7}{15}$, and $\frac{1}{2}$ as equivalent fractions having a least common denominator.

The least common denominator is 30.	$30 \div 3 = 10;\ \frac{2 \times 10}{3 \times 10} = \frac{20}{30}$ Ans
Divide 30 by each denominator.	$30 \div 15 = 2;\ \frac{7 \times 2}{15 \times 2} = \frac{14}{30}$ Ans
Multiply each term of the fraction by the value obtained.	$30 \div 2 = 15;\ \frac{1 \times 15}{2 \times 15} = \frac{15}{30}$ Ans

Example 2   Change $\frac{5}{8}$, $\frac{15}{32}$, $\frac{3}{4}$, and $\frac{9}{16}$ to equivalent fractions having a least common denominator.

The least common denominator is 32.

$32 \div 8 = 4; \dfrac{5 \times 4}{8 \times 4} = \dfrac{20}{32}$ Ans $32 \div 4 = 8; \dfrac{3 \times 8}{4 \times 88} = \dfrac{24}{32}$ Ans

$32 \div 32 = 1; \dfrac{15 \times 1}{32 \times 1} = \dfrac{15}{32}$ Ans $32 \div 16 = 2; \dfrac{9 \times 2}{16 \times 2} = \dfrac{18}{32}$ Ans

Adding Fractions

Procedure   To add fractions

- Express the fractions as equivalent fractions having the least common denominator.
- Add the numerators and write their sum over the least common denominator.
- Express an improper fraction as a mixed number when necessary and reduce the fractional part to lowest terms.

Example 1   Add $\dfrac{1}{2} + \dfrac{3}{5} + \dfrac{7}{10} + \dfrac{5}{6}$.

Express the fractions as equivalent fractions with 30 as the denominator.

Add the numerators and write their sum over the least common denominator, 30.

$$\frac{1}{2} = \frac{15}{30}$$
$$\frac{3}{5} = \frac{18}{30}$$
$$\frac{7}{10} = \frac{21}{30}$$

Express the fraction as a mixed number.

$$+ \frac{5}{6} = \frac{25}{30}$$
$$\overline{\quad\frac{79}{30}\quad} = 2\frac{19}{30} \quad \text{Ans}$$

Example 2   Determine the total length of the shaft shown. All dimensions are in inches.

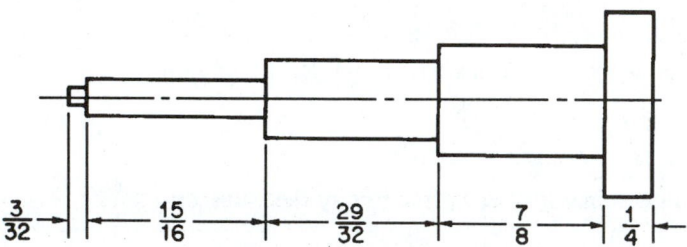

Express the fractions as equivalent fractions with 32 as the denominator.

Add the numerators and write their sum over the least common denominator, 32.

Express $\dfrac{98}{32}$ as a mixed number and reduce to lowest terms.

Total Length = $3\dfrac{1}{16}''$ Ans

$$\frac{3}{32} = \frac{3}{32}$$
$$\frac{15}{16} = \frac{30}{32}$$
$$\frac{29}{32} = \frac{29}{32}$$
$$\frac{7}{8} = \frac{28}{32}$$
$$+ \frac{1}{4} = \frac{8}{32}$$
$$\overline{\quad\frac{98}{32}\quad} = 3\frac{2}{32} = 3\frac{1}{16}$$

Adding Fractions, Mixed Numbers, and Whole Numbers

Procedure   To add fractions, mixed numbers, and whole numbers

- Add the whole numbers.
- Add the fractions.
- Combine whole number and fraction.

Example 1   Add $\frac{1}{3} + 7 + 3\frac{1}{2} + \frac{5}{12} + 2\frac{19}{24}$.

Express the fractional parts as equivalent
fractions with 24 as the denominator.

$$\frac{1}{3} = \frac{8}{24}$$

Add the whole numbers.

$$7 = 7$$

Add the fractions.

$$3\frac{1}{2} = 3\frac{12}{24}$$

Combine the whole number and the fraction.
Express the answer in lowest terms.

$$\frac{5}{12} = \frac{10}{24}$$

$$+ 2\frac{19}{24} = 2\frac{19}{24}$$

$$12\frac{49}{24} = 14\frac{1}{24} \quad \text{Ans}$$

Example 2   Find the distance between the two $\frac{1}{2}$-inch diameter holes in the plate
shown. All dimensions are in inches.

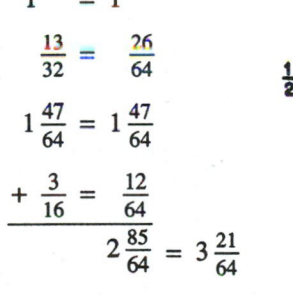

$$1 = 1$$
$$\frac{13}{32} = \frac{26}{64}$$
$$1\frac{47}{64} = 1\frac{47}{64}$$
$$+ \frac{3}{16} = \frac{12}{64}$$
$$2\frac{85}{64} = 3\frac{21}{64}$$

Distance $= 3\frac{21}{64}''$ Ans

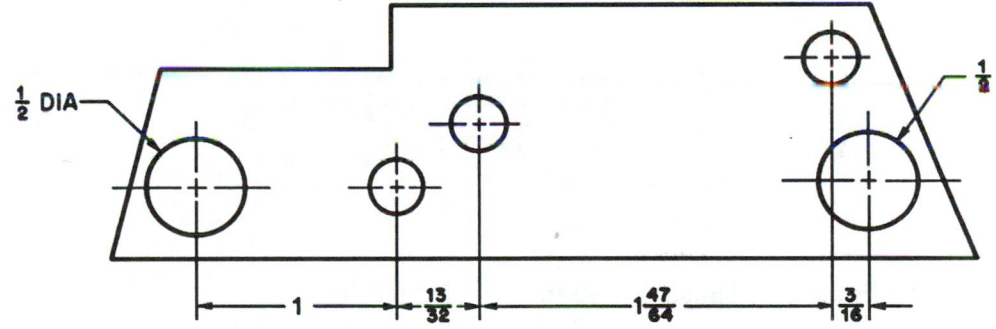

APPLICATION

Least Common Denominators

Determine the least common denominators of the following sets of fractions.

1. $\frac{2}{3}, \frac{1}{6}, \frac{5}{12}$ _____

2. $\frac{3}{5}, \frac{9}{10}, \frac{5}{6}$ _____

3. $\frac{5}{6}, \frac{7}{12}, \frac{3}{16}, \frac{19}{24}$ _____

4. $\frac{4}{5}, \frac{3}{4}, \frac{7}{10}, \frac{1}{2}$ _____

Equivalent Fractions with Least Common Denominators

Express these fractions as equivalent fractions having the least common denominator.

5. $\frac{1}{2}, \frac{3}{4}, \frac{5}{12}$ _____

6. $\frac{7}{16}, \frac{3}{8}, \frac{1}{2}$ _____

7. $\frac{9}{10}, \frac{1}{4}, \frac{3}{5}, \frac{1}{5}$ _____

8. $\frac{3}{16}, \frac{7}{32}, \frac{17}{64}, \frac{3}{4}$ _____

Adding Fractions

9. Determine the dimensions A, B, C, D, E, and F of this profile gage. All dimensions are in inches.

A = _____

B = _____

C = _____

D = _____

E = _____

F = _____

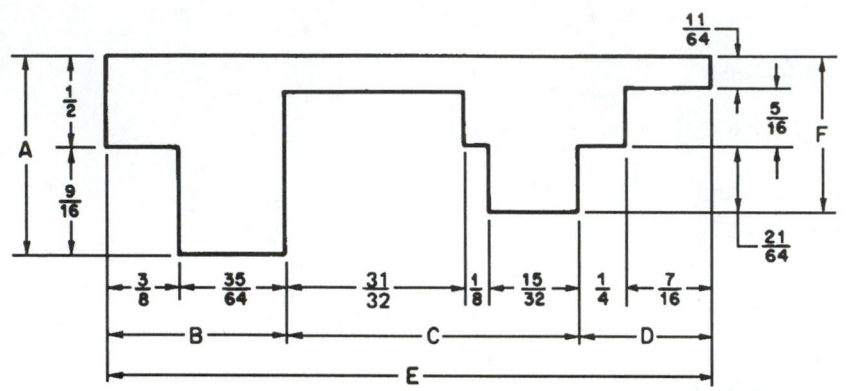

10. Determine the length, width, and height of this casting. All dimensions are in inches.

length = _____

width = _____

height = _____

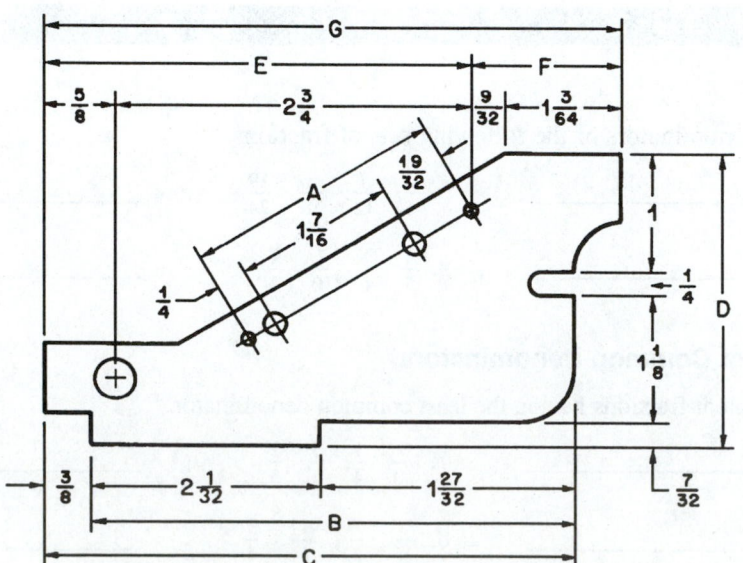

Adding Fractions, Mixed Numbers, and Whole Numbers

11. Determine dimensions A, B, C, D, E, F, and G of this plate. Reduce to lowest terms where necessary. All dimensions are in inches.

A = _____

B = _____

C = _____

D = _____

E = _____

F = _____

G = _____

12. Determine dimensions A, B, C, and D of this pin. All dimensions are in inches.

A = _____

B = _____

C = _____

D = _____

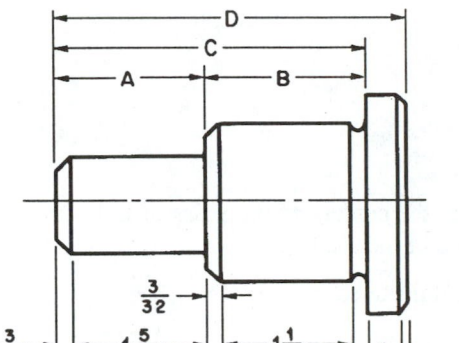

13. The operation sheet for machining an aluminum housing specifies 1 hour for facing, $2\frac{3}{4}$ hours for milling, $\frac{5}{6}$ hour for drilling, $\frac{3}{10}$ hour for tapping, and $\frac{2}{5}$ hour for setting up. What is the total time allotted for this job?

_____

UNIT 3 Subtraction of Common Fractions and Mixed Numbers

Objectives   After studying this unit you should be able to

- **Subtract fractions.**
- **Subtract mixed numbers.**

While making a part from a blueprint, a machinist often finds it necessary to express blueprint dimensions as working dimensions. Subtraction of fractions and mixed numbers is sometimes required in order to properly position a part on a machine, to establish hole locations, and to determine depths of cut.

Subtracting Fractions

Procedure   To subtract fractions

- Express the fractions as equivalent fractions having the least common denominator.
- Subtract the numerators.
- Write their difference over the least common denominator.
- Reduce the fraction to lowest terms.

Example 1   Subtract $\frac{3}{8}$ from $\frac{9}{16}$.

The least common denominator is 16. Express $\frac{3}{8}$ as 16ths.

$$\frac{9}{16} = \frac{9}{16}$$
$$-\frac{3}{8} = \frac{6}{16}$$

Subtract the numerators.
Write their difference over the least common denominator.

$$\frac{3}{16} \quad \text{Ans}$$

Example 2 Subtract $\frac{2}{5}$ from $\frac{3}{4}$.

$$\frac{3}{4} = \frac{15}{20}$$

$$-\frac{2}{5} = \frac{8}{20}$$
$$\frac{7}{20} \quad \text{Ans}$$

Example 3 Find the distances x and y between the centers of the pairs of holes in the strap shown. All dimensions are in inches.

To find distance x:

$$\frac{7}{8} = \frac{28}{32}$$

$$-\frac{11}{32} = \frac{11}{32}$$
$$\frac{17}{32}$$

$$x = \frac{17}{32}'' \quad \text{Ans}$$

To find distance y:

$$\frac{63}{64} = \frac{63}{64}$$

$$-\frac{1}{4} = \frac{16}{64}$$
$$\frac{47}{64}$$

$$y = \frac{47}{64}'' \quad \text{Ans}$$

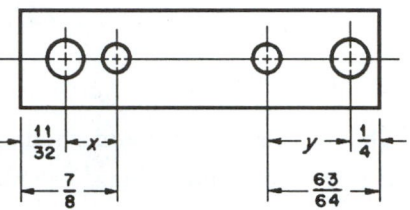

Subtracting Mixed Numbers

Procedure To subtract mixed numbers

- Subtract the whole numbers.
- Subtract the fractions.
- Combine whole number and fraction.

Example 1 Subtract $2\frac{1}{4}$ from $9\frac{3}{8}$.

Subtract the whole numbers.

$$9\frac{3}{8} = 9\frac{3}{8}$$

Subtract the fractions.
Combine.

$$-2\frac{1}{4} = 2\frac{2}{8}$$
$$7\frac{1}{8} \quad \text{Ans}$$

Example 2 Find the length of thread x of the bolt shown. All dimensions are in inches.

$$2\frac{7}{8} = 2\frac{28}{32}$$

$$-1\frac{3}{32} = 1\frac{3}{32}$$
$$1\frac{25}{32}$$

$$x = 1\frac{25}{32}'' \quad \text{Ans}$$

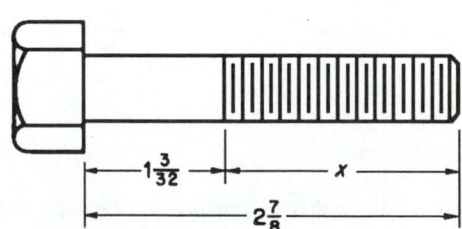

Example 3 Subtract $7\frac{15}{16}$ from $12\frac{5}{8}$.

$$12\frac{5}{8} = 12\frac{10}{16} = 11\frac{26}{16}$$

$$-7\frac{15}{16} = 7\frac{15}{16} = 7\frac{15}{16}$$
$$4\frac{11}{16} \quad \text{Ans}$$

➤ **Note:** Since $\frac{15}{16}$ cannot be subtracted from $\frac{10}{16}$, one unit of the whole number 12 is expressed as a fraction with the common denominator 16.

Example 4 Subtract $52\frac{31}{64}$ from 75.

$$75 \quad = 74\frac{64}{64}$$

$$\underline{- 52\frac{31}{64} = 52\frac{31}{64}}$$

$$22\frac{33}{64} \quad \text{Ans}$$

Example 5 Find dimension y of the counterbored block shown. All dimensions are in inches.

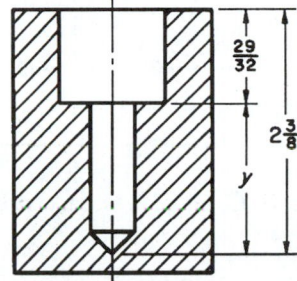

$$2\frac{3}{8} = 2\frac{12}{32} = 1\frac{44}{32}$$

$$\underline{-\frac{29}{32} = \frac{29}{32} = \frac{29}{32}}$$

$$1\frac{15}{32}$$

$$y = 1\frac{15}{32}'' \quad \text{Ans}$$

APPLICATION

Subtracting Fractions

1. Subtract each of the following fractions. Reduce to lowest terms where necessary.

   a. $\frac{5}{8} - \frac{9}{32}$ _____

   b. $\frac{7}{8} - \frac{5}{8}$ _____

   c. $\frac{9}{10} - \frac{19}{50}$ _____

   d. $\frac{5}{8} - \frac{9}{64}$ _____

   e. $\frac{9}{16} - \frac{13}{64}$ _____

   f. $\frac{19}{24} - \frac{3}{16}$ _____

2. Determine dimensions A, B, C, and D of this casting. All dimensions are in inches.

   A = _____
   B = _____
   C = _____
   D = _____

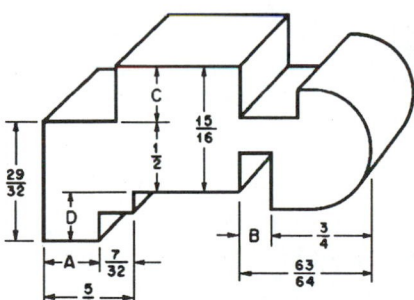

3. Determine dimensions A, B, C, D, E, and F of this drill jig. All dimensions are in inches.

   A = _____
   B = _____
   C = _____
   D = _____
   E = _____
   F = _____

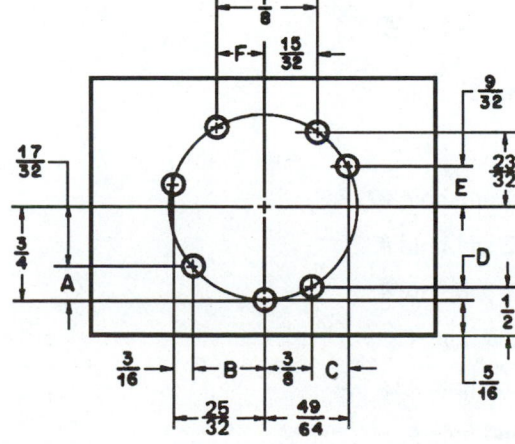

Subtracting Mixed Numbers

4. Determine dimensions A, B, C, D, E, F, and G of this tapered pin. All dimensions are in inches.

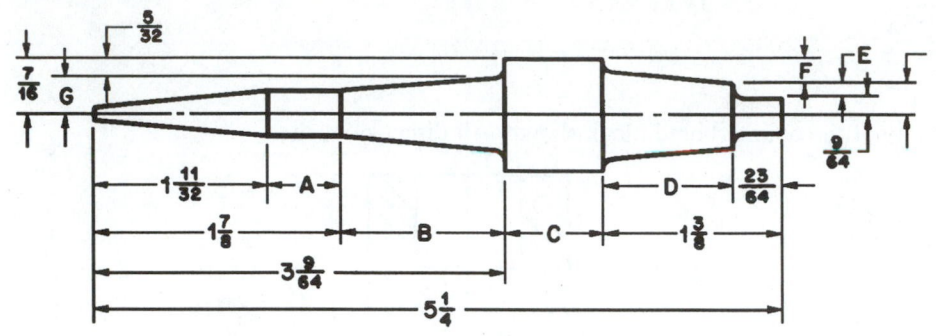

A = _____

B = _____

C = _____

D = _____

E = _____

F = _____

G = _____

5. Determine dimensions A, B, C, D, E, F, G, H, and I of this plate. All dimensions are in inches.

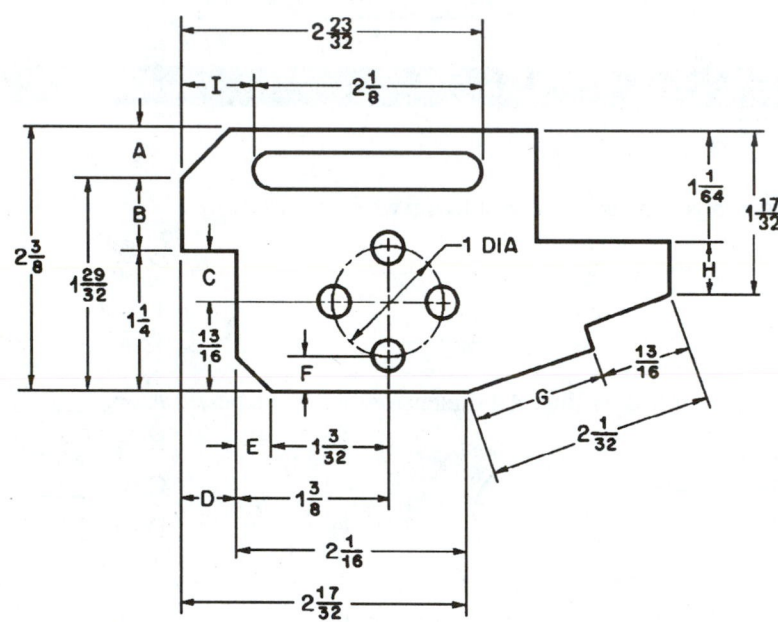

A = _____

B = _____

C = _____

D = _____

E = _____

F = _____

G = _____

H = _____

I = _____

6. Three holes are bored in a checking gage. The lower left edge of the gage is the reference point for the hole locations. Sketch the hole locations and determine the missing distances. From the reference point:

   Hole #1 is $1\frac{3}{32}''$ to the right, and $1\frac{5}{8}''$ up.

   Hole #2 is $2\frac{1}{64}''$ to the right, and $2\frac{3}{16}''$ up.

   Hole #3 is $3\frac{1}{4}''$ to the right, and $3\frac{1}{2}''$ up.

Determine:

a. The horizontal distance between hole #1 and hole #2.   _____

b. The horizontal distance between hole #2 and hole #3.   _____

c. The horizontal distance between hole #1 and hole #3.   _____

d. The vertical distance between hole #1 and hole #2.   _____

e. The vertical distance between hole #2 and hole #3.   _____

f. The vertical distance between hole #1 and hole #3.   _____

UNIT 4 Multiplication of Common Fractions and Mixed Numbers

Objectives   After studying this unit you should be able to

- **Multiply fractions.**
- **Multiply mixed numbers.**
- **Divide by common factors (cancellation).**

Multiplying Fractions

Procedure   To multiply two or more fractions

- Multiply the numerators and the denominators separately.
- Write the product of the numerators over the product of the denominators.
- Reduce the resulting fraction to lowest terms.

Example 1   Multiply $\frac{3}{4}$ by $\frac{8}{9}$.

Multiply the numerators. $\frac{3 \times 8}{4 \times 9} = \frac{24}{36} = \frac{24 \div 12}{36 \div 12} = \frac{2}{3}$ Ans

Multiply the denominators.

Write the product of the numerators over the product of the denominators.

Reduce the resulting fraction to lowest terms.

Example 2   Multiply $\frac{2}{3} \times \frac{5}{6} \times \frac{3}{10}$.

$\frac{2 \times 5 \times 3}{3 \times 6 \times 10} = \frac{30}{180} = \frac{30 \div 30}{180 \div 30} = \frac{1}{6}$ Ans

Example 3   Find the distance between centers of the first and last holes shown in this figure. All dimensions are in inches.

Multiply $6 \times \frac{7}{16} = \frac{6}{1} \times \frac{7}{16} = \frac{6 \times 7}{1 \times 16} = \frac{42}{16}$

Reduce $\frac{42}{16} = 2\frac{10}{16} = 2\frac{5}{8}$

Distance $= 2\frac{5}{8}''$ Ans

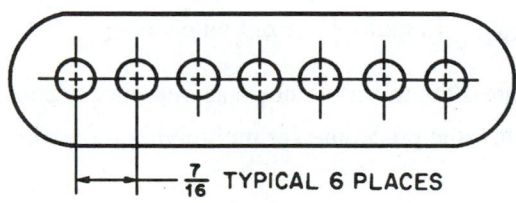

$\frac{7}{16}$ TYPICAL 6 PLACES

> **Note:**   The value of a number remains unchanged when the number is placed over a denominator of 1.

Dividing by Common Factors (Cancellation)

Problems involving multiplication of fractions are generally solved more quickly and easily if a numerator and denominator are divided by any common factors before the fractions are multiplied. This process of first dividing by common factors is commonly called *cancellation*.

Example 1   Multiply by cancellation method. $\frac{3}{4} \times \frac{8}{9}$

Divide by 3 which is the factor common to both the numerator 3 and the denominator 9.

$$\frac{3}{4} \times \frac{8}{9} = \frac{\overset{1}{\cancel{3}}}{\cancel{4}} \times \frac{\overset{2}{\cancel{8}}}{\cancel{9}} = \frac{1 \times 2}{1 \times 3} = \frac{2}{3} \quad \text{Ans}$$

$$3 \div 3 = 1$$
$$9 \div 3 = 3$$

Divide by 4 which is the factor common to both the denominator 4 and the numerator 8.

$$4 \div 4 = 1$$
$$8 \div 4 = 2$$

Multiply reduced fractions.

Example 2   Multiply $\frac{4}{7} \times \frac{5}{18} \times \frac{14}{15}$.

Divide 4 and 18 by 2.
Divide 7 and 14 by 7.
Divide 5 and 15 by 5.
Multiply.

$$\frac{\overset{2}{\cancel{4}}}{\cancel{7}} \times \frac{\overset{1}{\cancel{5}}}{\cancel{18}} \times \frac{\overset{2}{\cancel{14}}}{\cancel{15}} = \frac{2 \times 1 \times 2}{1 \times 9 \times 3} = \frac{4}{27} \quad \text{Ans}$$

Example 3   Multiply $\frac{5}{14} \times \frac{8}{9} \times \frac{7}{10}$.

Divide 5 and 10 by 5.
Divide 14 and 8 by 2.

$$\frac{\overset{1}{\cancel{5}}}{\cancel{14}} \times \frac{\overset{2}{\cancel{8}}}{9} \times \frac{\overset{1}{\cancel{7}}}{\cancel{10}} = \frac{1 \times 2 \times 1}{1 \times 9 \times 1} = \frac{2}{9} \quad \text{Ans}$$

The process is continued by dividing 7 and 7 by 7 and dividing 2 and 4 by 2.
Multiply.

Multiplying Mixed Numbers

Procedure   To multiply mixed numbers

- Express the mixed numbers as improper fractions.
- Follow the procedure for multiplying proper fractions.

Example 1   Multiply $2\frac{2}{5} \times 6\frac{7}{8}$.

Express $2\frac{2}{5}$ and $6\frac{7}{8}$ as improper fractions.
Divide 5 and 55 by 5.
Divide 12 and 8 by 4.

$$\frac{\overset{3}{\cancel{12}}}{\cancel{5}} \times \frac{\overset{11}{\cancel{55}}}{\cancel{8}} = \frac{3 \times 11}{1 \times 2} = \frac{33}{2} = 16\frac{1}{2} \quad \text{Ans}$$

Multiply and express the product as a mixed number.

Example 2 The block of steel shown is to be machined. The block measures $8\frac{3}{4}$ inches long, $4\frac{9}{16}$ inches wide, and $\frac{7}{8}$ inch thick. Find the volume of the block. All dimensions are in inches. (Volume = length × width × thickness.)

$$8\frac{3}{4} \times 4\frac{9}{16} \times \frac{7}{8} = \frac{35}{4} \times \frac{73}{16} \times \frac{7}{8} = \frac{35 \times 73 \times 7}{4 \times 16 \times 8}$$

$$= \frac{17885}{512} = 34\frac{477}{512}$$

Volume = $34\frac{477}{512}$ cubic inches Ans

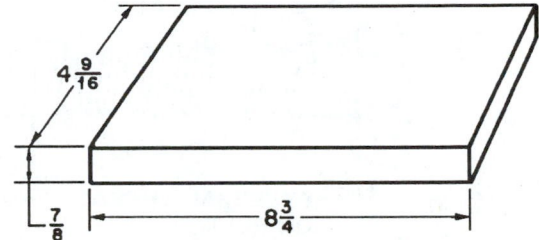

APPLICATION

Multiplying Fractions

1. Multiply these fractions. Reduce to lowest terms where necessary.

   a. $\frac{2}{3} \times \frac{1}{6}$ _____

   b. $\frac{1}{2} \times \frac{1}{4}$ _____

   c. $\frac{5}{8} \times \frac{13}{64}$ _____

   d. $\frac{3}{4} \times \frac{3}{5} \times \frac{2}{3}$ _____

   e. $7 \times \frac{9}{14} \times 3$ _____

   f. $\frac{7}{15} \times \frac{3}{8} \times \frac{5}{7}$ _____

2. Determine dimensions A, B, C, D, and E of the template shown. All dimensions are in inches.

   A = _____
   B = _____
   C = _____
   D = _____
   E = _____

3. A special washer-faced nut is shown. All dimensions are in inches.

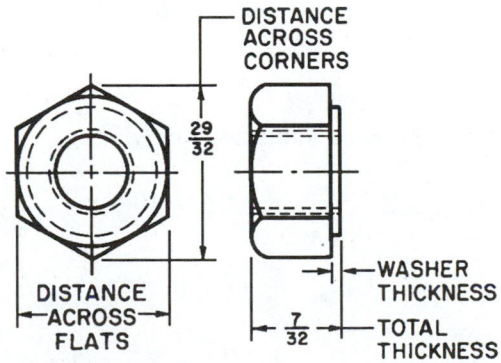

   a. Determine the distance across flats.

   Distance across flats = $\frac{55}{64}$ × Distance across corners _____

   b. Determine the washer thickness.

   Washer thickness = $\frac{1}{8}$ × Total thickness _____

4. The Unified Thread may have either a flat or rounded crest or root. If the sides of the Unified Thread are extended a sharp V-thread is formed. H is the height of a sharp V-thread. The pitch, P, is the distance between two adjacent threads.

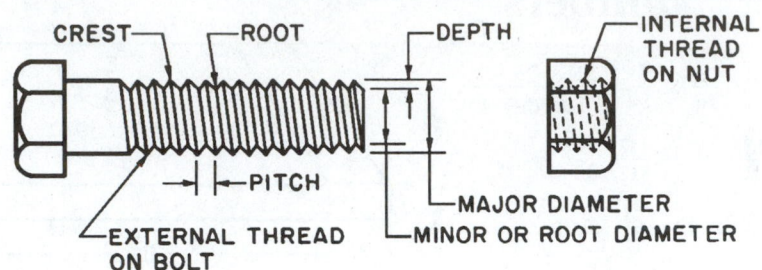

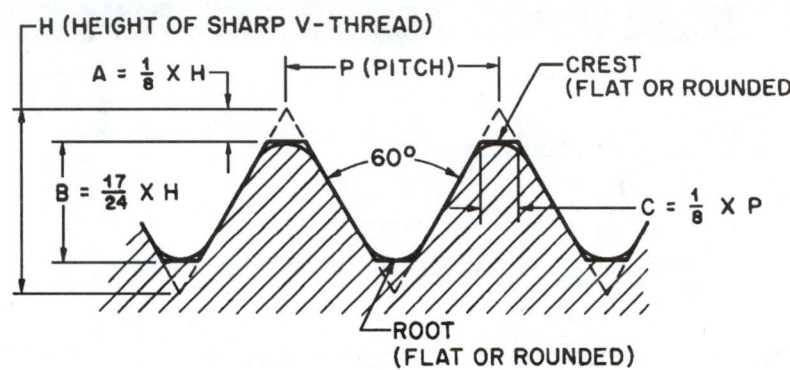

Find dimensions A, B, and C as indicated.

a. $H = \frac{7}{16}''$,   A = _____,   B = _____   f. $P = \frac{1}{4}''$,   C = _____

b. $H = \frac{3}{8}''$,   A = _____,   B = _____   g. $P = \frac{3}{32}''$,   C = _____

c. $H = \frac{15}{16}''$,   A = _____,   B = _____   h. $P = \frac{1}{20}''$,   C = _____

d. $H = \frac{21}{32}''$,   A = _____,   B = _____   i. $P = \frac{1}{28}''$,   C = _____

e. $H = \frac{3}{4}''$,   A = _____,   B = _____   j. $P = \frac{3}{16}''$,   C = _____

Multiplying Mixed Numbers

5. Multiply these mixed numbers. Reduce to lowest terms where necessary.

a. $1\frac{2}{3} \times 6\frac{3}{10}$ _____   d. $1\frac{2}{3} \times 10\frac{1}{4} \times \frac{3}{8}$ _____

b. $3\frac{5}{16} \times 7\frac{3}{4}$ _____   e. $2\frac{3}{32} \times 3 \times \frac{1}{8}$ _____

c. $4\frac{5}{8} \times 2\frac{1}{2}$ _____   f. $2\frac{2}{3} \times 2\frac{2}{3} \times 5\frac{1}{4}$ _____

6. How many inches of drill rod are required in order to make 20 drills each $3\frac{3}{16}''$ long? Allow $\frac{3}{32}''$ waste for each drill.   _____

UNIT 5 Division of Common Fractions and Mixed Numbers

Objectives After studying this unit you should be able to

- **Divide fractions.**
- **Divide mixed numbers.**

In machine technology, division of fractions and mixed numbers is used in determining production times and costs per machined unit, in calculating the pitch of screw threads, and in computing the number of parts that can be manufactured from a given amount of raw material.

Dividing Fractions as the Inverse of Multiplying Fractions

Division is the inverse of multiplication. Dividing by 2 is the same as multiplying by $\frac{1}{2}$.

$$5 \div 2 = 2\frac{1}{2}$$
$$5 \times \frac{1}{2} = 2\frac{1}{2}$$
$$5 \div 2 = 5 \times \frac{1}{2}$$

Two is the *inverse* of $\frac{1}{2}$, and $\frac{1}{2}$ is the *inverse* of 2. *Inverting* a fraction means turning the fraction upside down, such as, $\frac{1}{3}$ inverted is $\frac{3}{1}$, $\frac{8}{7}$ inverted is $\frac{7}{8}$, $\frac{63}{64}$ inverted is $\frac{64}{63}$, and $\frac{9}{16}$ inverted is $\frac{16}{9}$.

Procedure To divide fractions

- Invert the divisor.
- Change the division operation to a multiplication operation.
- Follow the procedure for multiplying fractions.

Example 1 Divide $\frac{5}{8}$ by $\frac{3}{4}$.

Invert the divisor.
Change the division operation to a multiplication operation.
Follow the procedure for multiplication.

$$\frac{5}{8} \div \frac{3}{4} = \frac{5}{\overset{}{\underset{2}{8}}} \times \frac{\overset{1}{4}}{3} = \frac{5}{6} \quad \text{Ans}$$

Example 2 The machine bolt shown has a pitch of $\frac{1}{16}$″. The pitch is the distance between 2 adjacent threads or the thickness of one thread. Find the number of threads in $\frac{7}{8}$″. All dimensions are in inches.

Divide $\frac{7}{8}$ by $\frac{1}{16}$.

$$\frac{7}{8} \div \frac{1}{16} = \frac{7}{\underset{1}{\cancel{8}}} \times \frac{\overset{2}{\cancel{16}}}{1} = 14 \quad \text{Ans}$$

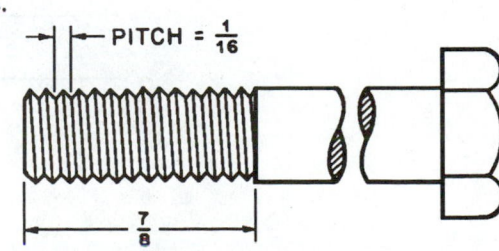

PITCH = $\frac{1}{16}$

$\frac{7}{8}$

Dividing Mixed Numbers

Procedure   To divide mixed numbers

- Express the mixed numbers as improper fractions.
- Follow the procedure for dividing fractions.

Example 1   Divide $7\frac{1}{2}$ by $2\frac{3}{8}$.

Express $7\frac{1}{2}$ and $2\frac{3}{8}$ as improper fractions. $7\frac{1}{2} \div 2\frac{3}{8} = \frac{15}{2} \div \frac{19}{8} =$

Invert the divisor.

Change the division operation to
a multiplication operation. $\frac{15}{\cancel{2}_1} \times \frac{\cancel{8}^4}{19} = \frac{60}{19} = 3\frac{3}{19}$ Ans

Multiply.

Example 2   A section of strip stock is shown with 5 equally spaced holes. Determine the distance between two consecutive holes. All dimensions are in inches.

➤ **Note:**   The number of spaces between the holes is one less than the number of holes.

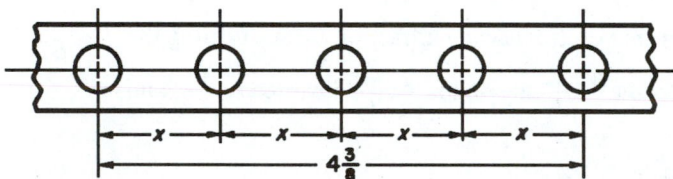

Express as improper fractions. $4\frac{3}{8} \div 4 = \frac{35}{8} \div \frac{4}{1} =$

Invert the divisor and multiply. $\frac{35}{8} \times \frac{1}{4} = \frac{35}{32} = 1\frac{3}{32}$

$$x = 1\frac{3}{32}''   \text{Ans}$$

APPLICATION

Inverting Fractions

Invert each of the following.

1. $\frac{7}{8}$ _____   3. $\frac{25}{8}$ _____

2. $\frac{1}{4}$ _____   4. 6 _____

Dividing Fractions

5. This casting shows seven tapped holes, A–G. The number of threads is determined by dividing the depth of thread by the thread pitch. Find the number of threads in each of the tapped holes. All dimensions are in inches.

A = _____

B = _____

C = _____

D = _____

E = _____

F = _____

G = _____

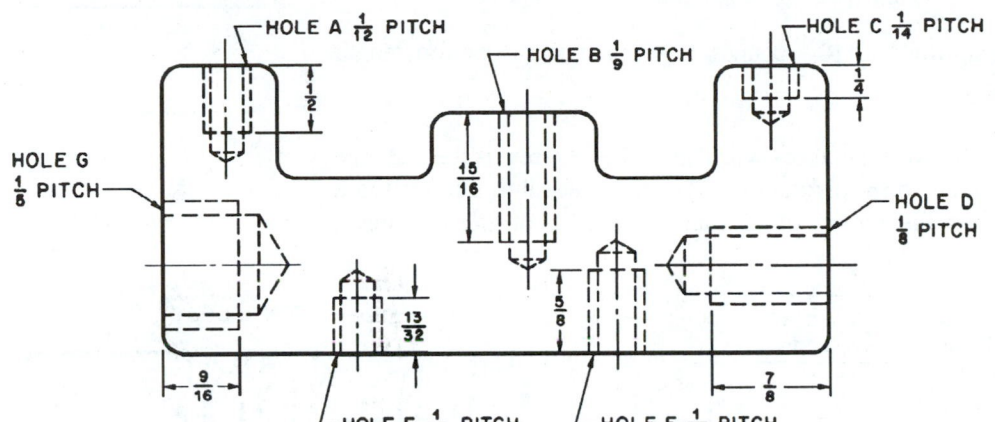

6. Bar stock is being cut on a lathe. The tool feeds (advances) $\frac{3}{64}$ inch each time the stock turns once (1 revolution). How many revolutions will the stock make when the tool advances $\frac{3}{4}$ inch?

_____

7. A groove $\frac{15}{16}$ inch deep is to be milled in a steel plate. How many cuts are required if each cut is $\frac{3}{16}$ inch deep?

_____

Dividing Mixed Numbers

8. This sheet metal section has 5 sets of drilled holes: A, B, C, D, and E. The holes within a set are equally spaced in the horizontal direction. Compute the horizontal distance between 2 consecutive holes for each set. All dimensions are in inches.

Set A = _____

Set B = _____

Set C = _____

Set D = _____

Set E = _____

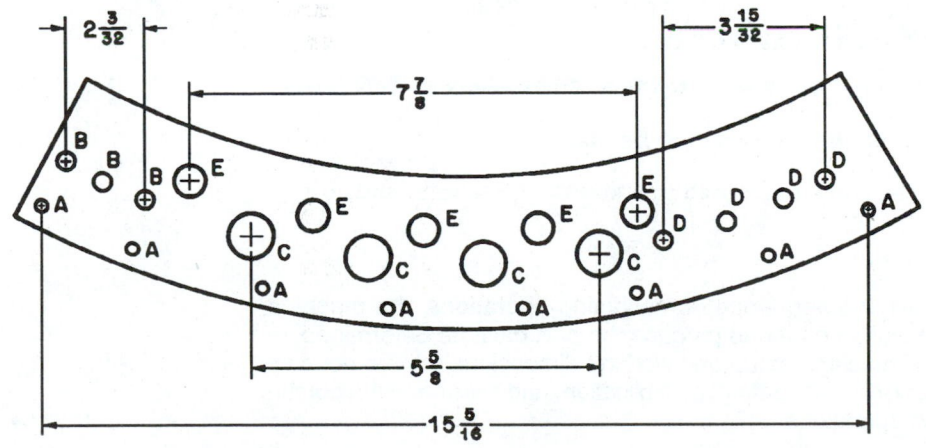

9. The feed on a boring mill is set for $\frac{1}{64}$ inch. How many revolutions does the work make when the tool advances $3\frac{3}{4}$ inches?

_____

10. How many complete pieces can be blanked from a strip of steel $27\frac{1}{4}$ feet long if each stamping requires $2\frac{3}{16}$ inches of material plus an allowance of $\frac{5}{16}$ inch at one end of the strip? (12 inches = 1 foot)

_____

11. A groove is milled the full length of a steel plate which is $3\frac{1}{4}$ feet long. This operation takes a total of $4\frac{1}{16}$ minutes. How many feet of steel are cut in one minute?   _____

12. How many binding posts can be cut from a brass rod $42\frac{1}{2}$ inches long if each post is $1\frac{7}{8}$ inches long? Allow $\frac{3}{32}$ inch waste for each cut.   _____

13. A bar of steel $23\frac{1}{4}$ feet long weighs $110\frac{1}{2}$ pounds. How much does a one-foot length of bar weigh?   _____

14. A single-threaded square thread screw is shown. The lead of a screw is the distance that the screw advances in one turn (revolution). The lead is equal to the pitch in a single-threaded screw. Given the number of turns and the amount of screw advance, determine the leads.

	Screw Advance	Number of Turns	Lead
a.	$2\frac{1}{4}''$	10	
b.	$7\frac{37}{64}''$	$24\frac{1}{4}$	
c.	$2\frac{7}{16}''$	$6\frac{1}{2}$	
d.	$1\frac{1}{2}''$	15	
e.	$6\frac{3}{10}''$	$12\frac{3}{5}$	

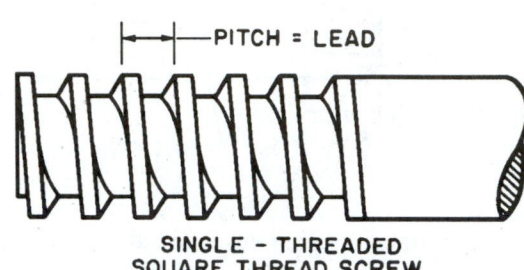

SINGLE – THREADED
SQUARE THREAD SCREW

UNIT 6 Combined Operations of Common Fractions and Mixed Numbers

Objectives   After studying this unit you should be able to

- Solve problems which involve combined operations of fractions and mixed numbers.
- Solve complex fractions.

Before a part is machined, the sequence of machining operations, the machine setup, and the working dimensions needed to produce the part must be determined. In actual practice, calculations of machine setup and working dimensions require not only the individual operations of addition, subtraction, multiplication, and division, but a combination of two or more of these operations.

Order of Operations for Combined Operations

Procedure

- **Do all the work in the parentheses first.** Parentheses are used to group numbers. In a problem expressed in fractional form, the numerator and the denominator are each considered as being enclosed in parentheses.

$$\frac{4\frac{3}{4} - \frac{1}{2}}{10 + 6\frac{5}{8}} = \left(4\frac{3}{4} - \frac{1}{2}\right) \div \left(10 + 6\frac{5}{8}\right)$$

If an expression contains parentheses with brackets, do the work within the innermost parentheses first.

- **Do multiplication and division next.** Perform multiplication and division in order from left to right.

- **Do addition and subtraction last.** Perform addition and subtraction in order from left to right.

Combining Addition and Subtraction

Example 1   Find the value of $3\frac{1}{2} - \frac{3}{8} + \frac{5}{16}$.

Subtract $\frac{3}{8}$ from $3\frac{1}{2}$. $3\frac{1}{2} - \frac{3}{8} = 3\frac{1}{8}$

Add $3\frac{1}{8}$ to $\frac{5}{16}$. $3\frac{1}{8} + \frac{5}{16} = 3\frac{7}{16}$ Ans

Example 2   Find x, the distance from the base of the plate to the center of hole #2. All dimensions are in inches.

$$x = \frac{9}{16}'' + 2\frac{1}{8}'' - \frac{13}{32}''$$

Add. $\frac{9}{16}'' + 2\frac{1}{8}'' = 2\frac{11}{16}''$

Subtract. $2\frac{11}{16}'' - \frac{13}{32}'' = 2\frac{9}{32}''$ Ans

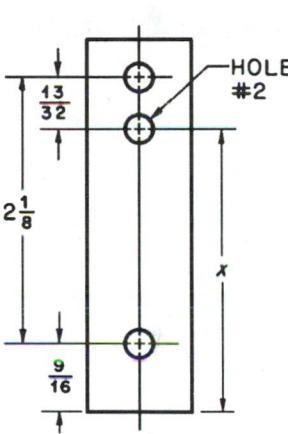

Combining Multiplication and Division

Example 1   Find the value of $\frac{2}{3} \times 8 \div 2\frac{1}{2}$.

Multiply. $\frac{2}{3} \times 8 = \frac{2 \times 8}{3 \times 1} = \frac{16}{3}$

Divide. $\frac{16}{3} \div 2\frac{1}{2} = \frac{16}{3} \times \frac{2}{5} = \frac{32}{15} = 2\frac{2}{15}$ Ans

Example 2   The stainless steel plate shown has grooves which are of uniform length and equally spaced within a distance of $33\frac{1}{2}$ inches. The time required to rough and finish mill a one-inch length of groove is $\frac{7}{10}$ minute. How many minutes are required to cut all the grooves? Disregard the time required to reposition the part. All dimensions are in inches.

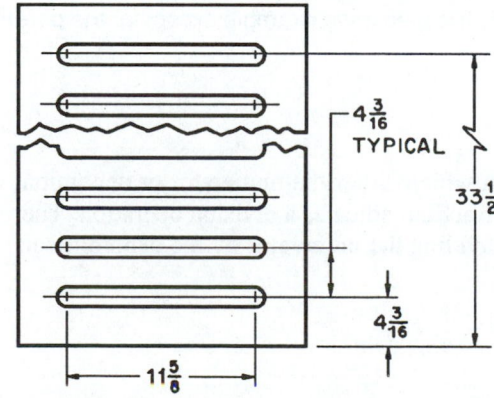

The number of grooves in $33\frac{1}{2}'' = 33\frac{1}{2} \div 4\frac{3}{16}$.

The time required to cut 1 groove $= \frac{7}{10} \times 11\frac{5}{8}$.

Total time equals the number of grooves multiplied by the time for each groove.

$$33\frac{1}{2} \div 4\frac{3}{16} \times \frac{7}{10} \times 11\frac{5}{8}$$

Divide. $33\frac{1}{2} \div 4\frac{3}{16} \times \frac{67}{2} \times \frac{16}{67} = 8$

Multiply. $8 \times \frac{7}{10} \times 11\frac{5}{8} = \frac{8}{1} \times \frac{7}{10} \times \frac{93}{8} = 65\frac{1}{10}$

Total Time $= 65\frac{1}{10}$ minutes   Ans

Combining Addition, Subtraction, Multiplication, and Division

Example 1   Find the value of $7\frac{5}{6} + 5\frac{1}{2} \div \frac{3}{4} - 10 \times \frac{7}{16}$.

First divide and multiply. $5\frac{1}{2} \div \frac{3}{4} = \frac{11}{2} \times \frac{4}{3} = 7\frac{1}{3}$

$$10 \times \frac{7}{16} = \frac{10}{1} \times \frac{7}{16} = 4\frac{3}{8}$$

Next add and subtract. $7\frac{5}{6} + 7\frac{1}{3} = 15\frac{1}{6}$

$$15\frac{1}{6} - 4\frac{3}{8} = 10\frac{19}{24}$$ Ans

Example 2   Find the value of $\left(7\frac{5}{6} + 5\frac{1}{2}\right) \div \frac{3}{4} - 10 \times \frac{7}{16}$.

First do the work in parentheses. $\left(7\frac{5}{6} + 5\frac{1}{2}\right) = 7\frac{5}{6} + 5\frac{3}{6} = 13\frac{1}{3}$

Next divide and multiply. $13\frac{1}{3} \div \frac{3}{4} = \frac{40}{3} \times \frac{4}{3} = \frac{160}{9} = 17\frac{7}{9}$

$$10 \times \frac{7}{16} = 4\frac{3}{8}$$

Then add and subtract. $17\frac{7}{9} - 4\frac{3}{8} = 13\frac{29}{72}$ Ans

➤ **Note:**   This example is the same as the preceding example except for the parentheses.

Complex Fractions

A *complex fraction* is an expression in which either the numerator or denominator or both are fractions or mixed numbers. A fraction indicates a division operation. Therefore, complex fractions can be solved by dividing the numerator by the denominator.

$$\frac{\frac{5}{9}}{\frac{1}{3}} = \frac{5}{9} \div \frac{1}{3}$$

Example Find the value of $\dfrac{5\frac{7}{8} + 2\frac{3}{4}}{3\frac{15}{16} - 1\frac{1}{8}}$.

➤ **Note:** The complete numerator is divided by the complete denominator. Therefore, parentheses are used to indicate that addition in the numerator and subtraction in the denominator must be performed before division.

$$\dfrac{5\frac{7}{8} + 2\frac{3}{4}}{3\frac{15}{16} - 1\frac{1}{8}} \quad \left(5\frac{7}{8} + 2\frac{3}{4}\right) \div \left(3\frac{15}{16} - 1\frac{1}{8}\right) = 8\frac{5}{8} \div 2\frac{13}{16} = 3\frac{1}{15} \quad \text{Ans}$$

APPLICATION

Order of Operations for Combined Operations

1. Solve the following examples of combined operations.

   a. $\frac{1}{2} + \frac{3}{16} - \frac{1}{4}$ _____

   b. $3\frac{7}{8} - 2\frac{3}{16} + \frac{3}{8}$ _____

   c. $\frac{3}{10} + 8\frac{2}{5} - 3\frac{1}{25}$ _____

   d. $27 - 2\frac{2}{3} + 4\frac{1}{6}$ _____

   e. $32\frac{1}{8} + 2\frac{3}{16} \times \frac{3}{4}$ _____

   f. $\frac{7}{9} \times \left(\frac{2}{3} + 3\frac{5}{6}\right)$ _____

   g. $12 - 4\frac{1}{2} + \frac{1}{2} + 2\frac{3}{4}$ _____

   h. $\left(16 - 4\frac{1}{2}\right) \div \frac{1}{2} + 5\frac{3}{8}$ _____

   i. $\left(16 - 4\frac{1}{2}\right) \div \left(\frac{1}{2} + 2\frac{1}{8}\right)$ _____

   j. $15\frac{1}{4} \times 1\frac{1}{3} + 2\frac{2}{3} + 4\frac{5}{6}$ _____

Complex Fractions

2. Find the value of the following complex fractions.

   a. $\dfrac{\frac{3}{4}}{\frac{1}{2}}$ _____

   b. $\dfrac{3\frac{7}{8}}{5}$ _____

   c. $\dfrac{\frac{15}{16}}{2\frac{1}{8}}$ _____

   d. $\dfrac{\frac{1}{3} + \frac{5}{6}}{3\frac{3}{4}}$ _____

   e. $\dfrac{6\frac{3}{4} - 2\frac{7}{8}}{3\frac{1}{2} + 1\frac{1}{16}}$ _____

   f. $\dfrac{10\frac{1}{2} \times \frac{1}{2}}{4 + 2\frac{1}{4}}$ _____

Related Problems

3. Refer to the shaft shown. Determine the missing dimensions in the table using the dimensions given. All dimensions are in inches.

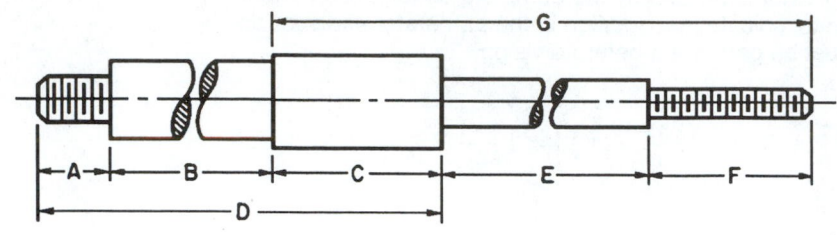

	A	B	C	D	E	F	G
a.	$\frac{1}{2}$		$1\frac{3}{8}$	$6\frac{3}{4}$		$\frac{15}{16}$	$7\frac{3}{8}$
b.		$3\frac{13}{16}$	$1\frac{5}{8}$	$5\frac{37}{64}$	$4\frac{3}{8}$	$\frac{3}{4}$	
c.	$\frac{7}{16}$	$4\frac{3}{32}$			$5\frac{1}{8}$	$\frac{27}{32}$	$7\frac{1}{32}$
d.	$\frac{5}{8}$		$1\frac{7}{16}$	$5\frac{31}{32}$		$\frac{7}{8}$	$7\frac{15}{16}$
e.		$3\frac{3}{4}$	$1\frac{11}{16}$	$6\frac{1}{32}$	$4\frac{61}{64}$	$\frac{25}{32}$	
f.	$\frac{11}{16}$	$4\frac{3}{16}$			$5\frac{3}{16}$	$\frac{7}{8}$	$7\frac{3}{64}$

4. The outside diameter of an aluminum tube is $3\frac{1}{16}$ inches. The wall thickness is $\frac{5}{32}$ inch. What is the inside diameter? _____

5. Four studs of the following lengths in inches are to be machined from bar stock: $1\frac{3}{4}''$, $1\frac{7}{8}''$, $2\frac{5}{16}''$, and $1\frac{11}{32}''$. Allow $\frac{1}{8}$ inch waste for each cut and $\frac{1}{32}$ inch on each end of each stud for facing. What is the total length of bar stock required? _____

6. Find dimensions A, B, C, and D of the idler bracket in the figure. All dimensions are in inches.

A = _____

B = _____

C = _____

D = _____

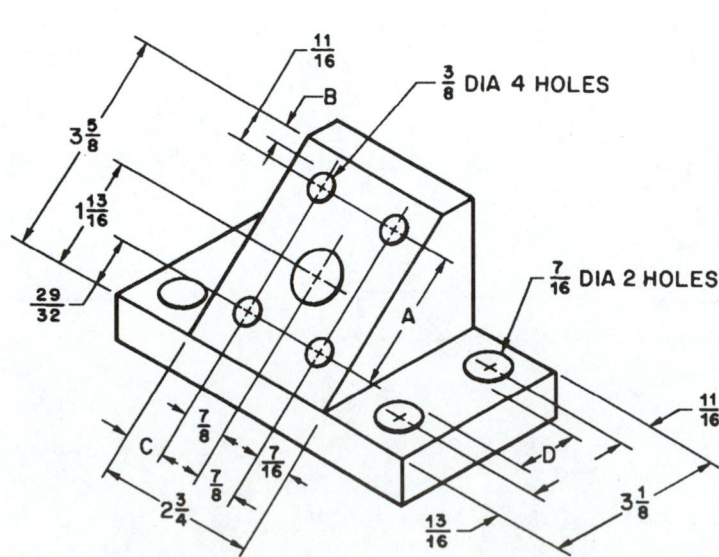

7. How long does it take to cut a distance of $1\frac{1}{4}$ feet along a shaft that turns 150 revolutions per minute with a tool feed of $\frac{1}{32}$ inch per revolution? _____

8. An angle iron $47\frac{1}{2}$ inches long has two drilled holes which are equally spaced from the center of the piece. The center distance between the two holes is $19\frac{7}{8}$ inches. What is the distance from each end of the piece to the closest hole? _____

9. A tube has an inside diameter of $\frac{3}{4}$ inch and a wall thickness of $\frac{1}{16}$ inch. The tube is to be fitted in a drilled hole in a block. What diameter hole should be drilled in the block to give $\frac{1}{64}$ inch total clearance? _____

10. Two views of a mounting block are shown. Determine dimensions A–G. All dimensions are in inches.

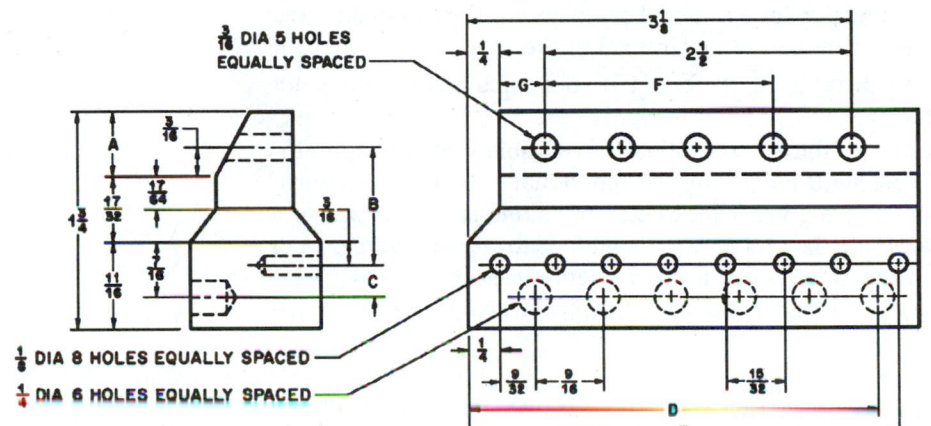

A = _____
B = _____
C = _____
D = _____
E = _____
F = _____
G = _____

11. The composition of an aluminum alloy by weight is $\frac{19}{20}$ aluminum and $\frac{1}{50}$ copper. The only other element in the alloy is magnesium. How many pounds of magnesium are required for casting 125 pounds of alloy? _____

12. Pieces of the following lengths are cut from a 15-inch steel bar: $2\frac{1}{2}''$, $1\frac{3}{4}''$, $1\frac{7}{8}''$, and $\frac{5}{16}''$. Allowing $\frac{1}{8}$-inch waste for each cut, what is the length of bar left after the pieces are cut? _____

UNIT 7 Computing with a Calculator: Fractions and Mixed Numbers

Objectives After studying this unit you should be able to

- Perform individual operations of addition, subtraction, multiplication, and division with fractions using a calculator.
- Perform combinations of operations with fractions using a calculator.

Fractions

The fraction key (a⬚c) is used when entering fractions and mixed numbers in a calculator. The answers to expressions entered as fractions will be given as fractions or mixed numbers with the fraction in lowest terms.

Enter the numerator, press ⬚ a⬚c ⬚, and enter the denominator. The fraction is displayed with the symbol ⌐ between the numerator and denominator.

Example   Enter $\frac{3}{4}$.

3 ⬚ a⬚c ⬚ 4, 3 ⌐ 4 is displayed.

Individual Arithmetic Operations: Fractions

The operations of addition, subtraction, multiplication, and division are performed with the four arithmetic keys and the equals key. The equals key completes all operations entered and readies the calculator for additional calculations. Certain makes and models of calculators have the execute key ⬚EXE⬚ instead of the equal key ⬚=⬚. If your calculator has the execute key, substitute ⬚EXE⬚ for ⬚=⬚. If your calculator has the enter key ⬚ENTER⬚, substitute ⬚ENTER⬚ for ⬚=⬚.

Examples of each of the four arithmetic operations of addition, subtraction, multiplication, and division are presented. Following the individual operation problems, combined operations expressions are given with calculator solutions. An answer to a problem should be checked by doing the problem a second time to ensure that improper data was not entered in its solution. Remember to clear or erase previous recorded data and calculations before doing a problem. Depending on the make and model of the calculator, press ⬚ AC ⬚ *once* or ⬚ON/c⬚ *twice.*

Example 1   Add. $\frac{3}{16} + \frac{19}{32}$

3 ⬚ a⬚c ⬚ 16 ⬚+⬚ 19 ⬚ a⬚c ⬚ 32 ⬚=⬚ 25 ⌐ 32, $\frac{25}{32}$ Ans

Example 2   Subtract. $\frac{7}{8} - \frac{5}{64}$

7 ⬚ a⬚c ⬚ 8 ⬚−⬚ 5 ⬚ a⬚c ⬚ 64 ⬚=⬚ 51 ⌐ 64, $\frac{51}{64}$ Ans

Example 3   Multiply. $\frac{3}{32} \times \frac{11}{16}$

3 ⬚ a⬚c ⬚ 32 ⬚x⬚ 11 ⬚ a⬚c ⬚ 16 ⬚=⬚ 33 ⌐ 512, $\frac{33}{512}$ Ans

Example 4   Divide. $\frac{5}{8} \div \frac{13}{15}$

5 ⬚ a⬚c ⬚ 8 ⬚÷⬚ 13 ⬚ a⬚c ⬚ 15 ⬚=⬚ 75 ⌐ 104, $\frac{75}{104}$ Ans

Mixed Numbers

Enter the whole number, press ⬚ a⬚c ⬚, enter the fraction numerator, press ⬚ a⬚c ⬚, and enter the denominator. Depending on the particular calculator, either the symbol _ or ⌐ is displayed between the whole number and fraction.

Example   Enter $15\frac{7}{16}$.

15 ⬚ a⬚c ⬚ 7 ⬚ a⬚c ⬚ 16

Either 15 _ 7 ⌐ 16 or 15 ⌐ 7 ⌐ 16 is displayed.

Individual Arithmetic Operations: Mixed Numbers

The following examples are of mixed numbers with individual arithmetic operations.

Example 1 Add. $7\frac{3}{64} + 23\frac{5}{8}$

7 [abc] 3 [abc] 64 [+] 23 [abc] 5 [abc] 8 [=] 30 ⌐ 43 ⌐ 64, $30\frac{43}{64}$ Ans

Example 2 Subtract. $43\frac{7}{8} - 36\frac{29}{32}$

43 [abc] 7 [abc] 8 [−] 36 [abc] 29 [abc] 32 [=] 6 ⌐ 31 ⌐ 32, $6\frac{31}{32}$ Ans

Example 3 Multiply. $38\frac{5}{6} \times 14\frac{13}{16}$

38 [abc] 5 [abc] 6 [x] 14 [abc] 13 [abc] 16 [=] 575 ⌐ 7 ⌐ 32, $575\frac{7}{32}$ Ans

Example 4 Divide. $159\frac{17}{64} \div 3\frac{7}{8}$

159 [abc] 17 [abc] 64 [÷] 3 [abc] 7 [abc] 8 [=] 41 ⌐ 25 ⌐ 248, $41\frac{25}{248}$ Ans

Practice Exercises, Individual Basic Operations with Fractions and Mixed Numbers

Evaluate the following expressions. The expressions are basic arithmetic operations. Remember to check your answers by doing each problem twice. The solutions to the problems directly follow the practice exercises. Compare your answers to the given solutions.

1. $\frac{5}{8} + \frac{11}{16}$

2. $\frac{31}{32} - \frac{7}{8}$

3. $\frac{9}{16} \times \frac{5}{8}$

4. $\frac{23}{25} \div \frac{4}{5}$

5. $85\frac{7}{64} + 107\frac{3}{4}$

6. $125\frac{7}{8} - 67\frac{63}{64}$

7. $62\frac{13}{16} \times 47\frac{1}{6}$

8. $785\frac{27}{32} \div 2\frac{3}{4}$

9. $\frac{59}{64} + 46\frac{27}{32}$

10. $37\frac{3}{8} - \frac{45}{64}$

Solutions to Practice Exercises, Individual Basic Operations with Fractions and Mixed Numbers

1. 5 [abc] 8 [+] 11 [abc] 16 [=] 1 ⌐ 5 ⌐ 16, $1\frac{5}{16}$ Ans

2. 31 [abc] 32 [−] 7 [abc] 8 [=] 3 ⌐ 32, $\frac{3}{32}$ Ans

3. 9 [abc] 16 [x] 5 [abc] 8 [=] 45 ⌐ 128, $\frac{45}{128}$ Ans

4. 23 [abc] 25 [÷] 4 [abc] 5 [=] 1 ⌐ 3 ⌐ 20, $1\frac{3}{20}$ Ans

5. 85 [abc] 7 [abc] 64 [+] 107 [abc] 3 [abc] 4 [=] 192 ⌐ 55 ⌐ 64, $192\frac{55}{64}$ Ans

6. 125 [abc] 7 [abc] 8 [−] 67 [abc] 63 [abc] 64 [=] 57 ⌐ 57 ⌐ 64, $57\frac{57}{64}$ Ans

7. 62 [abc] 13 [abc] 16 [x] 47 [abc] 1 [abc] 6 [=] 2962 ⌐ 21 ⌐

    32, $2,962\frac{21}{32}$ Ans

8. 785 $\boxed{a^{b}\!c}$ 27 $\boxed{a^{b}\!c}$ 32 $\boxed{+}$ 2 $\boxed{a^{b}\!c}$ 3 $\boxed{a^{b}\!c}$ 4 $\boxed{=}$ 285 _ 67 ⌐ 88,

    $285\frac{67}{88}$ Ans

9. 59 $\boxed{a^{b}\!c}$ 64 $\boxed{+}$ 46 $\boxed{a^{b}\!c}$ 27 $\boxed{a^{b}\!c}$ 32 $\boxed{=}$ 47 _ 49 ⌐ 64, $47\frac{49}{64}$ Ans

10. 37 $\boxed{a^{b}\!c}$ 3 $\boxed{a^{b}\!c}$ 8 $\boxed{-}$ 45 $\boxed{a^{b}\!c}$ 64 $\boxed{=}$ 36 _ 43 ⌐ 64, $36\frac{43}{64}$ Ans

Combined Operations

Because the following problems are combined operations expressions, your calculator must have algebraic logic to solve the problems shown. The expressions are solved by entering numbers and operations into the calculator in the same order as the expressions are written. Remember to check your answers by doing each problem twice.

Example 1   Evaluate. $275\frac{17}{32} + \frac{7}{8} \times 26\frac{3}{4}$

275 $\boxed{a^{b}\!c}$ 17 $\boxed{a^{b}\!c}$ 32 $\boxed{+}$ 7 $\boxed{a^{b}\!c}$ 8 $\boxed{\times}$ 26 $\boxed{a^{b}\!c}$ 3 $\boxed{a^{b}\!c}$ 4 $\boxed{=}$ 298 _ 15 ⌐ 16,

$298\frac{15}{16}$ Ans

Because the calculator has algebraic logic, the multiplication operation $\left(\frac{7}{8} \times 26\frac{3}{4}\right)$ was performed before the addition operation $\left(\text{adding } 275\frac{17}{32}\right)$ was performed.

Example 2   Evaluate. $\frac{35}{64} - \frac{5}{8} + 18 + 10\frac{2}{3}$

35 $\boxed{a^{b}\!c}$ 64 $\boxed{-}$ 5 $\boxed{a^{b}\!c}$ 8 $\boxed{+}$ 18 $\boxed{+}$ 10 $\boxed{a^{b}\!c}$ 2 $\boxed{a^{b}\!c}$ 3 $\boxed{=}$ 1 _ 39 ⌐ 64, $1\frac{39}{64}$ Ans

Example 3   Evaluate. $380\frac{29}{32} - \left(\frac{3}{16} + 9\frac{15}{64}\right) \times 12$

As previously discussed in Unit 6, operations enclosed within parentheses are done first. A calculator with algebraic logic performs the operations within parentheses before performing other operations in a combined operations expression. If an expression contains parentheses, enter the expression in the calculator in the order in which it is written. The parentheses keys must be used.

380 $\boxed{a^{b}\!c}$ 29 $\boxed{a^{b}\!c}$ 32 $\boxed{-}$ $\boxed{(}$ 3 $\boxed{a^{b}\!c}$ 16 $\boxed{+}$ 9 $\boxed{a^{b}\!c}$ 15 $\boxed{a^{b}\!c}$ 64 $\boxed{)}$ $\boxed{\times}$ 12 $\boxed{=}$

267 _ 27 ⌐ 32, $267\frac{27}{32}$ Ans

Example 4   Evaluate. $\dfrac{25\frac{47}{64} + 7 \times \frac{5}{8}}{\frac{3}{16} \times 2 + \frac{1}{8}}$

Recall that for a problem expressed in fractional form, the fraction bar is also used as a grouping symbol. The numerator and denominator are each considered as being enclosed in parentheses.

$\boxed{(}$ 25 $\boxed{a^{b}\!c}$ 47 $\boxed{a^{b}\!c}$ 64 $\boxed{+}$ 7 $\boxed{\times}$ 5 $\boxed{a^{b}\!c}$ 8 $\boxed{)}$ $\boxed{\div}$ $\boxed{(}$ 3 $\boxed{a^{b}\!c}$ 16 $\boxed{\times}$ 2 $\boxed{+}$ 1 $\boxed{a^{b}\!c}$

8 $\boxed{)}$ $\boxed{=}$ 60 _ 7 ⌐ 32, $60\frac{7}{32}$ Ans

The expression may also be evaluated by using the $\boxed{=}$ key to simplify the numerator without having to enclose the entire numerator in parentheses. However, parentheses must be used to enclose the denominator.

25 $\boxed{a^{b}\!c}$ 47 $\boxed{a^{b}\!c}$ 64 $\boxed{+}$ 7 $\boxed{\times}$ 5 $\boxed{a^{b}\!c}$ 8 $\boxed{=}$ $\boxed{\div}$ $\boxed{(}$ 3 $\boxed{a^{b}\!c}$ 16 $\boxed{\times}$ 2 $\boxed{+}$ 1 $\boxed{a^{b}\!c}$

8 $\boxed{)}$ $\boxed{=}$ 60 _ 7 ⌐ 32, $60\frac{7}{32}$ Ans

Practice Exercises, Combined Operations with Fractions and Mixed Numbers

Evaluate the following combined operations expressions. Remember to check you answers by doing each problem twice. The solutions to the problems directly follow the practice exercises. Compare your answers to the given solutions.

1. $\left(\frac{11}{16} + 12\frac{31}{32}\right) + \frac{1}{8}$

2. $\frac{108}{\frac{3}{8}} - 3\frac{5}{64}$

3. $\frac{43\frac{9}{10} - 17\frac{3}{5} + \frac{7}{20}}{5}$

4. $120\frac{13}{16} + 98\frac{5}{8} \times \left(6 - \frac{3}{4}\right)$

5. $\frac{56\frac{3}{4} + 20 \times \frac{7}{8}}{4 \times \frac{2}{3}}$

6. $\left(\frac{25}{32} - \frac{3}{4}\right) + \frac{1}{2} \times \frac{3}{4}$

7. $50 \times \left(28\frac{4}{5} - 17\frac{9}{10} + 27\right) \times \frac{3}{5}$

8. $\frac{40\frac{1}{2}}{1\frac{1}{8}} - \left(15\frac{5}{64} + 8\frac{29}{32}\right)$

9. $\frac{270 - 175\frac{1}{2} \times \frac{7}{8}}{\frac{1}{64} \times 128}$

Solutions to Practice Exercises, Combined Operations with Fractions and Mixed Numbers

1. $($ 11 [ab⁄c] 16 $+$ 12 [ab⁄c] 31 [ab⁄c] 32 $)$ $+$ 1 [ab⁄c] 8 $=$ 109 _ 1 ⌟
4, $109\frac{1}{4}$ Ans

or 11 [ab⁄c] 16 $+$ 12 [ab⁄c] 31 [ab⁄c] 32 $=$ $+$ 1 [ab⁄c] 8 $=$ 109 _ 1 ⌟
4, $109\frac{1}{4}$ Ans

2. 108 $\div$ 3 [ab⁄c] 8 $-$ 3 [ab⁄c] 5 [ab⁄c] 64 $=$ 284 _ 59 ⌟ 64, $284\frac{59}{64}$ Ans

3. $($ 43 [ab⁄c] 9 [ab⁄c] 10 $-$ 17 [ab⁄c] 3 [ab⁄c] 5 $+$ 7 [ab⁄c] 20 $)$ $+$
5 $=$ 5 _ 33 ⌟ 100, $5\frac{33}{100}$ Ans

or 43 [ab⁄c] 9 [ab⁄c] 10 $-$ 17 [ab⁄c] 3 [ab⁄c] 5 $+$ 7 [ab⁄c] 20 $=$ $+$ 5
$=$ 5 _ 33 ⌟ 100, $5\frac{33}{100}$ Ans

4. 120 [ab⁄c] 13 [ab⁄c] 16 $+$ 98 [ab⁄c] 5 [ab⁄c] 8 $\times$ $($ 6 $-$ 3 [ab⁄c] 4 $)$ $=$
638 _ 19 ⌟ 32, $638\frac{19}{32}$ Ans

5. $($ 56 [ab⁄c] 3 [ab⁄c] 4 $+$ 20 $\times$ 7 [ab⁄c] 8 $)$ $+$ $($ 4 $\times$ 2 [ab⁄c] 3 $)$ $=$
27 _ 27 ⌟ 32, $27\frac{27}{32}$ Ans

or 56 [ab⁄c] 3 [ab⁄c] 4 $+$ 20 $\times$ 7 [ab⁄c] 8 $=$ $+$ $($ 4 $\times$ 2 [ab⁄c] 3 $)$ $=$
27 _ 27 ⌟ 32, $27\frac{27}{32}$ Ans

6. $($ 25 [ab⁄c] 32 $-$ 3 [ab⁄c] 4 $)$ $+$ 1 [ab⁄c] 2 $\times$ 3 [ab⁄c] 4 $=$ 3 ⌟ 64,
$\frac{3}{64}$ Ans

or 25 [ab⁄c] 32 $-$ 3 [ab⁄c] 4 $=$ $+$ 1 [ab⁄c] 2 $\times$ 3 [ab⁄c] 4 $=$ 3 ⌟ 64,
$\frac{3}{64}$ Ans

7. 50 $\times$ $($ 28 [ab⁄c] 4 [ab⁄c] 5 $-$ 17 [ab⁄c] 9 [ab⁄c] 10 $+$ 27 $)$ $\times$ 3 [ab⁄c]
5 $=$ 1137, 1137   Ans

8. 40 $\boxed{a^{b/c}}$ 1 $\boxed{a^{b/c}}$ 2 $\boxed{+}$ 1 $\boxed{a^{b/c}}$ 1 $\boxed{a^{b/c}}$ 8 $\boxed{-}$ $\boxed{(}$ 5 $\boxed{a^{b/c}}$ 5 $\boxed{a^{b/c}}$ 64 $\boxed{+}$ 8 $\boxed{a^{b/c}}$ 29 $\boxed{a^{b/c}}$ 32 $\boxed{)}$ $\boxed{=}$ 12 _ 1 ⌐ 64, $12\frac{1}{64}$ Ans

9. $\boxed{(}$ 270 $\boxed{-}$ 175 $\boxed{a^{b/c}}$ 1 $\boxed{a^{b/c}}$ 2 $\boxed{\times}$ 7 $\boxed{a^{b/c}}$ 8 $\boxed{)}$ $\boxed{+}$ $\boxed{(}$ 1 $\boxed{a^{b/c}}$ 64 $\boxed{\times}$ 128 $\boxed{)}$ $\boxed{=}$ 58 _ 7 ⌐ 32, $58\frac{7}{32}$ Ans

or 270 $\boxed{-}$ 175 $\boxed{a^{b/c}}$ 1 $\boxed{a^{b/c}}$ 2 $\boxed{\times}$ 7 $\boxed{a^{b/c}}$ 8 $\boxed{=}$ $\boxed{+}$ $\boxed{(}$ 1 $\boxed{a^{b/c}}$ 64 $\boxed{\times}$ 128 $\boxed{)}$ $\boxed{=}$ 58 _ 7 ⌐ 32, $58\frac{7}{32}$ Ans

UNIT **8** Introduction to Decimal Fractions

Objectives **After studying this unit you should be able to**

- **Locate decimal fractions on a number line.**
- **Express common fractions having denominators of powers of ten as equivalent decimal fractions.**
- **Write decimal numbers in word form.**
- **Write numbers expressed in word form as decimal fractions.**

Most blueprints are dimensioned with decimal fractions rather than common fractions. The dials which are used in establishing machine settings and movement, in determining tool speeds and travel, and in measuring dimensions of parts are usually graduated in decimal units.

Explanation of Decimal Fractions

A decimal fraction is not written as a common fraction with a numerator and denominator. The denominator is omitted and replaced by a decimal point placed to the left of the numerator. *Decimal fractions* are equivalent to common fractions having denominators which are powers of 10, such as 10; 100; 1000; 10,000; 100,000; and 1,000,000. *Powers of 10* are numbers which are obtained by multiplying 10 by itself a certain number of times.

Meaning of Fractional Parts

The line segment shown is 1 unit long. It is divided into 10 equal smaller parts. The locations of common fractions and their decimal fraction equivalents are shown on the line.

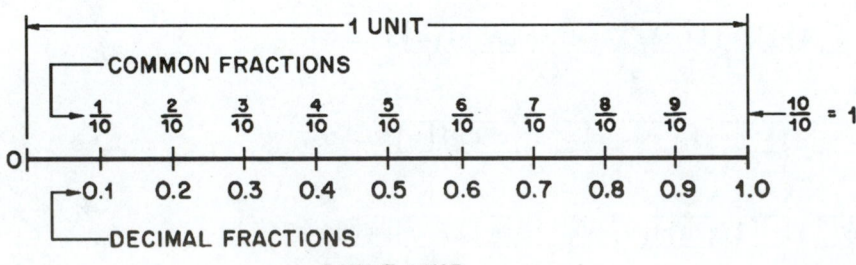

1 UNIT LINE

One of the ten equal small parts, $\frac{1}{10}$ (0.1) of the 1 unit line, is shown enlarged. The $\frac{1}{10}$ or 0.1 unit is divided into 10 equal smaller units. The locations of common fractions and their decimal fraction equivalents are shown on this line.

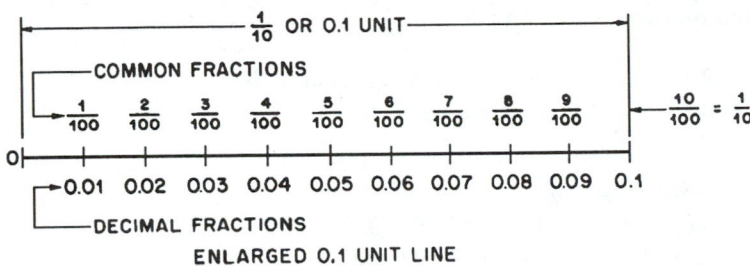

ENLARGED 0.1 UNIT LINE

If the $\frac{1}{100}$ (0.01) division is divided into 10 equal smaller parts, the resulting parts are $\frac{1}{1000}$ (0.001); $\frac{2}{1000}$ (0.002); $\frac{3}{1000}$ (0.003); ... $\frac{9}{1000}$ (0.009); $\frac{10}{1000} = \frac{1}{100}$ (0.01).

- Each time the decimal point is moved one place to the left, a value $\frac{1}{10}$ (0.1) times the previous value is obtained.
- Each time a decimal point is moved one place to the right, a value 10 times greater than the previous value is obtained.

Each time a decimal fraction is multiplied by 10 the decimal point is moved one place to the right. Each step in the following table shows both the decimal fraction and its equivalent common fraction.

Decimal Fraction	Common Fraction
$0.000003 \times 10 = 0.00003$	$3/1{,}000{,}000 \times 10 = 3/100{,}000$
$0.00003 \times 10 = 0.0003$	$3/100{,}000 \times 10 = 3/10{,}000$
$0.0003 \times 10 = 0.003$	$3/10{,}000 \times 10 = 3/1{,}000$
$0.003 \times 10 = 0.03$	$3/1000 \times 10 = 3/100$
$0.03 \times 10 = 0.3$	$3/100 \times 10 = 3/10$
$0.3 \times 10 = 3.$	$3/10 \times 10 = 3$

Reading and Writing Decimal Fractions

The following chart gives the names of the parts of a number with respect to the positions from the decimal point.

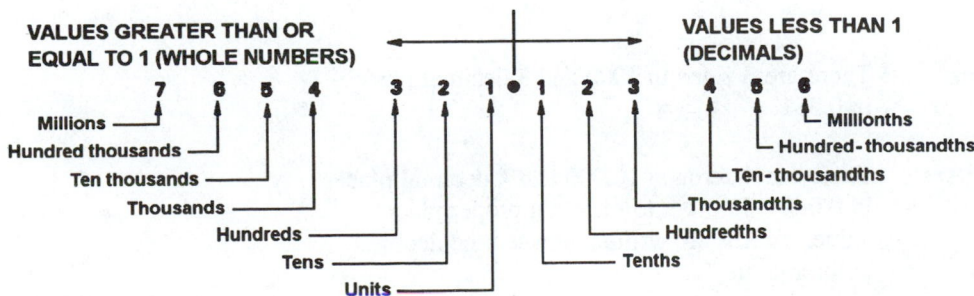

To read a decimal, read the number as a whole number. Then say the name of the decimal place of the last digit to the right.

Examples
1. 0.5 is read as five tenths.
2. 0.07 is read as seven hundredths.
3. 0.011 is read as eleven thousandths.

To write a decimal fraction from a word statement, write the number using a decimal point and zeros before the number as necessary for the given place value.

Examples
1. Two hundred nineteen ten-thousandths is written as 0.0219.
2. Forty-three hundred-thousandths is written as 0.00043.
3. Eight hundred seventeen millionths is written as 0.000817.

A number that consists of a whole number and a decimal fraction is called a *mixed decimal*. To read a mixed decimal, read the whole number, read the word *and* at the decimal point, and read the decimal.

Examples
1. 3.4 is read as three and four tenths.
2. 1.002 is read as one and two thousandths.
3. 16.0793 is read as sixteen and seven hundred ninety-three ten-thousandths.
4. 8.00032 is read as eight and thirty-two hundred-thousandths.

Simplified Method of Reading Decimal Fractions

Usually a simplified method of reading decimal fractions is used in the machine trades. This method is generally quicker, easier, and less likely to be misinterpreted. A tool-and-die maker reads 0.0265 inches as point zero, two, six, five inches. A machinist reads 4.172 millimeters as four, point one, seven, two millimeters.

Writing Decimal Fractions from Common Fractions Having Denominators Which Are Powers of Ten

A common fraction with a denominator which is a power of ten can be written as a decimal fraction. For a common fraction with a numerator smaller than the denominator, replace the denominator with a decimal point. The decimal point is placed to the left of the first digit of the numerator. There are as many decimal places as there are zeros in the denominator. When writing a decimal fraction it is advisable to place a zero to the left of the decimal point.

Examples

1. $\frac{9}{10} = 0.9$ Ans   There is 1 zero in 10 and 1 decimal place in 0.9.

2. $\frac{381}{1000} = 0.381$ Ans   There are 3 zeros in 1000 and 3 decimal places in 0.381.

3. $\frac{7}{10,000} = 0.0007$ Ans   There are 4 zeros in 10,000 and 4 decimal places in 0.0007. In order to maintain proper place value, 3 zeros are written between the decimal point and the 7.

APPLICATION

Meaning of Fractional Parts

1. Find the decimal value of each of the distances A, B, C, D, and E. Note the total unit value of the line.

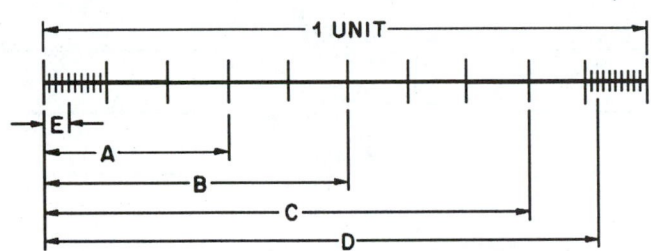

A = _____

B = _____

C = _____

D = _____

E = _____

2. Find the decimal value of each of the distances A, B, C, D, and E. Note the total unit value of the lines.

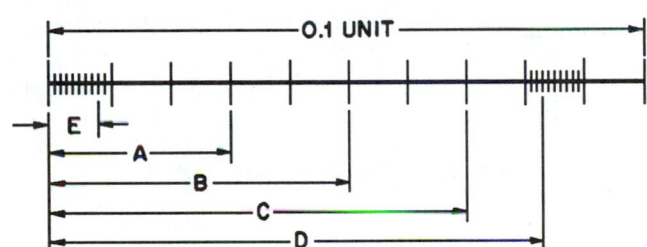

A = _____

B = _____

C = _____

D = _____

E = _____

3. Find the decimal value of each of the distances A, B, C, D, and E. Note the total unit value of the line.

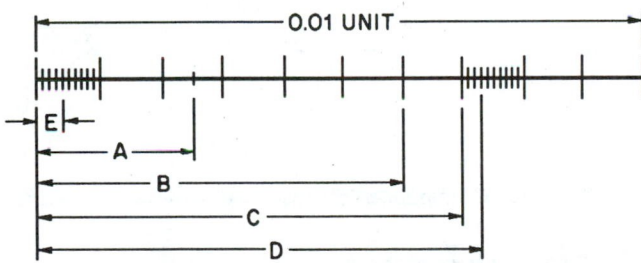

A = _____

B = _____

C = _____

D = _____

E = _____

In each of the following problems, the value on the left must be multiplied by one of the following numbers: 0.0001; 0.001; 0.01; 0.1; 10; 100; 1000; or 10,000 in order to obtain the value on the right of the equal sign. Determine the proper number.

4. $0.9 \times$ _____ = 0.0009

5. $0.7 \times$ _____ = 0.007

6. $0.03 \times$ _____ = 0.3

7. $0.0003 \times$ _____ = 0.003

8. $0.135 \times$ _____ = 0.00135

9. $4 \times$ _____ = 0.4

10. $0.0643 \times$ _____ = 0.000643

11. $0.0643 \times$ _____ = 6.43

12. $0.00643 \times$ _____ = 64.3

13. $643 \times$ _____ = 0.643

Reading and Writing Decimal Fractions

Write these numbers as words.

14. 0.064   _____

15. 0.007   _____

16. 0.132   _____

17. 0.0035   _____

18. 0.108   _____

19. 1.5   _____

20. 10.37   _____

21. 16.0007 _____

22. 4.0012   _____

23. 13.103   _____

Write these words as numbers.

24. eighty-four ten-thousandths   _____

25. three tenths   _____

26. forty-three and eight hundredths   _____

27. four and five hundred-thousandths   _____

28. thirty-five ten-thousandths   _____

29. ten and two tenths   _____

30. five and one ten-thousandth   _____

31. twenty and seventy-one hundredths   _____

Each of the following common fractions has a denominator which is a power of 10. Write the equivalent decimal fraction for each.

32. $\frac{9}{10}$ _____

33. $\frac{7}{10,000}$ _____

34. $\frac{17}{100}$ _____

35. $\frac{43}{100}$ _____

36. $\frac{61}{1000}$ _____

37. $\frac{999}{10,000}$ _____

38. $\frac{73}{1000}$ _____

39. $\frac{1973}{100,000}$ _____

40. $\frac{47,375}{100,000}$ _____

UNIT 9 Rounding Decimal Fractions and Equivalent Decimal and Common Fractions

Objectives   After studying this unit you should be able to

- **Round decimal fractions to any required number of places.**
- **Express common fractions as decimal fractions.**
- **Express decimal fractions as common fractions.**

When blueprint dimensions of a part are given in fractional units, a machinist is usually required to express these fractional values as decimal working dimensions. In computing material requirements and in determining stock waste and scrap allowances, it is sometimes more convenient to express decimal values as approximate fractional equivalents.

Rounding Decimal Fractions

When working with decimals, the computations and answers may contain more decimal places than are required. The number of decimal places needed depends on the degree of precision desired. The degree of precision depends on how the decimal value is going to be used. The tools, machines, equipment, and materials determine the degree of precision obtainable. For example, a length of 0.875376 inch cannot be cut on a milling machine. In cutting to the nearer thousandths of an inch, the machinist would consider 0.875376 inch as 0.875 inch. *Rounding a decimal* means expressing the decimal with a fewer number of decimal places.

Procedure   To round a decimal fraction

- Determine the number of decimal places required in an answer.

- If the digit directly following the last decimal place required is less than 5, drop all digits which follow the required number of decimal places.

- If the digit directly following the last decimal place required is 5 or larger, add one to the last required digit and drop all digits which follow the required number of decimal places.

Example 1   Round 0.873429 to three decimal places.

The digit following the third decimal place is 4.	0.873 ④ 29
Because 4 is less than 5, drop all digits after the third decimal place.	0.873 Ans

Example 2   Round 0.36845 to two decimal places.

The digit following the second decimal place is 8.	0.36 ⑧ 45
Because 8 is greater than 5, add 1 to the 6.	0.37 Ans

Example 3   Round 18.738257 to four decimal places.

The digit following the fourth decimal place is 5.	18.7382 ⑤ 7
Add 1 to the 2.	18.7383 Ans

Expressing Common Fractions as Decimal Fractions

A common fraction is an indicated division. For example, $\frac{3}{4}$ is the same as $3 \div 4$; $\frac{5}{16}$ is the same as $5 \div 16$; $\frac{99}{171}$ is the same as $99 \div 171$.

Because both the numerator and the denominator of a common fraction are whole numbers, expressing a common fraction as a decimal fraction requires division with whole numbers.

Procedure   To express a common fraction as a decimal fraction

- Divide the numerator by the denominator.

   A common fraction which divides evenly is expressed as an even or *terminating decimal*. A common fraction which will not divide evenly is expressed as a repeating or *nonterminating decimal.*

   The division should be carried out to one more place than the number of places required in the answer, then rounded one place.

Example 1 Express $\frac{2}{3}$ as a 4-place decimal.

Divide the numerator by the denominator.

After the 2, add one more zero than the required number of decimal places. (Add 5 zeros.)

$$\begin{array}{r} 0.66666 \\ 3\,)\overline{2.00000} \end{array}$$

Round 0.66666 to 4 places.

0.6667   Ans

Example 2 Express $\frac{5}{7}$ as a 2-place decimal.

Add 3 zeros after the 5.

$$\begin{array}{r} 0.714 \\ 7\,)\overline{5.000} \end{array}$$

Round to 2 places.

0.71   Ans

Expressing Decimal Fractions as Common Fractions

Procedure To express a decimal fraction as a common fraction

- Make the numerator of the common fraction the decimal with the decimal point omitted.

- The denominator is 1 followed by the same number of zeros as there are decimal places in the decimal fraction.

- Reduce to lowest terms.

Example 1 Express 0.375 as a common fraction.

The decimal 0.375 without the decimal point is the numerator. The numerator is 375.

$$0.375 = \frac{375}{1000} = \frac{3}{8} \quad \text{Ans}$$

The denominator is 1 with the same number of zeros as there are decimal places in the decimal fraction. There are three zeros. The denominator is 1000.

Reduce $\frac{375}{1000}$ to lowest terms.

Example 2 Express 0.27 as a common fraction.

The numerator is 27.

$$0.27 = \frac{27}{100} \quad \text{Ans}$$

The denominator is 100.

Example 3 Express 0.03125 as a common fraction.

The numerator is 3125.

The denominator is 100,000.

$$0.03125 = \frac{3125}{100,000} = \frac{1}{32} \quad \text{Ans}$$

Reduce $\frac{3125}{100,000}$ to lowest terms.

APPLICATION

Rounding Decimal Fractions

Round the following decimals to the indicated number of decimal places.

1. 0.63165 (3 places)   _____

2. 0.1247 (2 places)   _____

3. 0.23975 (3 places)   _____

4. 0.01723 (3 places)   _____

5. 0.03894 (2 places)   _____

6. 0.90039 (2 places)   _____

7. 0.72008 (4 places) _____ 9. 0.0003 (3 places) _____

8. 0.0006 (3 places) _____ 10. 0.099 (3 places) _____

Express Common Fractions as Decimal Fractions

Express the common fractions as decimal fractions. Express the answer to 4 decimal places.

11. $\frac{11}{16}$ _____ 15. $\frac{2}{3}$ _____ 19. $\frac{7}{32}$ _____

12. $\frac{7}{8}$ _____ 16. $\frac{10}{11}$ _____ 20. $\frac{1}{2}$ _____

13. $\frac{5}{8}$ _____ 17. $\frac{2}{25}$ _____ 21. $\frac{4}{7}$ _____

14. $\frac{3}{4}$ _____ 18. $\frac{47}{64}$ _____ 22. $\frac{3}{8}$ _____

Solve the following.

23. What decimal fraction of distance B is distance A? Express the answer to 4 decimal places. All dimensions are in inches. _____

24. Five pieces are cut from the length of round stock shown. After the pieces are cut, the remaining length is thrown away. What decimal fraction of the original length of round stock (17″) is the length which is thrown away? All dimensions are in inches. _____

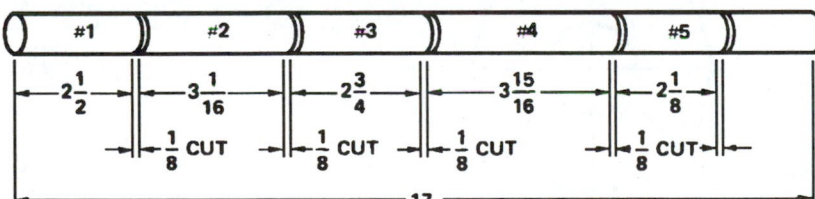

25. Dimensions in this figure are in feet and inches.

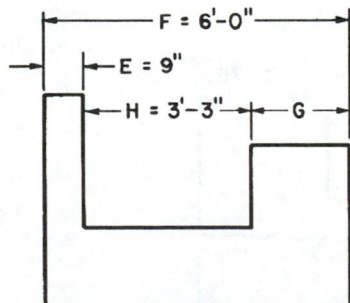

a. What decimal fraction of distance F is distance E? Note: Both the numerator and denominator of a common fraction must be in the same units before the value is expressed as a decimal fraction. Use 1 foot = 12 inches. _____

b. What decimal fraction of distance H is distance G? Express the answer to 4 decimal places. _____

Expressing Decimal Fractions as Common Fractions

Express the following decimal fractions as common fractions. Reduce to lowest terms.

26. 0.875 _____	33. 0.003 _____	40. 0.0005 _____
27. 0.125 _____	34. 0.008 _____	41. 0.03 _____
28. 0.4 _____	35. 0.502 _____	42. 0.09375 _____
29. 0.75 _____	36. 0.99 _____	43. 0.237 _____
30. 0.6 _____	37. 0.4375 _____	44. 0.45 _____
31. 0.6875 _____	38. 0.2113 _____	45. 0.045 _____
32. 0.67 _____	39. 0.8717 _____	46. 0.0045 _____

Solve the following.

47. What common fractional part of distance B is distance A? All dimensions are in inches.

_____

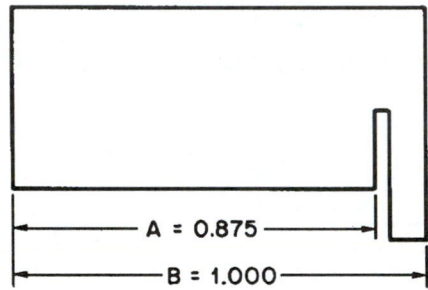

48. What common fractional part of diameter C is diameter D? All dimensions are in feet.

_____

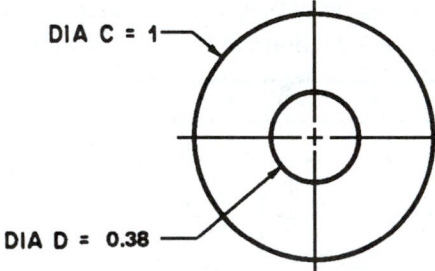

49. What common fractional part of distance A is each distance listed. All dimensions are in inches.

a. Distance B _____
b. Distance C _____
c. Distance D _____
d. Distance E _____
e. Distance F _____

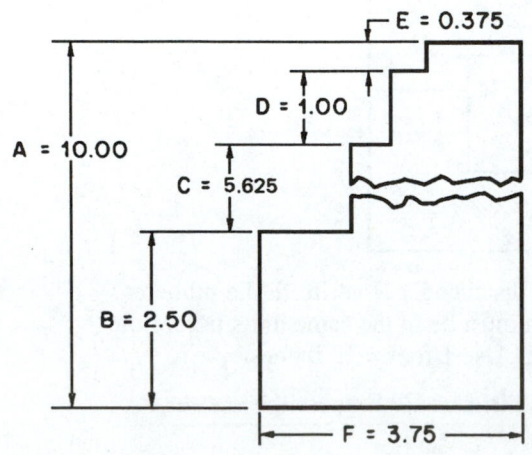

UNIT 10 Addition and Subtraction of Decimal Fractions

Objectives After studying this unit you should be able to

- **Add decimal fractions.**
- **Add combinations of decimals, mixed decimals, and whole numbers.**
- **Subtract decimal fractions.**
- **Subtract combinations of decimals, mixed decimals, and whole numbers.**

Adding and subtracting decimal fractions are required at various stages in the production of most products and parts. It is necessary to add and subtract decimals in order to estimate machining costs and production times, to compute stock allowances and tolerances, to determine locations and lengths of cuts, and to inspect finished parts.

Adding Decimal Fractions

Procedure To add decimal fractions

- Arrange the numbers so that the decimal points are directly under each other.
- Proceed with addition as with whole numbers.
- Place the decimal point in the sum directly under the other decimal points.

Example 1 Add. $7.35 + 114.075 + 0.3422 + 0.003 + 218.7$

➤ **Note:** To reduce the possibility of error, add zeros to decimals so that all the values have the same number of places to the right of the decimal point. Zeros added in this manner do not affect the value of the number.

Arrange the numbers so that the decimal points are directly under each other.

Proceed with addition as with whole numbers.

The decimal point of the sum is placed in the same position as the other decimal points.

$$
\begin{array}{r}
7.3500 \\
114.0750 \\
0.3422 \\
0.0030 \\
+ \ 218.7000 \\
\hline
340.4702 \quad \text{Ans}
\end{array}
$$

The decimal point location of a whole number is directly to the right of the last digit.

Example 2 Find the length x of the swivel bracket shown. All dimensions are in millimeters.

Add.
$$
\begin{array}{r}
8.78 \\
25.40 \\
12.80 \\
30.00 \\
3.90 \\
+ \ 9.25 \\
\hline
90.13
\end{array}
$$

$x = 90.13$ mm   Ans

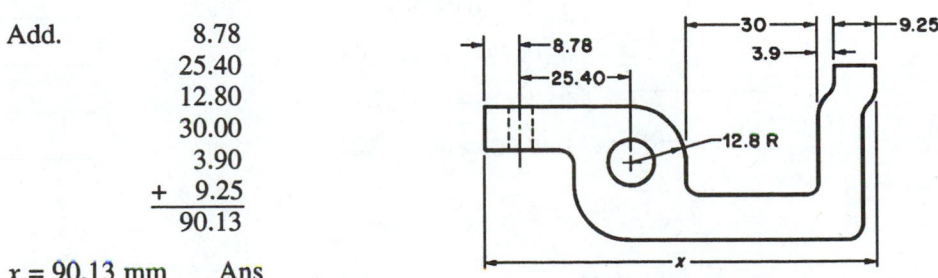

Subtracting Decimal Fractions

Procedure   To subtract decimal fractions

- Arrange the numbers so that the decimal points are directly under each other.
- Proceed with subtraction as with whole numbers.
- Place the decimal point in the difference under the other decimal points.

Example 1   Subtract 13.261 from 25.6.

Arrange the numbers so that the decimal points are directly under each other.

Add two zeros to 25.6 so that it has the same number of decimal places as 13.261. Place the decimal point of the answer in the same position as the other decimal points.

Subtract.

$$\begin{array}{r} 25.600 \\ -\ 13.261 \\ \hline 12.339 \end{array} \quad \text{Ans}$$

Example 2   Determine dimensions A, B, C, and D of the support bracket shown. All dimensions are given in inches.

Solve for A:
$A = 0.505 - 0.18$
$A = 0.325''$ Ans

$$\begin{array}{r} 0.505 \\ -\ 0.180 \\ \hline 0.325 \end{array}$$

Solve for B:
$B = 1.4 - 0.301$
$B = 1.099''$ Ans

$$\begin{array}{r} 1.400 \\ -\ 0.301 \\ \hline 1.099 \end{array}$$

Solve for C:
$C = 1.74 - 0.365$
$C = 1.375''$ Ans

$$\begin{array}{r} 1.740 \\ -\ 0.365 \\ \hline 1.375 \end{array}$$

Solve for D:
$D = 0.746 - 0.46$
$D = 0.286''$ Ans

$$\begin{array}{r} 0.746 \\ -\ 0.460 \\ \hline 0.286 \end{array}$$

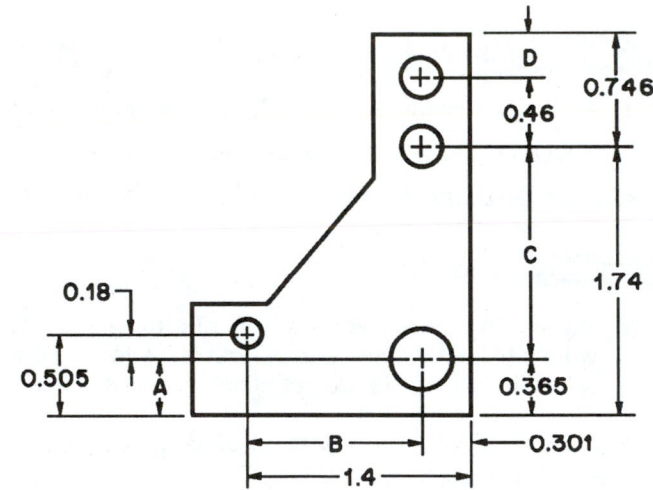

APPLICATION

Adding Decimal Fractions

1. Add the following numbers

   a. $0.375 + 10.4 + 5$ _____

   b. $0.003 + 0.13795$ _____

   c. $0.375 + 0.8 + 0.12$ _____

   d. $4.187 + 0.932 + 0.01$ _____

   e. $363.13 + 18.2 + 0.027$ _____

   f. $4 + 0.4 + 0.04 + 0.004$ _____

   g. $87 + 0.0239 + 7.23$ _____

   h. $0.0001 + 0.1 + 0.01$ _____

   i. $4.705 + 0.0937 + 0.98$ _____

   j. $0.063 + 4.9 + 324$ _____

2. Determine dimensions A, B, C, D, E, and F of the profile gage shown. All dimensions are in inches.

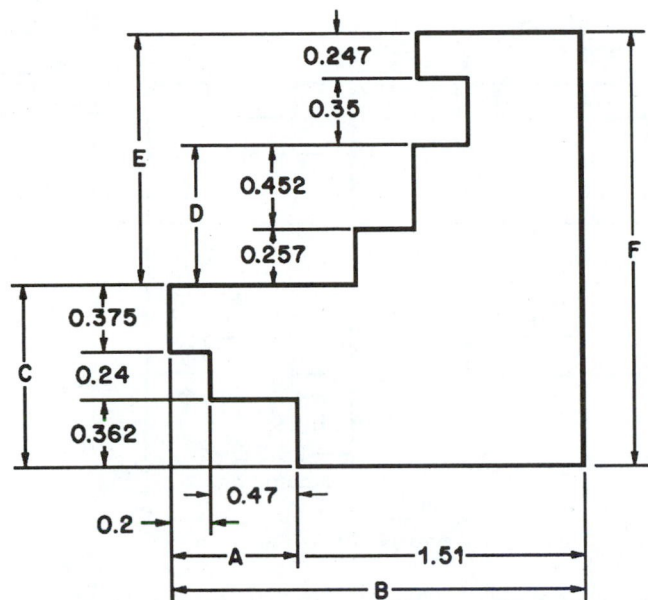

A = _____

B = _____

C = _____

D = _____

E = _____

F = _____

3. A sine plate is to be set to a desired angle by using size blocks of the following thicknesses: 3.000 inches, 0.500 inch, 0.250 inch. 0.125 inch, 0.100 inch, 0.1007 inch, and 0.1001 inch. Determine the total height that the sine plate is raised.

_____

4. Three cuts are required to turn a steel shaft. The depths of the cuts, in millimeters, are 6.25, 3.18, and 0.137. How much stock has been removed per side? Round answer to 2 decimal places.

_____

5. A thickness or feeler gauge is shown. Thickness gauges are widely used in manufacturing and machine service and repair occupations. Find the smallest combination of gauge leaves which total each of the following thicknesses: (More than one combination may total certain thicknesses.)

a. 0.014″   e. 0.011″

b. 0.033″   f. 0.042″

c. 0.021″   g. 0.029″

d. 0.038″   h. 0.049″

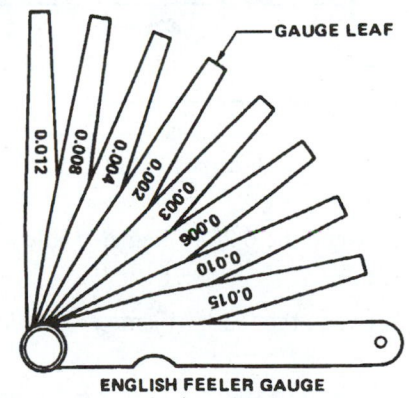

ENGLISH FEELER GAUGE

Subtracting Decimal Fractions

6. Subtract the following numbers. Where necessary, round answers to 3 decimal places.

a. $0.527 - 0.4136$ _____

b. $0.319 - 0.0127$ _____

c. $2.308 - 0.7859$ _____

d. $0.3 - 0.299$ _____

e. $0.4327 - 0.412$ _____

f. $23.062 - 0.973$ _____

g. $0.313 - 0.2323$ _____

h. $4.697 - 0.0002$ _____

i. $5.923 - 3.923$ _____

7. The front and right side views of a sliding shoe are shown. Determine dimensions A, B, C, D, E, and F. All dimensions are in millimeters.

A = _____

B = _____

C = _____

D = _____

E = _____

F = _____

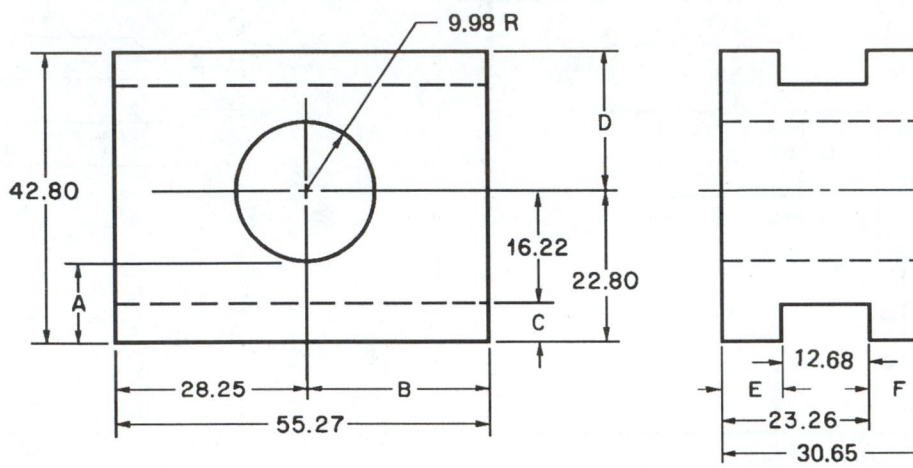

8. Refer to the plate shown and determine the following distances. All dimensions are in inches.

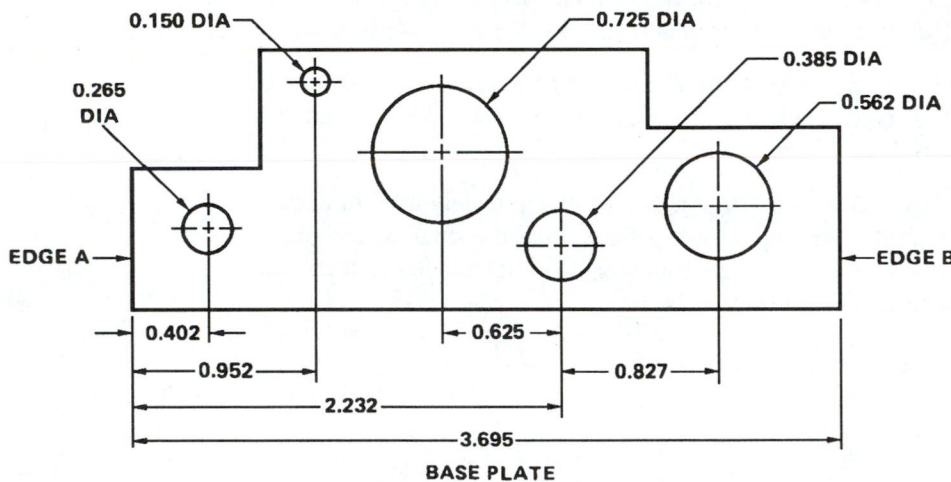

BASE PLATE

a. The center distance between the 0.265″ diameter hole and the 0.150″ diameter hole.

_____

b. The center distance between the 0.385″ diameter hole and the 0.150″ diameter hole.

_____

c. The distance between edge A and the center of the 0.725″ diameter hole.

_____

d. The distance between edge B and the center of the 0.385″ diameter hole.

_____

e. The distance between edge B and the center of the 0.562″ diameter hole.

_____

UNIT 11 Multiplication of Decimal Fractions

Objectives After studying this unit you should be able to

- **Multiply decimal fractions.**
- **Multiply combinations of decimals, mixed decimals, and whole numbers.**

A machinist must readily be able to multiply decimal fractions for computing machine feeds and speeds, for determining tapers, and for determining lengths and stock sizes. Multiplication of decimal fractions is also required in order to solve problems which involve geometry and trigonometry.

Multiplying Decimal Fractions

Procedure To multiply decimal fractions

- Multiply using the same procedure as with whole numbers.
- Beginning at the right of the product, point off the same number of decimal places as there are in the multiplicand and the multiplier combined.

Example 1 Multiply 50.123 by 0.87.

Multiply the same as with whole numbers.

Beginning at the right of the product, point off as many decimal places as there are in both the multiplicand and the multiplier.

Multiplicand $\rightarrow$	50.123 (3 places)
Multiplier $\rightarrow$	$\times 0.87$ (2 places)
	3 50861
	40 0984
Product $\rightarrow$	43.60701 (5 places) Ans

Example 2 Compute the lengths of thread on each end of this shaft. All dimensions are in inches.

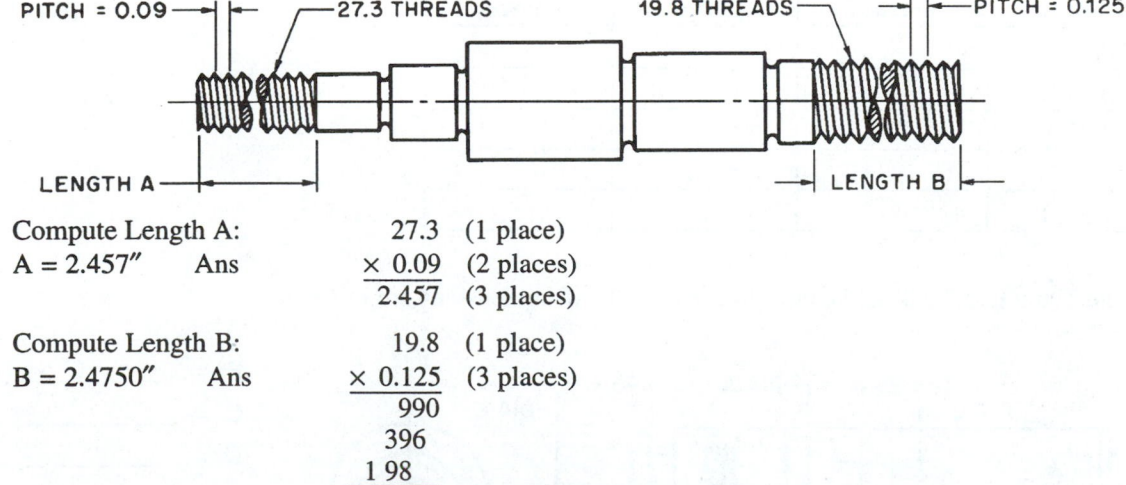

Compute Length A:

A = 2.457″   Ans

$$\begin{array}{r} 27.3 \ (1 \text{ place}) \\ \times \ 0.09 \ (2 \text{ places}) \\ \hline 2.457 \ (3 \text{ places}) \end{array}$$

Compute Length B:

B = 2.4750″   Ans

$$\begin{array}{r} 19.8 \ (1 \text{ place}) \\ \times \ 0.125 \ (3 \text{ places}) \\ \hline 990 \\ 396 \\ 1\ 98 \\ \hline 2.4750 \ (4 \text{ places}) \end{array}$$

When multiplying certain decimal fractions, the product has a smaller number of digits than the number of decimal places required. For these products add as many zeros to the left of the product as are necessary to give the required number of decimal places.

Example Multiply 0.0237 by 0.04. Round the answer to 5 decimal places.

The multiplicand, 0.0237, has four decimal places, and the multiplier, 0.04, has two decimal places. Therefore, the product must have six decimal places.

Add three zeros to the left of the product. Round 0.000948 to 5 places.

Multiply.

$$\begin{array}{r} 0.0237 \quad \text{(4 places)} \\ \times \quad 0.04 \quad \text{(2 places)} \\ \hline 0.000948 \quad \text{(6 places)} \end{array}$$

0.00095   Ans

APPLICATION

Multiplying Decimal Fractions

1. Multiply these numbers. Where necessary, round the answers to 4 decimal places.

   a. 4.693×0.012 _____   c. 40×0.15 _____

   b. 2.2×1.5 _____   d. 6.43×0.26 _____

2. A section of a spur gear is shown. Given the circular pitches for various gear sizes, determine the working depths, clearances, and tooth thicknesses. Round the answers to 4 decimal places.

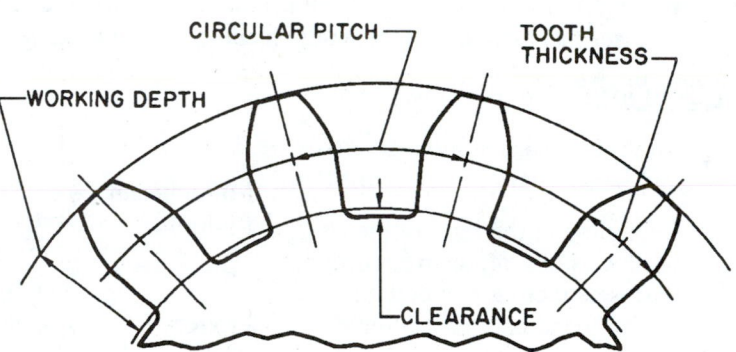

Working depth = 0.6366 × Circular Pitch
Clearance = 0.05 × Circular Pitch
Tooth thickness = 0.5 × Circular Pitch

	Circular Pitch (inches)	Working Depth (inches)	Clearance (inches)	Tooth Thickness (inches)
a.	0.3925			
b.	0.1582			
c.	0.8069			
d.	1.2378			
e.	1.5931			

3. Determine diameters A, B, C, D, and E of this shaft. All dimensions are in millimeters.

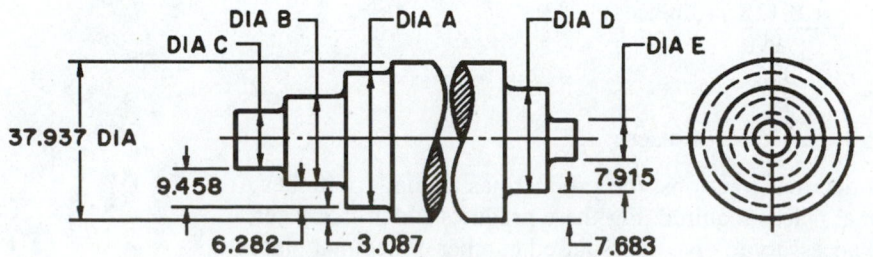

A = _____

B = _____

C = _____

D = _____

E = _____

4. Determine dimension *x* for each of these figures.

a. All dimensions are in inches.

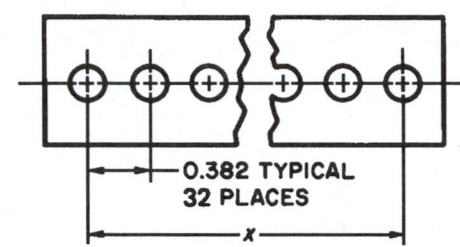

0.382 TYPICAL
32 PLACES

x

b. All dimensions are in millimeters.

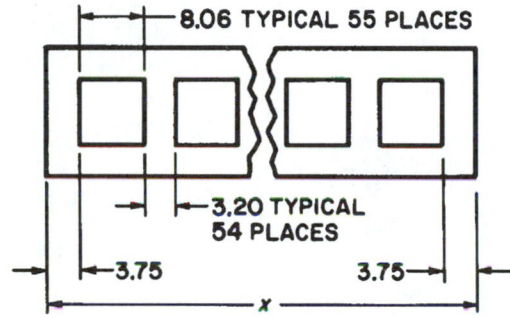

8.06 TYPICAL 55 PLACES

3.20 TYPICAL
54 PLACES

3.75   3.75

x

c. Round the answer to 3 decimal places. All dimensions are in inches.

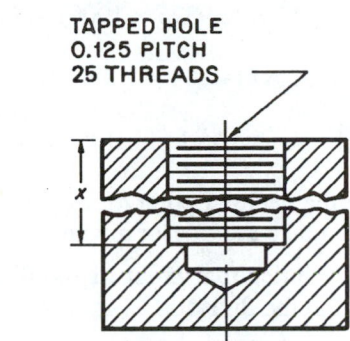

TAPPED HOLE
0.125 PITCH
25 THREADS

x

d. Round the answer to 3 decimal places. All dimensions are in inches.

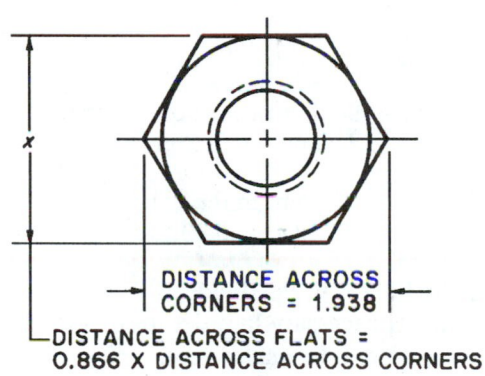

DISTANCE ACROSS
CORNERS = 1.938

DISTANCE ACROSS FLATS =
0.866 X DISTANCE ACROSS CORNERS

UNIT **12** Division of Decimal Fractions

Objectives   **After studying this unit you should be able to**

- **Divide decimal fractions.**
- **Divide decimal fractions with whole numbers.**
- **Divide decimal fractions with mixed decimals.**

Division with decimal fractions is used for computing the manufacturing cost and time per piece after total production costs and times have been determined. Division with decimal fractions is also required in order to compute thread pitches, gear tooth thicknesses and depths, cutting speeds, and depths of cut.

Dividing Decimal Fractions

Moving a decimal point to the right is equivalent to multiplying the decimal by a power of 10.

$0.237 \times 10 = 2.37$ $0.237 \times 1000 = 237.$

$0.237 \times 100 = 23.7$ $0.237 \times 10,000 = 2370.$

When dividing decimal fractions, the value of the answer (quotient) is not changed if the decimal points of both the divisor and the dividend are moved the same number of places to the right. It is the same as multiplying both divisor and dividend by the same number.

$$0.9375 \div 0.612 = (0.9375 \times 1000) \div (0.612 \times 1000) = 937.5 \div 612.$$

$$14.203 \div 6.87 = (14.203 \times 100) \div (6.87 \times 100) = 1420.3 \div 687.$$

Procedure To divide decimal fractions

- Move the decimal point of the divisor as many places to the right as are necessary to make the divisor a whole number.

- Move the decimal point of the dividend the same number of places as were moved in the divisor.

- Place the decimal point in the quotient directly above the decimal point in the dividend.

- Add zeros to the dividend if necessary.

- Divide as with whole numbers.

Example 1 Divide 0.643 by 0.28. Round the answer to 3 decimal places.

To make the divisor a whole number move the decimal point 2 places, 28.

The decimal point in the dividend is also moved 2 places, 64.3.

Add 3 zeros to the dividend. One extra place is necessary in order to round the answer to 3 decimal places.

Place the decimal point of the quotient directly above the decimal point of the dividend.

$$
\begin{array}{r}
2.2964 \approx 2.296 \quad \text{Ans} \\
28\,)\,\overline{64.3000} \\
\underline{56} \\
8\,3 \\
\underline{5\,6} \\
2\,70 \\
\underline{2\,52} \\
180 \\
\underline{168} \\
120 \\
\underline{112} \\
8
\end{array}
$$

Divide as with whole numbers.

Example 2 3.19 ÷ 0.072 (Round the answer to 2 decimal places.)

Move the decimal point 3 places in the divisor, and 3 places in the dividend.

Add 3 zeros to the dividend.

Place the decimal point of the quotient directly above the decimal point of the dividend.

Divide.

$$
\begin{array}{r}
44.305 \approx 44.31 \quad \text{Ans} \\
72\,)\,\overline{3190.000} \\
\underline{288} \\
310 \\
\underline{288} \\
22\,0 \\
\underline{21\,6} \\
400 \\
\underline{360} \\
40
\end{array}
$$

When dividing a decimal fraction or a mixed decimal by a whole number, it is not necessary to move the decimal point of either the divisor or the dividend. Add zeros to the right of the dividend, if necessary, to obtain the desired number of decimal places in the answer.

Examples

1. Divide 0.63 by 12 to 4 decimal places.

$$12 \overline{)\, 0.6300}^{\;0.0525} \quad \text{Ans}$$

2. Divide 33.97 by 5 to 3 decimal places.

$$5 \overline{)\, 33.970}^{\;6.794} \quad \text{Ans}$$

APPLICATION

Dividing Decimal Fractions

1. Divide the following numbers. Express the answers to the indicated number of decimal places.

   a. $0.69 \div 0.432$ (3 places) _____

   b. $0.92 \div 0.36$ (2 places) _____

   c. $0.001 \div 0.1$ (4 places) _____

   d. $10 \div 0.001$ (3 places) _____

   e. $1.023 \div 0.09$ (3 places) _____

   f. $\dfrac{16.3}{3.8}$ (2 places) _____

   g. $\dfrac{37}{0.273}$ (2 places) _____

   h. $\dfrac{0.005}{0.81}$ (4 places) _____

2. Rack sizes are given according to diametral pitch. Given 4 different diametral pitches, find the linear pitch and the whole depth of each rack to 4 decimal places. All dimensions are in inches.

   $$\text{Linear Pitch} = \frac{3.1416}{\text{Diametral Pitch}} \qquad \text{Whole Depth} = \frac{2.157}{\text{Diametral Pitch}}$$

	Diametral Pitch	Linear Pitch	Whole Depth
a.	6.75		
b.	2.75		
c.	7.25		
d.	16.125		

   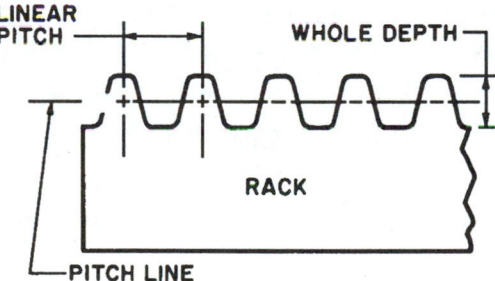

3. Four sets of equally spaced holes are shown in this machined plate. Determine dimensions A, B, C, and D to 2 decimal places. All dimensions are in millimeters.

   A = _____
   B = _____
   C = _____
   D = _____

   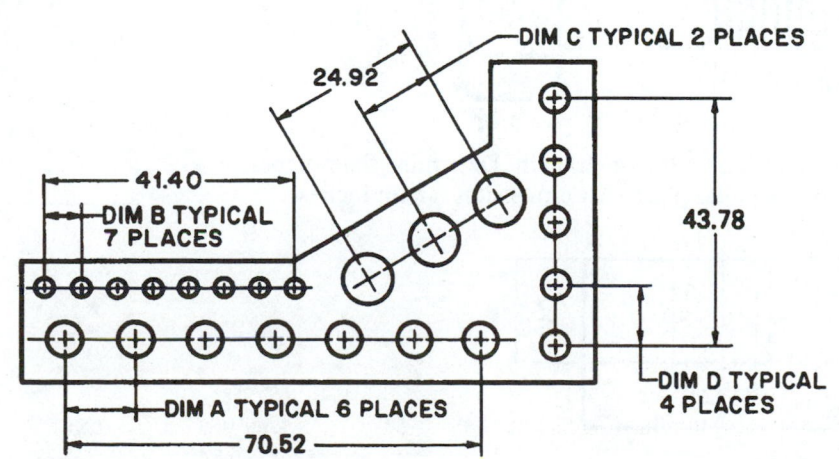

4. A cross-sectional view of a bevel gear is shown. Given the diametral pitch and the number of gear teeth, determine the pitch diameter, the addendum, and the dedendum. Round the answers to 4 decimal places.

$$\text{Pitch Diameter} = \frac{\text{Number of Teeth}}{\text{Diametral Pitch}}$$

$$\text{Dedendum} = \frac{1.1570}{\text{Diametral Pitch}}$$

$$\text{Addendum} = \frac{1}{\text{Diametral Pitch}}$$

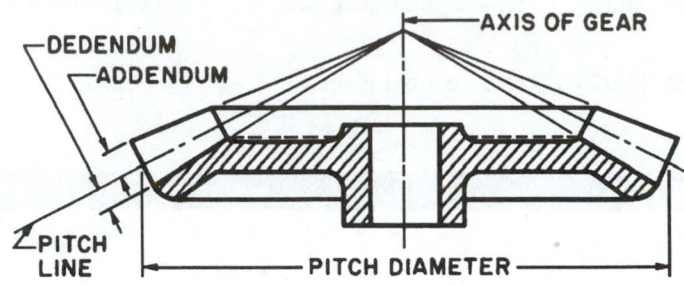

	Diametral Pitch	Number of Teeth	Pitch Diameter (inches)	Addendum (inches)	Dedendum (inches)
a.	4	45			
b.	6	75			
c.	8	44			
d.	3	54			

5. How many complete bushings each 14.60 millimeters long can be cut from a bar of bronze which is 473.75 millimeters long? Allow 3.12 millimeters waste for each piece. _____

6. A shaft is being cut in a lathe. The tool feeds (advances) 0.015 inch each time the shaft turns once (1 revolution). How many revolutions will the shaft turn when the tool advances 3.120 inches? Round the answer to 2 decimal places. _____

7. How much stock per stroke is removed by the wheel of a surface grinder if a depth of 4.725 millimeters is reached after 75 strokes? Round the answer to 3 decimal places. _____

8. An automatic screw machine is capable of producing one piece in 0.02 minute. How many pieces can be produced in 1.25 hours? _____

9. This bolt has 7.7 threads. Determine the pitch to 3 decimal places. All dimensions are in inches. _____

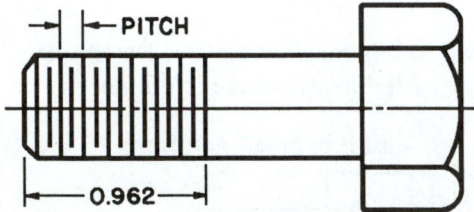

10. This block has a threaded hole with a 0.0625-inch pitch. Determine the number of threads for the given depth to 1 decimal place. All dimensions are in inches. _____

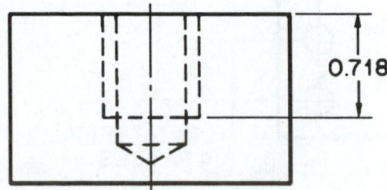

11. The length of a side of a square equals the distance from point A to point B divided by 1.4142. Determine the length of a side of this square plate to 2 decimal places. All dimensions are in millimeters.

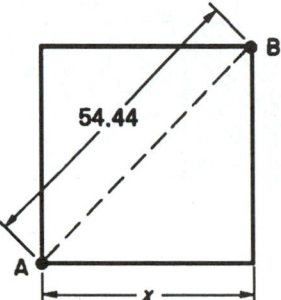

UNIT **13** Powers

Objectives After studying this unit you should be able to

- **Raise numbers to indicated powers.**

- **Solve problems which involve combinations of powers with other basic operations.**

Powers of numbers are used to compute areas of square plates and circular sections and to compute volumes of cubes, cylinders, and cones. Use of powers is particularly helpful in determining distances in problems which require applications of geometry and trigonometry.

Description of Powers

Two or more numbers multiplied to produce a given number are *factors* of the given number. Two factors of 8 are 2 and 4. The factors of 15 are 3 and 5. A *power* is the product of two or more equal factors. The third power of 5 is $5 \times 5 \times 5$ or 125. An *exponent* shows how many times a number is taken as a factor. It is written smaller than the number, above the number, and to the right of the number. The expression 3^2 means 3×3. The exponent 2 shows that 3 is taken as a factor twice. It is read as 3 to the second power or 3 squared.

Examples Find the indicated powers.

1. 2^5 Two to the fifth power means $2 \times 2 \times 2 \times 2 \times 2$ or 32.   Ans

2. 3^3 Three cubed means $3 \times 3 \times 3$ or 27.   Ans

3. 0.72^2 0.72 squared means 0.72×0.72 or 0.5184.   Ans

$A = s^2$ is called a *formula*. A formula is a short method of expressing an arithmetic relationship by the use of symbols. Known values may be substituted for the symbols and other values can be found.

Example 1   Determine the area of the square shown. The area of a square equals the length of a side squared. The answer is given in square units. All dimensions are in inches.

$$A = s^2$$

$$A = \left(\frac{7}{8}\text{ in}\right)^2$$

$$A = \frac{7}{8}\text{ in} \times \frac{7}{8}\text{ in}$$

$$A = \frac{49}{64}\text{ sq in} \quad \text{Ans}$$

SIDE = $\frac{7}{8}$

SIDE = $\frac{7}{8}$

Example 2   Find the volume of the cube shown. The volume of a cube equals the length of a side cubed. The answer is given in cubic units. All dimensions are in millimeters. Round answer to 1 decimal place.

$$V = s^3$$
$$V = (1.6\text{ mm})^3$$
$$V = 1.6\text{ mm} \times 1.6\text{ mm} \times 1.6\text{ mm}$$
$$V = 4.096\text{ mm}^3 \text{ or } 4.1\text{ mm}^3 \quad \text{Ans}$$

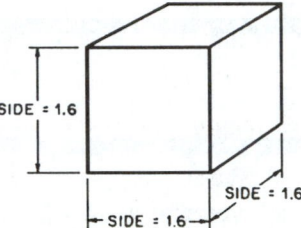

SIDE = 1.6

SIDE = 1.6

SIDE = 1.6

Use of Parentheses

In this example, only the numerator is squared.

$$\frac{2^2}{3} = \frac{2 \times 2}{3} = \frac{4}{3} = 1\frac{1}{3}$$

In this example, only the denominator is squared.

$$\frac{2}{3^2} = \frac{2}{3 \times 3} = \frac{2}{9}$$

Parentheses are used in grouping symbols. Parentheses indicate that both the numerator and the denominator of a fraction are raised to the given power.

$$\left(\frac{2}{3}\right)^2 = \frac{2^2}{3^2} = \frac{2 \times 2}{3 \times 3} = \frac{4}{9}$$

Procedure   To solve problems which involve operations within parentheses

- Perform the operations within the parentheses.
- Raise to the indicated power.

Examples

1. $(1.2 \times 0.6)^2 = 0.72^2 = 0.72 \times 0.72 = 0.5184 \quad \text{Ans}$
2. $(0.5 + 2.4)^2 = 2.9^2 = 2.9 \times 2.9 = 8.41 \quad \text{Ans}$
3. $(0.75 - 0.32)^2 = 0.43^2 = 0.43 \times 0.43 = 0.1849 \quad \text{Ans}$
4. $\left(\frac{14.4}{3.2}\right)^2 = 4.5^2 = 4.5 \times 4.5 = 20.25 \quad \text{Ans}$

When solving power problems which also require addition, subtraction, multiplication, or division, perform the power operation first.

Examples

1. $5 \times 3^2 - 12 = 5 \times 9 - 12 = 45 - 12 = 33$ Ans

2. $33.5 - 5.5^2 + 8.7 = 33.5 - 30.25 + 8.7 = 11.95$ Ans

3. $\frac{2.2^3 - 5.608}{1.4} = \frac{10.648 - 5.608}{1.4} = \frac{5.040}{1.4} = 3.6$ Ans

The symbol π (pi) represents a constant value used in mathematical relationships involving circles. Depending upon the specific problem to be solved, generally, the value of pi used is $3\frac{1}{7}$, 3.14, or 3.1416.

Example

Compute the volume of the cylinder shown to 2 decimal places. The answer is given in cubic units. All dimensions are in inches.

$V = \pi \times r^2 \times h$
$V = 3.14 \times (0.85 \text{ in})^2 \times 1.25 \text{ in}$
$V = 3.14 \times 0.7225 \text{ sq in} \times 1.25 \text{ in}$
$V = 2.8358 \text{ cu in}, 2.84 \text{ cu in}$ Ans

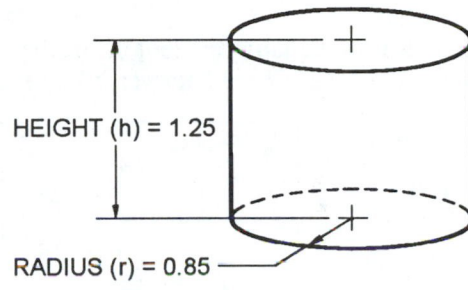

HEIGHT (h) = 1.25

RADIUS (r) = 0.85

Many problems require the application of the same formula more than once or the application of 2 different formulas in the solutions.

Example

Find the metal area of this square plate. Round the answer to 2 decimal places. All dimensions are in inches. $A = s^2$

The metal area equals the area of the large square minus the area of the removed square.

$A_1 = (5.250 \text{ in})^2$
$A_1 = 27.5625 \text{ sq in}$
$A_2 = (2.500 \text{ in})^2$
$A_2 = 6.2500 \text{ sq in}$
$A_3 = 27.5625 \text{ sq in} - 6.2500 \text{ sq in} = 21.31 \text{ sq in}$ Ans

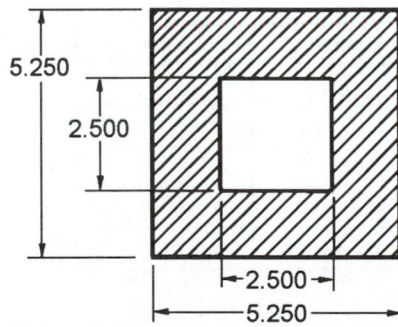

5.250

2.500

2.500

5.250

APPLICATION

Raising a Number to a Power

Raise the following numbers to the indicated power.

1. 3.4^3 _____

2. 1^8 _____

3. 100^4 _____

4. $\left(\frac{2}{3}\right)^3$ _____

5. $\frac{2}{3}^3$ _____

6. $\frac{3}{4}^3$ _____

7. $(0.3 \times 7)^2$ _____

8. $(20.7 + 7.2)^2$ _____

9. $\left(\frac{28.8}{7.2}\right)^3$ _____

Related Problems

In the following table the lengths of the sides of squares are given. Determine the areas of the squares. Round the answers to 2 decimal places where necessary.

	Side	Area			Side	Area
10.	1.25 in			15.	$\frac{3}{4}$ in	
11.	23.070 mm			16.	$\frac{7}{8}$ in	
12.	0.17 in			17.	$3\frac{3}{4}$ in	
13.	10.70 mm			18.	$\frac{13}{16}$ in	
14.	0.02 in			19.	$13\frac{3}{4}$ in	

$A = s^2$ where A = area
s = side

In the following table the lengths of the sides of cubes are given. Determine the volumes of the cubes. Round answers to 2 decimal places where necessary.

	Side	Volume			Side	Volume
20.	0.29 in			25.	$\frac{1}{3}$ in	
21.	20.60 mm			26.	$\frac{7}{8}$ in	
22.	3.930 in			27.	$1\frac{1}{2}$ in	
23.	14.00 mm			28.	$9\frac{1}{8}$ in	
24.	0.075 in			29.	$\frac{3}{4}$ in	

$V = s^3$ where V = volume
s = side

In the following table the radii of circles are given. Determine the areas of the circles. Round the answers to the nearest whole number.

	Radius	Area
30.	16.20 mm	
31.	15.60 mm	
32.	0.07 in	
33.	9.28 in	
34.	12.35 mm	

$A = \pi \times R^2$ where A = area
$\pi = 3.14$
R = radius

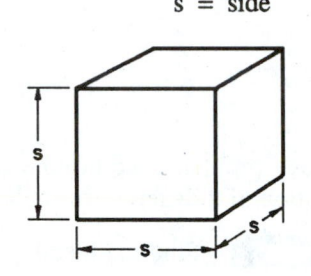

In the following table the diameters of spheres are given. Determine the volumes of the spheres. Round the answers to 1 decimal place where necessary.

	Diameter	Volume
35.	0.65 in	
36.	6.500 mm	
37.	0.75 in	
38.	10.80 mm	
39.	7.060 mm	

$V = \dfrac{\pi \times D^3}{6}$ where V = volume
$\pi = 3.14$
D = diameter

In the following table the radii and heights of cylinders are given. Determine the volumes of the cylinders. Round the answers to the nearest whole number.

	Radius	Height	Volume
40.	5.00 mm	3.20 mm	
41.	1.50 in	2.30 in	
42.	2.25 in	3.00 in	
43.	0.70 in	6.70 in	
44.	7.81 mm	6.72 mm	

$V = \pi \times R^2 \times H$ where V = volume
π = 3.14
R = radius
H = height

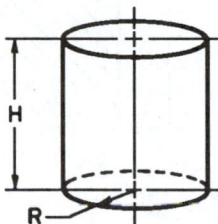

In the following table the diameters and heights of cones are given. Find the volumes of the cones. Round the answers to the nearest whole number.

	Diameter	Height	Volume
45.	3.20 in	4.00 in	
46.	3.00 in	5.00 in	
47.	10.60 mm	13.10 mm	
48.	9.90 mm	6.20 mm	
49.	0.37 in	0.96 in	

$V = 0.2618 \times D^2 \times H$ where V = volume
D = diameter
H = height

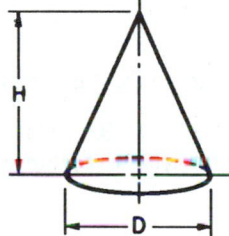

Solve the following problems. Use $\pi = 3.14$. Round answers to the nearest whole number.

50. Find the metal area of this washer. All dimensions are in millimeters.

$A = \pi \times R^2$

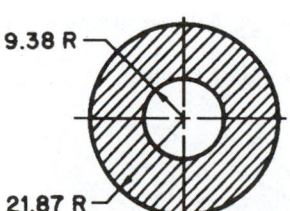

9.38 R

21.87 R

51. Find the metal area of this spacer. All dimensions are in millimeters.

Area of Square = s^2
Area of Circle = $\pi \times R^2$

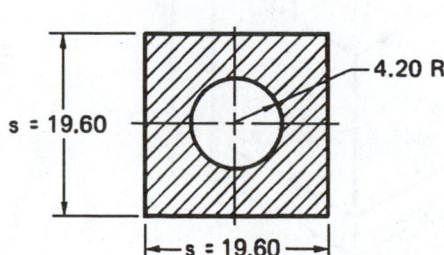

s = 19.60

4.20 R

s = 19.60

52. Find the area of this plate. All dimensions are in millimeters. _____

   ➤ **Hint:** The broken lines indicate one method of solution.

   $A = s^2$

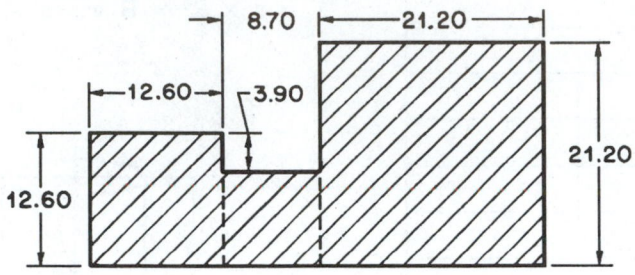

53. Find the metal volume of this bushing. All dimensions are in inches. _____

   $V = \pi \times R^2 \times H$

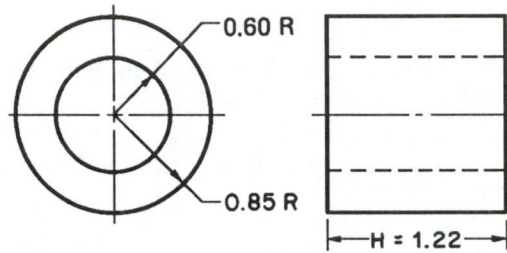

54. Find the volume of this pin. All dimensions are in inches. _____

   Volume of cylinder $= \pi \times R^2 \times H$
   Volume of cone $= 0.2618 \times D^2 \times H$

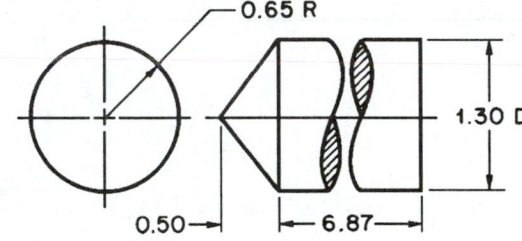

55. A materials estimator finds the weight of aluminum needed for the casting shown. Aluminum weighs 0.0975 pound per cubic inch. Find, to the nearer pound, the weight of aluminum required for 15 castings. All measurements are in inches. _____

   ➤ **Hint:** The broken line indicates a method of solution.

   $V = s^3$

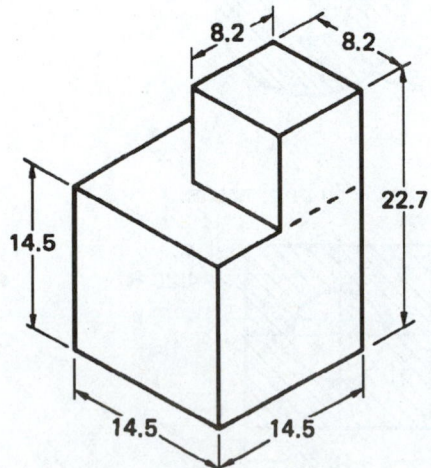

UNIT **14** Roots

Objectives **After studying this unit you should be able to**

- **Extract whole number roots.**
- **Determine square roots to any indicated number of decimal places.**
- **Solve problems which involve combinations of roots with other basic arithmetic operations.**

The operation of extracting roots of numbers is used to determine lengths of sides and heights of squares and cubes and radii of circular sections when areas and volumes are known. The machinist uses roots in computing distances between various parts of machined pieces from given blueprint dimensions.

Description of Roots

The *root* of a number is a quantity which is taken two or more times as an equal factor of the number. Determining a root is the opposite operation of determining a power. The *radical symbol* ($\sqrt{}$) is used to indicate a root of a number. The *index* is written smaller than the number, to the left and above the radical symbol. The index indicates the number of times that a root is to be taken as an equal factor to produce the given number. The index 2 is omitted for an indicated square root. For example, the square root of 9 is written $\sqrt{9}$. The expression $\sqrt{9}$ means to find the number which can be multiplied by itself and equal 9. Since $3 \times 3 = 9$, 3 is the square root of 9.

Examples   Find the indicated roots.

1. $\sqrt{36}$ Since $6 \times 6 = 36$, the square root of 36 is 6.   Ans

2. $\sqrt{144}$ Since $12 \times 12 = 144$, the square root of 144 is 12.   Ans

3. $\sqrt[3]{8}$ Since $2 \times 2 \times 2 = 8$, the cube root of 8 is 2.   Ans

4. $\sqrt[3]{125}$ Since $5 \times 5 \times 5 = 125$, the cube root of 125 is 5.   Ans

5. $\sqrt[4]{81}$ Since $3 \times 3 \times 3 \times 3 = 81$, the fourth root of 81 is 3.   Ans

Roots must be extracted in determining unknown dimensions represented in certain formulas.

Example 1   Compute the length of the side of the square shown. This square has an area of 25 square inches.

Since $A = s^2$, the length of a side of the square equals the square root of the area.

$s = \sqrt{A}$

$s = \sqrt{25 \text{ sq in}}$

$s = \sqrt{5 \text{ in} \times 5 \text{ in}}$

$s = 5 \text{ inches}$ Ans

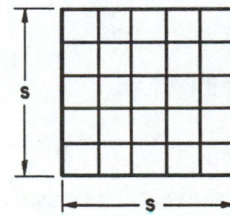

Example 2   Compute the length of the side of the cube shown. The volume of this cube equals 64 cubic inches.

$$s = \sqrt[3]{V}$$

$$s = \sqrt[3]{64 \ \text{cu in}}$$

$$s = \sqrt[3]{4 \ \text{in} \times 4 \ \text{in} \times 4 \ \text{in}}$$

$$s = 4 \ \text{inches} \qquad \text{Ans}$$

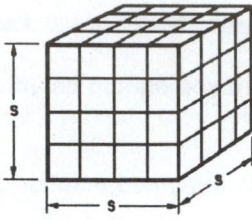

Roots of Fractions

In this example, only the root of the numerator is taken.

$$\frac{\sqrt{16}}{25} = \frac{\sqrt{4 \times 4}}{25} = \frac{4}{25}$$

In this example, only the root of the denominator is taken.

$$\frac{16}{\sqrt{25}} = \frac{16}{\sqrt{5 \times 5}} = \frac{16}{5} = 3\frac{1}{5}$$

A radical sign which encloses a fraction indicates that the roots of both the numerator and denominator are to be taken. The same answer is obtained by extracting both roots first and dividing second as by dividing first and extracting the root second.

Example   Find $\sqrt{\dfrac{36}{9}}$.

Method 1:   Extract both roots then divide.

$$\sqrt{\frac{36}{9}} = \frac{\sqrt{36}}{\sqrt{9}} = \frac{6}{3} = 2 \qquad \text{Ans}$$

Method 2:   Divide then extract the root.

$$\sqrt{\frac{36}{9}} = \sqrt{4} = 2 \qquad \text{Ans}$$

Expressions Enclosed Within the Radical Symbol

The radical symbol is a grouping symbol. An expression consisting of operations within the radical symbol is done using the order of operations.

Procedure   To solve problems which involve operations within the radical symbol

- Perform the operations within the radical symbol first using the order of operations.
- Then find the root.

Examples   Find the indicated roots.

1. $\sqrt{3 \times 12} = \sqrt{36} = \sqrt{6 \times 6} = 6 \qquad \text{Ans}$

2. $\sqrt{5 + 59} = \sqrt{64} = \sqrt{8 \times 8} = 8 \qquad \text{Ans}$

3. $\sqrt{128 - 7} = \sqrt{121} = \sqrt{11 \times 11} = 11 \qquad \text{Ans}$

Problems involving formulas may involve operations within a radical symbol.

Example   Compute the length of the chord of the circular segment shown. All dimensions are in inches.

$C = 2 \times \sqrt{H \times (2 \times R - H)}$

$C = 2 \times \sqrt{1.5 \times (2 \times 3.75 - 1.5)}$

$C = 2 \times \sqrt{1.5 \times 6}$

$C = 2 \times \sqrt{9}$

$C = 2 \times 3$

$C = 6$

Length of chord = 6 inches   Ans

$C = 2 \times \sqrt{H \times (2 \times R - H)}$

where C = length of chord

H = height of segment

R = radius of circle

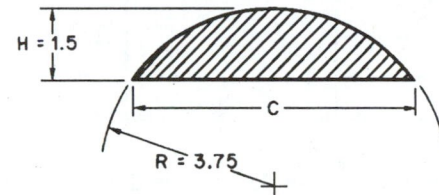

Roots That Are Not Whole Numbers

The root examples and exercises have all consisted of numbers that have whole number roots. These roots are relatively easy to determine by observation.

Most numbers do not have whole number roots. For example, $\sqrt{259} = 16.0935$ (rounded to 4 decimal places) and $\sqrt[3]{17.86} = 2.6139$ (rounded to 4 decimal places). The root of any positive number can easily be computed with a calculator. Calculator solutions to root expressions are given at the end of this unit on page 70.

APPLICATION

Radicals That Are Whole Numbers

The following problems have either whole number roots or numerators and denominators which have whole number roots. Determine these roots.

1. $\sqrt[3]{216}$ _____

2. $\sqrt{\frac{4}{9}}$ _____

3. $\frac{\sqrt{4}}{9}$ _____

4. $\frac{25}{\sqrt{36}}$ _____

5. $\sqrt{\frac{3}{4} \times \frac{3}{4}}$ _____

6. $\sqrt{0.5 \times 18}$ _____

7. $\sqrt{56.7 + 87.3}$ _____

8. $\sqrt{16.4 - 7.4}$ _____

9. $\sqrt[3]{\frac{428.8}{6.7}}$ _____

The following problems have whole number square roots. Solve for the missing values in the tables.

10. The areas of squares are given in the following table. Determine the lengths of the sides.

	Area (A)	Side (s)
a.	225 mm²	
b.	121 mm²	
c.	64 mm²	
d.	81 sq in	
e.	49 sq in	

$s = \sqrt{A}$

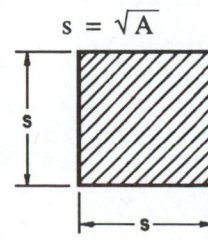

11. The volumes of cubes are given in the following table. Determine the lengths of the sides.

	Volume (V)	Side (s)
a.	216 mm³	
b.	64 cu in	
c.	512 cu in	
d.	1000 mm³	
e.	1 cu in	

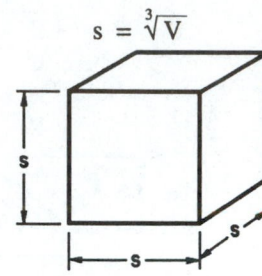

$$s = \sqrt[3]{V}$$

12. The areas of circles are given in this table. Determine the lengths of the radii. Use $\pi = 3.14$.

	Area (A)	Radius (R)
a.	50.24 sq in	
b.	12.56 sq in	
c.	314 mm²	
d.	28.26 sq in	
e.	153.86 mm²	

$$R = \sqrt{\frac{A}{\pi}}$$

13. The volumes of spheres are given in this table. Determine the lengths of the diameters.

	Volume (V)	Diameter (D)
a.	14.1372 cu in	
b.	113.0976 mm³	
c.	4.1888 cu in	
d.	0.5236 cu in	
e.	523.6 mm³	

$$D = \sqrt[3]{\frac{V}{0.5236}}$$

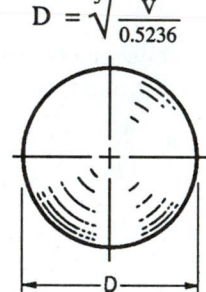

Radicals That Are Not Whole Numbers

The following problems have square roots that are not whole numbers. They require calculator computations. Refer to page 70 for calculator root solutions. Compute these roots to the indicated number of decimal places.

14. $\sqrt{15.63}$ (3 places) _____

15. $\sqrt{391}$ (2 places) _____

16. $\sqrt{\frac{3}{5}}$ (3 places) _____

17. $\sqrt{3\frac{1}{2}}$ (3 places) _____

18. $\sqrt{0.07 \times 28}$ (2 places) _____

19. $\sqrt{15.82 + 3.71}$ (2 places) _____

20. $\sqrt{178.5 - 163.7}$ (3 places) _____

21. $\sqrt{\frac{0.441}{60}}$ (4 places) _____

The following problems have roots that are not whole numbers. Solve for the missing values in the tables.

22. The volumes of cylinders and their heights are given in the following table. Find the lengths of the radii to 2 decimal places. Use $\pi = 3.14$.

	Volume (V)	Height (H)	Radius (R)
a.	249.896 mm³	7.00 mm	
b.	132.634 mm³	12.00 mm	
c.	14.00 cu in	29.00 in	
d.	10.00 cu in	28.00 in	

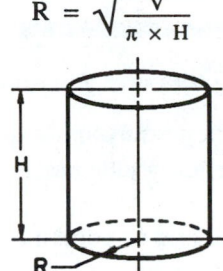

$$R = \sqrt{\frac{V}{\pi \times H}}$$

23. The volumes of cones and their heights are given in the following table. Compute the lengths of the diameters to 2 decimal places.

	Volume (V)	Height (H)	Diameter (D)
a.	116.328 mm³	8.00 mm	
b.	19.388 cu in	2.00 in	
c.	1257.6 mm³	10.00 mm	
d.	15 cu in	50.00 in	

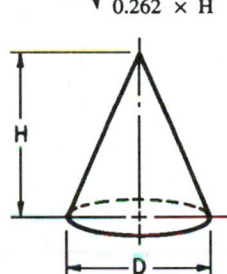

$$D = \sqrt{\frac{V}{0.262 \times H}}$$

Solve the following problems.

24. The pitch of broach teeth depends upon the length of cut, the depth of cut, and the material being broached.

$$\text{Minumum Pitch} = 3 \times \sqrt{L \times d \times F}$$

where
L = length of cut
d = depth of cut
F = a factor related to the type of material being broached

Find the minimum pitch, to 3 decimal places, for broaching cast iron where $L = 0.825''$, $d = 0.007''$, and $F = 5$.

_____

25. The dimensions of keys and keyways are determined in relation to the diameter of the shafts with which they are used.

$$D = \sqrt{\frac{L \times T}{0.3}}$$

where
D = shaft diameter
L = key length
T = key thickness

What is the shaft diameter that would be used with a key where $L = 2.70''$ and $T = 0.25''$?

_____

UNIT 15 Table of Decimal Equivalents and Combined Operations of Decimal Fractions

Objectives After studying this unit you should be able to

- Write decimal or fraction equivalents using a decimal equivalent table.
- Determine nearer fraction equivalents of decimals by using the decimal equivalent table.
- Solve problems consisting of combinations of operations by applying the order of operations.

Generally, fractional blueprint dimensions are given in multiples of 64ths of an inch. A machinist is often required to express these fractional dimensions as decimal equivalents for machine settings. When laying out parts such as castings that have ample stock allowances, it is sometimes convenient to use a fractional steel scale and to express decimal dimensions to the nearer equivalent fractions. The amount of computation and the chances of error can be reduced by using the decimal equivalent table.

Table of Decimal Equivalents

Using a decimal equivalent table saves time and reduces the chance of error. Decimal equivalent tables are widely used in the manufacturing industry. They are posted as large wall charts in work areas and are carried as pocket size cards. Skilled workers memorize many of the equivalents after using decimal equivalent tables.

The decimals listed in the table are given to six places. For actual on-the-job uses, a decimal is rounded to the degree of precision required for a particular application.

DECIMAL EQUIVALENT TABLE			
1/64—0.015625	17/64—0.256625	33/64—0.515625	49/64—0.765625
1/32———0.03125	9/32———0.28125	17/32———0.53125	25/32———0.78125
3/64—0.046875	19/64—0.296875	35/64—0.546875	51/64—0.796875
1/16———0.0625	5/16———0.3125	9/16———0.5625	13/16———0.8125
5/64—0.078125	21/64—0.328125	37/64—0.578125	53/64—0.828125
3/32———0.09375	11/32———0.34375	19/32———0.59375	27/32———0.84375
7/64—0.109375	23/64—0.359375	39/64—0.609375	55/64—0.859375
1/8———0.125	3/8———0.375	5/8———0.625	7/8———0.875
9/64—0.140625	25/64—0.390625	41/64—0.640625	57/64—0.890625
5/32———0.15625	13/32———0.40625	21/32———0.65625	29/32———0.90625
11/64—0.171875	27/64—0.421875	43/64—0.671875	59/64—0.921875
3/16———0.1875	7/16———0.4375	11/16———0.6875	15/16———0.9375
13/64—0.203125	29/64—0.453125	45/64—0.703125	61/64—0.953125
7/32———0.21875	15/32———0.46875	23/32———0.71875	31/32———0.96875
15/64—0.234375	31/64—0.484375	47/64—0.734375	63/64—0.984375
1/4———0.25	1/2———0.5	3/4———0.75	1———1.

The following examples illustrate the use of the decimal equivalent table.

Example 1   Find the decimal equivalent of $\frac{23}{32}''$

The decimal equivalent is shown directly
to the right of the common fraction.
$$\frac{23''}{32} = 0.71875'' \quad \text{Ans}$$

Example 2   Find the fractional equivalent of 0.3125″.

The fractional equivalent is shown
directly to the left of the decimal fraction.
$$0.3125'' = \frac{5}{16}'' \quad \text{Ans}$$

Example 3   Find the nearer fractional equivalents of the decimal dimensions given on the casting shown. All dimensions are in inches.

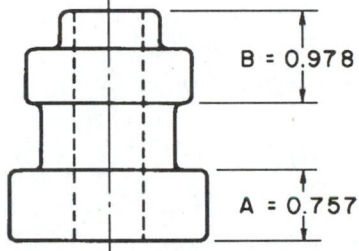

Compute dimension A.   The decimal 0.757 lies between 0.750 and 0.765625. The difference between 0.757 and 0.750 is 0.007. The difference between 0.757 and 0.765625 is 0.008625. Since 0.007 is less than 0.008625, the 0.750 value is closer to 0.757.

The nearer fractional equivalent of 0.750″ is $\frac{3}{4}''$ Ans

Compute dimension B.   The decimal 0.978 lies between 0.96875 and 0.984375. The difference between 0.978 and 0.96875 is 0.00925. The difference between 0.978 and 0.984375 is 0.006375. Since 0.006375 is less than 0.00925, the 0.984375 value is closer to 0.978.

The nearer fractional equivalent of 0.984375″ is $\frac{63}{64}''$ Ans

Combined Operations of Decimal Fractions

In the process of completing a job, a machinist must determine stock sizes, cutter sizes, feeds and speeds, and roughing allowances as well as cutting dimensions. Usually most and sometimes all of the fundamental operations of mathematics must be used for computations in the manufacture of a part.

Determination of powers and roots must also be considered in the order of operations. The following procedure incorporates all six fundamental operations.

Order of Operations

1. Do all operations within the grouping symbol first. Parentheses, the fraction bar and the radical symbol are used to group numbers. If an expression contains parentheses within parentheses or brackets do the work within the innermost parentheses first.

2. Do powers and roots next. The operations are performed in the order in which they occur. If a root consists of two or more operations within the radical symbol, perform all operations within the radical symbol, then extract the root.

3. Do multiplication and division next in the order in which they occur.

4. Do addition and subtraction last in the order in which they occur.

Example 1   Find the value of $7.875 + 3.2 \times 4.3 - 2.73$.

$$7.875 + 3.2 \times 4.3 - 2.73$$

Multiply. $3.2 \times 4.3 = 13.76$

$$7.875 + 13.76 - 2.73$$

Add. $7.875 + 13.76 = 21.635$

$$21.635 - 2.73$$

Subtract. $21.635 - 2.73 = 18.905$

18.905 Ans

Example 2   Find the value of $(27.34 - 4.82) \div (2.41 \times 1.78 + 7.89)$. Round the answer to 2 decimal places.

Perform operations within parentheses.

$$(27.34 - 4.82) \div (2.41 \times 1.78 + 7.89)$$

Subtract. $27.34 - 4.82 = 22.52$

$$22.52 \div (2.41 \times 1.78 + 7.89)$$

Multiply. $2.41 \times 1.78 = 4.2898$

$$22.52 \div (4.2893 + 7.89)$$

Add. $4.2898 + 7.89 = 12.1798$

$$22.52 \div 12.1798$$

Divide. $22.52 \div 12.1798 = 1.85$

1.85 Ans

Example 3   Find the value of $\dfrac{13.79 + (27.6 \times 0.3)^2}{\sqrt{23.04} + 0.875 - 3.76}$. Round the answer to 3 decimal places.

Grouping symbol operations are done first. Consider the numerator and the denominator as if each were within parentheses. All of the operations are performed in the numerator and in the denominator before the division is performed.

$$\frac{13.79 + (27.6 \times 0.3)^2}{\sqrt{23.04} + 0.875 - 3.76}$$

In the numerator:

$$[13.79 + (27.6 \times 0.3)^2] \div (\sqrt{23.04} + 0.875 - 3.76)$$

Multiply. $27.6 \times 0.3 = 8.28$

$$[13.79 + (8.28)^2] \div (\sqrt{23.04} + 0.875 - 3.76)$$

Square. $8.28^2 = 68.5584$

$$(13.79 + 68.5584) \div (\sqrt{23.04} + 0.875 - 3.76)$$

Add. $13.79 + 68.5584 = 82.3484$

In the denominator:

$$82.3484 \div (\sqrt{23.04} + 0.875 - 3.76)$$

Extract the square root. $\sqrt{23.04} = 4.8$

$$82.3484 \div (4.8 + 0.875 - 3.76)$$

Add. $4.8 + 0.875 = 5.675$

$$82.3484 \div (5.675 - 3.76)$$

Subtract. $5.675 - 3.76 = 1.915$

$$82.3484 \div 1.915$$

Divide. $82.3484 \div 1.915 = 43.002$

43.002 Ans

Example 4   Blanks in the shape of regular pentagons (5-sided figures) are punched from strip stock as shown. Determine the width of strip stock required, using the given dimensions and the formula for dimension R. Round the answer to 3 decimal places. All dimensions are in inches.

$$\text{Width} = R + 0.980 + 2 \times 0.125 \quad \text{where} \quad R = \sqrt{r^2 + s^2 + 4}$$

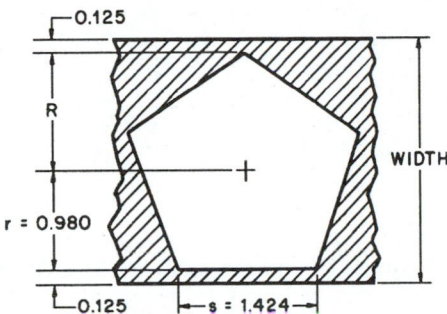

$$\sqrt{0.980^2 + 1.424^2 + 4} + 0.980 + 2 \times 0.125$$

Substitute the given values.

Compute the operations under the radical sign.

Square. $0.980^2 = 0.9604$

Square. $1.424^2 = 2.027776$

$$\sqrt{0.9604 + 2.027776 + 4} + 0.980 + 2 \times 0.125$$

Divide. $2.027776 \div 4 = 0.506944$

$$\sqrt{0.9604 + 0.506944} + 0.980 + 2 \times 0.125$$

Add. $\sqrt{0.9604 + 0.506944} = \sqrt{1.467344}$

$$\sqrt{1.467344} + 0.980 + 2 \times 0.125$$

Extract the square root. $\sqrt{1.467344} = 1.211$

$$1.211 + 0.980 + 2 \times 0.125$$

Multiply. $2 \times 0.125 = 0.250$

$$1.211 + 0.980 + 0.250$$

Add. $1.211 + 0.980 + 0.250 = 2.441$

Width = 2.441 inches   Ans

➤ **Note:** In solving expressions which consist of numerous multiplication and power operations, it is often necessary to carry out the work to two or three more decimal places than the number of decimal places required in the answer.

APPLICATION

Using the Decimal Equivalent Table

Find the fraction or decimal equivalents of these numbers using the decimal equivalent table.

1. $\frac{25}{32}$ _____

2. $\frac{7}{32}$ _____

3. $\frac{11}{32}$ _____

4. $\frac{13}{16}$ _____   6. 0.671875 _____   8. 0.28125 _____

5. $\frac{5}{64}$ _____   7. 0.3125 _____   9. 0.203125 _____

Find the nearer fraction equivalents of these decimals using the decimal equivalent table.

10. 0.541 _____   12. 0.465 _____   14. 0.209 _____

11. 0.762 _____   13. 0.498 _____   15. 0.805 _____

Combined Operations of Decimal Fractions

Solve these examples of combined operations. Round the answers to 2 decimal places where necessary.

16. $0.5231 + 10.375 + 4.32 \times 0.521$ _____

17. $81.07 + 12.1 + 2 \times 3.7$ _____

18. $\frac{56.050}{3.8} \times 0.875 - 3.92$ _____

19. $(24.78 - 19.32) \times 4.6$ _____

20. $(14.6 + 4 - 1.76)^2 \times 4.5$ _____

21. $27.16 + \sqrt{1.76 + 12.32}$ _____

22. $\left(\sqrt{3.98 + 0.87 \times 3.9}\right)^2$ _____

23. $(3.29 \times 1.7)^2 + (3.82 - 0.86)$ _____

24. $0.25 \times \left(\frac{\sqrt{64} \times 3.87}{8.32 \times 5.13}\right) + 18.3^2$ _____

25. $18.32 - \sqrt{\frac{7.86 \times 13.5}{3.5^2 - 0.52}} \times 0.7$ _____

Solve the following problems which require combined operations.

26. The figure shows the three-wire method of checking screw threads. With proper diameter wires and a micrometer, very accurate pitch diameter measurements can be made. Using the formula given, determine the micrometer dimension over wires of the American (National) Standard threads in the following table. Round the answer to 4 decimal places.

$$M = D - (1.5155 \times P) + (3 \times W)$$

	Major Diameter D (inches)	Pitch P (inches)	Wire Diameter W (inches)	Dimension Over Wires M (inches)
a.	0.8750	0.1250	0.0900	
b.	0.2500	0.0500	0.0350	
c.	0.6250	0.1000	0.0700	
d.	1.3750	0.16667	0.1500	
e.	2.5000	0.2500	0.1500	

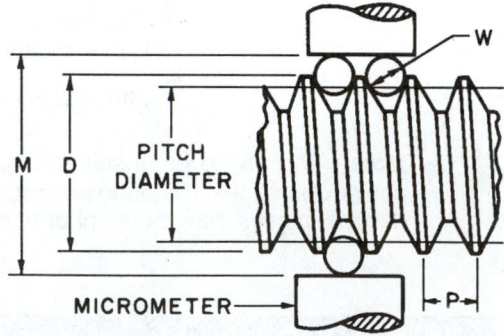

27. A bronze bushing with a diameter of 22.225 millimeters is to be pressed into a mounting plate. The assembly print calls for a bored hole in the plate to be 0.038 millimeter less in diameter than the bushing diameter. The hole diameter in the plate checks 22.103 millimeters. How much must the diameter of the plate hole be increased in order to meet the print specification? _____

28. A stamped sheet steel plate is shown. Compute dimensions A–F to 3 decimal places. All dimensions are in inches.

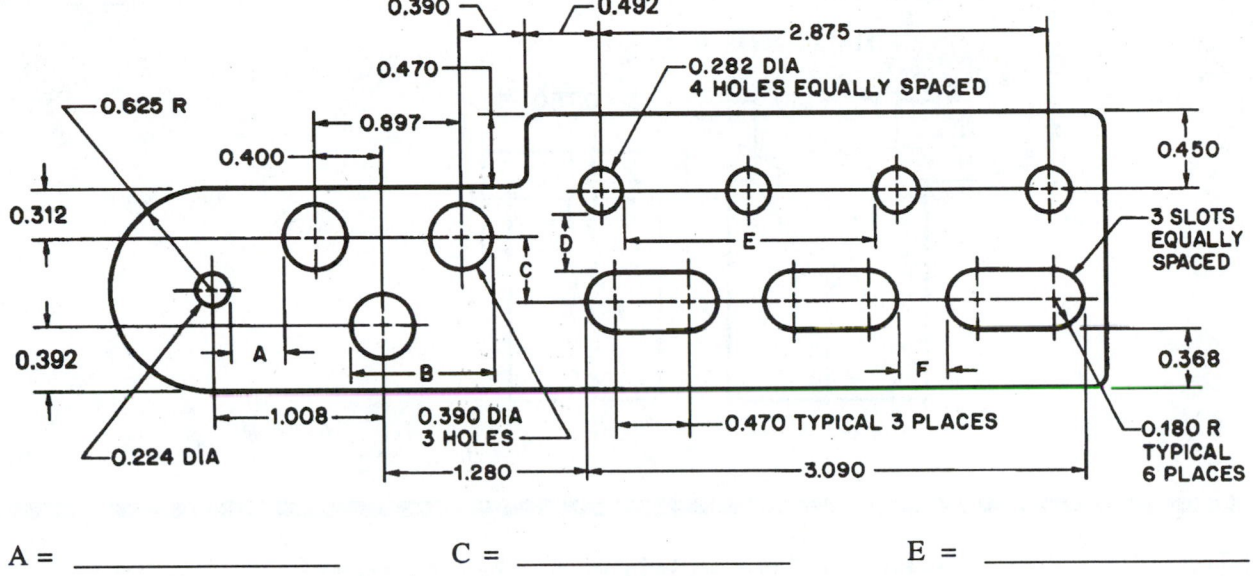

A = _____ C = _____ E = _____

B = _____ D = _____ F = _____

29. A flat is to be milled in three pieces of round stock each of a different diameter. The length of the flat is determined by the diameter of the stock and the depth of cut. The table gives the required length of flat and the stock diameter for each piece. Determine the depth of cut for each piece to 2 decimal places using this formula.

$$C = \frac{D}{2} - 0.5 \times \sqrt{4 \times \left(\frac{D}{2}\right)^2 - F^2}$$

	Diameter D	Length of Flat F	Depth of Cut C
a.	34.80 mm	30.50 mm	
b.	55.90 mm	40.60 mm	
c.	91.40 mm	43.40 mm	

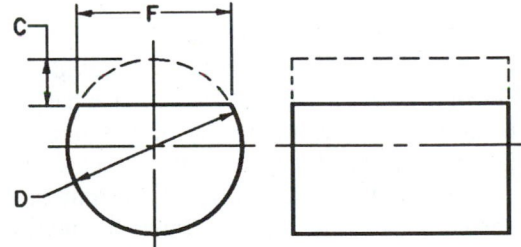

30. A groove is machined in a circular plate with a 41.36-millimeter diameter. Two milling cuts, one 6.30 millimeters deep and the other 3.15 millimeters, are made. A grinding operation then removes 0.40 millimeter. What is the distance from the center of the plate to the bottom of the groove? All dimensions are in millimeters. _____

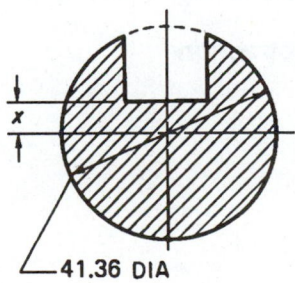

31. A 60° groove has been machined in a fixture. The groove is checked by placing a pin in the groove and indicating the distance between the top of the fixture and the top of the pin as shown. Compute distance H to 3 decimal places by using this formula. All dimensions are in inches.

$$H = 1.5 \times D - 0.866 \times W$$

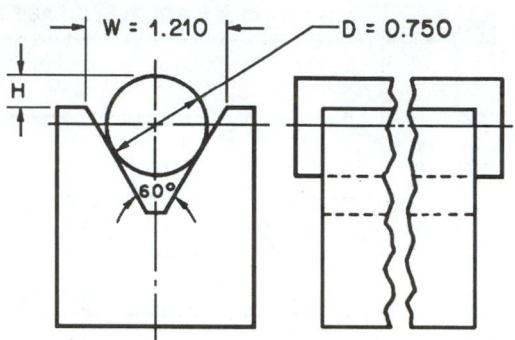

UNIT **16** Computing with a Calculator: Decimals

Objectives   After studying this unit you should be able to

- **Perform individual operations of addition, subtraction, multiplication, division, powers, and roots with decimals using a calculator.**
- **Perform combinations of operations with decimals using a calculator.**

Decimals

The decimal point key ($\boxed{\cdot}$) is used when entering decimal values in a calculator. When entering a decimal fraction in a calculator, the decimal point key is pressed at the position of the decimal point in the number. For example, to enter the number 0.732, first press $\boxed{\cdot}$ and then enter the digits. To enter the number 567.409, enter 567 $\boxed{\cdot}$ 409.

In calculator examples and illustrations of operations with decimals in this text, the decimal key $\boxed{\cdot}$ will *not* be shown to indicate the entering of a decimal point. Wherever the decimal point occurs in a number, it is understood that the decimal point key $\boxed{\cdot}$ is pressed.

Recall that your calculator must have algebraic logic to solve combined operations problems as they are shown in this text. Also recall the procedure for rounding numbers: Locate the digit in the number that gives the desired degree of precision; increase that digit by 1 if the digit immediately following is 5 or more; do not change the value of the digit if the digit immediately following is less than 5. Drop all digits that follow.

Decimals with Basic Operations of Addition, Subtraction, Multiplication, and Division

Example 1   Add.   19.37 + 123.9 + 7.04

19.36 $\boxed{+}$ 123.9 $\boxed{+}$ 7.04 $\boxed{=}$ 150.31   Ans

Example 2   Subtract.   2,876.78 − 405.052
2,876.78 $\boxed{-}$ 405.052 $\boxed{=}$ 2,471.728   Ans

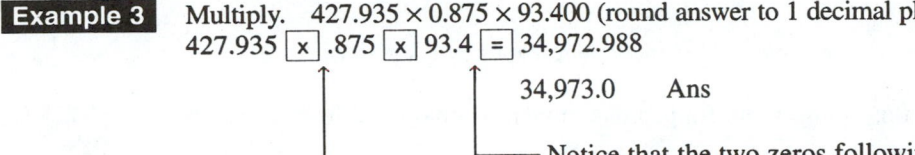

Example 3 Multiply. $427.935 \times 0.875 \times 93.400$ (round answer to 1 decimal place)

$427.935 \boxed{\times} .875 \boxed{\times} 93.4 \boxed{=} 34{,}972.988$

$34{,}973.0$ Ans

Notice that the two zeros following the 4 are not entered. The final zero or zeros to the right of the decimal point may be omitted.

Notice that the zero to the left of the decimal point is not entered. The leading zero is omitted.

Example 4 Divide. $813.7621 \div 6.466$ (round answer to 3 decimal places)

$813.7621 \boxed{\div} 6.466 \boxed{=} 125.85247$

125.852 Ans

Powers

Expressions involving powers and roots are readily computed with a scientific calculator. The *square* key is used to raise a number to the second power (to square a number). Depending on the calculator used, the square of a number is computed in one of the following ways:

Enter the number and press the square key ($\boxed{X^2}$).

Example To calculate 28.75^2, enter 28.75 and press $\boxed{X^2}$.

$28.75 \boxed{X^2} \rightarrow 826.5625$ Ans

➤ **Note:** Upon pressing $\boxed{X^2}$, the answer is displayed. It is not necessary to press $\boxed{=}$ with most calculators.

Or, enter the number, press the square key, $\boxed{X^2}$, and press $\boxed{EXE}$.

Example To calculate 28.75^2, enter 28.75, press $\boxed{X^2}$, and press $\boxed{EXE}$.

$28.75 \boxed{X^2} \boxed{EXE} 826.5625$ Ans

The universal power key ($\boxed{y^x}$), ($\boxed{x^y}$), or $\boxed{\wedge}$, depending on the calculator used, raises any positive number to a power. To raise a number to a power using the universal power key, do the following:

Enter the number to be raised to a power (y) or (x)

Press the universal power key $\boxed{y^x}$, $\boxed{x^y}$, or $\boxed{\wedge}$.

Enter the power (x) or (y).

Press the $\boxed{=}$ or $\boxed{EXE}$ key.

Examples

1. Calculate 15.72^3. Enter 15.72, press $\boxed{y^x}$, $\boxed{x^y}$, or $\boxed{\wedge}$, enter 3, and press $\boxed{=}$ or $\boxed{EXE}$.

$15.72 \boxed{y^x} 3 \boxed{=} 3884.7012$ Ans

2. Calculate 0.95^7

$.95 \boxed{y^x} 7 \boxed{=} 0.6983373$ Ans

Roots

To obtain the square root of any positive number, the square root key ($\boxed{\sqrt{x}}$) or ($\boxed{\sqrt{}}$) is used.

Depending on the calculator, the square root of a positive number is computed in one of the following ways.

1. Enter the number and press the square root key ($\boxed{\sqrt{x}}$).

Example Calculate $\sqrt{27.038}$. Enter 27.038 and press $\boxed{\sqrt{x}}$.

$$27.038 \ \boxed{\sqrt{x}} \rightarrow 5.199807689 \qquad \text{Ans}$$

2. Press the square root key ($\boxed{\sqrt{}}$), enter the number, and press $\boxed{\text{EXE}}$ or $\boxed{=}$.

➤ **Note:** The square root is a second function on certain calculators.

Example Calculate $\sqrt{27.038}$. Press $\boxed{\sqrt{}}$, enter 27.038, press $\boxed{\text{EXE}}$.

$$\boxed{\sqrt{}} \ 27.038 \ \boxed{\text{EXE}} \ 5.199807689 \qquad \text{Ans}$$

The root of any positive number can be computed with a calculator. Some calculators have a root key; with other calculators, roots are a second function.

Depending on the calculator, root calculations are generally performed as follows.

1. Procedure for calculators that have the root key $\boxed{\sqrt[x]{}}$. Enter the root to be taken, press $\boxed{\sqrt[x]{}}$, enter the number whose root is to be taken, press $\boxed{\text{EXE}}$ or $\boxed{=}$.

Example Calculate $\sqrt[5]{475.19}$. Enter 5, press $\boxed{\sqrt[x]{}}$, enter 475.19, press $\boxed{\text{EXE}}$ or $\boxed{=}$.

➤ **Note:** Where $\boxed{\sqrt[x]{}}$ is a second function, press $\boxed{\text{SHIFT}}$ before pressing $\boxed{\sqrt[x]{}}$.

$$5 \ \boxed{\sqrt[x]{}} \ 475.19 \ \boxed{\text{EXE}} \ 3.430626662 \qquad \text{Ans}$$

2. Procedure for calculators that do not have the root key $\boxed{\sqrt[x]{}}$ and roots are second functions. The procedures vary somewhat depending on the calculator used. Procedure for calculators with $\boxed{x\sqrt{y}}$ as a second function. Enter the number you want to find the root for, press $\boxed{\text{2nd}}$, press $\boxed{x\sqrt{y}}$, enter the root to be taken, press $\boxed{=}$.

Example Calculate $\sqrt[5]{475.19}$. Enter 475.19, press $\boxed{\text{2nd}}$, press $\boxed{x\sqrt{y}}$, enter 5, press $\boxed{=}$.

$$475.19 \ \boxed{\text{2nd}} \ \boxed{x\sqrt{y}} \ 5 \ \boxed{=} \ 3.430626662 \qquad \text{Ans}$$

or procedure for calculators with $\boxed{x^{1/y}}$ as a second function.

Enter the number for which you are taking the root, press $\boxed{\text{SHIFT}}$, press $\boxed{x^{1/y}}$, enter the root to be taken, press $\boxed{=}$.

Example Calculate $\sqrt[5]{475.19}$. Enter 475.19, press $\boxed{\text{SHIFT}}$, press $\boxed{x^{1/y}}$, enter 5, press $\boxed{=}$.

$$475.19 \ \boxed{\text{SHIFT}} \ \boxed{x^{1/y}} \ 5 \ \boxed{=} \ 3.4306267 \qquad \text{Ans}$$

Practice Exercises, Individual Basic Operations

Evaluate the following expressions. The expressions are basic arithmetic operations including powers and roots. Remember to check your answers by doing each problem twice. The solutions to the problems directly follow the practice exercises. Compare your answers to the given solutions. Round each answer to the indicated number of decimal places.

1. 276.84 + 312.094 (2 places)

2. 16.09 + 0.311 + 5.516 (1 place)

3. 6,704.568 − 4,989.07 (2 places)

4. 0.9244 − 0.0822 (3 places)

5. 43.4967 × 6.0913 (4 places)

6. 8.503 × 0.779 × 13.248 (3 places)

7. 54.419 ÷ 6.7 (1 place)

8. 0.9316 ÷ 0.0877 (4 places)

9. 36.22^2 (2 places)

10. 7.063^5 (1 place)

11. $\sqrt{28.73721}$ (4 places)

12. $\sqrt[5]{1,068.470}$ (3 places)

Solutions to Individual Basic Operations

1. 276.84 $\boxed{+}$ 312.094 $\boxed{=}$ 588.934, 588.93   Ans

2. 16.09 $\boxed{+}$ .311 $\boxed{+}$ 5.516 $\boxed{=}$ 21.917, 21.9   Ans

3. 6704.568 $\boxed{-}$ 4989.07 $\boxed{=}$ 1715.498, 1,715.50   Ans

4. .9244 $\boxed{-}$ .0822 $\boxed{=}$ 0.8422, 0.842   Ans

5. 43.4967 $\boxed{\times}$ 6.0913 $\boxed{=}$ 264.95145, 264.9515   Ans

6. 8.503 $\boxed{\times}$ .779 $\boxed{\times}$ 13.248 $\boxed{=}$ 87.752593, 87.753   Ans

7. 54.419 $\boxed{\div}$ 6.7 $\boxed{=}$ 8.1222388, 8.12   Ans

8. .9316 $\boxed{\div}$ .0877 $\boxed{=}$ 10.622577, 10.6226   Ans

9. 36.22 $\boxed{x^2}$ → 1311.8884, 1311.89   Ans

10. 7.063 $\boxed{y^x}$ 5 $\boxed{=}$ 17577.052, 17,577.1   Ans

11. 28.73721 $\boxed{\sqrt{x}}$ → 5.3607098, 5.3607   Ans

12. 5 $\boxed{\sqrt[x]{\ }}$ 1068.47 $\boxed{EXE}$ 4.03415394   Ans

    or 1068.47 $\boxed{2nd}$ $\boxed{\sqrt[x]{y}}$ 5 $\boxed{=}$ 4.03415394   Ans

    or 1068.47 $\boxed{SHIFT}$ $\boxed{x^{\frac{1}{y}}}$ 5 $\boxed{=}$ 4.03415394   Ans

Combined Operations

Because the following problems are combined operations expressions, your calculator must have algebraic logic to solve the problems shown. The expressions are solved by entering numbers and operations into the calculator in the same order as the expressions are written.

Examples

1. Evaluate. 30.75 + 15 ÷ 4.02 (round answer to 2 decimal places)

    30.75 $\boxed{+}$ 15 $\boxed{\div}$ 4.02 $\boxed{=}$ 34.481343, 34.48   Ans

2. Evaluate. $51.073 - \dfrac{4}{0.091} + 33.151 \times 2.707$ (round answer to 2 decimal places)

    51.073 $\boxed{-}$ 4 $\boxed{\div}$ .091 $\boxed{+}$ 33.151 $\boxed{\times}$ 2.707 $\boxed{=}$ 96.856713, 96.86   Ans

3. Evaluate. 46.23 + (5 + 6.92) × (56.07 − 38.5)

As previously discussed in the order of operations, operations enclosed within parentheses are performed first. A calculator with algebraic logic performs the operations within parentheses before performing other operations in a combined operations expression. If an expression contains parentheses, enter the expression into the calculator in the order in which it is written. The parentheses keys $[\,(\,]$ and $[\,)\,]$ must be used.

46.23 $[+]$ $[\,(\,]$ 5 $[+]$ 6.92 $[\,)\,]$ $[\times]$ $[\,(\,]$ 56.07 $[-]$ 38.5 $[\,)\,]$ $[=]$ 255.6644   Ans

4. Evaluate. $\dfrac{13.463\ +\ 9.864\ \times\ 6.921}{4.373\ +\ 2.446}$ (round answer to 3 decimal places)

Recall that for problems expressed in fractional form, the fraction bar is also used as a grouping symbol. The numerator and denominator are each considered as being enclosed in parentheses.

$$(13.463\ +\ 9.864\ \times\ 6.921)\ \div\ (4.373\ +\ 2.446)$$

$[\,(\,]$ 13.463 $[+]$ 9.864 $[\times]$ 6.921 $[\,)\,]$ $[\div]$ $[\,(\,]$ 4.373 $[+]$ 2.446 $[\,)\,]$ $[=]$ 11.985884, 11.986   Ans

The expression may also be evaluated by using the $[=]$ key to simplify the numerator without having to enclose the entire numerator in parentheses. However, parentheses must be used to enclose the denominator.

13.463 $[+]$ 9.864 $[\times]$ 6.921 $[=]$ $[\div]$ $[\,(\,]$ 4.373 $[+]$ 2.446 $[\,)\,]$ $[=]$ 11.985884, 11.986   Ans

5. Evaluate. $\dfrac{100.32\ -\ (16.87\ +\ 13)}{111.36\ -\ 78.47}$ (round answer to 2 decimal places)

$$\dfrac{100.32\ -\ (16.87\ +\ 13)}{111.36\ -\ 78.47}\ =\ (100.32\ -\ (16.87\ +\ 13))\ \div\ (11.36\ -\ 78.47)$$

Observe these parentheses

To be sure that the complete numerator is evaluated before dividing by the denominator, enclose the complete numerator within parentheses. This is an example of an expression containing parentheses within parentheses.

$[\,(\,]$ 100.32 $[-]$ $[\,(\,]$ 16.87 $[+]$ 13 $[\,)\,]$ $[\,)\,]$ $[\div]$ $[\,(\,]$ 111.36 $[-]$ 78.47 $[\,)\,]$ $[=]$ 2.1419884, 2.14   Ans

Using the $[=]$ key to simplify the numerator:

100.32 $[-]$ $[\,(\,]$ 16.87 $[+]$ 13 $[\,)\,]$ $[=]$ $[\div]$ $[\,(\,]$ 111.36 $[-]$ 78.47 $[\,)\,]$ $[=]$ 2.1419884, 2.14   Ans

6. Evaluate. $\dfrac{873.03\ +\ 12.12^{\,3}\ \times\ 41}{\sqrt{16.43}\ -\ 266.76\ +\ 107.88}$ (round answer to 2 decimal places)

$[\,(\,]$ 873.03 $[+]$ 12.12 $[y^x]$ 3 $[\times]$ 41 $[\,)\,]$ $[\div]$ $[\,(\,]$ 16.43 $[\sqrt{x}]$ $[-]$ 266.76 $[+]$ 107.88 $[\,)\,]$ $[=]$ 46732.658, 46,732.66   Ans

Using the $[=]$ key to simplify the numerator:

873.03 $[+]$ 12.12 $[y^x]$ 3 $[\times]$ 41 $[=]$ $[\div]$ $[\,(\,]$ 16.43 $[\sqrt{x}]$ $[-]$ 266.76 $[+]$ 107.88 $[\,)\,]$ $[=]$ 46732.658, 46,732.66   Ans

Practice Exercises, Combined Operations

Evaluate the following combined operations expressions. Remember to check your answers by doing each problem twice. The solutions to the problems directly follow the practice exercises. Compare your answers to the given solutions. Round each answer to the indicated number of decimal places.

1. $503.97 - 487.09 \times 0.777 + 65.14$ (2 places)

2. $27.028 + \dfrac{5}{6.331} - 5.875 \times 1.088$ (3 places)

3. $23.073 \times (0.046 + 5.934 - 3.049) - 17.071$ (3 places)

4. $30.180 \times (0.531 + 12.939 - 2.056) - 60.709$ (3 places)

5. $\dfrac{643.72 - 18.192 \times 0.783}{470.07 - 88.33}$ (2 places)

6. $\dfrac{793.32 - 2.67 \times 0.55}{107.9 + 88.93}$ (1 place)

7. $2{,}446 + 8.917^3 \times 5.095$ (3 places)

8. $679.07 + (36 + 19.973 - 0.887)^2 \times 2.05$ (1 place)

9. $43.71 - \sqrt{256.33 - 107} + 17.59$ (2 places)

10. $\dfrac{\sqrt[5]{14.773} + 93.977 \times \sqrt[3]{282.608}}{3.033}$ (3 places)

11. $\dfrac{1{,}202.03 + \sqrt[3]{706.8 - 44.317}}{(14.03 \times 0.54 - 2.08)^2} - 2.63$ (1 place)

Solutions to Practice Exercises, Combined Operations

1. $503.97 \boxed{-} 487.09 \boxed{\text{x}} .777 \boxed{+} 65.14 \boxed{=} 190.64107$, 190.64   Ans

2. $27.028 \boxed{+} 5 \boxed{+} 6.331 \boxed{-} 5.875 \boxed{\text{x}} 1.088 \boxed{=} 21.425765$, 21.426   Ans

3. $23.073 \boxed{\text{x}} \boxed{(} .046 \boxed{+} 5.934 \boxed{-} 3.049 \boxed{)} \boxed{-} 17.071 \boxed{=} 50.555963$,
    50.556   Ans

4. $30.180 \boxed{\text{x}} \boxed{(} .531 \boxed{+} 12.939 \boxed{-} 2.056 \boxed{)} \boxed{-} 60.709 \boxed{=} 283.76552$,
    283.766   Ans

5. $\boxed{(} 634.72 \boxed{-} 18.192 \boxed{\text{x}} .783 \boxed{)} \boxed{+} \boxed{(} 470.07 \boxed{-} 88.33 \boxed{)} \boxed{=} 1.6489644$,
    1.65   Ans

    or $643.72 \boxed{-} 18.192 \boxed{\text{x}} .783 \boxed{=} \boxed{+} \boxed{(} 470.07 \boxed{-} 88.33 \boxed{)} \boxed{=} 1.6489644$,
    1.65   Ans

6. $\boxed{(} 793.32 \boxed{-} 2.67 \boxed{\text{x}} .55 \boxed{)} \boxed{+} \boxed{(} 107.9 \boxed{+} 88.93 \boxed{)} \boxed{=} 4.0230224$,
    4.0   Ans

    or $793.32 \boxed{-} 2.67 \boxed{\text{x}} .55 \boxed{=} \boxed{+} \boxed{(} 107.9 \boxed{+} 88.93 \boxed{)} \boxed{=} 4.0230224$,
    4.0   Ans

7. $2446 \boxed{+} 8.917 \boxed{y^x} 3 \boxed{\text{x}} 5.095 \boxed{=} 6058.4387$, 6,058.4387   Ans

8. $679.07 \boxed{+} \boxed{(} 36 \boxed{+} 19.973 \boxed{-} .887 \boxed{)} \boxed{X^2} \boxed{\text{x}} 2.05 \boxed{=} 6899.7282$,
    6,899.7   Ans

9. $43.71 \boxed{-} \boxed{(} 256.33 \boxed{-} 107 \boxed{)} \boxed{\sqrt{x}} \boxed{+} 17.59 \boxed{=} 49.079935$, 49.08   Ans

    or $43.71 \boxed{-} \boxed{\sqrt{}} \boxed{(} 256.33 - 107 \boxed{)} \boxed{+} 17.59 \boxed{\text{EXE}} 49.079935$,
    49.08   Ans

10. $\boxed{(}$ 5 $\boxed{\sqrt[x]{}}$ 14.773 $\boxed{+}$ 93.977 $\boxed{x}$ 3 $\boxed{\sqrt[x]{}}$ 282.608 $\boxed{)}$ $\boxed{+}$ 3.033 $\boxed{\text{EXE}}$
    203.89927, 203.899   Ans

    or $\boxed{(}$ 14.773 $\boxed{\text{2nd}}$ $\boxed{x\sqrt{y}}$ 5 $\boxed{+}$ 93.977 $\boxed{x}$ 282.608 $\boxed{\text{2nd}}$ $\boxed{x\sqrt{y}}$ 3 $\boxed{)}$ $\boxed{+}$
    3.033 $\boxed{=}$ 203.89927, 203.899   Ans

11. $\boxed{(}$ 1202.03 $\boxed{\div}$ 3 $\boxed{\sqrt[x]{}}$ $\boxed{(}$ 706.8 $\boxed{-}$ 44.317 $\boxed{)}$ $\boxed{)}$ $\boxed{+}$ $\boxed{(}$ 14.03 $\boxed{x}$ .54 $\boxed{-}$
    2.08 $\boxed{)}$ $\boxed{X^2}$ $\boxed{-}$ 2.63 $\boxed{\text{EXE}}$ 1.9345574, 1.9   Ans

    or $\boxed{(}$ 1202.03 $\boxed{\div}$ $\boxed{(}$ 706.8 $-$ 44.317 $\boxed{)}$ $\boxed{\text{2nd}}$ $\boxed{x\sqrt{y}}$ 3 $\boxed{)}$ $\boxed{)}$ $\boxed{+}$ $\boxed{(}$ 14.03
    $\boxed{x}$ .54 $\boxed{-}$ 2.08 $\boxed{)}$ $\boxed{X^2}$ $\boxed{-}$ 2.63 $\boxed{=}$ 1.9345574, 1.9   Ans

UNIT 17 Introduction to Percents

Objectives   After studying this unit you should be able to

- **Express decimal fractions and common fractions as percents.**
- **Express percents as decimal fractions and common fractions.**

Percents are widely used in both business and nonbusiness fields. Merchandise selling prices and discounts, wage deductions, and equipment depreciation are determined by percentages.

In manufacturing technology percentage concepts have many applications, such as expressing production increases or decreases, power inputs and outputs, quality control product rejections, and material allowances for waste and nonconforming parts.

Definition of Percent

The *percent* (%) indicates the number of hundredths of a whole. The square shown to the right is divided into 100 equal parts. The whole (large square) contains 100 small parts, or 100 percent of the small squares. Each small square is one part of 100 parts or $\frac{1}{100}$ of the large square. Therefore, each small square is $\frac{1}{100}$ of 100 percent or 1 percent.

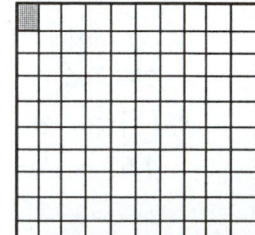

1 part of 100 parts

$$\frac{1}{100} = 0.01 = 1\%$$

Example   What percent of the square shown to the right is shaded?

The large square is divided into 4 equal smaller squares. Three of the smaller squares are shaded.

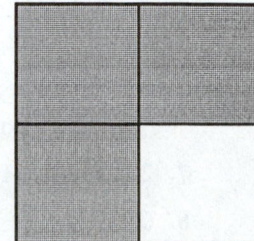

3 parts of 4 parts

$$\frac{3}{4} = 0.75 = 75\%   \text{Ans}$$

Expressing Decimal Fractions as Percents

A decimal fraction can be expressed as a percent by moving the decimal point two places to the right and inserting the percent symbol. Moving the decimal point two places to the right is actually multiplying by 100.

Examples

1. Express 0.0152 as a percent.

   Move the decimal point 2 places
   to the right. $0.01\ 52 = 1.52\%$ Ans

   Insert the percent symbol.

2. Express 3.876 as a percent.

   Move the decimal point 2 places
   to the right. $3.87\ 6 = 387.6\%$ Ans

   Insert the percent symbol.

Expressing Common Fractions and Mixed Numbers as Percents

To express a common fraction as a percent, first express the common fraction as a decimal fraction. Then express the decimal fraction as a percent. If necessary to round, the decimal fraction must be two more decimal places than the desired number of places for the percent.

Examples

1. Express $\frac{7}{8}$ as a percent.

   Express $\frac{7}{8}$ as a decimal fraction. $\frac{7}{8} = 0.875$

   Express 0.875 as a percent. $0.875 = 87.5\%$ Ans

2. Express $5\frac{2}{3}$ as a percent to
   1 decimal place.

   Express $5\frac{2}{3}$ as a decimal fraction. $5\frac{2}{3} = 5.667$

   Express 5.667 as a percent. $5.667 = 566.7\%$ Ans

   $5\ \boxed{+}\ 2\ \boxed{+}\ 3\ \boxed{=}\ \boxed{\times}\ 100\ \boxed{=}\ 566.6666667,\ 566.7$ Ans

Expressing Percents as Decimal Fractions

Expressing a percent as a decimal fraction can be done by dropping the percent symbol and moving the decimal point two places to the left. Moving the decimal point two places to the left is actually dividing by 100.

Examples

1. Express $38\frac{16}{21}\%$ as a decimal fraction. Round the answer to 4 decimal places.

   Express $38\frac{16}{21}\%$ as 38.76% $38\frac{16}{21}\% = 38.76\% = 0.3876$ Ans

Drop the percent symbol and move the decimal point 2 places to the left.

$38\ \boxed{+}\ 16\ \boxed{+}\ 21\ \boxed{=}\ \boxed{+}\ 100\ \boxed{=}\ 0.387619047,\ 0.3876$ Ans
or $38\ \boxed{a^{b}\!/\!c}\ 16\ \boxed{a^{b}\!/\!c}\ 21\ \boxed{+}\ 100\ \boxed{=}\ 0.387619047,\ 0.3876$ Ans

Express each percent as a decimal fraction. Round the answers to 3 decimal places.

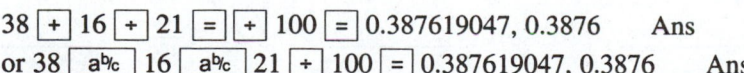

1. 0.48%   0.005   Ans

2. $15\frac{3}{4}\%$ 0.158   Ans

Expressing Percents as Common Fractions

A percent is expressed as a fraction by first finding the equivalent decimal fraction. The decimal fraction is then expressed as a common fraction.

Examples

1. Express 37.5% as a common fraction.

   Express 37.5% as a decimal fraction. $37.5\% = 0.375$

   Express 0.375 as a common fraction. $0.375 = \dfrac{375}{1,000} = \dfrac{3}{8}$ Ans

2. 375 $\boxed{a^{b/c}}$ 1000 $\boxed{=}$ 3 $\lrcorner$ 8, $\dfrac{3}{8}$ Ans

Express each percent as a common fraction.

1. 10% $10\% = 0.10 = \dfrac{10}{100} = \dfrac{1}{10}$ Ans

2. 0.5% $0.5\% = 0.005 = \dfrac{5}{1,000} = \dfrac{1}{200}$ Ans

3. $222\dfrac{1}{2}\%$ $222\dfrac{1}{2}\% = 222.5\% = 2.225 = 2\dfrac{225}{1,000} = 2\dfrac{9}{40}$ Ans

   2 $\boxed{a^{b/c}}$ 225 $\boxed{a^{b/c}}$ 1000 $\boxed{=}$ 2 _ 9 $\lrcorner$ 40, $2\dfrac{9}{40}$ Ans

APPLICATION

Determining Percents

Determine the percent of each figure that is shaded.

1.   2.   3.   4.

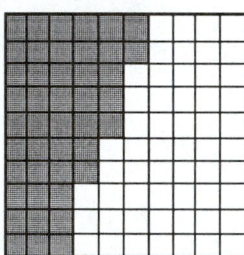
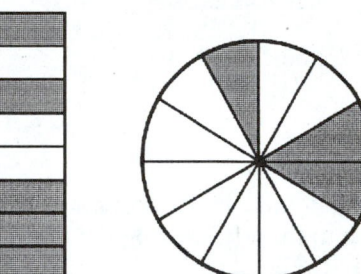
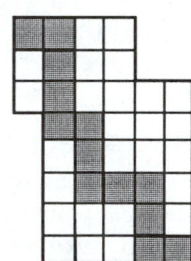

1. _____
2. _____
3. _____
4. _____

Expressing Decimals and Fractions as Percents

Express each value as a percent.

5. 0.35 _____   10. 1.33 _____   15. $\dfrac{1}{4}$ _____   20. $\dfrac{1}{250}$ _____

6. 0.96 _____   11. 2.076 _____   16. $\dfrac{21}{80}$ _____   21. $1\dfrac{59}{100}$ _____

7. 0.04 _____   12. 0.0639 _____   17. $\dfrac{3}{20}$ _____   22. $2\dfrac{7}{25}$ _____

8. 0.062 _____   13. 0.0002 _____   18. $\dfrac{37}{50}$ _____   23. $14\dfrac{5}{8}$ _____

9. 0.008 _____   14. 3.005 _____   19. $\dfrac{17}{32}$ _____   24. $3\dfrac{1}{200}$ _____

Expressing Percents as Decimals

Express each percent as a decimal fraction or mixed decimal.

25. 82% _____   29. 27.76% _____   33. 4.73% _____   37. $2\frac{3}{8}$% _____

26. 19% _____   30. 103% _____   34. $12\frac{1}{2}$% _____   38. 0.05% _____

27. 3% _____   31. 224.9% _____   35. $\frac{3}{4}$% _____   39. $37\frac{1}{4}$% _____

28. 2.6% _____   32. 0.6% _____   36. 0.1% _____   40. $205\frac{1}{10}$% _____

Expressing Percents as Fractions

Express each percent as a common fraction or mixed number.

41. 50% _____   44. 4% _____   47. 190% _____   50. 100.1% _____

42. 25% _____   45. 16% _____   48. 0.2% _____   51. 0.9% _____

43. 62.5% _____   46. 275% _____   49. 1.8% _____   52. 0.05% _____

UNIT **18** Basic Calculations of Percentages, Percents, and Rates

Objectives   After studying this unit you should be able to

- Determine the percentage, given the base and rate.
- Determine the percent (rate), given the percentage and base.
- Determine the base, given the rate and percentage.

Types of Simple Percentage Problems

A simple percentage problem has three parts. The parts are the rate, the base, and the percentage. In the problem 10% of $80 = $8, the rate is 10%, the base is $80, and the percentage is $8. The *rate* is the percent. The *base* is the number of which the rate or percent is taken. It is the whole or a quantity equal to 100%. The *percentage* is the quantity of the percent of the base. In solving problems, the rate, percentage, and base must be identified.

In solving percentage problems, the words *is* and *of* are often helpful in identifying the three parts. The word *is* generally relates to the rate or percentage and the word *of* generally relates to the base.

Examples

1. What *is* 25% *of* 120? *Is* relates to 25% (the rate) and *of* relates to 120 (the base).

2. What percent *of* 48 *is* 12? *Is* relates to 12 (the percentage) and *of* relates to 48 (the base).

3. 60 *is* 30% *of* what number? *Is* relates to 60 (the percentage) and 30% (the rate). *Of* relates to "what number" (the base).

There are three types of simple percentage problems. The type used depends on which two quantities are given and which quantity must be found. The three types are as follows:

- Finding the percentage, given the rate (percent) and the base.

  A problem of this type is, "What is 15% of 384?"

  The formula for finding the percentage, given the rate and the base is:

$$\text{Percentage} = \text{Rate} \times \text{Base}$$

$$P = R \times B$$

If the rate is less than 100%, the percentage is less than the base.
If the rate is greater than 100%, the percentage is greater than the base.

- Finding the rate (percent), given the base and the percentage.

  A problem of this type is, "What percent of 48 is 12?"

  The rearranged formula for finding the rate, given the base and the percentage is:

$$\text{Rate} = \frac{\text{Percentage}}{\text{Base}}$$

$$R = \frac{P}{B}$$

If the percentage is less than the base, the rate is less than 100%.
If the percentage is greater than the base, the rate is greater than 100%.

- Finding the base, given the rate (percent) and the percentage.

  A problem of this type is, "Fifty is 30% of what number?"

  The rearranged formula for finding the base, given the rate and the percentage is:

$$\text{Base} = \frac{\text{Percentage}}{\text{Rate}}$$

$$B = \frac{P}{R}$$

Practical applications involve numbers that have units or names of quantities called *denominate numbers*. The base and the percentage have the same unit or denomination. For example, if the base unit is expressed in inches, the percentage is expressed in inches. The rate is not a denominate number; it does not have a unit or denomination. Rate is the part to be taken of the whole quantity, the base.

Finding the Percentage, Given the Base and Rate

In some problems, the base and rate are given and the percentage must be found. First, express the rate (percent) as an equivalent decimal fraction. Then solve with the formula, Percentage = Rate × Base.

Examples

1. What is 15% of 60?

   The rate is 15%.
   The base is 60. It is the number of which the rate is taken—the whole or a quantity equal to 100%.
   The percentage is to be found. It is the quantity of the percent of the base.
   Express the rate, 15%, as an equivalent decimal fraction. 15% = 0.15

   Find the percentage:

   Percentage = Rate × Base, Percentage = 0.15 × 60 = 9   Ans

2. Find $56\frac{9}{25}\%$ of \$183.76.

The rate is $56\frac{9}{25}\%$.

The base is \$183.76.

The percentage is to be found.

Express the rate, $56\frac{9}{25}\%$ as an equivalent decimal fraction. $56\frac{9}{25}\% = 0.5636$

Find the percentage:

Percentage = Rate × Base, Percentage = 0.5636 × \$183.87 = \$103.57   Ans

$56\ \boxed{+}\ 9\ \boxed{+}\ 25\ \boxed{=}\ \boxed{+}\ 100\ \boxed{x}\ 183.76\ \boxed{=}\ 103.567136, \$103.57$ Ans

or $56\ \boxed{a^{b}\!/\!c}\ 9\ \boxed{a^{b}\!/\!c}\ 25\ \boxed{+}\ 100\ \boxed{x}\ 183.76\ \boxed{=}\ 103.567136, \$103.57$ Ans

Finding the Percent (Rate), Given the Base and Percentage

In some problems, the base and percentage are given, and the percent (rate) must be found. Solve with the formula

$$\text{Rate} = \frac{\text{Percentage}}{\text{Base}}$$

Examples

1. What percent of 12.87 is 9.620? Round the answer to 1 decimal place.

   Since a percent of 12.87 is to be taken, the base or whole quantity equal to 100% is 12.87.

   The percentage or quantity of the percent of the base is 9.620.

   The rate is to be found.

   Since the percentage, 9.620, is less than the base, 12.87, the rate must be less than 100%.

   Solve with the formula:

   $$\text{Rate} = \frac{\text{Percentage}}{\text{Base}}, \quad \text{Rate} = \frac{9.620}{12.87} = 0.74747$$

   Express 0.74747 as a percent. $0.74747 = 74.7\%$ Ans (rounded)

   $9.62\ \boxed{+}\ 12.87\ \boxed{x}\ 100\ \boxed{=}\ 74.74747475, 74.7\%$ Ans (rounded)

2. What percent of 9.620 is 12.87? Round the answer to 1 decimal place.

   Notice that although the numbers are the same as in Example 1, the base and percentage are reversed.

   Since a percent of 9.620 is to be taken, the base or whole quantity equal to 100% is 9.620.

   The percentage or quantity of the percent of the base is 12.87.

   Since the percentage, 12.87, is greater than the base, 9.620, the rate must be greater than 100%.

   $$\text{Rate} = \frac{\text{Percentage}}{\text{Base}}, \quad \text{Rate} = \frac{12.87}{9.620} = 1.33784$$

   Express 1.33784 as a percent. $1.33784 = 133.8\%$ Ans (rounded)

   $12.87\ \boxed{+}\ 9.62\ \boxed{x}\ 100\ \boxed{=}\ 133.7837838, 133.8\%$ Ans (rounded)

Finding the Base, Given the Percent (Rate) and the Percentage

In some problems, the percent (rate) and the percentage are given, and the base must be found. First, express the percent as its decimal fraction equivalent. Then solve with the formula

$$\text{Base} = \frac{\text{Percentage}}{\text{Rate}}$$

Examples

1. 816 is 68% of what number?

   The rate is 68%.

   Since 816 is the quantity of the percent of the base, the percentage is 816; 816 is 68% of the base.

   The base to be found is the whole quantity equal to 100%.

   Since the rate, 68% is less than 100%, the base must be greater than the percentage.

   Express 68% as an equivalent decimal fraction. 68% = 0.68

   Solve with the formula:

   $$\text{Base} = \frac{\text{Percentage}}{\text{Rate}}, \quad \text{Base} = \frac{816}{0.68} = 1,200 \qquad \text{Ans}$$

2. $149.50 is $115\frac{2}{3}\%$ of what value?

   The rate is $115\frac{2}{3}\%$.

   Since $149.50 is the quantity of the percent of the base, the percentage is $149.50; $149.50 is $115\frac{2}{3}\%$ of the base.

   The base to be found is the whole quantity equal to 100%. Since the rate, $115\frac{2}{3}\%$ is greater than 100%, the percentage must be greater than the base.

   Express $115\frac{2}{3}\%$ as an equivalent decimal fraction. $115\frac{2}{3}\% = 115.67\%$
   = 1.1567

   Solve with the formula:

   $$\text{Base} = \frac{\text{Percentage}}{\text{Rate}}, \quad \text{Base} = \frac{\$149.50}{1.1567} = \$129.25 \qquad \text{Ans}$$

      149.5 $+$ $($ 115 $+$ 2 $\div$ 3 $)$ $\times$ 100 $=$ 129.2507205, $129.25   Ans

APPLICATION

Finding Percentage

Find each percentage. Round the answers to 2 decimal places when necessary.

1. 20% of 80   _____

2. 2.15% of 80   _____

3. 60% of 200   _____

4. 15.23% of 150   _____

5. 25% of 312.6   _____

6. 7% of 140.34   _____

7. 156% of 65   _____

8. 0.8% of 214   _____

9. 12.7% of 295 _____

10. 122% of 1.68 _____

11. 140% of 280 _____

12. 1.8% of 1240 _____

13. 39% of 18.3 _____

14. 0.42% of 50 _____

15. 0.03% of 424.6 _____

16. $8\frac{1}{2}$% of 375 _____

17. $\frac{3}{4}$% of 132 _____

18. 296.5% of 81 _____

19. $15\frac{1}{4}$% of $35\frac{1}{4}$ _____

20. $\frac{17}{50}$% of $139\frac{3}{10}$ _____

Finding Percent (Rate)

Find each percent (rate). Round the answers to 2 decimal places when necessary.

21. What percent of 8 is 4? _____

22. What percent of 20.7 is 5.6? _____

23. What percent of 100 is 37? _____

24. What percent of 84.37 is 70.93? _____

25. What percent of 70.93 is 84.37? _____

26. What percent of 258 is 97? _____

27. What percent of 132.7 is 206.3? _____

28. What percent of 19.5 is 5.5? _____

29. What percent of 1.25 is 0.5? _____

30. What percent of 0.5 is 1.25? _____

31. What percent of $6\frac{1}{2}$ is 2? _____

32. What percent of 134 is $156\frac{3}{4}$? _____

33. What percent of $\frac{7}{8}$ is $\frac{3}{8}$? _____

34. What percent of $\frac{3}{8}$ is $\frac{7}{8}$? _____

35. What percent of 3.08 is 4.76? _____

36. What percent of 0.65 is 0.09? _____

37. What percent of $12\frac{1}{4}$ is 3? _____

38. What percent of 312 is 400.9? _____

39. What percent of $\frac{3}{4}$ is $\frac{3}{8}$? _____

40. What percent of $13\frac{4}{5}$ is $6\frac{3}{10}$? _____

Finding Base

Find each base. Round the answers to 2 decimal places when necessary.

41. 15 is 10% of what number? _____

42. 25 is 80% of what number? _____

43. 80 is 25% of what number? _____

44. 3.8 is 95.3% of what number? _____

45. 13.6 is 8% of what number? _____

46. 123.86 is 88.7% of what number? _____

47. 203 is 110% of what number? _____

48. $44\frac{1}{3}$ is 60% of what number? _____

49. $7\frac{1}{2}$ is 180% of what number? _____

50. 10 is $6\frac{1}{4}$ % of what number? _____

51. 190.75 is 70.5% of what number? _____

52. 6.6 is 3.3% of what number? _____

53. 88 is 205% of what number? _____

54. 1.3 is 0.9% of what number? _____

55. $\frac{7}{8}$ is 175% of what number? _____

56. $\frac{1}{10}$ is $1\frac{1}{5}$ % of what number? _____

59. $20\frac{1}{2}$ is 71% of what number? _____

57. 9.3 is 238.6% of what number? _____

60. $\frac{3}{4}$ is 123% of what number? _____

58. 0.84 is 2.04% of what number? _____

Finding Percentage, Percent, or Base

Find each percentage, percent (rate), or base. Round the answers to 2 decimal places when necessary.

61. What percent of 24 is 18? _____

67. 72.4% of 212.7 is _____.

62. What is 30% of 50? _____

68. What percent of 228 is 256? _____

63. What is 123.8% of 12.6? _____

69. 51.03 is 88% of what number? _____

64. 73 is 82% of what number? _____

70. 36.5 is _____% of 27.6.

65. What percent of $10\frac{1}{2}$ is 2? _____

71. $2\frac{1}{4}$% of 150 is _____.

66. _____ is 48% of 94.82?

72. _____ is 18% of 120.66?

UNIT **19** Percent Practical Applications

Objectives   After studying this unit you should be able to

- Solve simple percentage practical applications in which two of the three parts are given.
- Solve more complex percentage practical applications in which two of the three parts are not directly given.

Identifying Rate, Base, and Percentage in Various Types of Practical Applications

In solving simple problems, generally, there is no difficulty in identifying the rate or percent. A common mistake is to incorrectly identify the percentage and the base. There is sometimes confusion as to whether a value is a percentage or a base; the base and percentage are incorrectly interchanged.

The following statements summarize the information that was given when each of the three types of problems was discussed and solved. A review of the statements should be helpful in identifying the rate, percentage, and base.

- The rate (percent) is the part taken of the whole quantity (base).
- The base is the whole quantity or a quantity that is equal to 100%. It is the quantity of which the rate is taken.
- The percentage is the quantity of the percent that is taken of the base. It is the quantity equal to the percent that is taken of the whole.

- If the rate is 100%, the percentage and the base are the same quantity.
  If the rate is less than 100%, the percentage is less than the base.
  If the rate is greater than 100%, the percentage is greater than the base.

- In practical applications, the percentage and the base have the same unit or denomination. The rate does not have a unit or denomination.

- The word *is* generally relates to the rate or percentage, and the word *of* generally relates to the base.

Finding Percentage in Practical Applications

Example A production run of steel pins is estimated as $3,275. Material cost is estimated as 35% of the total cost. What is the estimated material cost to the nearest dollar?

Think the problem through to determine what is given and what is to be found.
The rate is 35%.
The base is $3,275. It is the total cost or the whole quantity.
The percentage, which is the material cost, is to be found.
35% = 0.35

Percentage = Rate × Base, Percentage = 0.35 × $3,275 = $1,146   Ans

Finding Percent (Rate) in Practical Applications

Example An inspector rejects 23 out of a total production of 630 electrical switches. What percent of the total production is rejected? Round the answer to 1 decimal place.

Think the problem through to determine what is given and what is to be found.
Since a percent of the total production of 630 switches is to be found, the base or whole quantity equal to 100% is 630 switches.
The percentage or quantity of the percent of the base is 23 switches.
The rate is to be found.

$$\text{Rate} = \frac{\text{Percentage}}{\text{Base}}, \quad \text{Rate} = \frac{23 \text{ switches}}{630 \text{ switches}} = 0.03651$$

Express 0.03651 as a percent. 0.03651 = 3.7%   Ans (rounded)

 23 ÷ 630 × 100 = 3.650793651, 3.7%   Ans (rounded)

Finding the Base in Practical Applications

Example A motor is said to be 80% efficient if the output (power delivered) is 80% of the input (power received). How many horsepower does a motor receive if it is 80% efficient with a 6.20 horsepower (hp) output?

Think the problem through to determine what is given and what is to be found.
The rate is 80%.
Since the output of 6.20 hp is the quantity of the percent of the base, the percentage is 6.20 hp (6.20 hp is 80% of the base).
The base to be found is the input; the whole quantity equal to 100%.
Express 80% as an equivalent decimal fraction. 80% = 0.80

$$\text{Base} = \frac{\text{Percentage}}{\text{Rate}}, \quad \text{Base} = \frac{6.20 \text{ hp}}{0.80} = 7.75 \text{ hp}   \text{Ans}$$

More Complex Percentage Practical Applications

In certain percentage problems, two of the three parts are not directly given. One or more additional operations may be required in setting up and solving a problem. Examples of these types of problems follow.

Examples

1. By replacing high-speed cutters with carbide cutters, a machinist increases production by 35%. Using carbide cutters, 270 pieces per day are produced. How many pieces per day were produced with high-speed steel cutters?

   Think the problem through. The base (100%) is the daily production using high-speed steel cutters. Since the base is increased by 35%, the carbide cutter production of 270 pieces is 100% + 35% or 135% of the base. Therefore, the rate is 135% and the percentage is 270. The base is to be found.

   Express 135% as an equivalent decimal fraction. $135\% = 1.35$

   $$\text{Base} = \frac{\text{Percentage}}{\text{Rate}}, \quad \text{Base} = \frac{270 \text{ pieces per day}}{1.35} = 200 \text{ pieces per day} \quad \text{Ans}$$

2. A mechanic purchases a set of socket wrenches for $54.94. The purchase price is 33% less than the list price. What is the list price?

   Think the problem through. The base (100%) is the list price. Since the base is decreased by 33%, the purchase price, $54.94, is 100% − 33% or 67% of the base. Therefore, the rate is 67% and the percentage is $54.94. The base is to be found.

   Express 67% as its equivalent decimal fraction. $67\% = 0.67$

   $$\text{Base} = \frac{\text{Percentage}}{\text{Rate}}, \quad \text{Base} = \frac{\$54.94}{0.67} = \$82 \quad \text{Ans}$$

3. An aluminum bar measures 137.168 millimeters before it is heated. When heated, the bar measures 137.195 millimeters. What is the percent increase in length? Express the answers to 2 decimal places.

   Think the problem through. The base (100%) is the bar length before heating, 137.168 millimeters. The increase in length is 137.195 millimeters − 137.168 millimeters or 0.027 millimeter. Therefore, the percentage is 0.027 millimeter, and the base is 137.168 millimeters. The rate (percent) is to be found.

   $$\text{Rate} = \frac{\text{Percentage}}{\text{Base}}, \quad \text{Rate} = \frac{0.027 \text{ mm}}{137.168 \text{ mm}} = 0.0001968$$

   Express 0.0001968 as a percent. $0.0001968 = 0.01968\%$

   0.02% Ans (rounded)

 $137.195 \boxed{-} 137.168 \boxed{=} \boxed{\div} 137.168 \boxed{\times} 100 \boxed{=} 0.019683891$

   0.02% Ans (rounded)

APPLICATION

Finding Percentage, Percent, and Base in Practical Applications

Solve the following problems.

1. The total amount of time required to machine a part is 12.5 hours. Milling machine operations take 7.0 hours. What percent of the total time is spent on the milling machine?

2. A casting, when first poured, is 17.875 centimeters long. The casting shrinks 0.188 centimeter as it cools. What is the percent shrinkage? Round the answer to 2 decimal places.

_____

3. A machine operator completes a job in 80% of the estimated time. The estimated time is $8\frac{1}{2}$ hours. How long does the job actually take?

_____

4. A machine is sold for 42% of the original cost. If the original cost is $9,255.00, find the selling price of the used machine.

_____

5. On a production run, 8% of the units manufactured are rejected. If 120 units are rejected how many total units are produced?

_____

6. An engine loses 4.2 horsepower through friction. The power loss is 6% of the total rated horsepower. What is the total horsepower rating?

_____

7. A small manufacturing plant employs 130 persons. On certain days, 16 employees are absent. What percent of the total number of employees are absent? Round the answer to the nearest whole percent.

_____

8. This year's earnings of a company are 140% of last year's earnings. The company earned $910,000 this year. How much did the company earn last year?

_____

9. In three hours 73.50 feet of railing are fabricated. This is 28% of a total order. How many feet of railing are ordered?

_____

10. How many pounds of manganese bronze can be made with 955.0 pounds of copper if the manganese bronze is to contain 58% copper by weight? Round the answer to the nearest whole pound.

_____

11. An alloy of manganese bronze is made up by weight of 58% copper, 40% tin, 1.5% manganese, and 0.5% other materials. How many pounds of each metal are there in 1,250 pounds of alloy? Round the answers to the nearest whole pound.

_____

12. A manufacturer estimates the following percent costs to produce a product: labor, 38%; materials, 45%; overhead, 17%. The total cost of production is $120,000. Determine each of the dollar costs.

_____

13. An iron casting shrinks $\frac{1}{8}$ inch per foot. What is the percent shrinkage? Round the answer to the nearest whole percent.

_____

14. A hot brass casting when first poured in a mold is 9.25 inches long. The shrinkage is 1.38%. What is the length of the casting when cooled? Round the answer to 2 decimal places.

_____

15. The following table shows the number of pieces manufactured in three consecutive days. The numbers of defective pieces are shown as rework and scrap for each day. Determine the percents of rework and scrap for each day. Round the answers to 1 decimal place.

_____

Date	Number of Pieces Manufactured	Number of Defective Pieces		% Defective Pieces	
		Rework	Scrap	Rework	Scrap
9/16	1,650	44	59		
9/17	1,596	29	48		
9/18	1,685	52	34		

16. The power output of a machine is equal to the product of the power input and the percent efficiency. Power Output = Power Input × Percent Efficiency. What is the output of a machine with an 8.0 horsepower motor running at full capacity and at 82% efficiency? Round the answer to 1 decimal place.

17. Material cost for a job is $1,260. The cost is 38.6% of the total cost. What is the total cost? Round the answer to the nearest dollar.

18. A machinist's weekly gross income is $745. The following percent deductions are made from the gross income:

> Federal Income Tax, 14.20%
> State Income Tax, 4.50%
> Social Security, 7.60%

Determine the net income (take-home pay) after the deductions are made.

19. In the heat treatment of steel, a rough approximation of temperatures can be made by observing the color of the heated steel. At approximately 1300°F (degrees Fahrenheit) the steel is dark red. What percent increase in temperature from the 1300°F must be made for the heated steel to turn to each of the following colors? Round the answers to the nearest whole percent.

    a. Dull cherry-red at approximately 1470°F

    b. Orange-yellow at approximately 2200°F

    c. Brilliant white at approximately 2730°F

20. The following table lists the percent of carbon by weight for various types of carbon steel tools. Determine the amount of carbon needed to produce 2.60 tons of carbon steel required in the production of each type of tool. Round the answers to the nearest pound.

Type of Machinist's Tool	Percent Carbon	Type of Machinist's Tool	Percent Carbon
1. Twist Drill	1.15	4. Ordinary File	1.25
2. Wrench	0.75	5. Machinist's Hammer	0.95
3. Threading Die	1.05	6. Chuck Jaw	0.85

21. A machine shop has 2,840 castings in stock at the beginning of the month. At the end of the first week, 28.0% of the stock is used. At the end of the second week, 50.0% of the stock remaining is used. How many castings remain in stock at the end of the second week?

22. It is estimated that 125 metres of channel iron are required for a job. Channel iron is ordered, including an additional 20% allowance for scrap and waste. Actually, 175 metres of channel iron are used for the job. The amount actually used is what percent more than the estimated amount? Round the answer to the nearest whole percent.

23. An alloy of red brass is composed of 85% copper, 5% tin, 6% lead, and zinc. Find the number of pounds of zinc required to make 450 pounds of alloy.

24. The day shift of a manufacturing firm produces 6% defective pieces out of a total production of 1,638 pieces. The night shift produces $4\frac{1}{2}$% of defective pieces out of a total of 1,454 pieces. How many more acceptable pieces are produced by the day shift than by the night shift?

25. The following table shows the number of pieces of a product produced each day during one week. Also shown are the number of pieces rejected each day by the quality control department. Find the percent rejection for the week's production. Round the answer to 1 decimal place.

	MON.	TUES.	WED.	THUR.	FRI.
Number of Pieces Produced	735	763	786	733	748
Number of Pieces Rejected	36	43	52	47	31

26. A manufacturer estimates that 15,500 pieces per day could be produced with the installation of new machinery. The machines now used produce 11,000 pieces per day. What percent increase in production would be gained by replacing the present machinery with new machinery? Round the answer to the nearest whole percent.

27. The average percent defective product of a manufacturing plant is 1.20%. On a particular day 50 pieces were rejected out of a total daily production of 2,730 pieces. What is the percent increase of defective pieces for the day above the average percent defective? Round the answer to 2 decimal places.

28. Before machining, a steel forging weighs 7.8 pounds. A milling operation removes 1.5 pounds, drilling removes 0.7 pound, and grinding removes 0.5 pound. What percent of the original weight of the forging is the final machined forging? Round the answer to 1 decimal place.

29. A machine is 85% efficient and loses 1.3 horsepower through its drivetrain. Determine the horsepower input of the machine. Round the answer to 1 decimal place.

30. The cost of one dozen cutters is listed as $525. A multiple discount of 12% and 8% is applied to the purchase. Determine the net (selling) price of the cutters.

   ➤ **Note:** With multiple discounts, the first discount is subtracted from the list price. The second discount is subtracted from the price computed after the first discount was subtracted.

31. A manufacturer's production this week is 3,620 pieces. This is 13.5% greater than last week's production. Find last week's production. Round the answer to the nearest whole piece.

32. Two machines are used to produce the same product. One machine has a capability of producing 750 pieces per 8-hour shift. It is operating at 80% of its capability. The second machine has a capability of producing 900 pieces per 8-hour shift. It is operating at 75% of its capability. Find the total number of pieces produced per hour with both machines operating. Round the answer to the nearest whole piece.

33. Allowing for scrap, a firm produced 1,890 pieces. The number produced is 8% more than the number of pieces required for the order. How many pieces does the order call for? Round the answer to the nearest whole piece.

34. A manufacturing company receives $122,000 upon the completion of a job. Total expenses for the job are $110,400. What percent of the job is profit? Round the answer to 1 decimal place.

35. Manufacturing costs consist of labor costs, material costs, and overhead. Refer to the following table. What percent of the total manufacturing cost for each of Jobs 1, 2, and 3 is each manufacturing cost? Round the answer to the nearest whole percent.

1. _____

2. _____

3. _____

	MANUFACTURING COSTS		
Job	Labor Costs	Material Costs	Overhead Costs
1	$1,890	$ 875	$1,240
2	$ 930	$1,060	$ 880
3	$2,490	$1,870	$1,600

UNIT 20   Achievement Review— Section One

Objective

You should be able to solve the exercises and problems in this Achievement Review by applying the principles and methods covered in units 1–19.

1. Express each of the following fractions as equivalent fractions as indicated.

   a. $\frac{3}{8} = \frac{?}{32}$ _____

   b. $\frac{7}{10} = \frac{?}{100}$ _____

   c. $\frac{1}{4} = \frac{?}{64}$ _____

   d. $\frac{9}{16} = \frac{?}{128}$ _____

2. Express each of the following mixed numbers as improper fractions.

   a. $3\frac{1}{5}$ _____

   b. $2\frac{9}{10}$ _____

   c. $5\frac{3}{4}$ _____

   d. $13\frac{3}{8}$ _____

   e. $6\frac{9}{32}$ _____

3. Express each of the following improper fractions as mixed numbers.

   a. $\frac{5}{2}$ _____

   b. $\frac{21}{5}$ _____

   c. $\frac{75}{4}$ _____

   d. $\frac{115}{32}$ _____

   e. $\frac{329}{64}$ _____

4. Express each of the following fractions as a fraction in lowest terms.

   a. $\frac{8}{16}$ _____

   b. $\frac{12}{100}$ _____

   c. $\frac{30}{32}$ _____

   d. $\frac{18}{64}$ _____

   e. $\frac{28}{128}$ _____

5. Express the fractions in each of the following sets as equivalent fractions having the least common denominator.

   a. $\frac{1}{4}, \frac{3}{16}, \frac{9}{32}$ _____

   b. $\frac{7}{16}, \frac{5}{32}, \frac{9}{64}$ _____

   c. $\frac{7}{10}, \frac{3}{4}, \frac{9}{25}, \frac{13}{20}$ _____

6. Add or subtract each of the following values. Express the answers in lowest terms.

a. $\frac{1}{8} + \frac{5}{8}$ _____

b. $\frac{7}{16} + \frac{15}{16}$ _____

c. $\frac{5}{8} + \frac{13}{32}$ _____

d. $3\frac{7}{10} + \frac{49}{100}$ _____

e. $\frac{9}{32} + \frac{1}{4} + \frac{21}{64}$ _____

f. $\frac{11}{16} - \frac{7}{16}$ _____

g. $\frac{17}{20} - \frac{3}{5}$ _____

h. $\frac{49}{64} - \frac{3}{8}$ _____

i. $6 - \frac{11}{16}$ _____

j. $13\frac{1}{8} - 9\frac{7}{32}$ _____

7. Multiply or divide each of the following values. Express the answers in lowest terms.

a. $\frac{1}{2} \times \frac{5}{8}$ _____

b. $\frac{3}{4} \times \frac{4}{5} \times \frac{2}{3}$ _____

c. $5\frac{7}{32} \times \frac{3}{8}$ _____

d. $3\frac{1}{10} \times 8\frac{1}{4}$ _____

e. $\frac{3}{16} \times 20 \times 5\frac{1}{2}$ _____

f. $\frac{3}{10} \div \frac{2}{5}$ _____

g. $\frac{14}{15} \div \frac{7}{25}$ _____

h. $16 \div \frac{1}{3}$ _____

i. $2\frac{17}{32} \div \frac{9}{24}$ _____

j. $2\frac{29}{32} \div 8\frac{3}{4}$ _____

8. Perform each of the indicated combined operations.

a. $\frac{3}{4} + \frac{5}{16} - \frac{3}{8}$ _____

b. $20\frac{1}{2} + 3\frac{3}{8} \times \frac{1}{8}$ _____

c. $\frac{3}{5} \times \frac{7}{8} + 3\frac{3}{4}$ _____

d. $\left(18 - 5\frac{3}{4}\right) \div \frac{1}{2} + 3\frac{7}{8}$ _____

e. $\dfrac{16 - 4\frac{1}{2}}{\frac{1}{2} + 2\frac{1}{8}}$ _____

f. $\dfrac{5\frac{1}{4} \times \frac{1}{2}}{6 \div 3\frac{3}{4}}$ _____

9. How many complete pieces can be blanked from a strip of aluminum 72 inches long if each stamping requires $1\frac{3}{8}$ inches of material plus an allowance of $\frac{3}{4}$ inch at one end of the strip? _____

10. How many inches of bar stock are needed to make 30 spacers each $1\frac{3}{16}$ inches long? Allow $\frac{1}{8}$ inch waste for each spacer. _____

11. A shaft is turned at 200 revolutions per minute with a tool feed of $\frac{1}{32}$ inch per revolution. How many minutes does it take to cut a distance of 50 inches along the shaft? _____

12. A shop order calls for 1800 steel pins each $1\frac{5}{8}$ inches long. If $\frac{3}{16}$ inch is allowed for cutting off and facing each pin, how many complete 10-foot lengths of stock are needed for the order? (12 inches = 1 foot) _____

13. Compute dimensions A, B, C, D, and E of the support bracket shown. All dimensions are given in inches.

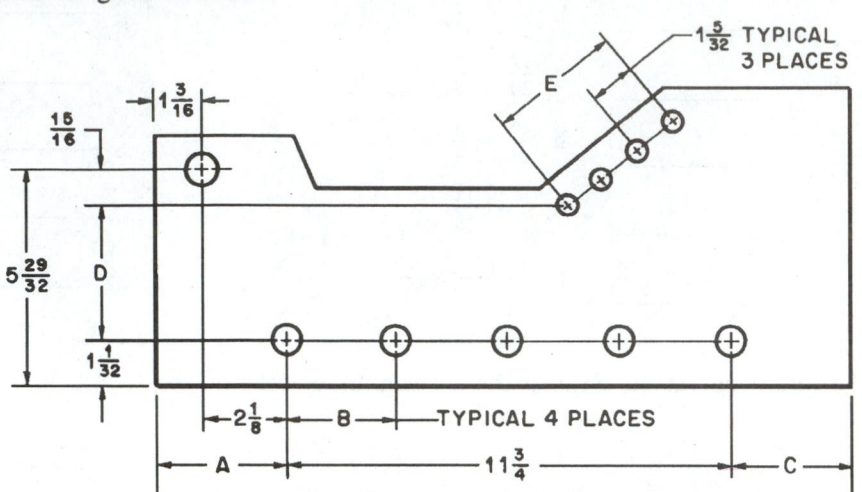

A = _____

B = _____

C = _____

D = _____

E = _____

14. Write each of the following numbers as words.

    a. 0.6 _____

    b. 0.74 _____

    c. 0.147 _____

    d. 0.0086 _____

    e. 4.208 _____

    f. 16.0419 _____

15. Write each of the following words as decimal fractions or mixed decimals.

    a. three tenths _____

    b. twenty-six thousandths _____

    c. nine and twenty-six thousandths _____

    d. five and eighty-one ten-thousandths _____

16. Round each of the following numbers to the indicated number of decimal places.

    a. 0.596 (2 places) _____

    b. 5.0463 (3 places) _____

    c. 0.80729 (4 places) _____

    d. 7.0005 (3 places) _____

17. Express each of the following common fractions as decimal fractions. Where necessary, round the answers to 3 decimal places.

    a. $\frac{3}{4}$ _____

    b. $\frac{7}{8}$ _____

    c. $\frac{2}{3}$ _____

    d. $\frac{2}{25}$ _____

    e. $\frac{13}{20}$ _____

18. Express each of the following decimal fractions as common fractions in lowest terms.

    a. 0.7 _____

    b. 0.525 _____

    c. 0.007 _____

    d. 0.915 _____

    e. 0.0075 _____

19. Add or subtract each of the following values.

    a. 0.875 + 0.712 _____

    b. 5.004 + 0.92 + 0.5034 _____

    c. 0.006 + 12.3 + 0.0009 _____

    d. 2.99 + 6.015 + 0.1003 _____

    e. 23 + 0.0007 + 0.007 + 0.4 _____

    f. 0.879 − 0.523 _____

    g. 0.1863 − 0.0419 _____

    h. 5.400 − 5.399 _____

    i. 0.009 − 0.0068 _____

    j. 14.001 − 13.999 _____

20. Multiply or divide each of the following values. Round the answer to 4 decimal places where necessary.

a. 0.923×0.6 _____ f. $0.85 \div 0.39$ _____

b. 3.63×2.30 _____ g. $0.100 \div 0.01$ _____

c. 4.81×0.07 _____ h. $4.016 \div 0.03$ _____

d. 0.005×0.180 _____ i. $123 \div 0.665$ _____

e. 12.123×0.001 _____ j. $0.0098 \div 5.036$ _____

21. Raise each of the following values to the indicated powers.

a. 2.6^2 _____ c. 0.006^2 _____ e. $\left(\dfrac{20.8}{6.5}\right)^3$ _____

b. 0.50^3 _____ d. $\left(\dfrac{3}{5}\right)^2$ _____

22. Determine the whole number roots of each of the following values as indicated.

a. $\sqrt{49}$ _____ d. $\sqrt{46.83 + 17.17}$ _____

b. $\sqrt[3]{64}$ _____ e. $\sqrt{39.2 \times 1.25}$ _____

c. $\sqrt{\dfrac{36}{81}}$ _____

23. Determine the square roots of each of the following values to the indicated number of decimal places.

a. $\sqrt{379}$ (2 places) _____ c. $\sqrt{\dfrac{2}{5}}$ (3 places) _____

b. $\sqrt{0.8736}$ (3 places) _____ d. $\sqrt{93.876 - 47.904}$ (3 places) _____

24. Find the decimal or fraction equivalents of each of the following numbers using the decimal equivalent table.

a. $\dfrac{5}{8}$ _____ c. $\dfrac{21}{64}$ _____ e. 0.671875 _____

b. $\dfrac{17}{32}$ _____ d. 0.65625 _____

25. Determine the nearer fractional equivalents of each of the following decimals using the decimal equivalent table.

a. 0.465 _____ b. 0.769 _____ c. 0.038 _____ d. 0.961 _____

26. Solve each of the following combined operations expressions. Round answers to 2 decimal places.

a. $0.4321 + 10.870 + 3.43 \times 0.93$ _____ c. $35.98 + \sqrt{6.35 - 4.81}$ _____

b. $(12.60 \div 3 - 0.98)^2 \times 3.60$ _____ d. $6 \times \left(\dfrac{\sqrt{81} \times 4.03}{3.30 \times 2.75}\right) - 1.7^2$ _____

27. The basic form of an ISO Metric Thread is shown. Given a thread pitch of 1.5 millimeters, compute thread dimensions A, B, C, D, E, and F to 3 decimal places.

A = _____

B = _____

C = _____

D = _____

E = _____

F = _____

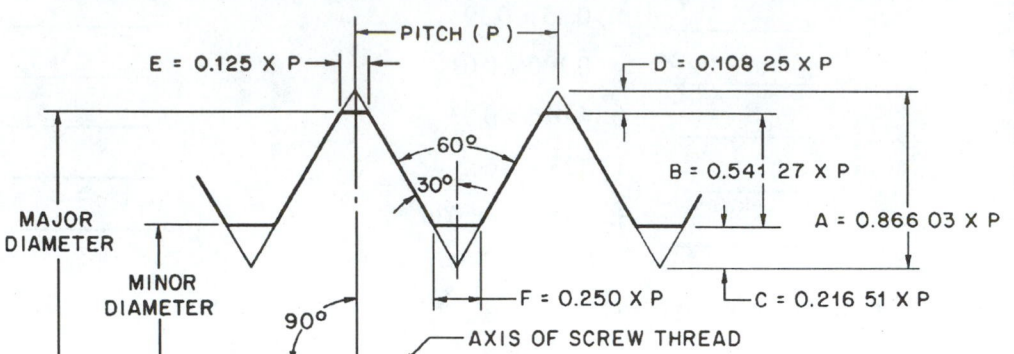

28. A combination of three gage blocks is selected to provide a total thickness of 0.4573 inch. Two blocks 0.250 inch and 0.118 inch thick are selected. What is the required thickness of the third block?

_____

29. A piece of round stock is being turned to a 17.86-millimeter diameter. A machinist measures the diameter of the piece as 18.10 millimeters. What depth of cut should be made to turn the piece to the required diameter?

_____

30. A plate 57.20 millimeters thick is to be machined to a thickness of 44.10 millimeters. The plate is to be rough cut with the last cut a finish cut 0.30 millimeter deep. If each rough cut is 3.20 millimeters deep, how many rough cuts are required?

_____

31. A shaft is turned in a lathe at 120 revolutions per minute. The cutting tool advances 0.030 inch per revolution. How long is the length of cut along the shaft at the end of 3.50 minutes?

_____

32. Express each value as a percent.

a. 0.72 _____   b. 2.037 _____   c. $\frac{1}{25}$ _____   d. 0.0003 _____

33. Express each percent as a decimal fraction or mixed decimal.

a. 19% _____   b. 0.7% _____   c. $\frac{3}{4}$% _____   d. $310\frac{3}{10}$% _____

34. Express each percent as a common fraction or mixed number.

a. 30% _____   b. 140% _____   c. 12.5% _____   d. 0.65% _____

35. Find each percentage. Round the answers to 2 decimal places when necessary.

a. 15% of 60 _____   f. 130% of 212 _____

b. 3% of 42.3 _____   g. $12\frac{1}{2}$% of 32 _____

c. 72.8% of 120 _____   h. $\frac{1}{4}$% of 627.3 _____

d. 0.7% of 812 _____

e. 42.6% of 53.76 _____

36. Find each percent (rate). Round the answers to 2 decimal places when necessary.

    a. What percent of 10 is 2? _____

    b. What percent of 2 is 10? _____

    c. What percent of 88.7 is 21.9? _____

    d. What percent of 275 is 108? _____

    e. What percent of 2.84 is 0.8? _____

    f. What percent of $12\frac{1}{4}$ is 3? _____

    g. What percent of 312 is 400.9? _____

37. Find each base. Round the answers to 2 decimal places when necessary.

    a. 20 is 60% of what number? _____

    b. 4.1 is 24.9% of what number? _____

    c. 340 is 152% of what number? _____

    d. 44.08 is 73.5% of what number? _____

    e. 9.3 is 238.6% of what number? _____

    f. 0.84 is 2.04% of what number? _____

    g. $\frac{3}{4}$ is 123% of what number? _____

38. Find each percentage, percent (rate), or base. Round the answers to 2 decimal places when necessary.

    a. What percent of 24 is 18? _____

    b. What is 30% of 50? _____

    c. What is 123.8% of 12.6? _____

    d. 73 is 82% of what number? _____

    e. What percent of $10\frac{1}{2}$ is 2? _____

    f. __?__ is 48% of 94.82. _____

    g. 72.4% of 212.7 is __?__ . _____

    h. What percent of 228 is 256? _____

    i. 51.08 is 88% of what number? _____

    j. 36.5 is __?__ % of 27.6. _____

    k. $2\frac{1}{4}$% of 150 is __?__ . _____

    l. __?__ is 18% of 120.66. _____

39. The carbon content of machine steel for gauges usually ranges from 0.15% to 0.25%. Round the answers for a and b to 2 decimal places.

    a. What is the minimum weight of carbon in 250 kilograms of machine steel? _____

    b. What is the maximum weight of carbon in 250 kilograms of machine steel? _____

40. A piece of machinery is purchased for $8,792. In one year the machine depreciates 14.5%. By how many dollars does the machine depreciate in one year? Round the answer to the nearest dollar. _____

41. Engine pistons and cylinder heads are made of an aluminum casting alloy that contains 4% silicon, 1.5% magnesium, and 2% nickel. Round the answers to the nearest tenth kilogram.

    a. How many kilograms of silicon are needed to produce 575 kilograms of alloy? _____

    b. How many kilograms of magnesium are needed to produce 575 kilograms of alloy? _____

    c. How many kilograms of nickel are needed to produce 575 kilograms of alloy? _____

42. Before starting two jobs, a shop has an inventory of eighteen 15.0-foot lengths of flat stock. The first job requires 30% of the inventory. The second job requires 25% of the inventory remaining after the first job. How many feet of flat stock remain in inventory at the end of the second job? Round the answer to the nearest whole foot. _____

43. An alloy of stainless steel contains 73.6% iron, 18% chromium, 8% nickel, 0.1% carbon, and sulfur. How many pounds of sulfur are required to make 5,800 pounds of stainless steel? Round the answer to the nearest whole pound.

   _____

44. Two machines together produce a total of 2,015 pieces. Machine A operates for $6\frac{1}{2}$ hours and produces an average of 170 pieces per hour. Machine B operates for 7 hours. What percent of the average hourly production of Machine A is the average hourly production of Machine B? Round the answer to the nearest whole percent.

   _____

Section Two
Linear Measurement: English and Metric

UNIT 21 English and Metric Units of Measure

Objectives After studying this unit you should be able to

- Express English lengths as larger or smaller English linear units.
- Express metric lengths as larger or smaller metric linear units.
- Express metric length units as English length units.
- Express English length units as metric length units.

The ability to measure with tools and instruments and to compute measurements is a basic requirement in the machine trades. The units of measure used in the United States are established and maintained by the Bureau of Standards. These units of measure are based on international standards. Industrial standards of measure are determined by the American National Standards Institute (ANSI). This institute, with the cooperation of other similar organizations throughout the world, establishes and maintains industrial standards of measure, specifications and practice.

In 1960, a modernized system of metrics called the International System of Units (SI) was established by international agreement. All but a few countries are now converting from nonmetric systems or are revising their version of the metric system to SI standards. Presently, the United States uses both the English and the SI metric systems of measure. The use of the metric system in this country is continually increasing. It is important that the machine craftsperson be able to compute and measure with metric units as well as with English units.

Both the English and metric systems include all types of units of measure, such as length, area, volume, and capacity. In the machine trades, linear or length measure is used most often.

Measurement Definitions

Measurement is the comparison of a quantity with a standard unit. A *linear measurement* is a means of expressing the distance between two points; it is the measurement of lengths. A linear measurement has two parts: a unit of length and a multiplier.

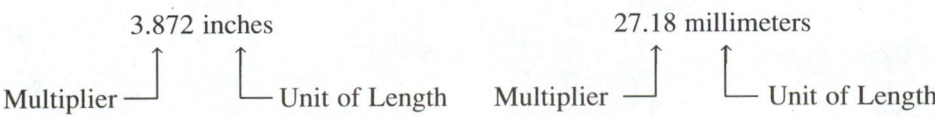

The measurements 3.872 inches and 27.18 millimeters are examples of denominate numbers. A *denominate number* is a number that specifies a unit of measure.

English Units of Linear Measure

The yard is the standard unit of linear measure in the English system. From the yard, other units such as the inch and foot are established. The smallest unit is the inch. Common English units of length with their symbols are shown in this table.

ENGLISH UNITS OF LINEAR MEASURE
1 yard (yd) = 3 feet (ft)
1 yard (yd) = 36 inches (in)
1 foot (ft) = 12 inches (in)
1 mile (mi) = 1760 yards (yd)
1 mile (mi) = 5280 feet (ft)

In the machine trades, English linear units other than the inch are seldom used. English measure dimensions on engineering drawings are given in inches. Although English linear units other than the inch are rarely required for on-the-job applications, you should be able to use any units in the system.

Expressing Larger English Units of Linear Measure as Smaller Units

Procedure   To express a larger unit of length as a smaller unit of length

- Multiply the given length by the number of smaller units contained in one of the larger units.

Example 1   Express $2\frac{1}{2}$ feet as inches.

Since 12 inches equal 1 foot,
multiply $2\frac{1}{2}$ by 12.

$$2\frac{1}{2} \times 12 = 30$$

$$2\frac{1}{2} \text{ feet} = 30 \text{ inches}   \text{Ans}$$

Example 2   How many inches are in 0.25 yard?

Since 36 inches equal 1 yard,
multiply 0.25 by 36.

$$0.25 \times 36 = 9$$

$$0.25 \text{ yard} = 9 \text{ inches}   \text{Ans}$$

Expressing Smaller English Units of Linear Measure as Larger Units

Procedure   To express a smaller unit of length as a larger unit of length

- Divide the given length by the number of smaller units contained in one of the larger units.

Example 1   Express 67.2 inches as feet.

Since 12 inches equal 1 foot,
divide 67.2 by 12.

$$67.2 \div 12 = 5.6$$

$$67.2 \text{ inches} = 5.6 \text{ feet}   \text{Ans}$$

Example 2   How many yards are in 122.4 inches?

Since 36 inches equal 1 yard,
divide 122.4 by 36.

$$122.4 \div 36 = 3.4$$
$$122.4 \text{ inches} = 3.4 \text{ yards} \quad \text{Ans}$$

Metric Units of Linear Measure

An advantage of the metric system is that it allows easy and fast computations. Since metric units are based on powers of ten, computations are simplified. To express a metric unit as a smaller or larger unit, all that is required is to move the decimal point a certain number of places to the left or right.

The metric system does not require difficult conversions as with the English system. For example, it is easier to remember that 1000 meters equal 1 kilometer than to remember that 1720 years equal 1 mile. The meter is the standard unit of linear measure in the metric system. Other linear metric units are based on the meter.

You will observe that in various technical publications and materials the spellings of metric units end in *er* or *re*. For example, units are expressed as *meter* or *metre* and *millimeter* or *millimetre*. Generally, in the manufacturing industry, the *er* spelling is used. Therefore, throughout this book *er* spellings of metric units are used.

Metric measure dimensions on engineering drawings are given in millimeters. In the machine trades, metric linear units other than the millimeter are seldom used. However, you should be able to use any units in the system. Metric units of length with their symbols are shown in this table. Observe that each unit is ten times greater than the unit directly above it.

METRIC UNITS OF LINEAR MEASURE	
1 millimeter (mm) = 0.001 meter (m)	1000 millimeters (mm) = 1 meter (m)
1 centimeter (cm) = 0.01 meter (m)	100 centimeters (cm) = 1 meter (m)
1 decimeter (dm) = 0.1 meter (m)	10 decimeters (dm) = 1 meter (m)
1 meter (m) = 1 meter (m)	1 meter (m) = 1 meter (m)
1 dekameter (dam) = 10 meters (m)	0.1 dekameter (dam) = 1 meter (m)
1 hectometer (hm) = 100 meters (m)	0.01 hectometer (hm) = 1 meter (m)
1 kilometer (km) = 1000 meters (m)	0.001 kilometer (km) = 1 meter (m)

The following metric power of ten prefixes are based on the meter.

milli means one thousandth (0.001)
centi means one hundredth (0.01)
deci means one tenth (0.1)

deka means ten (10)
hecto means hundred (100)
kilo means thousand (1000)

The most frequently used metric units of length are the kilometer (km), meter (m), centimeter (cm), and millimeter (mm). In actual applications, the dekameter (dam) and hectometer (hm) are not used. The decimeter (dm) is seldom used.

Writing Metric Quantities

Periods are *not* used after the unit symbols. For example, write 1 mm, *not* 1 m.m. or 1mm., when expressing the millimeter as a symbol. A comma is *not* used to separate digits in groups of three. Many countries use commas for decimal markers. To avoid confusion, a space is left between groups of three digits counting from the decimal point. If there are only four digits to the left or right of the decimal point, the space is optional.

Examples   1. Write 11 240 mm, *not* 11,240 mm.
2. Write 0.869 54 mm, *not* 0.86954 mm.

Expressing Equivalent Units Within the Metric System

To express a given unit of length as a larger unit, move the decimal point a certain number of places to the left. To express a given unit of length as a smaller unit, move the decimal point a certain number of places to the right. The procedure of moving decimal points is shown in the following examples. Refer to the table of metric units of linear measure.

Example 1 Express 72 millimeters (mm) as centimeters (cm).

Since a centimeter is the next larger unit to a millimeter, move the decimal point 1 place to the left. (In moving the decimal point 1 place to the left, you are actually dividing by 10.)

7 2 .

72 mm = 7.2 cm   Ans

Example 2 Express 0.96 centimeter (cm) as millimeters (mm).

Since a millimeter is the next smaller unit to a centimeter, move the decimal point 1 place to the right. (In moving the decimal point 1 place to the right, you are actually multiplying by 10.)

0 . 9 6

0.96 cm = 9.6 mm   Ans

Example 3 Express 0.245 meter (m) as millimeters (mm).

Since a millimeter is three smaller units from a meter, move the decimal point 3 places to the right. (In moving the decimal point 3 places to the right, you are actually multiplying by 10^3 or 1000.)

0 . 2 4 5

0.245 m = 245 mm   Ans

Example 4 Add.   0.3 meter (m) + 12.6 centimeters (cm) + 76 millimeters (mm) Express the answer in millimeters.

Express each value in millimeters.

$$
\begin{aligned}
0.3 \text{ m} &= 300 \text{ mm} \\
12.6 \text{ cm} &= 126 \text{ mm} \\
76 \text{ mm} &= 76 \text{ mm}
\end{aligned}
$$

Add.
300 $\boxed{+}$ 126 $\boxed{+}$ 76 $\boxed{=}$ 502, 502 mm   Ans

Metric-English Linear Equivalents (Conversion Factors)

Since both the English and metric systems are used in this country, it is sometimes necessary to express equivalents between systems. Dimensioning an engineering drawing with both English and metric dimensions is called dual dimensioning. Since dual dimensioning tends to clutter a drawing and introduces additional opportunities for error, many companies do not use the system. Instead, some companies use metric dimensions only, with an inch-millimeter conversion table on or attached to the print. However, certain companies use dual dimensioning; it can be a practical method for industries which have plants in foreign countries. Examples of two types of dual dimensioning are shown.

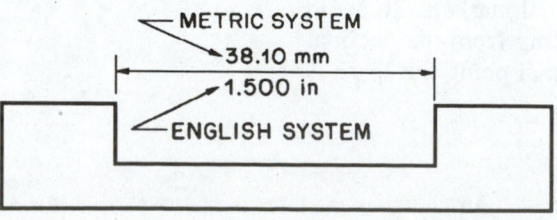
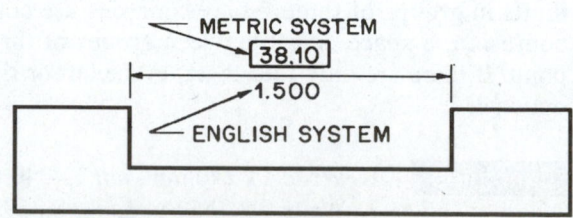

The commonly used equivalent factors of linear measure are shown in this table. Equivalent factors are commonly called conversion factors.

METRIC-ENGLISH LINEAR EQUIVALENTS (CONVERSION FACTORS)	
Metric to English Units	**English to Metric Units**
1 millimeter (mm) = 0.03937 inch (in)	1 inch (in) = 25.4 millimeters (mm)
1 centimeter (cm) = 0.3937 inch (in)	1 inch (in) = 2.54 centimeters (cm)
1 meter (m) = 39.37 inches (in)	1 foot (ft) = 0.3048 meter (m)
1 meter (m) = 3.2808 feet (ft)	1 yard (yd) = 0.9144 meter (m)
1 kilometer (km) = 0.6214 mile (mi)	1 mile (mi) = 1.609 kilometers (km)

Metric-English linear equivalents other than millimeter-inch equivalents are seldom used in the machine trades. However, you should be able to express any unit in one measuring system as a unit in the other system. The relationship between English decimal inch units and metric millimeter units is shown by comparing these scales.

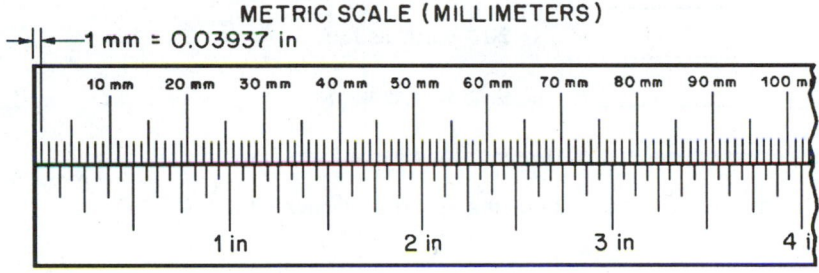

METRIC SCALE (MILLIMETERS)

DECIMAL INCH SCALE

Procedure   To express a unit in one system as an equivalent unit in the other system

- Multiply the given measurement by the appropriate conversion factor in the Metric-English Linear Equivalent Table.

Examples

1. Express 12.700 inches as millimeters.

   Since 1 in = 25.4 mm, 12.700 × 25.4 mm = 322.58 mm   Ans

2. Express 6.780 centimeters as inches. Round the answer to 3 decimal places.

   Since 1 cm = 0.3937 in, 6.78 × 0.3937 in = 2.669 in   Ans

3. This template is dimensioned in millimeters. Determine, in inches, the total length of the template. Round the answer to 3 decimal places.

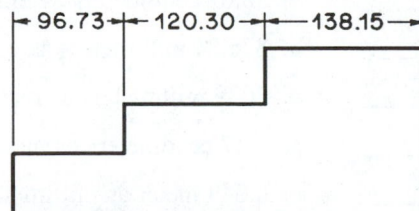

Add the dimensions in millimeters as they are given and express the sum in inches. Since 1 mm = 0.03937 in, multiply the sum in inches by 0.03937.

96.73 ⊞ 120.30 ⊞ 138.15 ⊟ ⊠ .03937 in = 13.983 in   Ans

or ⦅ 96.73 ⊞ 120.30 ⊞ 138.15 ⦆ ⊠ .03937 ⊟ 13.983437,
   13.983 in   Ans

APPLICATION

English Units of Linear Measure

1. Express each of the following lengths as indicated.

   a. 96 inches as feet   _____   k. 51 feet as yards   _____

   b. 123 inches as feet   _____   l. $\frac{1}{3}$ yard as inches   _____

   c. $3\frac{1}{2}$ feet as inches   _____   m. 258 inches as feet   _____

   d. 0.4 yard as inches   _____   n. $7\frac{2}{3}$ feet as inches   _____

   e. $1\frac{1}{4}$ yards as inches   _____   o. 0.20 yard as inches   _____

   f. 144 inches as yards   _____   p. 140.25 feet as yards   _____

   g. 75 inches as feet   _____   q. 333 inches as yards   _____

   h. 8 yards as feet   _____   r. 186 inches as feet   _____

   i. 4.2 yards as feet   _____   s. $20\frac{2}{3}$ yards as feet   _____

   j. 27 feet as yards   _____   t. 9.25 feet as inches   _____

2. A $3\frac{1}{2}$-inch diameter milling cutter revolving at 120.0 revolutions per minute has a cutting speed of 120.0 feet per minute. What is the cutting speed in inches per minute?

   _____

3. How many complete 6-foot lengths of round stock should be ordered to make 230 pieces each 1.300 inches long? Allow $1\frac{1}{2}$ lengths of stock for cutoff and scrap.

   _____

4. Pieces each 3.25 inches long are to be cut from lengths of bar stock. Allowing 0.10 inch for cutoff per piece, how many complete pieces can be cut from twelve 8-foot lengths of stock?

   _____

Metric Units of Linear Measure

5. Express each of the following lengths as indicated.

   a. 2.9 centimeters as millimeters   _____   k. 0.0086 meter as millimeters   _____

   b. 15.78 centimeters as millimeters   _____   l. 1.046 meters as centimeters   _____

   c. 219.75 millimeters as centimeters   _____   m. 30.03 centimeters as millimeters _____

   d. 97.83 millimeters as centimeters   _____   n. 876.84 millimeters as centimeters _____

   e. 0.97 meter as centimeters   _____   o. 2039 millimeters as meters   _____

   f. 0.17 meter as millimeters   _____   p. 3.47 centimeters as meters   _____

   g. 153 millimeters as meters   _____   q. 0.049 meter as millimeters   _____

   h. 673 centimeters as meters   _____   r. 7.321 meters as centimeters   _____

   i. 0.93 millimeter as centimeters   _____   s. 6.377 centimeters as millimeters _____

   j. 0.08 centimeter as millimeters   _____   t. 0.934 meter as millimeters   _____

6. Perform the indicated operations. Express the answer in the indicated unit.

   a. 25.73 mm + 7.6 cm = ? mm _____

   b. 3.7 m + 98 cm = ? m _____

   c. 59.6 cm − 63.7 mm = ? cm _____

   d. 184.8 mm − 12.3 cm = ? mm _____

   e. 1.06 m − 43.7 cm = ? cm _____

   f. 0.793 m − 523.8 mm = ? mm _____

   g. 214 mm + 87.6 cm + 0.9 m = ? m _____

   h. 0.056 m + 4.93 cm + 57.3 mm = ? mm _____

   i. 54.4 mm + 5.05 cm + 204.3 mm = ? mm _____

   j. 3.927 m − 812 mm = ? m _____

7. An aluminum slab 0.082 meter thick is machined with three equal cuts, each cut is 10 millimeters deep. Determine the finished thickness of the slab in millimeters. _____

8. A piece of sheet metal is 1.12 meters wide. Strips each 3.4 centimeters wide are cut. Allow 3 millimeters for cutting each strip.

   a. Determine the number of complete strips cut. _____

   b. Determine the width of the waste strip in millimeters. _____

Metric-English Linear Equivalents

9. Express each of the following English units of length as the indicated metric unit of length. Where necessary round the answer to 3 decimal places.

   a. 37.000 millimeters as inches _____

   b. 126.800 millimeters as inches _____

   c. 17.300 centimeters as inches _____

   d. 0.840 centimeter as inches _____

   e. 2.400 meters as inches _____

   f. 0.090 meter as inches _____

   g. 8.000 meters as feet _____

   h. 10.200 meters as feet _____

   i. 736.00 millimeters as inches _____

   j. 34.050 millimeters as inches _____

   k. 56.300 centimeters as inches _____

   l. 2.000 meters as yards _____

   m. 45.000 centimeters as feet _____

   n. 78.000 millimeters as feet _____

10. Express each of the following metric units of length as the indicated English unit of length. Where necessary, round the answer to 2 decimal places.

    a. 4.000 inches as millimeters _____

    b. 0.360 inch as millimeters _____

    c. 34.00 inches as centimeters _____

    d. 20.85 inches as centimeters _____

    e. 6.00 feet as meters _____

    f. 0.75 foot as meters _____

    g. 3.50 yards as meters _____

    h. 1.30 yards as meters _____

    i. 2.368 inches as millimeters _____

    j. 0.73 inch as centimeters _____

    k. 216.00 inches as meters _____

    l. $\frac{1}{2}$ inch as millimeters _____

    m. $3\frac{1}{4}$ inches as centimeters _____

    n. $75\frac{3}{8}$ inches as meters _____

11. Determine the total length of stock in inches required to make 35 bushings, each 34.2 mm long. Allow $\frac{3}{16}''$ waste for each bushing. Round the answer to 1 decimal place. _____

12. The part shown is to be made in a machine shop using decimal-inch machinery and tools. All dimensions are in millimeters. Express each of the dimensions, A–L, in inches to 3 decimal places.

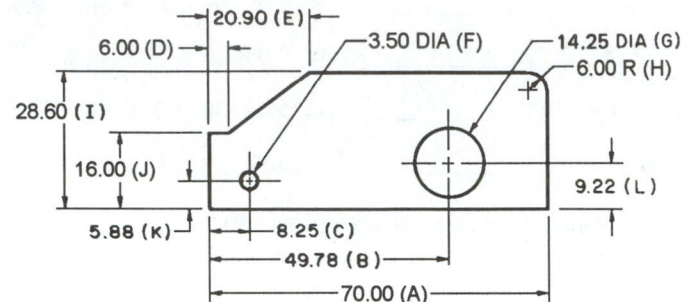

A = _____	E = _____	I = _____
B = _____	F = _____	J = _____
C = _____	G = _____	K = _____
D = _____	H = _____	L = _____

13. The shaft shown is dimensioned in inches. Express each dimension, A–J, in millimeters and round each dimension to 2 decimal places.

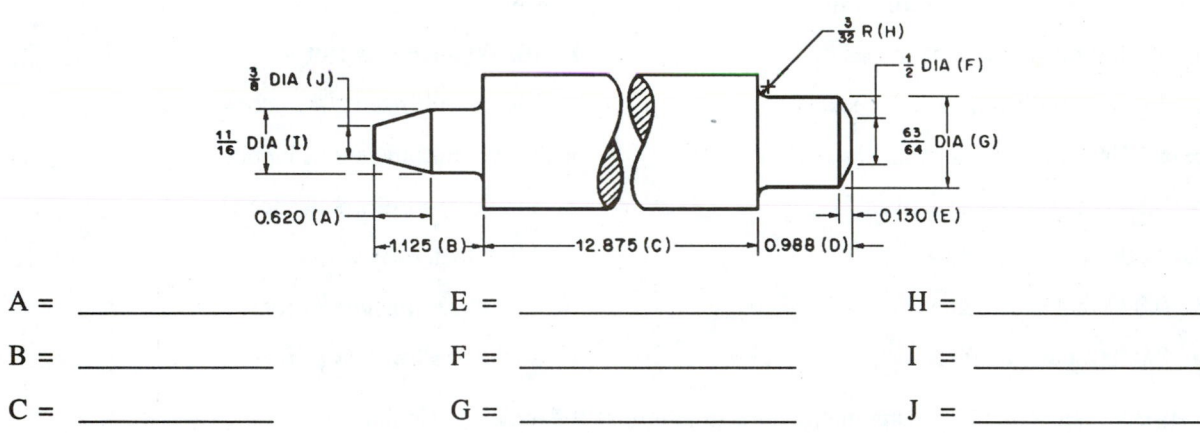

A = _____	E = _____	H = _____
B = _____	F = _____	I = _____
C = _____	G = _____	J = _____
D = _____		

UNIT 22 Degree of Precision, Greatest Possible Error, Absolute Error, and Relative Error

Objectives After studying this unit you should be able to

- Determine the degree of precision of any given number.
- Compute the greatest possible error of English and metric length units.
- Compute absolute error and relative error.

Degree of Precision

The cost of producing a part increases with the degree of precision called for; therefore, no greater degree of precision than is actually required should be specified on a drawing. The degree of precision specified for a particular machining operation dictates the type of machine, the machine setup, and the measuring instrument used for that operation.

The *exact* length of an object cannot be measured. All measurements are approximations. By increasing the number of graduations on a measuring instrument, the degree of precision is increased. Increasing the number of graduations enables the user to get closer to the *exact* length. The precision of a measurement depends on the measuring instrument used. The degree of precision of a measuring instrument depends on the smallest graduated unit of the instrument.

Machinists often work to 0.001-inch or 0.02-millimeter precision. In the manufacture of certain products, very precise measurements to 0.00001 inch or 0.0003 millimeter and 0.000001 inch or 0.000 03 millimeter are sometimes required.

Various measuring instruments have different limitations on the degree of precision possible. The accuracy achieved in measurement does not only depend on the limitations of the measuring instrument. Accuracy can also be affected by errors of measurement. Errors can be caused by defects in the measuring instruments and by environmental changes such as differences in temperature. Perhaps the greatest cause of error is the inaccuracy of the person using the measuring instrument.

Limitations of Measuring Instruments

Following are the limitations on the degree of precision possible of some commonly used manufacturing measuring instruments.

Steel rules: $\frac{1}{64}''$ (fractional-inch); 0.01″ (decimal-inch); 0.5 mm (metric).

Micrometers: 0.001″ (decimal-inch) and 0.0001″ (with vernier scale); 0.01 mm (metric) and 0.002 mm (with vernier scale).

Vernier and dial calipers: 0.001″ (decimal-inch); 0.02 mm (metric).

Dial indicators (comparison measurement): Graduations as small as 0.00005″ (decimal-inch); 0.002 mm (metric).

Precision gage blocks (comparison measurement): Accurate to 0.000002″ (decimal-inch); 0.000 06 mm (metric). The degree of precision of measurement is only as precise as the measuring instrument that is used with the blocks.

High amplification comparators (mechanical, optical, pneumatic, electronic): Graduations as small as 0.000001″.

Degree of Precision of Numbers

The degree of precision of a number depends upon the unit of measurement. The degree of precision of a number increases as the number of decimal places increases.

Example 1   The degree of precision of 2″ is to the nearer inch as shown in A. The range of values includes all numbers equal to or greater than 1.5″ or less than 2.5″.

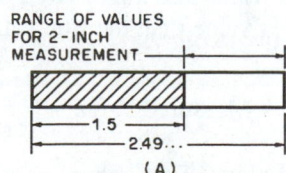

(A)

Example 2   The degree of precision of 2.0″ is to the nearer 10th of an inch as shown in B. The range of values includes all numbers equal to or greater than 1.95″ and less than 2.05″.

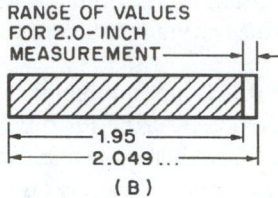

Example 3   The degree of precision of 2.00″ is to the nearer 100th of an inch as shown in C. The range of values includes all numbers equal to or greater than 1.995″ and less than 2.005″.

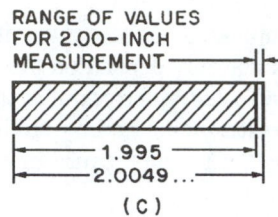

Example 4   The degree of precision of 2.000″ is to the nearer 1000th of an inch. The range of values includes all numbers equal to or greater than .9995″ and less than 2.0005″.

Greatest Possible Error

The *greatest possible error* of a measurement is one-half the smallest graduated unit of the measurement used to make the measurement. Therefore, the greatest possible error is equal to $\frac{1}{2}$ or 0.5 of the precision.

Examples

1. A machinist reads a measurement of 36 millimeters on a steel rule. The smallest graduation on the rule used is 1 millimeter, therefore, the precision is 1 millimeter. Since the greatest possible error is one-half of the smallest graduated unit, the greatest possible error is 0.5 × 1 mm or 0.5 mm. The actual length measured is between 35 mm − 0.5 mm and 36 mm + 0.5 mm or between 35.5 mm and 36.5 mm.

2. A tool and die maker reads a measurement of 0.4754 inch on a vernier scale micrometer. The smallest graduation on the micrometer is 0.0001 inch, therefore the precision is 0.0001 inch. The greatest possible error is 0.5 × 0.0001″ or 0.00005″. The actual length measured is between 0.4754″ − 0.00005″ and 0.4754″ + 0.00005″ or between 0.47535″ and 0.47545″.

Absolute Error and Relative Error

Absolute error and relative error are commonly used to express the amount of error between an actual or true value and a measured value.

Absolute error is the difference between a true value and a measured value. Since the measured value can be either a smaller or larger value than the true value, subtract the smaller value from the larger value.

$$\text{Absolute Error} = \text{True Value} - \text{Measured Value}$$
or

$$\text{Absolute Error} = \text{Measured Value} - \text{True Value}$$

Relative error is the ratio of the absolute error to the true value. It is expressed as a percent.

$$\text{Relative Error} = \frac{\text{Absolute Error}}{\text{True Value}} \times 100$$

Examples

1. The actual or true value of the diameter of a shaft is 1.7056 inches. The shaft is measured as 1.7040 inches. Compute the absolute and relative error.

   The true value is larger than the measured value, therefore:

   Absolute Error = True Value − Measured Value

   Absolute Error = 1.7056 in − 1.7040 in = 0.0016 in   Ans

   $$\text{Relative Error} = \frac{\text{Absolute Error}}{\text{True Value}} \times 100$$

   $$\text{Relative Error} = \frac{0.0016\,\text{in}}{1.7056\,\text{in}} \times 100 = 0.094\% \quad \text{Ans (rounded)}$$

 .0016 $+$ 1.7056 $\times$ 100 $=$ 0.09380863, 0.094%   Ans

2. An inspector measured a taper angle as 3.01 degrees. The true value of the angle is 2.98 degrees. Compute the absolute and relative error.

   The measured value is larger than the true value, therefore:

   Absolute Error = Measured Value − True Value

   Absolute Error = 3.01° − 2.98° = 0.03°   Ans

   $$\text{Relative Error} = \frac{\text{Absolute Error}}{\text{True Value}} \times 100$$

   $$\text{Relative Error} = \frac{0.03°}{2.98°} \times 100 = 1.0\% \quad \text{Ans (rounded)}$$

APPLICATION

Degree of Precision

For each measurement find

a. the degree of precision.
b. the value which is equal to or less than the range of values.
c. the value which is greater than the range of values.

1. 4.3″   a. _____   b. _____   c. _____

2. 1.62″   a. _____   b. _____   c. _____

3. 4.078″   a. _____   b. _____   c. _____

4. 6.07″   a. _____   b. _____   c. _____

5. 15.885″   a. _____   b. _____   c. _____

6. 9.1837″   a. _____   b. _____   c. _____

7. 11.003″   a. _____   b. _____   c. _____

8. 36.0″   a. _____   b. _____   c. _____

9. 7.01″   a. _____   b. _____   c. _____

10. 23.00″   a. _____   b. _____   c. _____

11. 6.1″   a. _____   b. _____   c. _____

12. 14.01070″   a. _____   b. _____   c. _____

13. 26.87 mm   a. _____   b. _____   c. _____

14. 15.4 mm   a. _____   b. _____   c. _____

15. 117.06 mm   a. _____   b. _____   c. _____

16. 0.976 mm   a. _____   b. _____   c. _____

17. 48.01 mm   a. _____   b. _____   c. _____

18. 104.799 mm   a. _____   b. _____   c. _____

19. 7.00 mm   a. _____   b. _____   c. _____

20. 34.082 5 mm   a. _____   b. _____   c. _____

21. 8.001 mm   a. _____   b. _____   c. _____

22. 14.0000 mm   a. _____   b. _____   c. _____

Greatest Possible Error

For each of the exercises in the following tables, the measurement made and the smallest graduation of the measuring instrument are given. Determine the greatest possible error and the smallest and largest possible actual length for each.

	ENGLISH SYSTEM				
			Greatest Possible Error (inches)	Actual Length	
	Measurement Made (inches)	Smallest Graduation of Measuring Instrument Used (inches)		Smallest Possible (inches)	Largest Possible (inches)
23.	5.30	0.05 (steel rule)			
24.	15.68	0.02 (steel rule)			
25.	0.753	0.001 (vernier caliper)			
26.	0.226	0.001 (micrometer)			
27.	0.9369	0.0001 (vernier micrometer)			
28.	$3\frac{5}{8}$	$\frac{1}{64}$ (steel rule)			

	METRIC SYSTEM				
			Greatest Possible Error (millimeters)	Actual Length	
	Measurement Made (millimeters)	Smallest Graduation of Measuring Instrument Used (millimeters)		Smallest Possible (millimeters)	Largest Possible (millimeters)
29.	64	1 (steel rule)			
30.	105	0.5 (steel rule)			
31.	98.5	0.5 (steel rule)			
32.	53.38	0.02 (vernier caliper)			
33.	13.37	0.01 (micrometer)			
34.	12.778	0.002 (vernier micrometer)			

Absolute and Relative Error

For each of the values in the following table, the true value and measured value are given. Determine the absolute and relative error of each. Where necessary, round the answers to 3 decimal places.

	True Value	Measured Value		True Value	Measured Value
35.	38.720 in	38.700 in	41.	1.050 mm	1.020 mm
36.	0.530 mm	0.520 mm	42.	0.9347 in	0.9341 in
37.	12.700°	12.900°	43.	1.005°	1.015°
38.	0.485 in	0.482 in	44.	27.200 in	26.900 in
39.	23.860 mm	24.000 mm	45.	18.276 in	18.302 in
40.	6.056°	6.100°	46.	0.983 mm	1.000 mm

UNIT 23 Tolerance, Clearance, and Interference

Objectives   After studying this unit you should be able to

- Compute total tolerances and maximum and minimum limits of dimensions.
- Compute maximum and minimum clearances of mating parts.
- Compute maximum and minimum interferences of mating parts.
- Express unilateral tolerances as bilateral tolerances.

Tolerance

Tolerance is the amount of variation permitted on the dimensions or surfaces of manufactured parts. *Limits* are the extreme permissible dimensions of a part. Tolerance is equal to the difference between the maximum and minimum limits of any specified dimension of a part.

Example   The maximum limit of a hole diameter is 0.878 inch and the minimum limit is 0.872 inch. Find the tolerance.

The tolerance is $0.878'' - 0.872'' = 0.006''$ Ans

A *basic dimension* is the standard size from which the maximum and minimum limits are determined. Usually tolerances are given in such a way as to show the amount of variation and in which direction from the basic dimension these variations can occur. *Unilateral tolerance* means that the total tolerance is taken in one direction from the basic dimension. *Bilateral tolerance* means that the tolerance is divided partly plus (+) or above and partly minus (−) or below the basic dimension. A *mean dimension* is a value which is midway between the maximum and minimum limits. Where bilateral tolerances are used with equal plus and minus tolerances, the mean dimension is equal to the basic dimension.

Example 1 The part shown is dimensioned with a unilateral tolerance. The dimensions are given in inches. The basic dimension is 3.7500″. The total tolerance is a minus (−) tolerance. Find the maximum permissible dimension (maximum limit) and minimum permissible dimension (minimum limit).

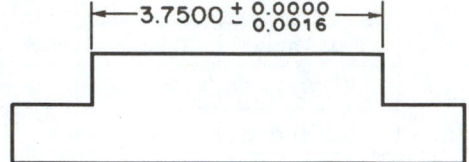

Maximum Limit: 3.7500″ + 0.0000″ = 3.7500″   Ans

Minimum Limit: 3.7500″ − 0.0016″ = 3.7484″   Ans

Example 2 The part shown is dimensioned with a bilateral tolerance. The dimensions are given in millimeters. The basic dimension is 62.79 mm. The tolerance is given in two directions, plus (+) and minus (−). Find the maximum limit and the minimum limit.

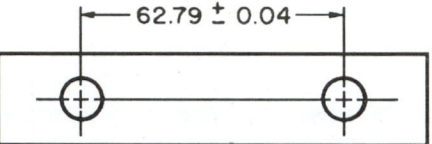

Maximum Limit: 62.79 mm + 0.04 mm = 62.83 mm   Ans

Minimum Limit: 62.79 mm − 0.04 mm = 62.75 mm   Ans

Example 3 What is the mean dimension of a part if the maximum dimension (maximum limit) is 46.35 millimeters and the minimum dimension (minimum limit) is 46.27 millimeters?

Subtract.   46.35 mm − 46.27 mm = 0.08 mm

Divide.   0.08 mm ÷ 2 = 0.04 mm

Subtract.   46.35 mm − 0.04 mm = 46.31 mm   Ans

➤ **Note:** The mean dimension is midway between 46.35 mm and 46.27 mm.

$$46.31 \text{ mm} \begin{array}{l} + \ 0.04 \text{ mm} = 46.35 \text{ mm (Max. limit)} \\ - \ 0.04 \text{ mm} = 46.27 \text{ mm (Min. limit)} \end{array}$$

Expressing Unilateral Tolerance as Bilateral Tolerance

In the actual processing of parts, given unilateral tolerances are sometimes changed to bilateral tolerances. A machinist may prefer to work to a mean dimension and take equal plus and minus tolerances while machining a part. The following example shows the procedure for expressing a unilateral tolerance as a bilateral tolerance.

Example The part shown is dimensioned with unilateral tolerances. Express the unilateral tolerance as a bilateral tolerance. Dimensions are given in inches.

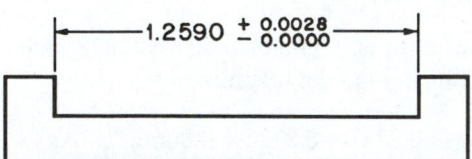

Divide the total tolerance by 2.
0.0028″ ÷ 2 = 0.0014″

Determine the mean dimension.
1.2590″ + 0.0014″ = 1.2604″

Show as a bilateral tolerance.
1.2604″ ± 0.0014″   Ans

Fits of Mating Parts

Fits between mating parts, such as between shafts and holes, have wide application in the manufacturing industry. The tolerances applied to each of the mating parts determines the relative looseness or tightness of fit between parts.

When one part is to move within another there is a *clearance* between the parts. A shaft made to turn in a bushing is an example of a clearance fit. The shaft diameter is less than the bushing hole diameter. When one part is made to be forced into the other there is *interference* between parts. A pin pressed into a hole is an example of an interference fit. The pin diameter is greater than the hole diameter.

Allowance is the intentional difference in the dimensions of mating parts which provides for different classes of fits. *Allowance* is the minimum clearance or the maximum interference which is intended between mating parts. Allowance represents the condition of the tightest permissible fit.

Example 1   A mating shaft and a hole with a clearance fit dimensioned with bilateral tolerances is shown. All dimensions are in inches. Determine the following:

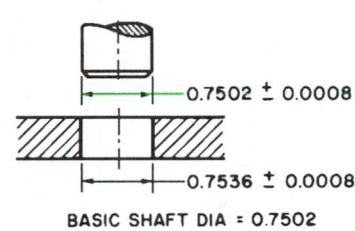

BASIC SHAFT DIA = 0.7502
BASIC HOLE DIA = 0.7536

a. Maximum shaft diameter $0.7502'' + 0.0008'' = 0.7510''$ Ans

b. Minimum shaft diameter $0.7502'' - 0.0008'' = 0.7494''$ Ans

c. Maximum hole diameter $0.7536'' + 0.0008'' = 0.7544''$ Ans

d. Minimum hole diameter $0.7536'' - 0.0008'' = 0.7528''$ Ans

e. Maximum clearance equals
    maximum hole diameter minus
    minimum shaft diameter $0.7544'' - 0.7494'' = 0.0050''$ Ans

f. Minimum clearance equals
    minimum hole diameter minus
    maximum shaft diameter $0.7528'' - 0.7510'' = 0.0018''$ Ans

Since allowance is defined as the minimum clearance, the allowance = 0.0018″   Ans

Example 2   A pin which is to be pressed into a hole is shown. This is an example of an interference fit dimensioned with unilateral tolerances. Dimensions are in millimeters. Determine the following:

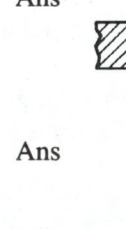

BASIC PIN DIA = 13.860
BASIC HOLE DIA = 13.855

a. Maximum pin diameter   13.860 mm + 0.010 mm = 13.870 mm   Ans

b. Minimum pin diameter   13.860 mm − 0.000 mm = 13.860 mm   Ans

c. Maximum hole diameter   13.855 mm + 0.000 mm = 13.855 mm   Ans

d. Minimum hole diameter   13.855 mm − 0.010 mm = 13.845 mm   Ans

e. Minimum interference
    equals minimum pin
    diameter minus maximum
    hole diameter   13.860 mm − 13.855 mm = 0.005 mm   Ans

f. Maximum interference
    equals maximum pin
    diameter minus minimum
    hole diameter   13.870 mm − 13.845 mm = 0.025 mm   Ans

Since allowance is defined as the maximum interference, the allowance = 0.025 mm   Ans

APPLICATION

Tolerance, Maximum and Minimum Limits

Refer to the following tables and determine the tolerance, maximum limit, or minimum limit as required for each problem.

1. English System

	Tolerance	Maximum Limit	Minimum Limit
a.		$5\frac{7}{16}''$	$5\frac{13}{32}''$
b.		$7'\ 9\frac{1}{16}''$	$7'\ 8\frac{15}{16}''$
c.	0.03″	16.76″	
d.	0.007″		0.904″
e.		1.7001″	1.6998″
f.	0.004″		10.999″

2. Metric System

	Tolerance	Maximum Limit	Minimum Limit
a.		50.7 mm	49.9 mm
b.		26.8 cm	26.6 cm
c.	0.04 mm		258.03 mm
d.	0.12 mm	79.65 mm	
e.	0.006 cm		12.731 cm
f.		4.01 mm	3.98 mm

Unilateral and Bilateral Tolerance

3. Refer to this figure. Dimension A with its tolerance is given in each of the following problems. Determine the maximum dimension (maximum limit) and the minimum dimension (minimum limit) for each.

a. Dimension A = 4.640″ $^{+0.003''}_{-0.000''}$

maximum _____ minimum _____

b. Dimension A = 5.927″ $^{+0.005''}_{-0.000''}$

maximum _____ minimum _____

c. Dimension A = 2.004″ $^{+0.000''}_{-0.004''}$

maximum _____ minimum _____

d. Dimension A = 4.6729″ $^{+0.0000''}_{-0.0012''}$

maximum _____ minimum _____

e. Dimension A = 1.0875″ $^{+0.0009''}_{-0.0000''}$

maximum _____ minimum _____

f. Dimension A = 28.16 mm $^{+0.00\text{ mm}}_{-0.06\text{ mm}}$

maximum _____ minimum _____

g. Dimension A = 43.94 mm $^{+0.04\text{ mm}}_{-0.00\text{ mm}}$

maximum _____ minimum _____

h. Dimension A = 118.66 mm $^{+0.07\text{ mm}}_{-0.00\text{ mm}}$

maximum _____ minimum _____

i. Dimension A = 73.398 mm $^{+0.000\text{ mm}}_{-0.012\text{ mm}}$

maximum _____ minimum _____

j. Dimension A = 45.106 mm $^{+0.009\text{ mm}}_{-0.000\text{ mm}}$

maximum _____ minimum _____

4. The following dimensions are given with bilateral tolerances. For each value determine the maximum dimension (maximum limit) and the minimum dimension (minimum limit).

a. 2.812″ ± 0.006″
maximum _____ minimum _____

b. 3.003″ ± 0.004″
maximum _____ minimum _____

c. 3.971″ ± 0.010″
maximum _____ minimum _____

d. 4.0562″ ± 0.0012″
maximum _____ minimum _____

e. 1.3799″ ± 0.0009″
maximum _____ minimum _____

f. 2.0000″ ± 0.0007″
maximum _____ minimum _____

g. 43.46 mm ± 0.05 mm
    maximum _____ minimum _____

h. 107.07 mm ± 0.08 mm
    maximum _____ minimum _____

i. 62.04 mm ± 0.10 mm
    maximum _____ minimum _____

j. 10.203 mm ± 0.024 mm
    maximum _____ minimum _____

k. 289.005 mm ± 0.007 mm
    maximum _____ minimum _____

l. 66.761 mm ± 0.015 mm
    maximum _____ minimum _____

5. Express each of the following unilateral tolerances as bilateral tolerances having equal plus and minus values.

a. $0.938''$ $^{+0.010''}_{-0.000''}$ _____

b. $1.686''$ $^{+0.002''}_{-0.000''}$ _____

c. $3.000''$ $^{+0.000''}_{-0.004''}$ _____

d. $0.073''$ $^{+0.000''}_{-0.008''}$ _____

e. $4.1873''$ $^{+0.0014''}_{-0.0000''}$ _____

f. $1.0021''$ $^{+0.0000''}_{-0.0074''}$ _____

g. $0.0010''$ $^{+0.0000''}_{-0.0008''}$ _____

h. $8.4649''$ $^{+0.0022''}_{-0.0000''}$ _____

i. 44.30 mm $^{+0.02\ mm}_{-0.00\ mm}$ _____

j. 10.06 mm $^{+0.00\ mm}_{-0.08\ mm}$ _____

k. 64.89 mm $^{+0.06\ mm}_{-0.00\ mm}$ _____

l. 37.988 mm $^{+0.056\ mm}_{-0.000\ mm}$ _____

m. 125.00 mm $^{+0.000\ mm}_{-0.017\ mm}$ _____

n. 43.091 mm $^{+0.000\ mm}_{-0.026\ mm}$ _____

o. 98.879 mm $^{+0.009\ mm}_{-0.000\ mm}$ _____

Fits of Mating Parts

The following problems require computations with both clearance fits and interference fits between mating parts. Find the missing values in the following tables.

6. Refer to the figure to determine the values in the table. The answer to the first problem is given. Allowance is equal to the minimum clearance. All dimensions are in inches.

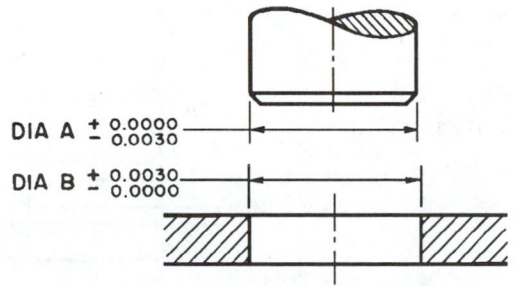

DIA A $\pm\ ^{0.0000}_{0.0030}$

DIA B $\pm\ ^{0.0030}_{0.0000}$

		Basic Dimension	Maximum Diameter (Max. Limit)	Minimum Diameter (Min. Limit)	Maximum Clearance	Minimum Clearance (Allowance)
a.	DIA A	1.4580	1.4580	1.4550	0.0090	0.0030
	DIA B	1.4610	1.4640	1.4610		
b.	DIA A	0.6345				
	DIA B	0.6365				
c.	DIA A	2.1053				
	DIA B	2.1078				

7. Refer to the figure to determine the values in the table. Allowance is equal to the maximum interference. All dimensions are in millimeters.

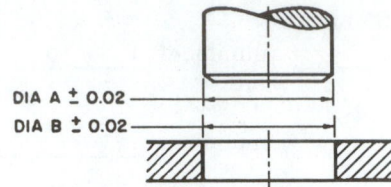

DIA A ± 0.02
DIA B ± 0.02

		Basic Dimension	Maximum Diameter (Max. Limit)	Minimum Diameter (Min. Limit)	Maximum Interference (Allowance)	Minimum Interference
a.	DIA A	20.73				
	DIA B	20.68				
b.	DIA A	32.07				
	DIA B	32.01				
c.	DIA A	12.72				
	DIA B	12.65				

8. Refer to the figure to determine the values in the table. Allowance is equal to minimum clearance. All dimensions are in inches.

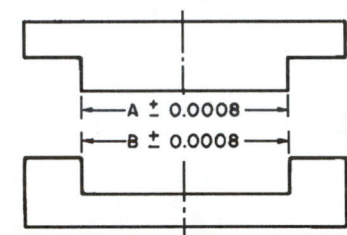

A ± 0.0008
B ± 0.0008

		Basic Dimension	Maximum Dimension (Max. Limit)	Minimum Dimension (Min. Limit)	Maximum Clearance	Minimum Clearance (Allowance)
a.	DIM A	0.9995				
	DIM B	1.0020				
b.	DIM A	2.0334				
	DIM B	2.0360				
c.	DIM A	1.4392				
	DIM B	1.4412				

9. Refer to the figure to determine the values in the table. Allowance is equal to the maximum interference. All dimensions are in millimeters.

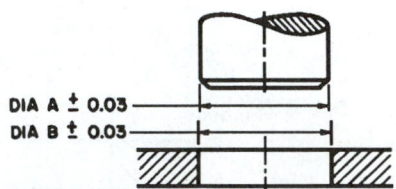

DIA A ± 0.03
DIA B ± 0.03

		Basic Dimension	Maximum Diameter (Max. Limit)	Minimum Diameter (Min. Limit)	Maximum Interference (Allowance)	Minimum Interference
a.	DIA A	87.58				
	DIA B	87.50				
b.	DIA A	9.94				
	DIA B	9.85				
c.	DIA A	130.03				
	DIA B	129.96				

Related Problems

10. Spacers are manufactured to the mean dimension and tolerance shown in this figure. An inspector measures 10 spacers and records the following thicknesses:

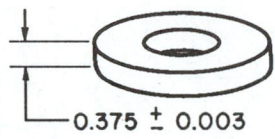

0.375 ± 0.003

0.372″	0.379″	0.370″	0.377″	0.371″
0.376″	0.375″	0.373″	0.378″	0.380″

Which spacers are defective (above the maximum limit or below the minimum limit)? All dimensions are in inches.

_____

11. A tool-and-die maker grinds a pin to an 18.25-millimeter diameter as shown. The pin is to be pressed (an interference fit) in a hole. The minimum interference allowed is 0.03 millimeter. The maximum interference allowed is 0.07 millimeter. Determine the mean diameter of the hole. All dimensions are in millimeters.

_____

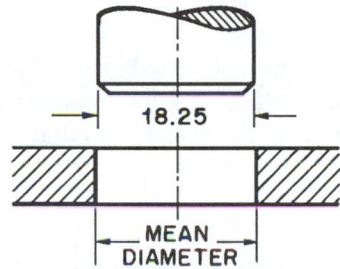

18.25

MEAN DIAMETER

12. A piece is to be cut to the dimensions and tolerances shown. Determine the maximum permissible value of length A. All dimensions are in inches.

_____

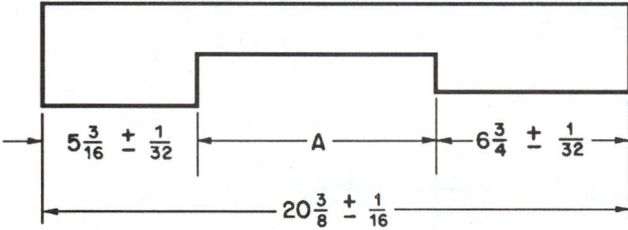

$5\frac{3}{16} \pm \frac{1}{32}$ A $6\frac{3}{4} \pm \frac{1}{32}$

$20\frac{3}{8} \pm \frac{1}{16}$

13. Determine the maximum and minimum permissible wall thickness of the steel sleeve shown. All dimensions are in millimeters.

maximum _____

minimum _____

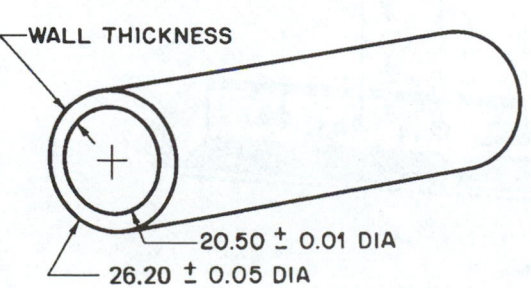

WALL THICKNESS

20.50 ± 0.01 DIA

26.20 ± 0.05 DIA

14. Mating parts are shown. The pins in the top piece fit into the holes in the bottom piece. All dimensions are in inches. Determine the following:

a. The mean pin diameters. _____

b. The mean hole diameters. _____

c. The maximum dimension A. _____

d. The minimum dimension A. _____

e. The maximum dimension B. _____

f. The minimum dimension B. _____

g. The maximum total clearance between dimension C and dimension D. _____

h. The minimum total clearance between dimension C and dimension D. _____

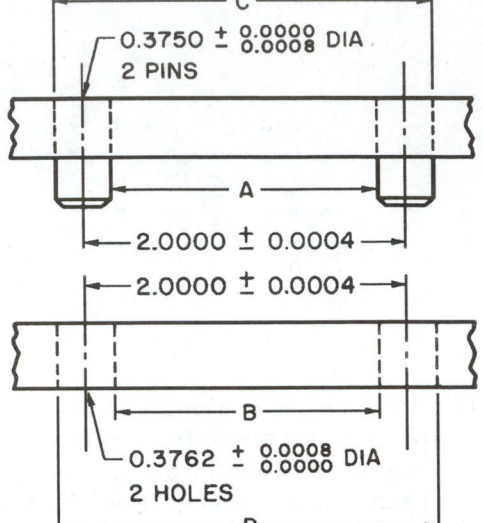

15. Drawing (A) gives the locations with tolerances of 6 holes that are to be drilled in a length of angle iron. A machinist drills the holes then checks them for proper locations from edge A. The actual locations of the drilled holes are shown in drawing (B). Which holes are drilled out of tolerance (located incorrectly)?

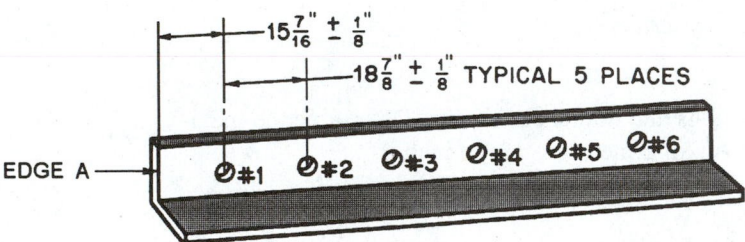

(A)   Specifications for locations of holes

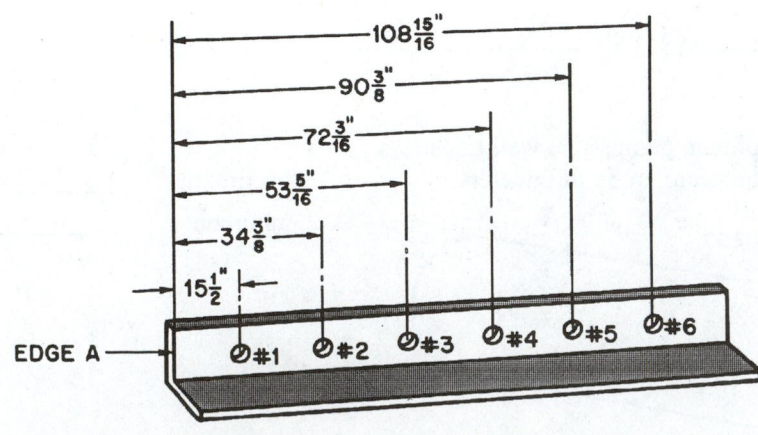

(B)   Actual locations of drilled holes

UNIT 24 English and Metric Steel Rules

Objectives   **After studying this unit you should be able to**

- **Read measurements on fractional-inch and decimal-inch steel rules.**
- **Measure lengths using fractional-inch and decimal-inch scales.**
- **Read measurements on metric steel rules.**
- **Measure lengths using metric scales**

Steel rules are widely used for machine shop applications which do not require a high degree of precision. The steel rule is often the most practical measuring instrument to use for checking dimensions where stock allowances for finishing are provided. Steel rules are also used for locating roughing cuts on machined pieces and for determining the approximate locations of parts for machine setups. Steel rules used in the machine shop are generally six inches long, although rules anywhere from a fraction of an inch to several inches in length are also used.

Correct Procedure in the Use of Steel Rules

The end of a rule receives more wear than the rest of the rule. Therefore, the end should not be used as a reference point unless it is used with a knee (a straight block).

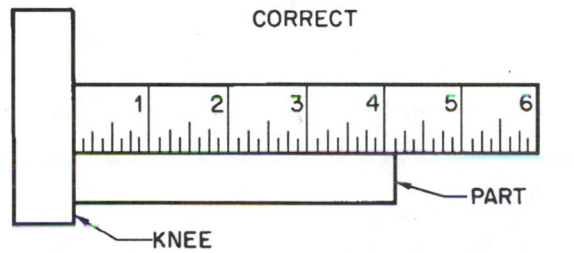
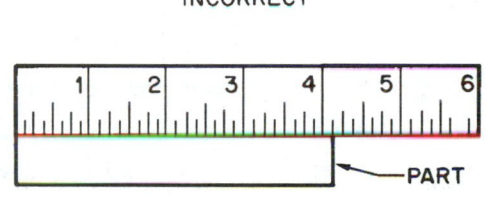

If a knee is not used, the 1-inch graduation of English measure rules should be used as a reference point as shown. The 1 inch must be subtracted from the measurement obtained. For metric measure rules, use the 10-millimeter graduation as the reference point. The 10 millimeters must be subtracted from the measurement obtained.

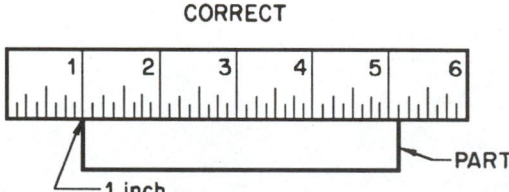

The scale edge of the rule should be put on the part to be measured. Following the correct procedure eliminates parallax error (error caused by the scale and the part being in different planes).

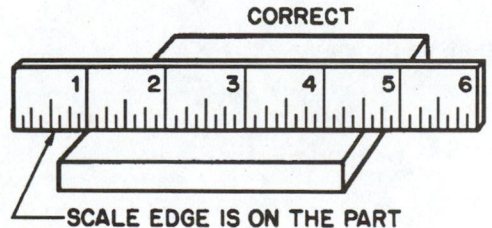
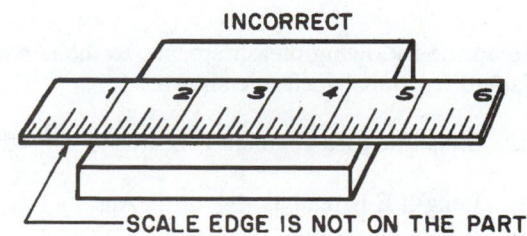

Reading Fractional-Inch Rules

The smallest division of fractional rules is $\frac{1}{64}$ inch. An enlarged fractional-inch rule is shown. The top scale is graduated in 64ths of an inch and the bottom scale in 32nds of an inch. The staggered graduations are halves, quarters, eighths, sixteenths, and thirty-seconds.

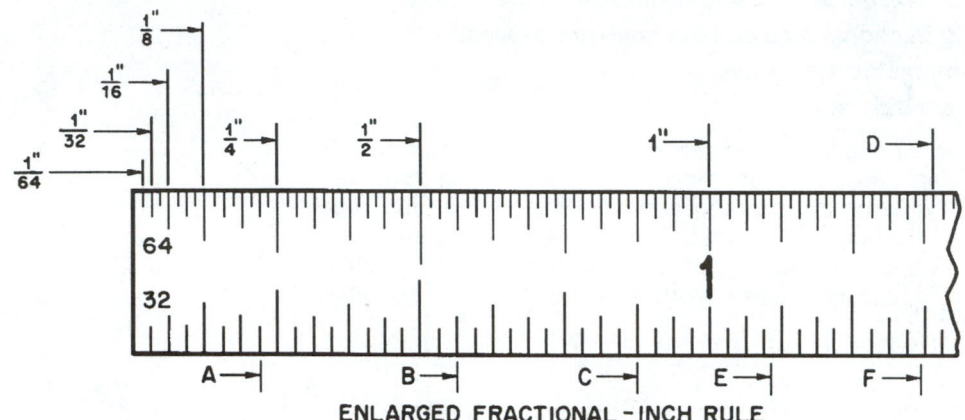

ENLARGED FRACTIONAL – INCH RULE

Measurements can read by noting the last complete inch unit and counting the number of fractional units past the inch unit. Generally, a short-cut method of reading measurements is used as illustrated by the following examples.

Examples   Read the following measurements on the enlarged fractional-inch rule shown.

1. Length A: Subtract one $\frac{1}{32}''$ graduation from $\frac{1}{4}''$.

$$\frac{1''}{4} - \frac{1''}{32} = \frac{8''}{32} - \frac{1''}{32} = \frac{7''}{32} \qquad \text{Ans}$$

2. Length B: Add one $\frac{1}{16}''$ graduation to $\frac{1}{2}''$.

$$\frac{1''}{2} + \frac{1''}{16} = \frac{8''}{16} + \frac{1''}{16} = \frac{9''}{16} \qquad \text{Ans}$$

3. Length C: Subtract one $\frac{1}{8}''$ graduation from $1''$.

$$1'' - \frac{1''}{8} = \frac{8''}{8} - \frac{1''}{8} = \frac{7''}{8} \qquad \text{Ans}$$

4. Length D: Add one $\frac{1}{64}''$ graduation to $1\frac{3}{8}''$.

$$1\frac{3''}{8} + \frac{1''}{64} = 1\frac{24''}{64} + \frac{1''}{64} = 1\frac{25''}{64} \qquad \text{Ans}$$

Often the edge of an object being measured does not fall exactly on a rule graduation. In these cases, read the measurement to the nearer rule graduation.

Examples   Read the following measurements, to the nearer graduation, on the enlarged fractional-inch rule shown.

1. Length E: Since the measurement is nearer to $1\frac{3}{32}''$ than $1\frac{1}{8}''$,

Length E is read as $1\frac{3}{32}''$.   Ans

2. Length F: Since the measurement is nearer to $1\frac{3}{8}''$ than $1\frac{11}{32}''$,

Length F is read as $1\frac{3}{8}''$.   Ans

Reading Decimal-Inch Rules

An enlarged decimal-inch rule is shown. The top scale is graduated in hundredths of an inch (0.01"). The bottom scale is graduated in fiftieths of an inch (0.02"). The staggered graduations are halves, tenths, and fiftieths.

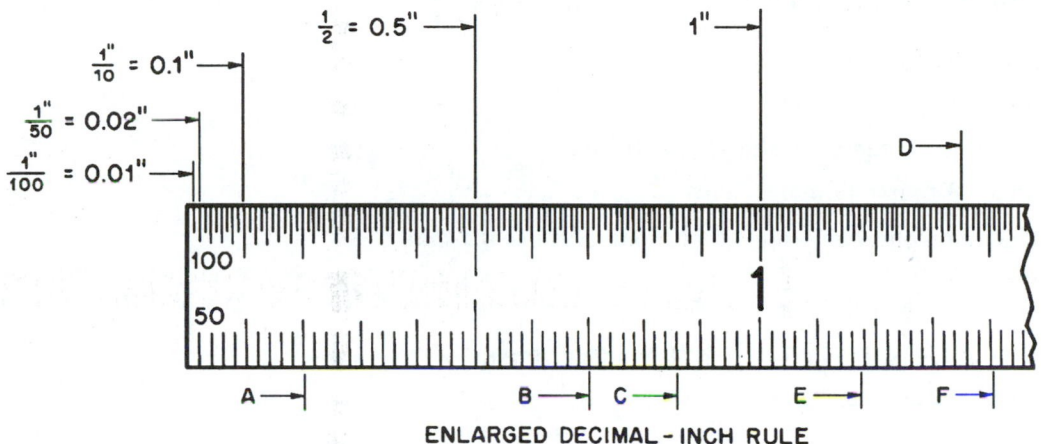

ENLARGED DECIMAL - INCH RULE

Examples   Read the following measurements on the enlarged decimal-inch rule shown.

1. Length A: Count two 0.1" graduations.
   $2 \times 0.1'' = 0.2''$ Ans

2. Length B: Add two 0.1" graduations to the 0.5".
   $0.5'' + 0.2'' = 0.7''$ Ans

3. Length C: Add three 0.02" graduations to 0.8".
   $0.8'' + 0.06'' = 0.86''$ Ans

4. Length D: Add 1", plus three 0.1" graduations, plus five 0.01" graduations.
   $1'' + 0.3'' + 0.05'' = 1.35''$ Ans

5. Length E: Since the measurement is nearer to 1.18" than 1.16",
   Length E is read as 1.18".   Ans

6. Length F: Since the measurement is nearer 1.40" than 1.42",
   Length F is read as 1.40".   Ans

Reading a Metric Rule

An enlarged metric rule is shown. The top scale is graduated in half millimeters (0.5 mm). The bottom scale is graduated in millimeters (1 mm).

The following examples show the method of reading measurements with a metric rule with 0.5 mm and 1 mm scales.

Examples   Read the following measurements on the enlarged metric rule shown.

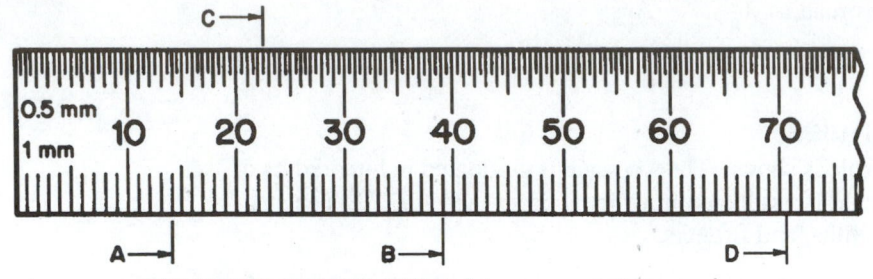

ENLARGED METRIC RULE (1 mm and 0.5 mm)

1. Length A: Add four 1-mm graduations to 10 mm.

   10 mm + 4 mm = 14 mm   Ans

2. Length B: Subtract one 1-mm graduation from 40 mm.

   40 mm − 1 mm = 39 mm   Ans

3. Length C: Add 20 mm, plus two 1-mm graduations, plus one 0.5-mm graduation.

   20 mm + 2 mm + 0.5 mm = 22.5 mm   Ans

4. Length D: Since the measurement is nearer 71 mm than 70 mm, Length D is read as 71 mm.   Ans

APPLICATION

Fractional-Inch Steel Rules

1. Read measurements a–p on the enlarged fractional-inch rule shown.

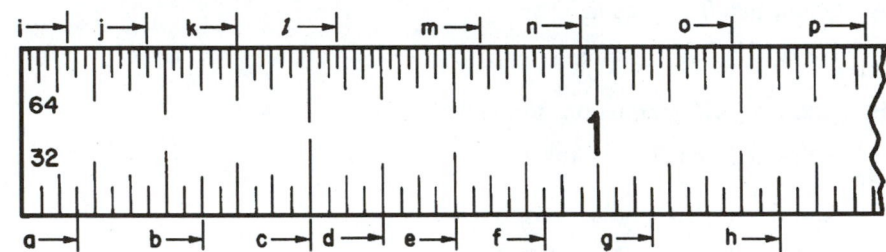

a. _____	e. _____	h. _____	k. _____	n. _____
b. _____	f. _____	i. _____	l. _____	o. _____
c. _____	g. _____	j. _____	m._____	p. _____
d. _____				

2. Measure the length of each of the following line segments to the nearer $\frac{1}{16}''$.

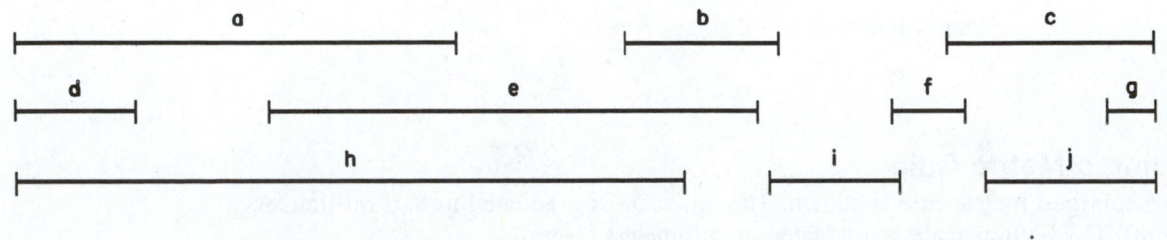

| a. _____ | c. _____ | e. _____ | g. _____ | i. _____ |
| b. _____ | d. _____ | f. _____ | h. _____ | j. _____ |

3. Measure the lengths of dimensions a–n of the template shown to the nearer $\frac{1}{32}''$.

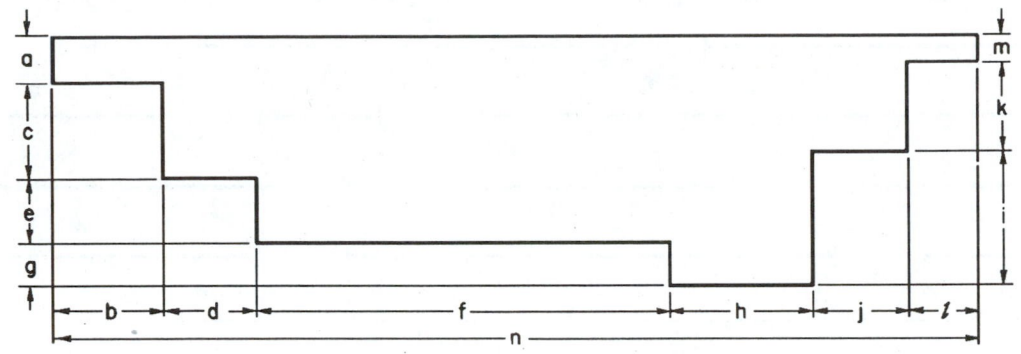

a. _____   d. _____   g. _____   j. _____   m. _____

b. _____   e. _____   h. _____   k. _____   n. _____

c. _____   f. _____   i. _____   l. _____

4. Measure the length of each of the following line segments to the nearer $\frac{1}{64}''$.

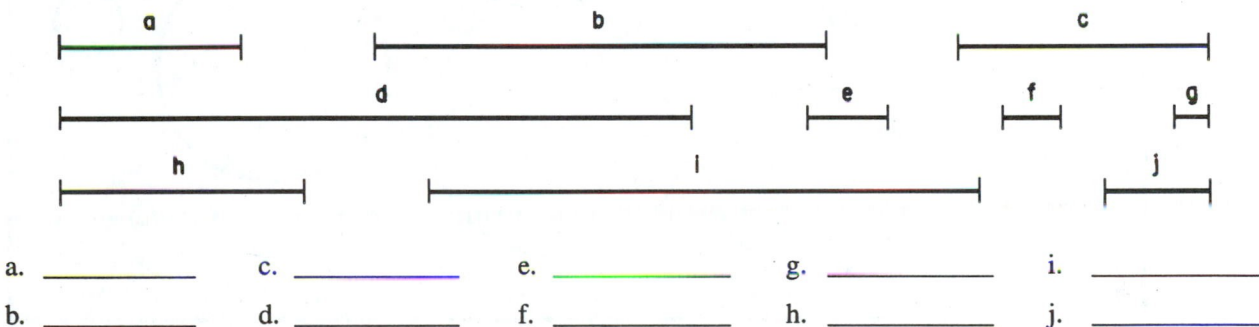

a. _____   c. _____   e. _____   g. _____   i. _____

b. _____   d. _____   f. _____   h. _____   j. _____

Decimal-Inch Steel Rules

5. Read measurements a–p on the enlarged decimal-inch rule shown.

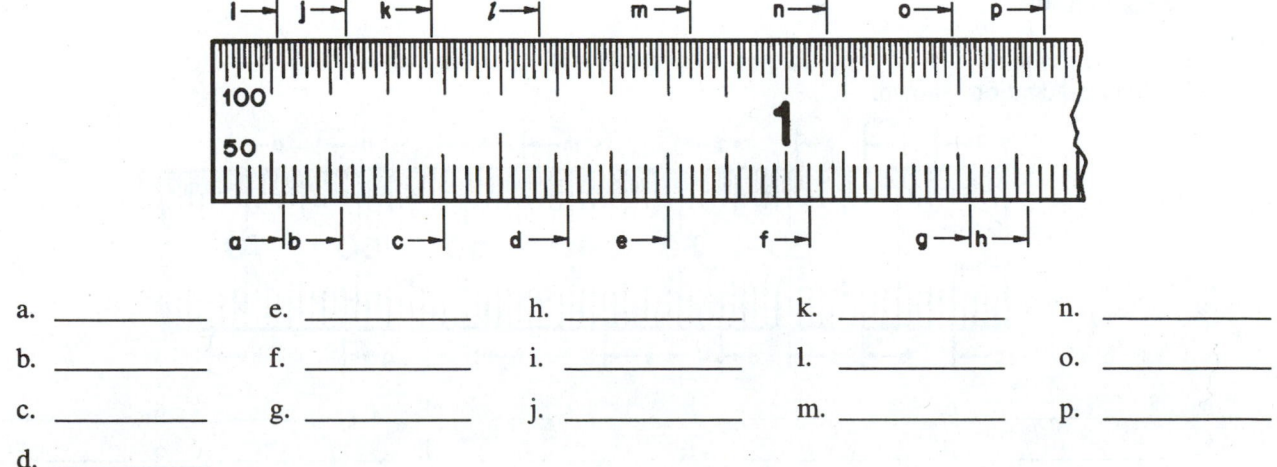

a. _____   e. _____   h. _____   k. _____   n. _____

b. _____   f. _____   i. _____   l. _____   o. _____

c. _____   g. _____   j. _____   m. _____   p. _____

d. _____

6. Measure the length of each of the following line segments to the nearer fiftieth of an inch (0.02″).

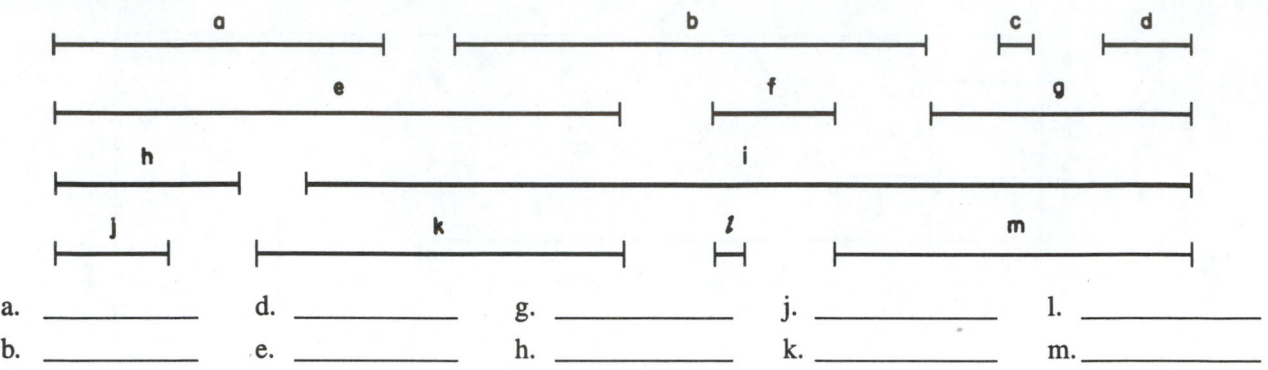

a. _____	d. _____	g. _____	j. _____	l. _____
b. _____	e. _____	h. _____	k. _____	m._____
c. _____	f. _____	i. _____		

7. Measure the diameters of the holes in the plate shown to the nearer fiftieth of an inch (0.02″).

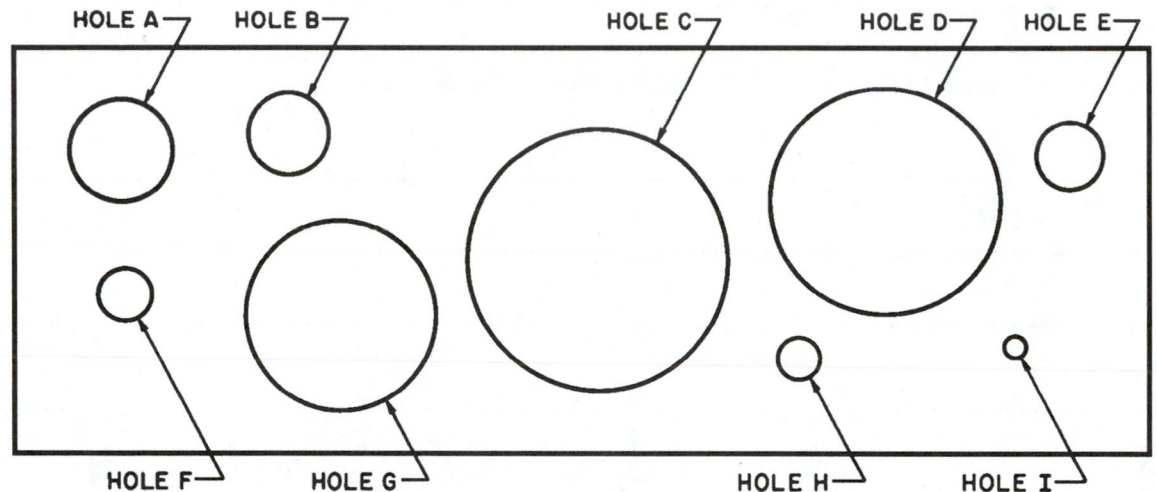

➤ **Note:** Measure to the inside of the hole line thickness.

| A = _____ | C = _____ | E = _____ | G = _____ | I = _____ |
| B = _____ | D = _____ | F = _____ | H = _____ | |

Metric Steel Rules

8. Read measurements a–p on the enlarged metric rule with 1-millimeter and 0.5-millimeter graduations shown.

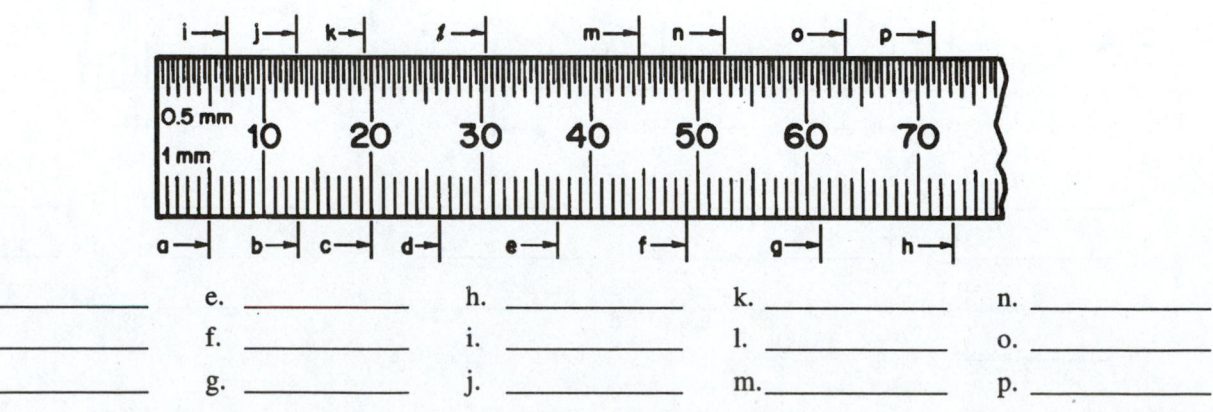

a. _____	e. _____	h. _____	k. _____	n. _____
b. _____	f. _____	i. _____	l. _____	o. _____
c. _____	g. _____	j. _____	m._____	p. _____
d. _____				

9. Measure the length of each of the following line segments to the nearer whole millimeter.

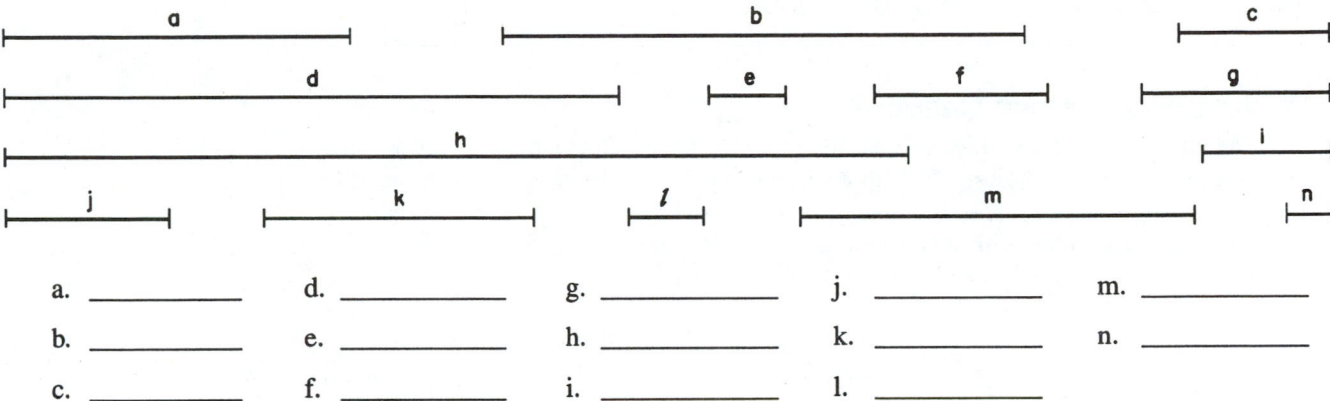

a. _____   d. _____   g. _____   j. _____   m. _____

b. _____   e. _____   h. _____   k. _____   n. _____

c. _____   f. _____   i. _____   l. _____

10. Measure dimensions a–k on the pattern shown to the nearer whole millimeter.

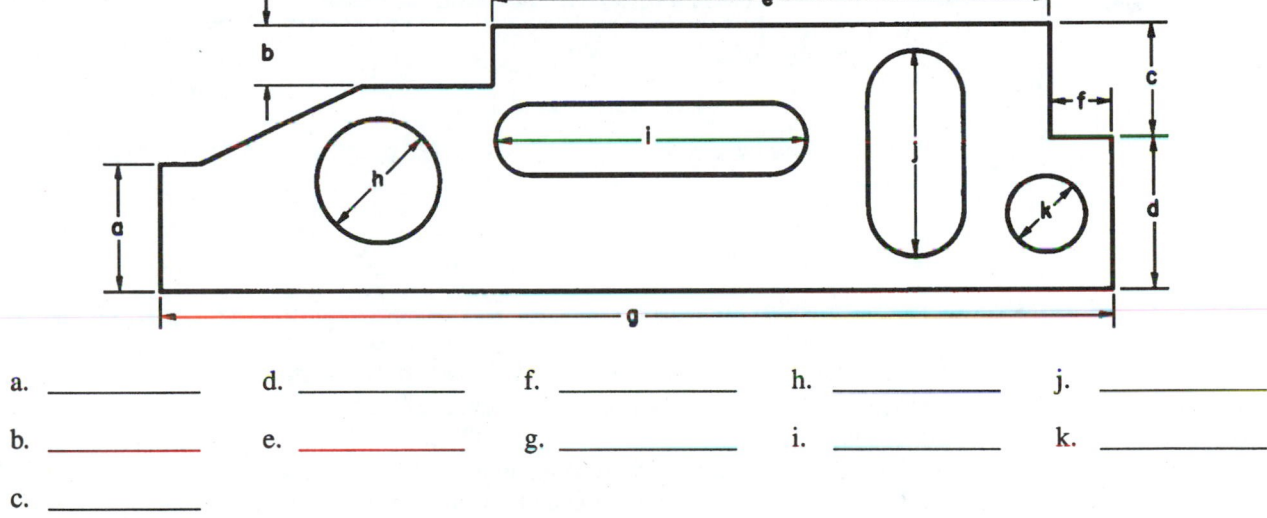

a. _____   d. _____   f. _____   h. _____   j. _____

b. _____   e. _____   g. _____   i. _____   k. _____

c. _____

UNIT 25 English Vernier Calipers and Height Gages

Objectives   After studying this unit you should be able to

- **Read measurements set on a decimal-inch vernier caliper.**
- **Set given measurements on a decimal-inch vernier caliper.**
- **Read measurements set on a decimal-inch vernier height gage.**
- **Set given measurements on a decimal-inch vernier height gage.**

Decimal-inch vernier calipers are used in machine shop applications when the degree of precision to thousandths of an inch is adequate. They are used for measuring lengths of parts, distances between holes, and both inside and outside diameters of cylinders.

Vernier height gages are widely used on surface plates and on machine tables. The height gage with an indicator attachment is used for checking locations of surfaces and holes. The height gage with a scriber attachment is used to mark reference lines, locations, and stock allowances on castings and forgings.

Decimal-Inch Vernier Caliper

The basic parts of a vernier caliper are a main scale which is similar to a steel rule with a fixed jaw and a sliding jaw with a vernier scale. The vernier scale slides parallel to the main scale and provides a degree of precision to 0.001″. Calipers are available in a wide range of lengths with different types of jaws and scale graduations. A vernier caliper which is commonly used in machine shops is shown.

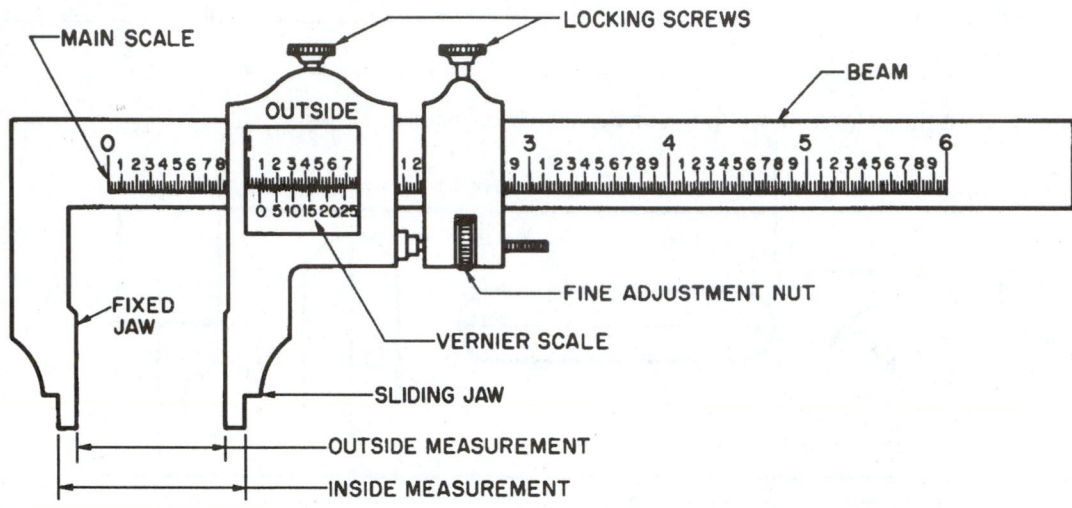

The main scale is divided into inches and the inches are divided into 10 divisions each equal to 0.1″. The 0.1″ divisions are divided into 4 parts each equal to 0.025″. The vernier scale consists of 25 divisions.

A vernier scale is shown. The vernier scale has 25 divisions in a length equal to a length on the main scale that has 24 divisions. The difference between a main scale division and a vernier division is $\frac{1}{25}$ of 0.025″ or 0.001″.

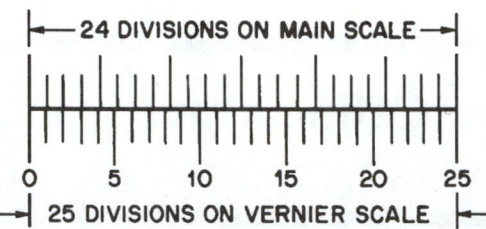

Reading and Setting a Measurement on a Decimal-Inch Vernier Caliper

A measurement is read by adding the thousandths reading on the vernier scale to the reading from the main scale.

Procedure   To read a measurement on a decimal-inch vernier caliper

- Read the number of 1″ graduations, 0.1″ graduations, and 0.025″ graduations on the main scale that are left of the zero graduation on the vernier scale.

- On the vernier scale, find the graduation that most closely coincides with a graduation on the main scale. Add this vernier reading which indicates the number of 0.001″ graduations to the main scale reading.

Setting a given measurement is the reverse procedure of reading a measurement on the vernier caliper.

Example 1   Read the measurement set on this vernier caliper.

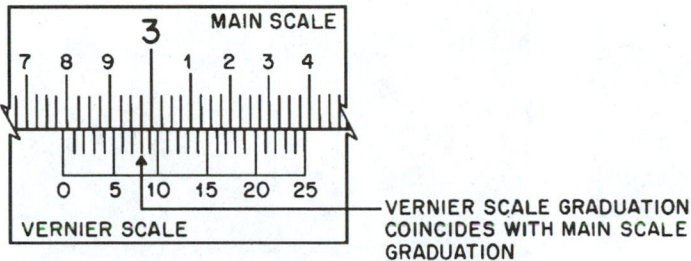

In reference to the zero division on the vernier scale read two 1″ divisions, seven 0.1″ divisions, and three 0.025″ divisions on the main scale.

$$(2'' + 0.7'' + 0.075'' = 2.775'')$$

Observe which vernier scale graduation most closely coincides with a main scale graduation. The eight vernier scale graduation coincides; therefore, 0.008″ is added to 2.775″.

Measurement: 2.775″ + 0.008″ = 2.783″   Ans

Example 2   Set 1.237″ on a vernier caliper.

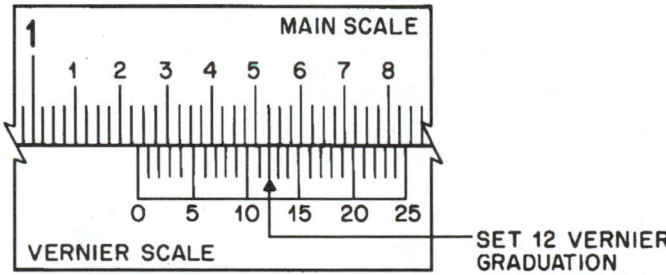

Move the vernier zero graduation to 1″ + 0.2″ + 0.025″ on the main scale.

An additional 0.012″ (1.237″ − 1.225″) is set by adjusting the sliding jaw until the 12 graduation on the vernier scale coincides with a graduation on the main scale.

The 1.237-inch setting is shown.

The accuracy of measurement obtainable with a vernier caliper depends on the user's ability to align the caliper with the part which is being measured and the user's "feel" when measuring. The line of measurement must be parallel to the beam of the caliper and lie in the same plane as the caliper. Care must be used to prevent too loose or too tight a caliper setting.

The front side of the English vernier caliper (25 divisions) is used for outside measurements as shown on the left. The reverse or back side is used for inside measurements as shown on the right.

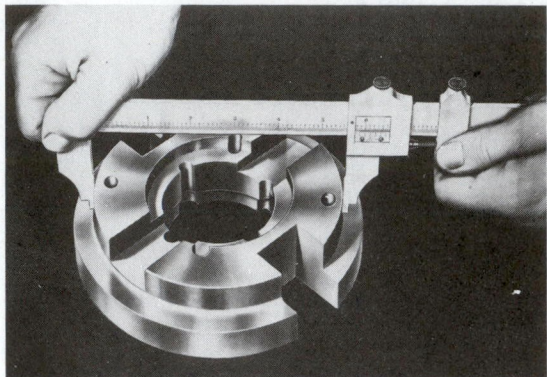

Measuring an outside diameter   The measurement is read on the front side of the caliper.   (The L.S. Starrett Company)

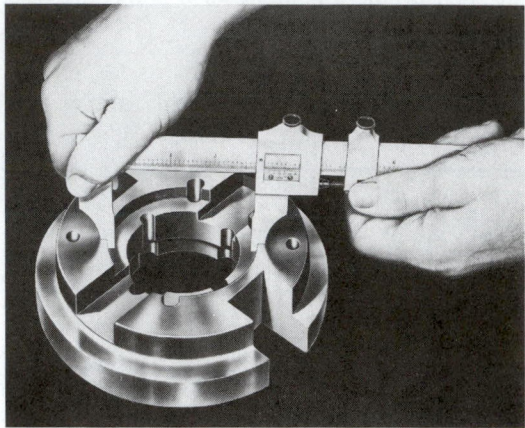

Measuring an inside diameter   The measurement is read on the back side of the caliper.   (The L.S. Starrett Company)

Decimal-Inch Vernier Height Gage

The vernier height gage and vernier caliper are similar in operation. The height gage also has a sliding jaw; the fixed jaw is the surface plate with which the height gage is usually used. The gage can be used with a scriber, a depth gage attachment, or an indicator. The indicator is the most widely used and, generally, the most accurate attachment. The parts of a vernier height gage are shown.

Measurements on the vernier height gage are read and set using the same procedure as with the vernier caliper.

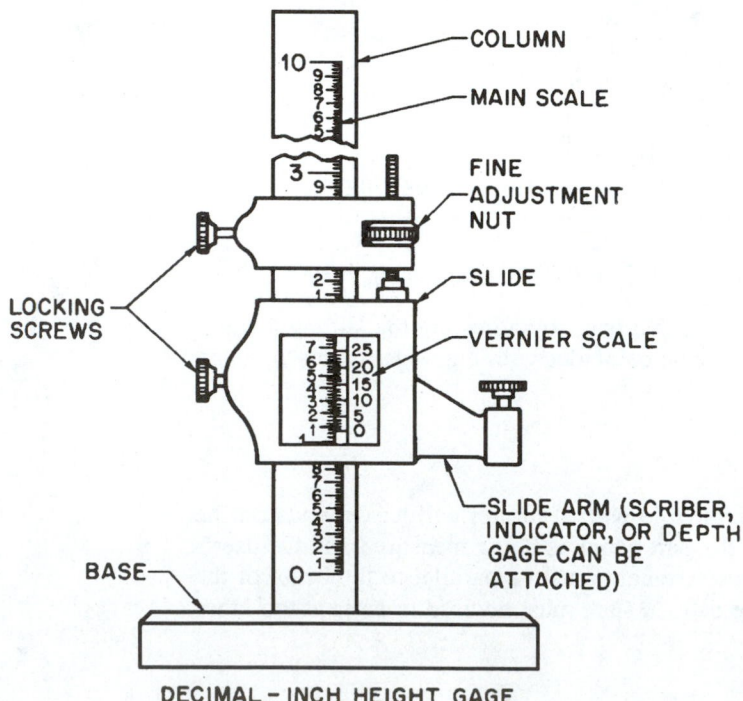

DECIMAL – INCH HEIGHT GAGE

Example 1   Read the measurement set on this vernier height gage.

In reference to the zero division on the vernier scale read 5″, four 0.1″ divisions, and two 0.025″ divisions on the main scale. (5″ + 0.4″ + 0.050″ = 5.450″)

Observe which vernier scale graduation most closely coincides with the main scale graduation. The twenty-first vernier scale graduation coincides; therefore, 0.021″ is added to 5.450″.

Measurement = 5.450″ + 0.021″ = 5.471″   Ans

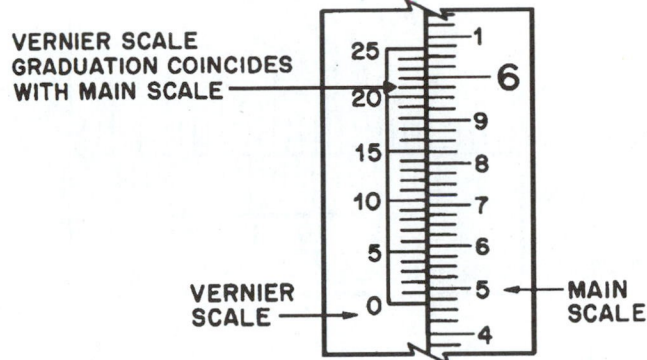

Example 2   Set 8.398″ on a vernier height gage.

Move the vernier zero graduation to 8″ + 0.3″ + 0.075″ = 8.375″.

An additional 0.023″ (8.398″ − 8.375″) is set by turning the fine adjustment screw until the 23 graduation on the vernier scale coincides with a graduation on the main scale.

The 8.398-inch setting is shown.

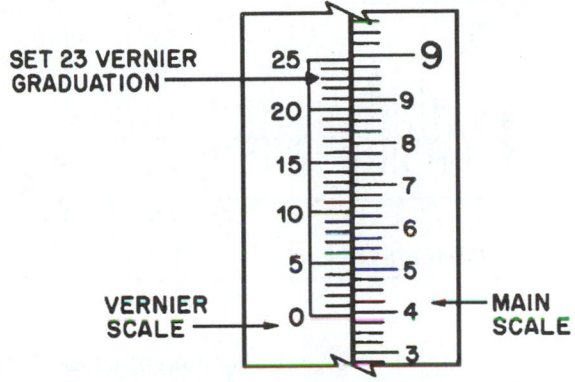

APPLICATION

Decimal-Inch Vernier Caliper

1. Read the decimal-inch vernier caliper measurements a–h for the following settings.

a. _____

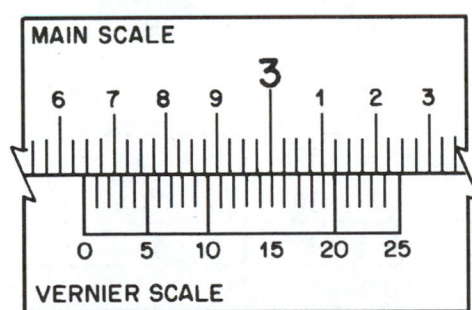

c. _____

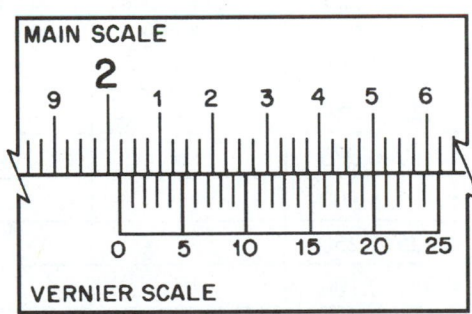

b. _____

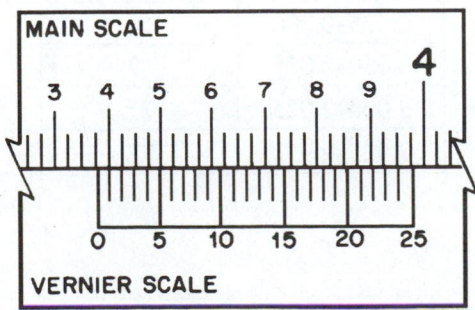

d. _____

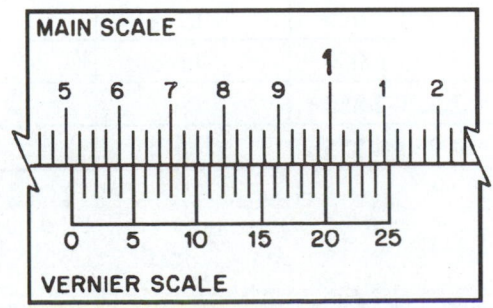

e. _____

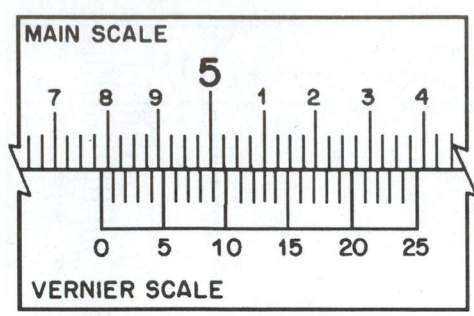

g. _____

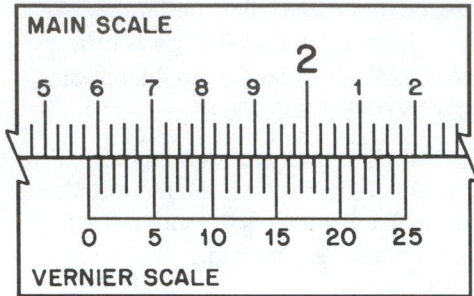

f. _____

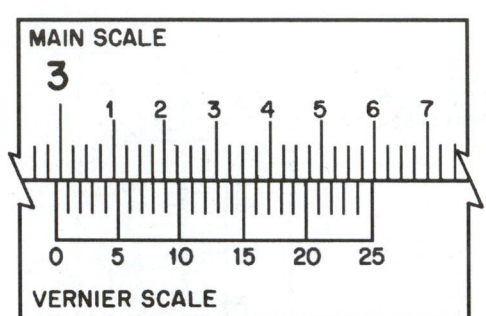

h. _____

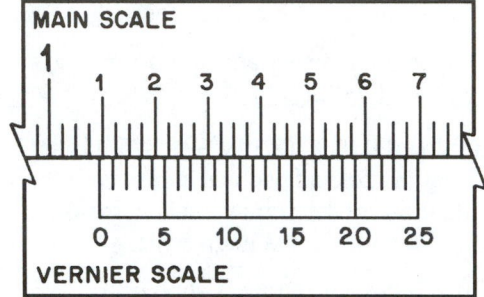

2. The following tables give the position of the zero graduation on the vernier scale in reference to the main scale and the vernier scale graduation that coincides with a main scale graduation. Determine the vernier caliper settings. The answer to the first problem is given.

	Zero Vernier Graduation Lies Between These Main Scale Graduations (inches)	Vernier Graduation That Coincides with a Main Scale Graduation	Vernier Caliper Setting (inches)		Zero Vernier Graduation Lies Between These Main Scale Graduations (inches)	Vernier Graduation That Coincides with a Main Scale Graduation	Vernier Caliper Setting (inches)
a.	1.875–1.900	19	1.894	m.	0.000–0.025	5	
b.	3.025–3.050	21		n.	0.825–0.850	17	
c.	0.050–0.075	6		o.	3.550–3.575	23	
d.	5.775–5.800	11		p.	5.075–5.100	20	
e.	1.225–1.250	7		q.	3.325–3.350	15	
f.	0.075–0.100	16		r.	2.075–2.100	6	
g.	3.000–3.025	4		s.	4.400–4.425	10	
h.	2.650–2.675	9		t.	1.025–1.050	13	
i.	1.000–1.025	13		u.	0.675–0.700	18	
j.	5.975–6.000	18		v.	0.050–0.075	2	
k.	2.825–2.850	8		w.	3.000–3.025	21	
l.	4.950–4.975	1		x.	2.925–2.950	22	

3. Refer to the following sentence and to the following given vernier caliper settings to find values A, B, and C. "The zero vernier scale graduation lies between A and B on the main scale and the vernier graduation C coincides with the main scale graduation." The answer to the first problem is given.

	Vernier Caliper Setting (inches)	A (inches)	B (inches)	C
a.	3.242	3.225	3.250	17
b.	2.877			
c.	4.839			
d.	0.611			
e.	4.369			
f.	0.084			
g.	7.857			

	Vernier Caliper Setting (inches)	A (inches)	B (inches)	C
h.	1.646			
i.	4.034			
j.	0.022			
k.	3.333			
l.	5.999			
m.	0.278			
n.	0.965			

4. The distance between the centers of two holes can be checked with a vernier caliper. The position of the caliper in measuring the inside distance between two holes is shown. To determine the setting on the caliper, subtract the radius of each hole (one-half the diameter) from the center distance. The following problems give the hole diameters and the distances between centers. For each determine (1) the main scale setting and (2) the vernier scale setting. All dimensions are in inches.

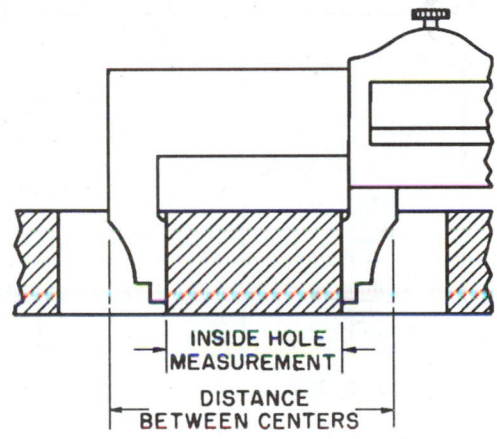

INSIDE HOLE MEASUREMENT

DISTANCE BETWEEN CENTERS

a. 2 HOLES 0.232 DIA

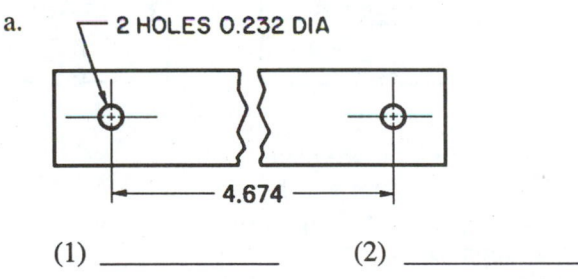

4.674

(1) _____ (2) _____

b. 2 HOLES 0.186 DIA

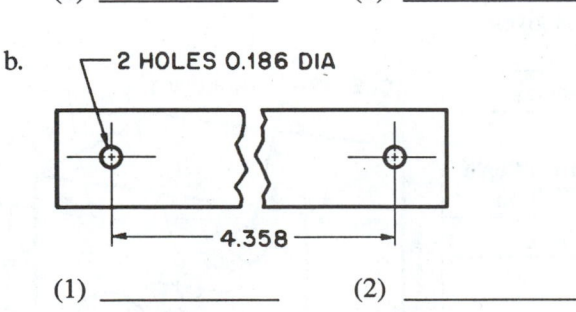

4.358

(1) _____ (2) _____

c. 0.123 DIA   0.137 DIA

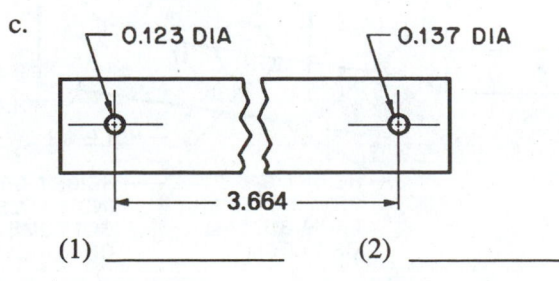

3.664

(1) _____ (2) _____

d. **Note:** Hole tolerances are shown. Maximum and minimum vernier scale settings are required.

0.750 ± 0.004 DIA   0.478 ± 0.002 DIA

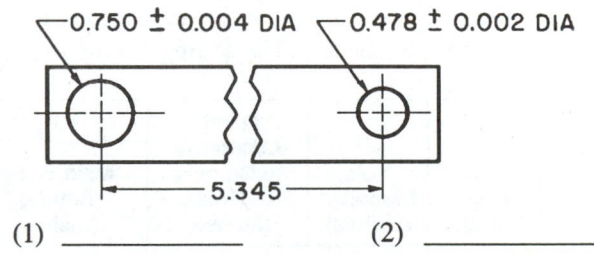

5.345

(1) _____ (2) _____

e. **Note:** Hole tolerances and center distance tolerances are shown. Maximum and minimum vernier scale settings are required.

0.375 ± 0.003 DIA   0.327 ± 0.005 DIA

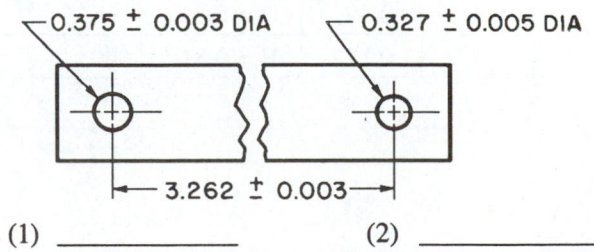

3.262 ± 0.003

(1) _____ (2) _____

Decimal-Inch Height Gage

5. Read height gage measurement a–h for the following settings.

a. _____   c. _____   e. _____   g. _____

b. _____   d. _____   f. _____   h. _____

6. The hole locations of this block are checked by placing the block on a surface plate and indicating the bottom of each hole using a height gage with an indicator attachment. Determine the height gage settings from the bottom of the part to the bottom of the holes. Assume that the actual hole diameters and locations are the same as the given dimensions. The setting for the first problem is given.

Hole Number	Hole Diameter (inches)	Given Locations to Centers of Holes (inches)	Height Gage Settings	
			Main Scale Setting (inches)	Vernier Scale Setting
1	0.376	A = 0.640	0.450–0.475	2
2	0.258	B = 1.008		
3	0.188	C = 0.514		
4	0.496	D = 0.312		
5	0.127	E = 0.810		

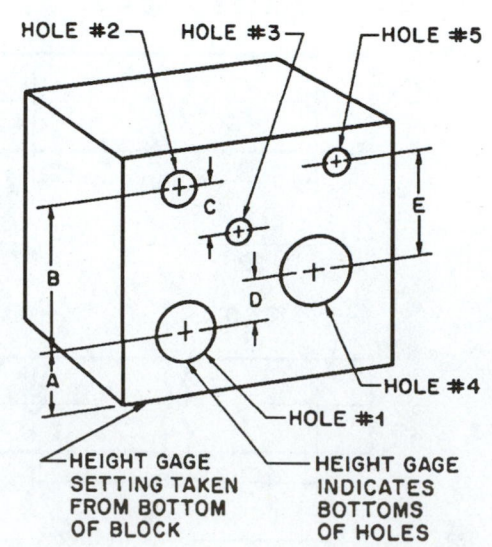

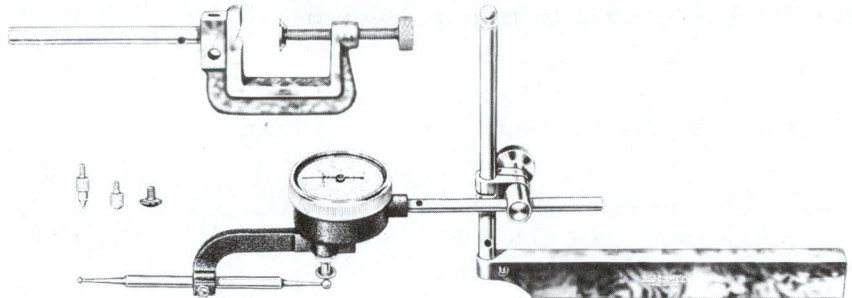

Dial indicator attachment   (Courtesy of L.S. Starrett Company)

UNIT 26 Metric Vernier Calipers and Height Gages

Objectives   After studying this unit you should be able to

- **Read measurements set on a metric vernier caliper.**
- **Set given measurements on a metric vernier caliper.**
- **Read measurements set on a metric vernier height gage.**
- **Set given measurements on a metric vernier height gage.**

Metric vernier calipers and height gages are used in machine shop applications when the degree of precision to 0.02 millimeter is adequate.

Metric Vernier Caliper

The same principles are used in reading and setting metric measure vernier calipers as those in decimal-inch measure. The main scale is divided in millimeter divisions. Each millimeter division is divided in half or 0.5-millimeter divisions. Every tenth millimeter graduation is numbered in sequence as 10 mm, 20 mm, 30 mm, etc. The vernier scale has 25 divisions; each division is $\frac{1}{25}$ of 0.5 millimeter or 0.02 millimeter.

Reading and Setting Measurements on a Metric Vernier Caliper

A measurement is read by adding the 0.02-millimeter reading on the vernier scale to the reading from the main scale.

Procedure   To read a measurement on a metric vernier caliper

- Read the number of millimeter divisions and 0.5 millimeter divisions on the main scale that are to the left of the zero graduation on the vernier scale.
- On the vernier scale, find the graduation that most closely coincides with a graduation on the main scale. Multiply the graduation by 0.02 millimeter and add the value obtained to the main scale reading.

Setting a given measurement is the reverse procedure of reading a measurement on the vernier caliper.

Example 1 Read the measurement set on the metric scales shown.

To the left of the zero division on the vernier scale read 21 millimeter divisions and one 0.5-millimeter division on the main scale. (21 mm + 0.5 mm = 21.5 mm)

Observe which vernier scale graduation most closely coincides with a main scale graduation. The sixth vernier scale graduation coincides. Since each vernier scale graduation represents 0.02 mm, multiply 6 times 0.02 mm = 0.12 mm. Add 0.12 mm to 21.5 mm.

Measurement: 0.12 mm + 21.5 mm = 21.62 mm   Ans

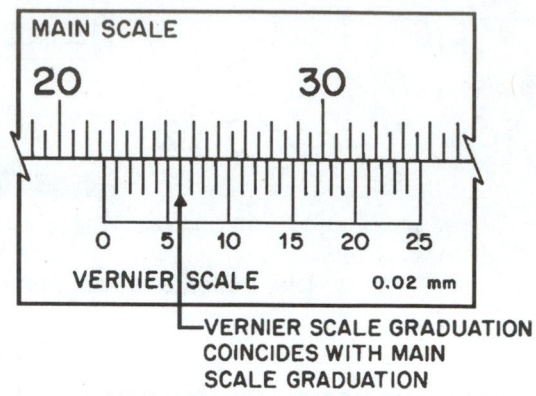

Example 2 Set 50.96 millimeters on a vernier caliper.

Move the vernier scale zero graduation to 50 millimeters plus 0.5 millimeter on the main scale. (50 mm + 0.5 mm = 50.5 mm)

An additional 0.46 millimeter (50.96 mm − 50.5 mm) is set by adjusting the sliding jaw. Since each vernier scale graduation represents 0.02 mm, divide 0.46 mm by 0.02 mm; 0.46 mm ÷ 0.02 mm = 23. Adjust the sliding jaw until the 23 graduation on the vernier scale coincides with a graduation on the main scale.

The 50.96-millimeter setting is shown.

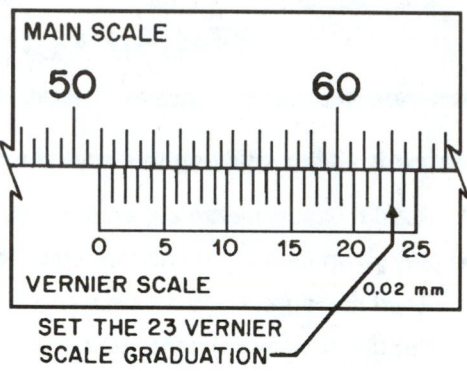

Metric Vernier Height Gage

As with the English vernier height gage and caliper, the metric vernier height gage and metric vernier caliper are similar in operation.

Reading and Setting Measurements on a Metric Vernier Height Gage

Measurements on the metric vernier height gage are read and set using the same procedure as with the metric vernier caliper.

Example 1 Read the measurement set on the metric vernier height gage scales shown.

Below the zero division on the vernier scale read 70 millimeter divisions and one 0.5 millimeter division on the main scale. (70 mm + 0.5 mm = 70.5 mm)

Observe which vernier scale graduation most closely coincides with a main scale graduation. The eighth vernier scale graduation coincides. Since each vernier scale graduation represents 0.02 mm, multiply 8 times 0.02 mm = 0.16 mm. Add 0.16 mm to 70.5 mm.

Measurement: 0.16 mm + 70.5 mm = 70.66 mm   Ans

Example 2 Set 42.74 millimeters on a vernier height gage.

Move the vernier scale zero graduation to 42 millimeters plus 0.5 millimeter on the main scale. (42 mm + 0.5 mm = 42.5 mm)

An additional 0.24 millimeter (42.74 mm − 42.5 mm) is set by carefully adjusting the sliding jaw. Since each vernier scale graduation represents 0.02 mm, divide 0.24 mm by 0.02 mm; 0.24 mm ÷ 0.02 mm = 12. Adjust the sliding jaw until the 12 graduation on the vernier scale coincides with a graduation on the main scale.

The 42.74-millimeter setting is shown.

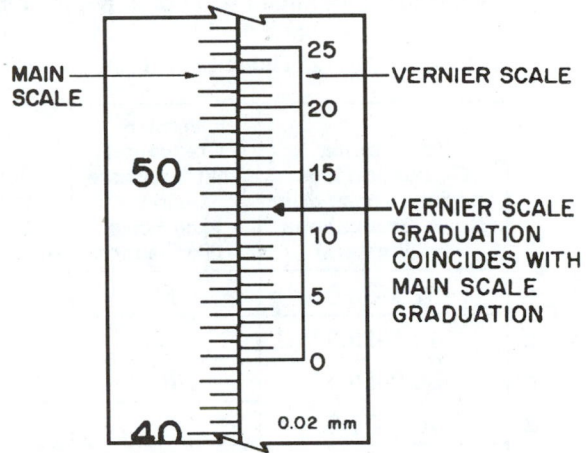

APPLICATION

Metric Vernier Caliper

1. Read the metric vernier caliper measurements for the following settings.

a. _____

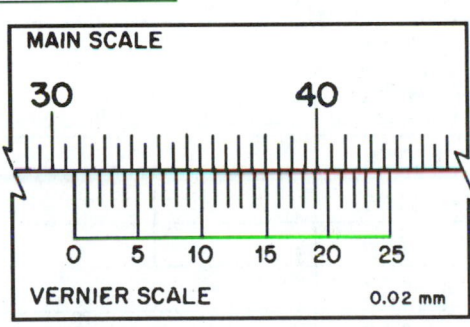

d. _____

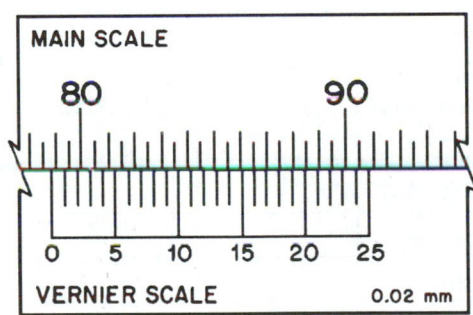

b. _____

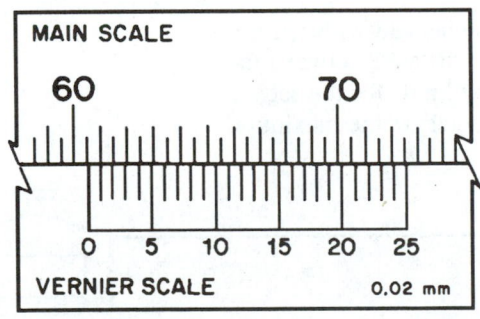

e. _____

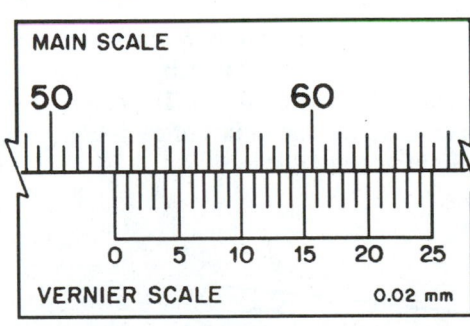

c. _____

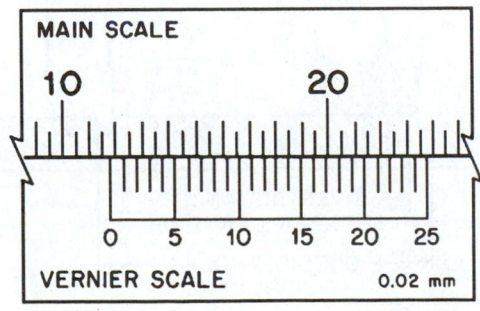

f. _____

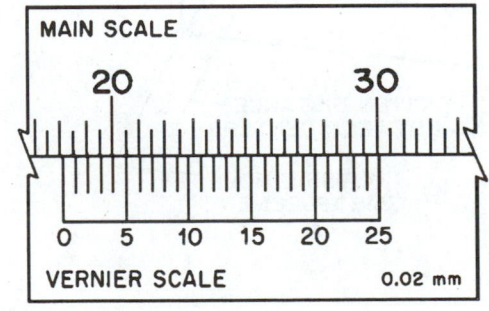

2. The following table gives the position of the zero graduation on the vernier scale in reference to the main scale of a metric vernier caliper. Also listed is the vernier scale graduation that coincides with a main scale graduation. Determine each vernier caliper setting. The answer to the first exercise is given.

	Zero Vernier Graduation Lies Between These Main Scale Graduations (millimeters)	Vernier Graduation That Coincides with a Main Scale Graduation	Vernier Caliper Setting (millimeters)		Zero Vernier Graduation Lies Between These Main Scale Graduations (millimeters)	Vernier Graduation That Coincides with a Main Scale Graduation	Vernier Caliper Setting (millimeters)
a.	52.5–53.0	14	52.78	g.	48.0–48.5	23	
b.	14.5–15.0	2		h.	77.5–78.0	21	
c.	86.0–86.5	16		i.	16.5–17.0	1	
d.	70.5–71.0	21		j.	98.0–98.5	11	
e.	24.0–24.5	8		k.	41.0–41.5	17	
f.	39.5–40.0	13		l.	56.5–57.0	20	

3. Refer to the following sentence and to the given vernier caliper settings in the following table to determine the values A, B, and C. "The zero vernier scale graduation lies between A and B on the main scale, and the vernier graduation C coincides with a main scale graduation." The answer to the first problem is given.

	Vernier Caliper Setting (millimeters)	A (millimeters)	B (millimeters)	C		Vernier Caliper Setting (millimeters)	A (millimeters)	B (millimeters)	C
a.	37.68	37.5	38.0	9	f.	10.38			
b.	18.56				g.	43.06			
c.	42.04				h.	77.40			
d.	88.82				i.	81.22			
e.	56.68				j.	93.98			

4. The distance between the two holes in the object shown is checked with vernier calipers. The position of the caliper in measuring the outside distance between the two holes in the part is shown. To determine the high and low limits for the setting on the caliper, add the radius of each hole (one-half the diameter) to the maximum or minimum center distance.

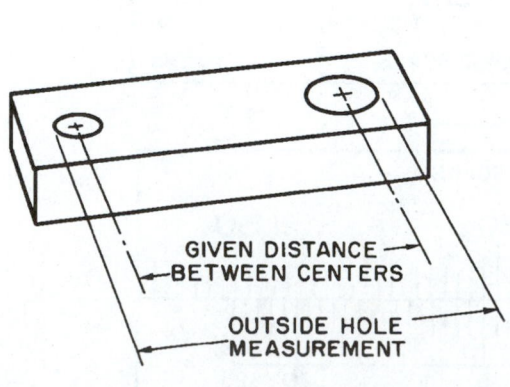

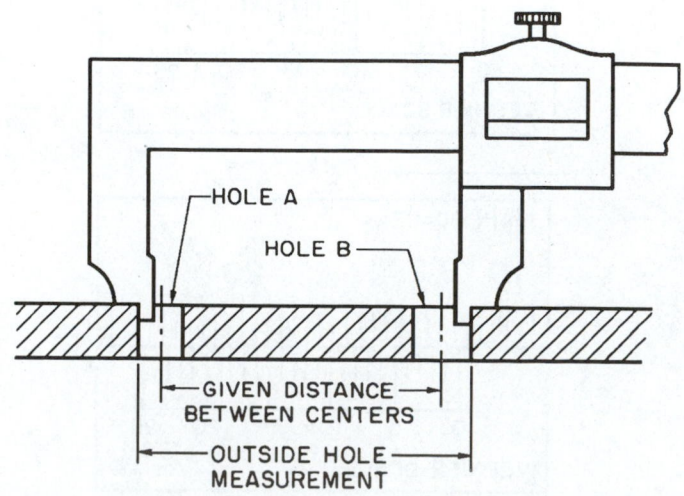

Refer to the data in the following table. Determine the vernier caliper scale settings for each problem. The answer to the first problem is given. All dimensions are in millimeters.

	ACTUAL HOLE DIAMETERS			VERNIER CALIPER SCALE SETTINGS		
	Hole A	Hole B	Given Distance Between Hole Centers	Main Scale Setting	High Limit Vernier Scale Setting	Low Limit Vernier Scale Setting
a.	10.52	12.86	56.92 ± 0.07	68.5–69.0	9	2
b.	14.10	17.18	72.08 ± 0.08			
c.	16.54	19.06	95.36 ± 0.04			
d.	9.98	14.80	44.41 ± 0.06			
e.	8.40	11.66	67.33 ± 0.10			
f.	17.46	21.82	86.57 ± 0.12			

Metric Height Gage

5. Read the metric vernier height gage measurements for the following settings.

a. _____   c. _____   e. _____   g. _____

b. _____   d. _____   f. _____   h. _____

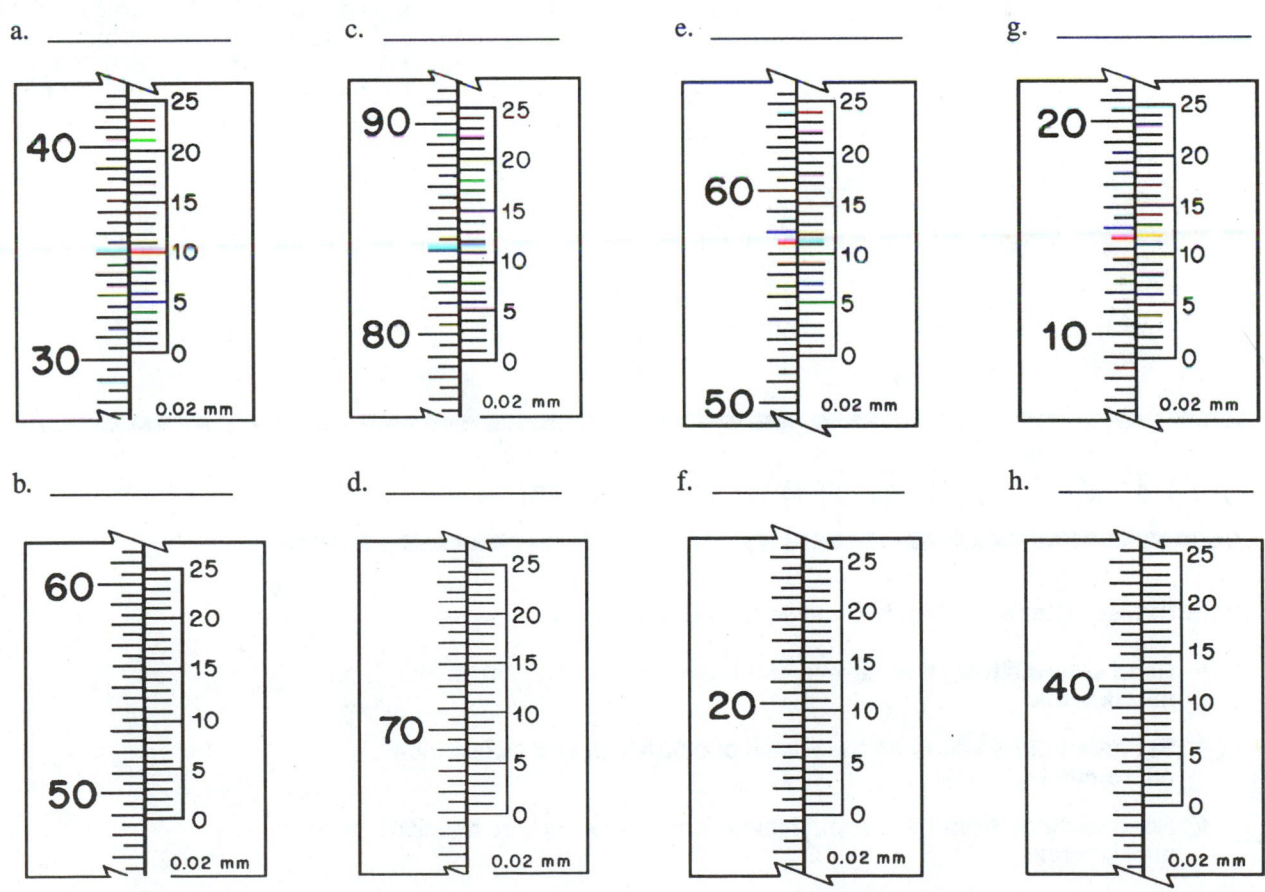

6. The hole locations of this block are checked by placing the block on a surface plate and indicating the bottom of each hole using a height gage with an indicator attachment. Determine the height gage settings from the bottom of the block to the bottom of the holes. Assume that the actual hole diameters and locations are the same as the given dimensions. The setting for the first hole is given.

Hole Number	Hole Diameter (millimeters)	Given Locations to Centers of Holes (millimeters)	HEIGHT GAGE SETTINGS	
			Main Scale Setting (millimeters)	Vernier Scale Setting
1	12.32	A = 15.78	9.5–10.0	6
2	6.38	B = 25.75		
3	4.50	C = 12.46		
4	14.76	D = 8.04		
5	5.84	E = 21.44		

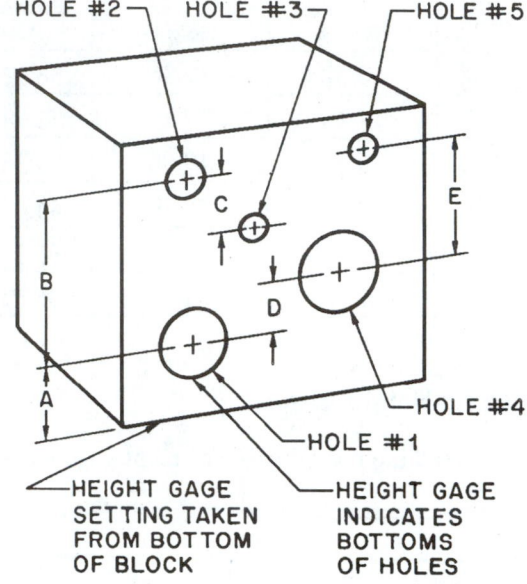

HEIGHT GAGE SETTING TAKEN FROM BOTTOM OF BLOCK

HEIGHT GAGE INDICATES BOTTOMS OF HOLES

UNIT 27 English Micrometers

Objectives   After studying this unit you should be able to

- **Read settings from the barrel and thimble scales of a 0.001-inch micrometer.**

- **Set given dimensions on the scales of 0.001-inch and 0.0001-inch micrometers.**

- **Read settings from the barrel, thimble, and vernier scales of 0.0001-inch micrometers.**

Micrometers are basic measuring instruments used by machinists in the processing and checking of parts. Micrometers are available in a wide range of sizes and types. Outside micrometers are used to measure dimensions between parallel surfaces of parts and outside diameters of cylinders. Other types, such as depth micrometers, screw thread micrometers, disc and blade micrometers, bench micrometers, and inside micrometers, also have wide application in the machine shop. A few of the many types of micrometers are shown.

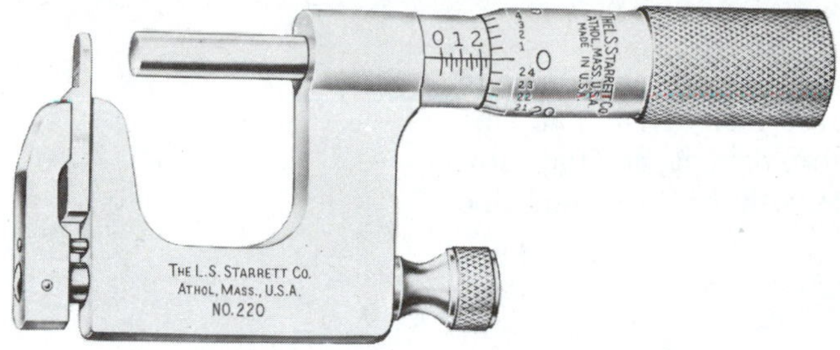

Anvil Micrometer
(The L.S. Starrett Company)

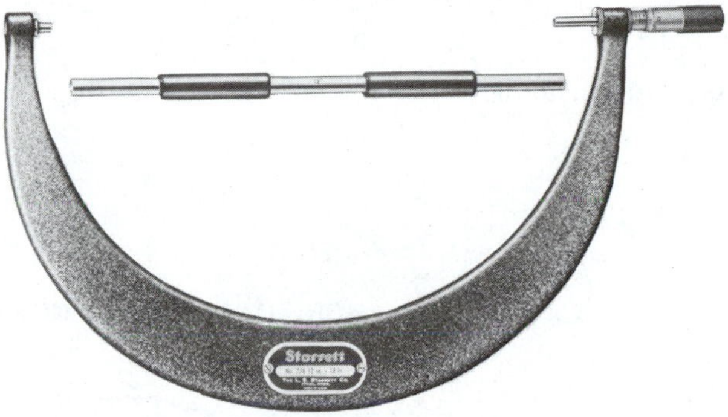

Bow Micrometer
(The L.S. Starrett Company)

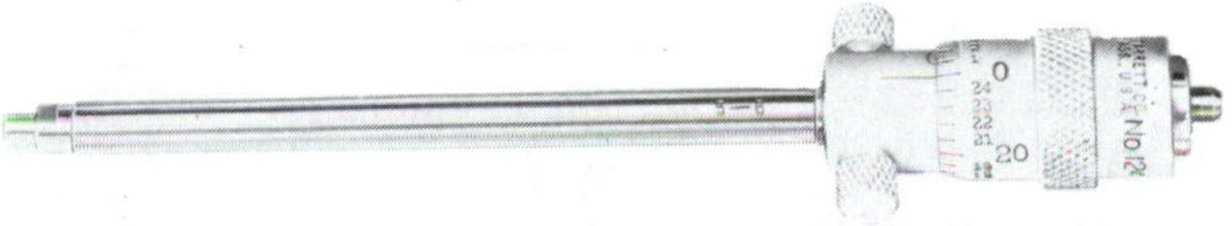

Inside Micrometer
(The L.S. Starrett Company)

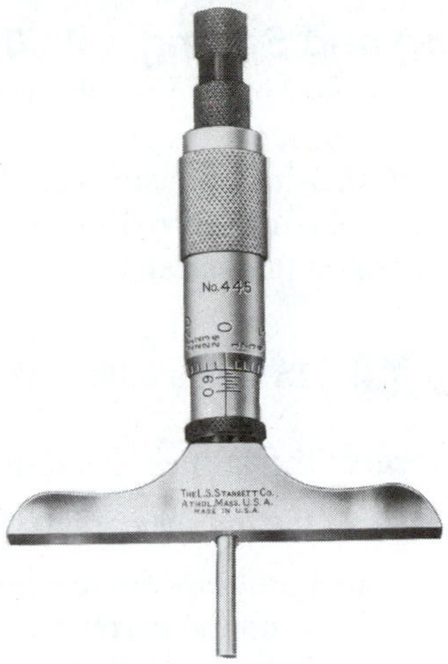

Micrometer Depth Gage
(The L.S. Starrett Company)

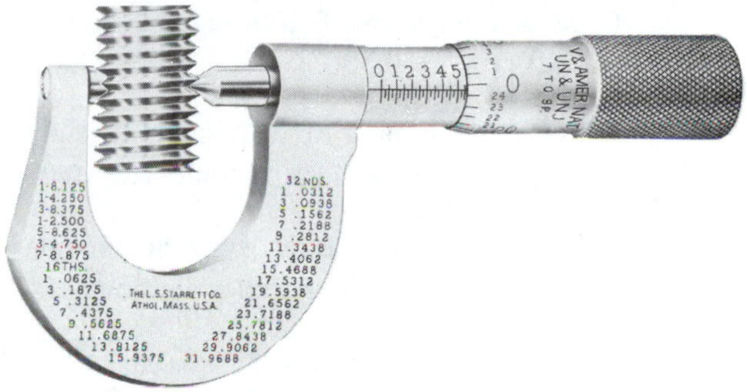

Screw Thread Micrometer
(The L.S. Starrett Company)

The 0.001-Inch Micrometer

A 0.001-inch outside micrometer is shown with its principal parts labeled.

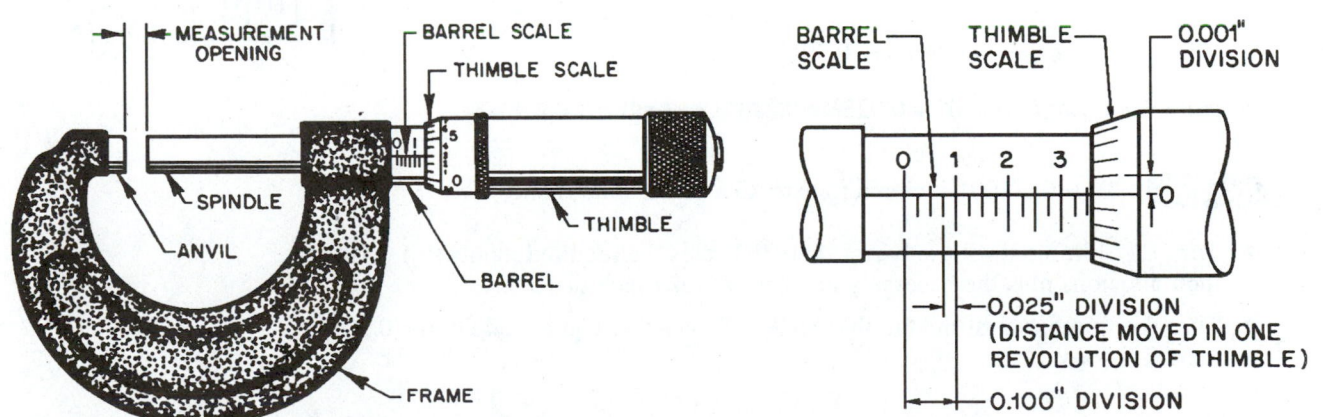

The part to be measured is placed between the anvil and the spindle. The barrel of a micrometer consists of a scale which is one inch long. The one-inch length is divided into ten divisions each equal to 0.100 inch. The 0.100-inch divisions are further divided in four divisions each equal to 0.025 inch.

The thimble has a scale which is divided into twenty-five parts. One revolution of the thimble moves 0.025 inch on the barrel scale. Therefore, a movement of one graduation on the thimble equals $\frac{1}{25}$ of 0.025 inch or 0.001 inch along the barrel.

Reading and Setting a 0.001-Inch Micrometer

A micrometer is read by observing the position of the bevel edge of the thimble in reference to the scale on the barrel. Observe the greatest 0.100-inch division and the number of 0.025-inch divisions on the barrel scale. To this barrel reading, add the number of the 0.001-inch divisions on the thimble that coincide with the horizontal line (reading line) on the barrel scale.

Procedure To read a 0.001-inch micrometer

- Observe the greatest 0.100-inch division on the barrel scale.

- Observe the number of 0.025-inch divisions on the barrel scale.

- Add the thimble scale reading (0.001-inch division) that coincides with the horizontal line on the barrel scale.

Example 1 Read the micrometer setting shown.

Observe the greatest 0.100-inch division on the barrel scale. (three 0.100″ = 0.300″)

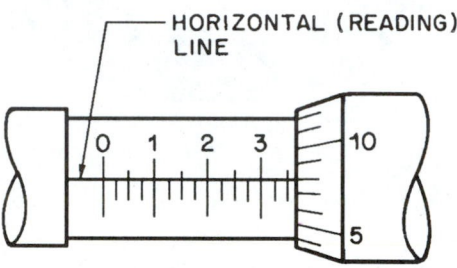

Observe the number of 0.025-inch divisions between the 0.300-inch mark and the thimble. (two 0.025″ = 0.050″)

Add the thimble scale reading that coincides with the horizontal line on the barrel scale. (eight 0.001″ = 0.008″)

Micrometer reading: 0.300″ + 0.050″ + 0.008″ = 0.358″   Ans

Example 2 Read the micrometer setting shown.

On the barrel scale, two 0.100″ = 0.200″.

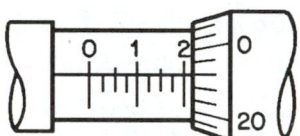

On the barrel scale, zero 0.025″ = 0″.

On the thimble scale, twenty-three 0.001″ = 0.023″.

Micrometer reading: 0.200″ + 0.023″ = 0.223″   Ans

Procedure To set a 0.001-inch micrometer to a given dimension

- Turn the thimble until the barrel scale indicates the required number of 0.100-inch divisions plus the necessary number of 0.025-inch divisions.

- Turn the thimble until the thimble scale indicates the required additional 0.001-inch divisions.

Example 1 Set 0.949 inch on a micrometer.

Turn the thimble to nine 0.100-inch divisions plus one 0.025-inch division on the barrel scale. (9 × 0.100″ + 0.025″ = 0.925″)

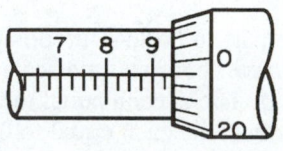

Turn the thimble an additional twenty-four 0.001-inch thimble scale divisions. (0.949″ − 0.925″ = 0.024″)

The 0.949-inch setting is shown.

Example 2 Set 0.520 inch on a micrometer.

Turn the thimble to five 0.100-inch divisions on the barrel scale.
$(5 \times 0.100'' = 0.500'')$

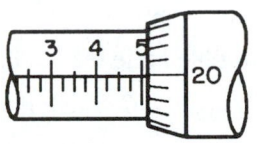

Turn the thimble an additional twenty 0.001-inch divisions.
$(0.520'' - 0.500'' = 0.020'')$

The 0.520-inch setting is shown.

The Vernier (0.0001-Inch) Micrometer

The addition of a vernier scale on the barrel of a 0.001-inch micrometer increases the degree of precision of the instrument to 0.0001 inch. The barrel scale and thimble scale of a vernier micrometer are identical to that of a 0.001-inch micrometer. The figure shows the relative positions of the barrel scale, thimble scale, and vernier scale of a 0.0001-inch vernier micrometer.

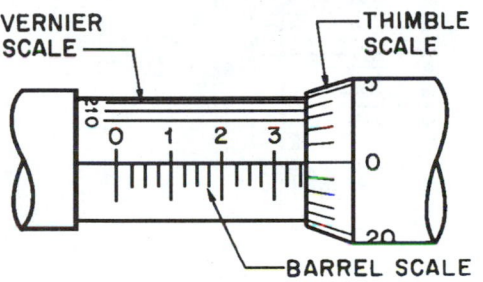
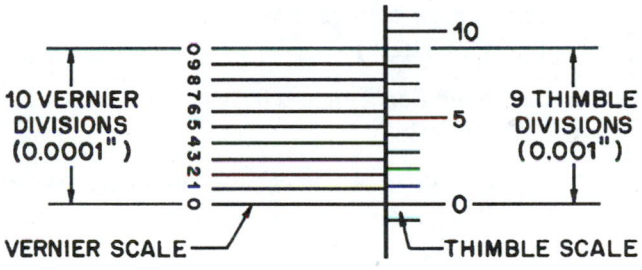

The vernier scale consists of ten divisions. Ten vernier divisions on the circumference of the barrel are equal in length to nine divisions of the thimble scale. The difference between one vernier division and one thimble division is 0.0001-inch. A flattened view of a vernier and a thimble scale is shown.

Reading and Setting the Vernier (0.0001-Inch) Micrometer

Reading a vernier micrometer is the same as reading a 0.001-inch micrometer except for the addition of reading the vernier scale. A particular vernier graduation coincides with a thimble scale graduation. This vernier graduation gives the number of 0.0001-inch divisions that are added to the barrel and thimble scale readings.

Example 1 Read the vernier micrometer setting shown in this flattened view.

Read the barrel scale reading. Three 0.100″ divisions plus three 0.025″ divisions = 0.375″.

Read the thimble scale. The reading is between the 0.009″ and 0.010″ divisions, therefore, the thimble reading is 0.009″.

Read the vernier scale. The 0.0004″ division of the vernier scale coincides with a thimble division.

Vernier micrometer reading:
$0.375'' + 0.009'' + 0.0004'' = 0.3844''$ Ans

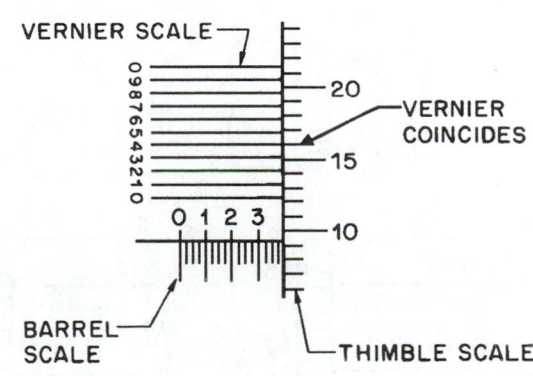

Example 2 Read the vernier micrometer setting shown in this flattened view.

On the barrel scale read 0.200″

On the thimble scale read 0.020″

On the vernier scale read 0.0008″

Vernier micrometer reading:
0.200″ + 0.020″ + 0.0008″ = 0.2208″   Ans

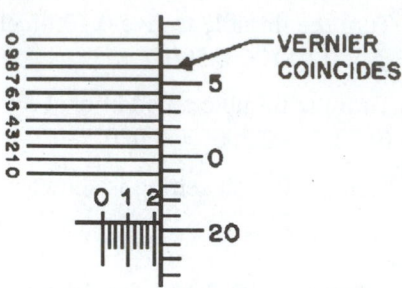

Setting a vernier (0.0001-inch) micrometer is the same as setting a 0.001-inch micrometer except for the addition of setting the vernier scale.

Example   Set 0.2336 inch on a vernier micrometer.

Turn the thimble to two 0.100-inch divisions plus one 0.025-inch division on the barrel scale.
(2 × 0.100″ + 0.025″ = 0.225″)

Turn the thimble an additional eight 0.001-inch divisions.
(0.2336″ − 0.225″ = 0.0086″)

Turn the thimble carefully until a graduation on the thimble scale coincides with the 0.0006-inch division on the vernier scale.
(0.2336″ − 0.233″ = 0.0006″)

The 0.2336-inch setting is shown.

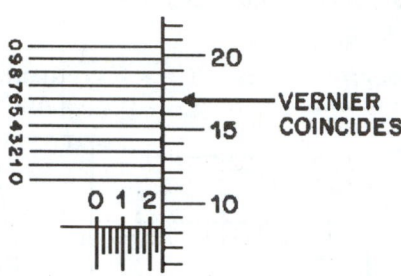

APPLICATION

0.001-Inch Micrometer

Read the settings on the following 0.001-inch micrometer scales.

1. _____

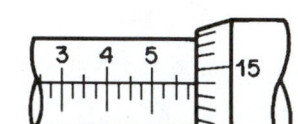

4. _____

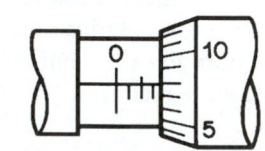

7. _____

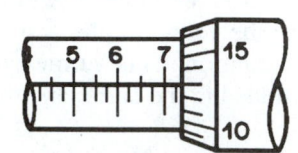

10. _____

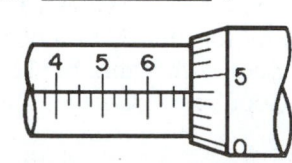

2. _____

5. _____

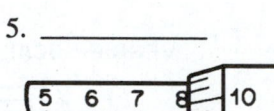

8. _____

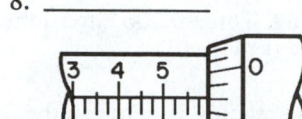

11. _____

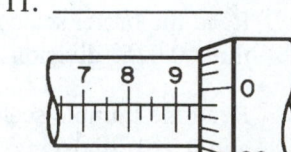

3. _____

6. _____

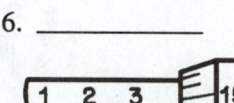

9. _____

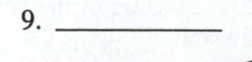

12. _____

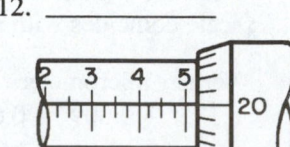

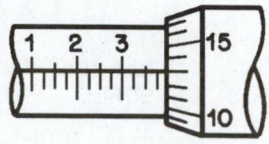

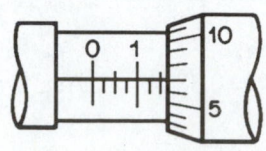

Given the following barrel scale and thimble scale settings of a 0.001-inch micrometer, determine the readings in the tables. The answer to the first problem is given.

	Barrel Scale Setting is Between: (inches)	Thimble Scale Setting (inches)	Micrometer Reading (inches)
13.	0.425–0.450	0.016	0.441
14.	0.075–0.100	0.009	
15.	0.150–0.175	0.003	
16.	0.875–0.900	0.012	
17.	0.400–0.425	0.024	

	Barrel Scale Setting is Between: (inches)	Thimble Scale Setting (inches)	Micrometer Reading (inches)
18.	0.000–0.025	0.023	
19.	0.025–0.050	0.013	
20.	0.750–0.775	0.017	
21.	0.975–1.000	0.008	
22.	0.625–0.650	0.016	

Given the following 0.001-inch micrometer readings, determine the barrel scale and thimble scale settings. The answer to the first problem is given.

	Micrometer Reading (inches)	Barrel Scale Setting is Between: (inches)	Thimble Scale Setting (inches)
23.	0.387	0.375–0.400	0.012
24.	0.841		
25.	0.973		
26.	0.002		
27.	0.079		

	Micrometer Reading (inches)	Barrel Scale Setting is Between: (inches)	Thimble Scale Setting (inches)
28.	0.998		
29.	0.038		
30.	0.281		
31.	0.427		
32.	0.666		

The Vernier (0.0001-Inch) Micrometer

Read the settings on the following 0.0001-inch micrometer scales. The vernier, thimble, and barrel scales are shown in flattened views.

33. _____

35. _____

37. _____

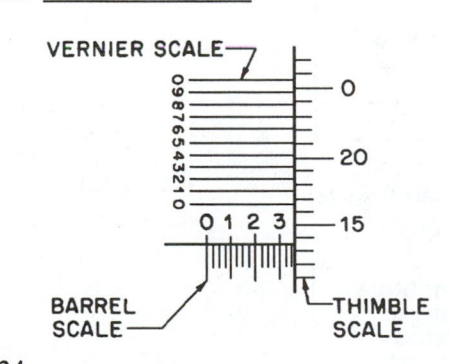

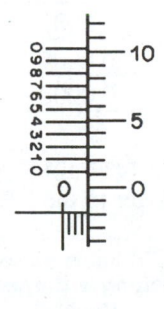

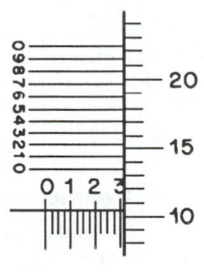

34. _____

36. _____

38. _____

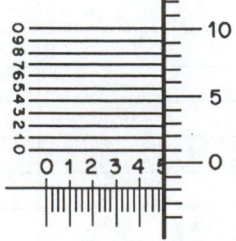

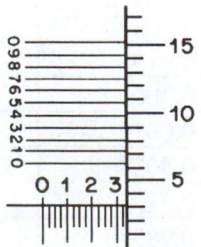

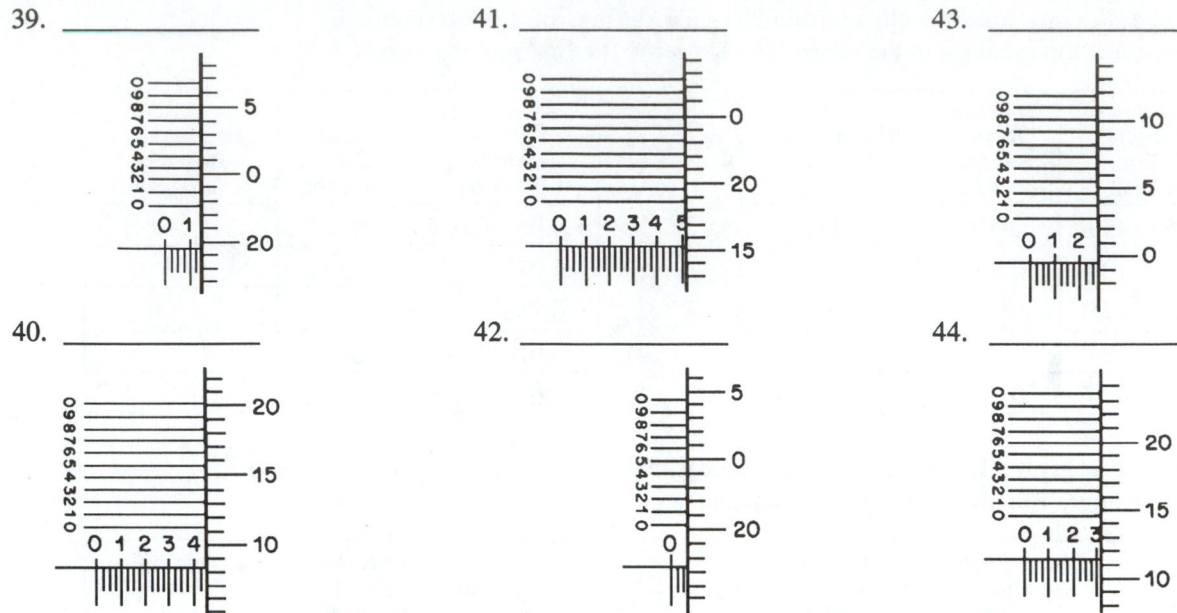

39. _____ 41. _____ 43. _____

40. _____ 42. _____ 44. _____

Given the following barrel scale, thimble scale, and vernier scale settings of a 0.0001-inch micrometer, determine the micrometer readings in these tables. The answer to the first problem is given.

	Barrel Scale Setting is Between: (inches)	Thimble Scale Setting is Between: (inches)	Vernier Scale Setting (inches)	Micrometer Reading (inches)
45.	0.375–0.400	0.017–0.018	0.0008	0.3928
46.	0.125–0.150	0.008–0.009	0.0003	
47.	0.950–0.975	0.021–0.022	0.0007	
48.	0.075–0.100	0.011–0.012	0.0005	
49.	0.300–0.325	0.000–0.001	0.0004	
50.	0.625–0.650	0.021–0.022	0.0002	
51.	0.000–0.025	0.000–0.001	0.0009	
52.	0.275–0.300	0.020–0.021	0.0007	
53.	0.850–0.875	0.009–0.010	0.0004	
54.	0.225–0.250	0.014–0.015	0.0008	

Given the following 0.0001-inch micrometer readings, determine the barrel scale, thimble scale, and vernier scale settings. The answer to the first problem is given.

	Micrometer Reading (inches)	Barrel Scale Setting is Between: (inches)	Thimble Scale Setting is Between: (inches)	Vernier Scale Setting (inches)
55.	0.7846	0.775–0.800	0.009–0.010	0.0006
56.	0.1035			
57.	0.0083			
58.	0.9898			
59.	0.3001			
60.	0.0012			
61.	0.8008			
62.	0.3135			
63.	0.9894			
64.	0.0479			

UNIT **28** Metric Micrometers

Objectives   After studying this unit you should be able to

- Read settings from the barrel and thimble scales of a 0.010-millimeter micrometer.
- Set given dimensions on the scales of 0.01-millimeter and 0.002-millimeter micrometers.
- Read settings from the barrel, thimble, and vernier scales of 0.002-millimeter micrometers.

The 0.01-Millimeter Micrometer

The construction, parts, and operation of a 0.01-millimeter micrometer are basically the same as a 0.001-inch micrometer. A 0.01-millimeter outside micrometer is shown.

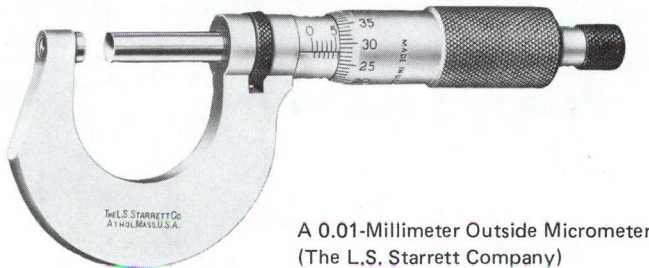

A 0.01-Millimeter Outside Micrometer
(The L.S. Starrett Company)

The barrel of a 0.01-millimeter micrometer consists of a scale which is 25 millimeters long. Refer to the barrel and thimble scales shown. The 25-millimeter barrel scale length is divided into 25 divisions each equal to 1 millimeter. Every fifth millimeter is numbered from 0 to 25 (0, 5, 10, 15, 20, 25). On the lower part of the barrel scale each millimeter is divided in half (0.5 mm).

The thimble has a scale which is divided into 50 parts. One revolution of the thimble moves 0.5 millimeter on the barrel scale. Therefore, a movement of one graduation on the thimble equals $\frac{1}{50}$ of 0.5 millimeter or 0.01 millimeter along the barrel.

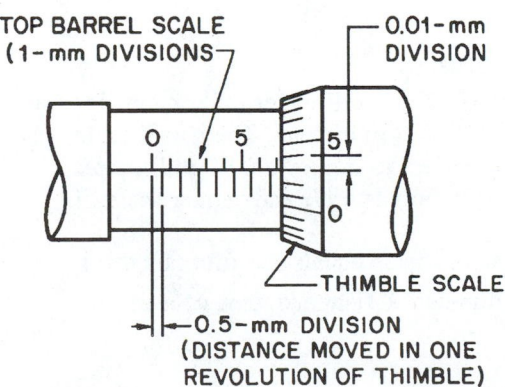

Reading and Setting a 0.01-Millimeter Micrometer

Procedure   To read a 0.01-millimeter micrometer

- Observe the number of 1-millimeter divisions on the barrel scale.

- Observe the number of 0.5-millimeter divisions (either 0 or 1) on the lower part of the barrel scale.
- Add the thimble scale reading (0.01-millimeter division) that coincides with the horizontal line (reading line) on the barrel side.

Example 1   Read the micrometer setting shown.

Observe the number of 1-millimeter divisions on the barrel scale. Read as 4 millimeters.

Observe the number of 0.5-millimeter divisions on the lower barrel scale.

Read as zero 0.5-millimeter divisions.

Add the thimble scale reading that coincides with the horizontal line on the barrel scale. Read as thirty-three 0.01-millimeter divisions.

Micrometer reading: 4 mm + 0.33 mm = 4.33 mm   Ans

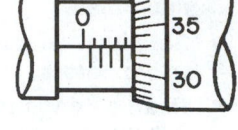

Example 2   Read this micrometer setting.

On the barrel scale read 17 millimeters.

On the lower barrel scale read one 0.5 millimeter.

On the thimble scale read twenty-six 0.01 millimeter.

Micrometer reading: 17 mm + 0.5 mm + 0.26 mm = 17.76 mm   Ans

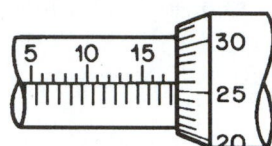

Procedure   To set a 0.01-millimeter micrometer

- Turn the thimble until the scale indicates the required number of 1-millimeter divisions plus the necessary number of 0.5-millimeter divisions.
- Turn the thimble until the thimble scale indicates the required additional 0.01-millimeter divisions.

Example   Set 14.94 millimeters on a micrometer.

Turn the thimble to fourteen 1-millimeter divisions plus one 0.5-millimeter division on the barrel scale. (14 mm + 0.5 mm = 14.5 mm)

Turn the thimble an additional forty-four 0.01-millimeter thimble scale divisions. (14.94 mm − 14.5 mm = 0.44 mm)

The 14.94-millimeter setting is shown.

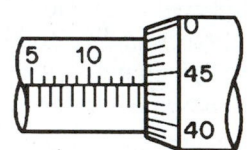

The Vernier (0.002-Millimeter) Micrometer

The addition of a vernier scale on the barrel of a 0.01-millimeter micrometer increases the degree of precision of the instrument to 0.002 millimeter. The barrel scale and thimble scale of a vernier micrometer are identical to that of a 0.01-millimeter micrometer. The relative positions of the barrel scale, thimble scale and vernier scale of a vernier micrometer are shown.

The vernier scale consists of five divisions. Each division equals one-fifth of a thimble division or $\frac{1}{5}$ of 0.01 millimeter or 0.002 millimeter. A flattened view of a vernier and thimble scale is shown.

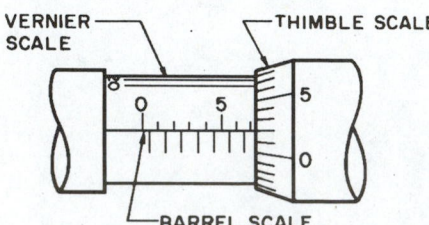

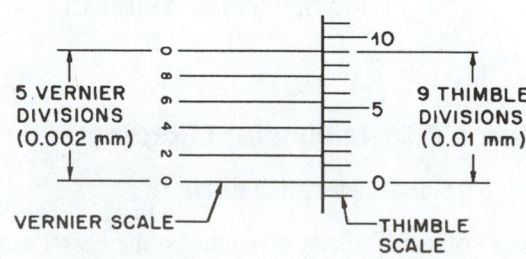

Reading and Setting a Vernier (0.002-Millimeter) Micrometer

Reading a vernier (0.002-millimeter) micrometer is the same as reading a 0.01-millimeter micrometer except for the addition of reading the vernier scale. Observe which division on the vernier scale coincides with a division on the thimble scale. If the vernier division which coincides is marked 2, add 0.002 millimeter to the barrel and thimble scale reading. Add 0.004 millimeter for a coinciding vernier division marked 4, add 0.006 millimeter for a division marked 6, and add 0.008 millimeter for a division marked 8.

Example 1   A flattened view of a vernier micrometer is shown. Read this setting.

On the barrel scale read 6 millimeters.

On the lower barrel scale read zero 0.5 millimeter.

On the thimble scale read the number of 0.01-millimeter divisions. The reading is between the twenty-six and twenty-seven 0.01-millimeter divisions. Read as 0.26 millimeter.

Read the vernier scale. The 4 on the vernier scale coincides with a thimble scale division. Add 0.004 millimeter to the barrel and thimble scale readings.

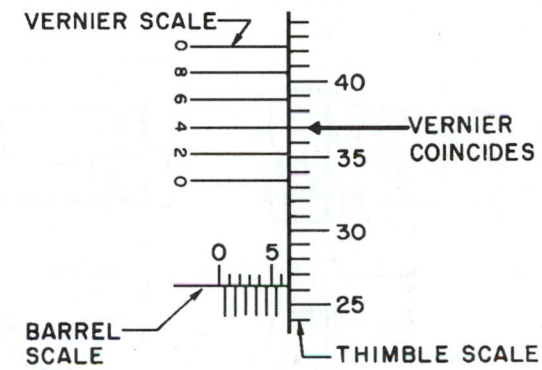

The vernier micrometer reading:
6 mm + 0.26 mm + 0.004 mm = 6.264 mm   Ans

Example 2   A flattened view of a vernier micrometer is shown. Read this setting.

On the barrel scale read 9 millimeters.

On the lower barrel scale read one 0.5 millimeter.

On the thimble scale read the number of 0.01-millimeter divisions. The reading is between the forty-three and forty-four 0.01-millimeter divisions. Read as 0.43 millimeter.

Read the vernier scale. The 8 on the vernier scale coincides with a thimble scale division. Add 0.008 millimeter to the barrel and thimble scale readings.

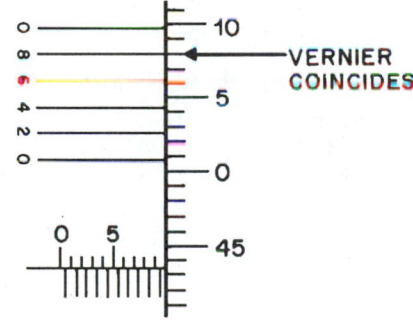

The vernier micrometer reading:
9 mm + 0.5 mm + 0.43 mm + 0.008 mm = 9.938 mm   Ans

Setting a 0.002-millimeter vernier micrometer is the same as setting a 0.01-millimeter micrometer except for the addition of setting the vernier scale.

Example   Set 1.862 millimeters on a vernier micrometer.

Turn the thimble to one 1 millimeter-division plus one 0.5 millimeter-division on the barrel scale.
(1 mm + 0.5 mm = 1.5 mm)

Turn the thimble an additional thirty-six 0.01-millimeter thimble scale divisions.
(1.862 mm – 1.5 mm = 0.362 mm)

Turn the thimble carefully until a graduation on the thimble coincides with the 0.002 millimeter division on the vernier scale.
(0.362 mm – 0.36 mm = 0.002 mm)

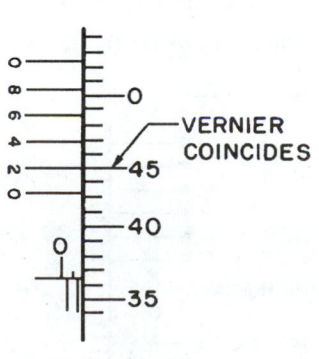

The 1.862-millimeter setting is shown.

APPLICATION

0.01-Millimeter Micrometer

Read the settings on the following 0.01-millimeter scales.

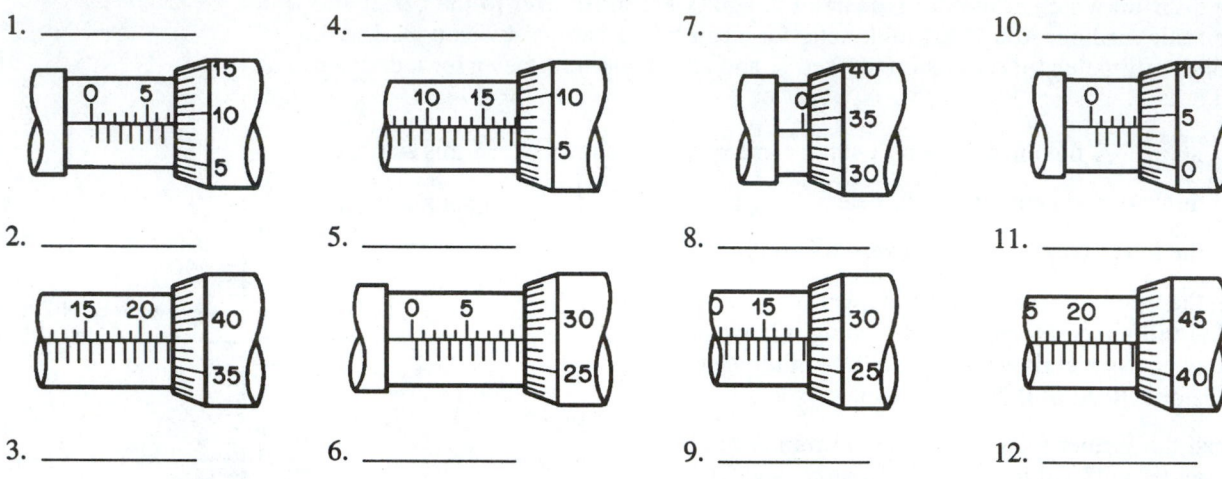

1. _____ 4. _____ 7. _____ 10. _____

2. _____ 5. _____ 8. _____ 11. _____

3. _____ 6. _____ 9. _____ 12. _____

Given the micrometer readings in the following table, determine the barrel scale and thimble scale settings. The answer to the first problem is given.

	Micrometer Reading (millimeters)	Barrel Scale Setting is Between: (millimeters)	Thimble Scale Setting (millimeters)		Micrometer Reading (millimeters)	Barrel Scale Setting is Between: (millimeters)	Thimble Scale Setting (millimeters)
13.	12.86	12.5–13.0	0.36	20.	7.56		
14.	10.34			21.	8.44		
15.	15.08			22.	19.72		
16.	3.92			23.	23.08		
17.	0.78			24.	4.78		
18.	7.06			25.	21.82		
19.	18.12			26.	13.90		

Vernier (0.002-Millimeter) Micrometer

Read the settings on the following 0.002-millimeter vernier micrometer scales.

27. _____   28. _____   29. _____

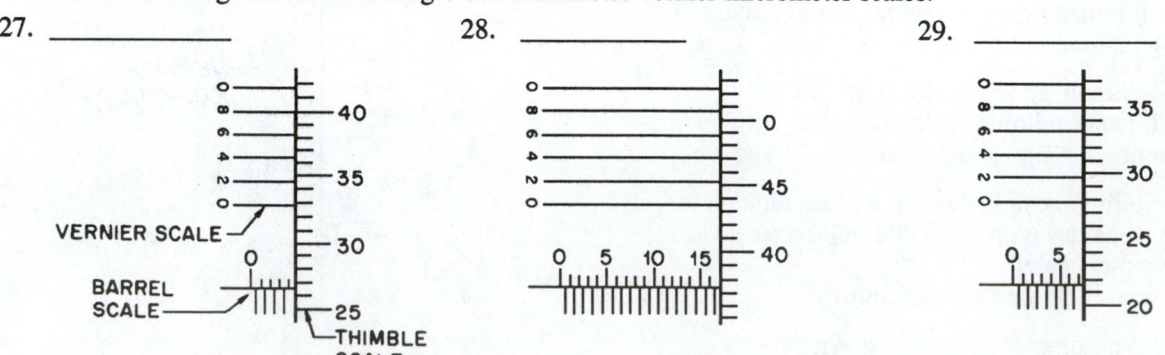

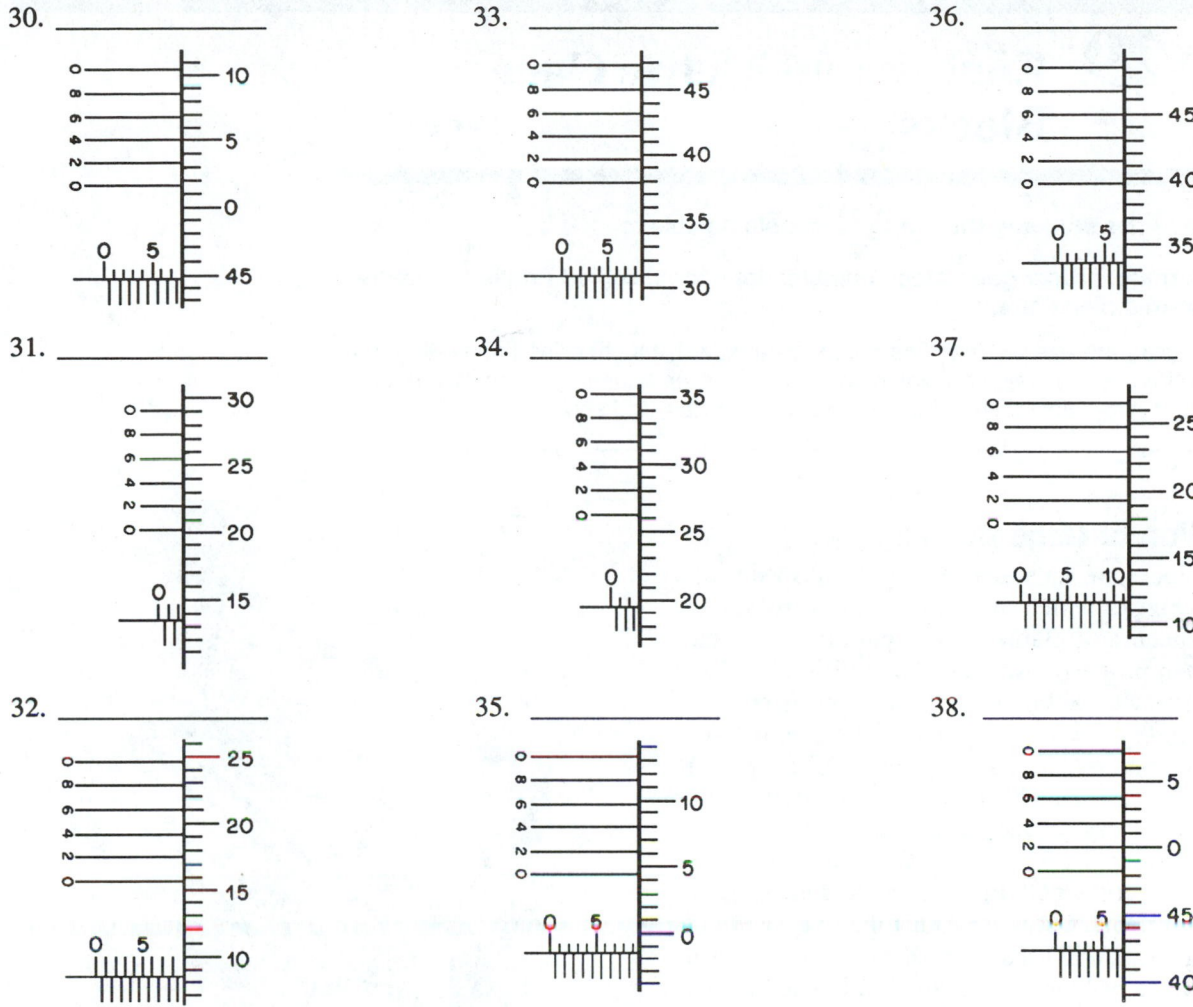

30. _____

31. _____

32. _____

33. _____

34. _____

35. _____

36. _____

37. _____

38. _____

Given the following 0.002-millimeter vernier micrometer readings in the table, determine the barrel scale, thimble scale, and vernier scale settings. The answer to the first problem is given.

	Micrometer Reading (millimeters)	Barrel Scale Setting is Between: (millimeters)	Thimble Scale Setting is Between: (millimeters)	Vernier Scale Setting (millimeters)
39.	14.874	14.5–5.0	0.37–0.38	0.004
40.	21.258			
41.	9.238			
42.	11.862			
43.	4.056			
44.	8.768			
45.	7.004			
46.	19.192			
47.	5.708			
48.	13.998			
49.	9.234			
50.	0.756			
51.	14.582			
52.	7.766			

UNIT 29 English and Metric Gage Blocks

Objective After studying this unit you should be able to

- **Determine proper gage block combinations for specified English or metric system dimensions.**

Gage blocks are used in machine shops as standards for checking and setting (calibration) of micrometers, calipers, dial indicators, and other measuring instruments. Other applications of gage blocks are for layout, machine setups, and surface plate inspection.

Description of Gage Blocks

Gage blocks are square or rectangular shaped hardened steel blocks which are manufactured to a high degree of accuracy, flatness, and parallelism. Gage blocks, when properly used, provide millionths of an inch accuracy with millionths of an inch precision.

By *wringing* blocks (slipping blocks one over the other using light pressure), a combination of the proper blocks can be achieved which provides a desired length. Wringing the blocks produces a very thin air gap that is similar to liquid film in holding the blocks together. There are a variety of both English unit and metric gage block sets available. These tables list the thicknesses of blocks of a frequently used English gage block set and the thicknesses of blocks of a commonly used metric gage block set.

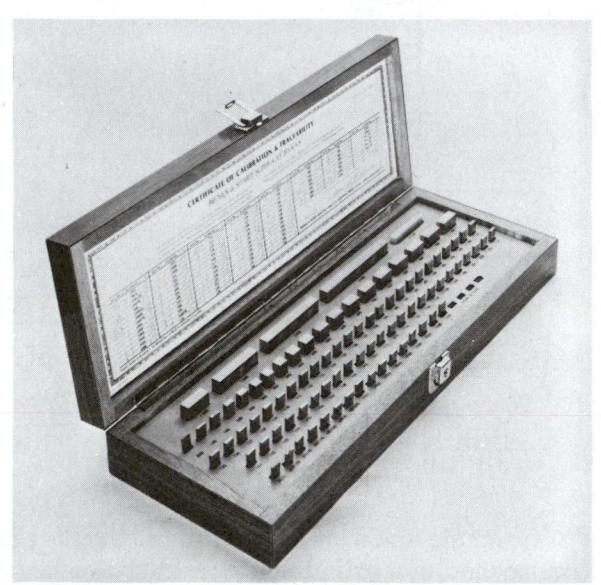

A complete set of gage blocks (Brown & Sharpe Mfg. Co.)

BLOCK THICKNESSES OF AN ENGLISH GAGE BLOCK SET*

9 Blocks 0.000 1″ Series

0.100 1	0.100 2	0.100 3	0.100 4	0.100 5	0.100 6	0.100 7	0.100 8	0.100 9

49 Blocks 0.001″ Series

0.101	0.102	0.103	0.104	0.105	0.106	0.107	0.108	0.109
0.110	0.111	0.112	0.113	0.114	0.115	0.116	0.117	0.118
0.119	0.120	0.121	0.122	0.123	0.124	0.125	0.126	0.127
0.128	0.129	0.130	0.131	0.132	0.133	0.134	0.135	0.136
0.137	0.138	0.139	0.140	0.141	0.142	0.143	0.144	0.145
0.146	0.147	0.148	0.149					

19 Blocks 0.050″ Series

0.050	0.100	0.150	0.200	0.250	0.300	0.350	0.400	0.450
0.500	0.550	0.600	0.650	0.700	0.750	0.800	0.850	0.900
0.950								

4 Blocks 1.000″ Series

1.000	2.000	3.000	4.000

*All thicknesses are in inches

BLOCK THICKNESSES OF A METRIC GAGE BLOCK SET*								
9 Blocks 0.001 mm Series								
1.001	1.002	1.003	1.004	1.005	1.006	1.007	1.008	1.009
9 Blocks 0.01 mm Series								
1.01	1.02	1.03	1.04	1.05	1.06	1.07	1.08	1.09
9 Blocks 0.1 mm Series								
1.1	1.2	1.3	1.4	1.5	1.6	1.7	1.8	1.9
9 Blocks 1 mm Series								
1	2	3	4	5	6	7	8	9
9 Blocks 10 mm Series								
10	20	30	40	50	60	70	80	90

*All thicknesses are in millimeters

Determining Gage Block Combinations

Usually there is more than one combination of blocks which will give a desired length. The most efficient procedure for determining block combinations is to eliminate digits of the desired measurement from right to left. This procedure saves time, minimizes the number of blocks, and reduces the chances of error. The following examples show how to apply the procedure in determining block combinations.

Example 1 Determine a combination of gage blocks for 2.9468 inches. Refer to the gage block sizes given in the Table of Block Thicknesses of an English Gage Block Set. All dimensions are in inches.

Choose the block which eliminates the last digit to the right, the 8. Choose the 0.1008″ block. Subtract. (2.9468″ − 0.1008″ = 2.846″)

Eliminate the last digit, 6, of 2.846″. Choose the 0.146″ block which eliminates the 4 as well as the 6. Subtract. (2.846″ − 0.146″ = 2.700″)

Eliminate the last non-zero digit, 7, of 2.700″. Choose the 0.700″ block. Subtract. (2.700″ − 0.700″ = 2.000″)

The 2.000″ block completes the required dimension as shown.

Check. Add the blocks chosen.
0.1008″ + 0.146″ + 0.700″ + 2.000″ = 2.9468″

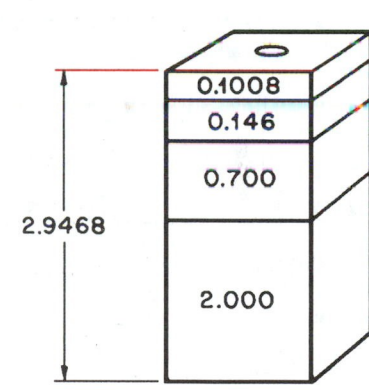

Example 2 Determine a combination of gage blocks for 10.2843 inches. Refer to the gage block sizes given in the Table of Block Thicknesses for an English Gage Block Set. All dimensions are in inches.

Eliminate the 3. Choose the 0.1003″ block.
Subtract. (10.2843″ − 0.1003″ = 10.184″)

Eliminate the 4. Choose the 0.134″ block.
Subtract. (10.184″ − 0.134″ = 10.050″)

Eliminate the 5. Choose the 0.050″ block.
Subtract. (10.050″ − 0.050″ = 10.000″)

The 1.000″, 2.000″, 3.000″ and 4.000″ blocks complete the required dimensions as shown.

Check. (0.1003″ + 0.134″ + 0.050″ + 1.000″ + 2.000″ + 3.000″ + 4.000″ = 10.2843″)

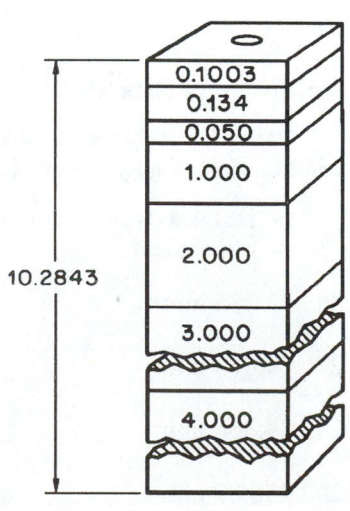

Example 3 Determine a combination of gage blocks for 157.372 millimeters. Refer to the gage block sizes given in the Table of Block Thicknesses for a Metric Gage Block Set. All dimensions are in millimeters.

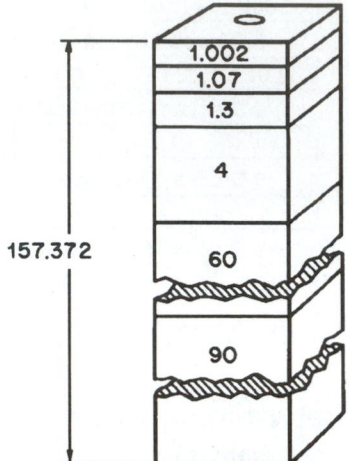

Eliminate the 2. Choose the 1.002 mm block.
Subtract.   (157.372 mm − 1.002 mm = 156.37 mm)

Eliminate the 7. Choose the 1.07 mm block.
Subtract.   (156.37 mm − 1.07 mm = 155.3 mm)

Eliminate the 3. Choose the 1.3 mm block.
Subtract.   (155.3 mm − 1.3 mm = 154 mm)

Eliminate the 4. Choose the 4 block.
Subtract.   (154 mm − 4 mm = 150 mm)

The 60 and 90 block complete the required dimension as shown.

Check.   (1.002 mm + 1.07 mm + 1.3 mm + 4 mm + 60 mm
+ 90 mm = 157.372 mm)

APPLICATION

English Gage Blocks

Using the Table of Block Thicknesses for an English Gage Block Set, determine a combination of gage blocks for each of the following dimensions.

> **Note:** Usually more than one combination of blocks will give the desired dimension.

1. 3.8638″ _____	10. 9.050″ _____	18. 0.6754″ _____
2. 1.8702″ _____	11. 4.8757″ _____	19. 7.7777″ _____
3. 3.1222″ _____	12. 1.0001″ _____	20. 10.0101″ _____
4. 0.6333″ _____	13. 0.2621″ _____	21. 9.4346″ _____
5. 0.2759″ _____	14. 2.7311″ _____	22. 4.8208″ _____
6. 5.8002″ _____	15. 5.090″ _____	23. 6.003″ _____
7. 7.973″ _____	16. 6.0807″ _____	24. 10.0021″ _____
8. 0.9999″ _____	17. 2.9789″ _____	25. 0.6998″ _____
9. 10.250″ _____		

Metric Gage Blocks

Using the Table of Block Thicknesses for a Metric Gage Block Set, determine a combination of gage blocks for each of the following dimensions.

> **Note:** Usually more than one combination of blocks will give the desired dimension.

26. 43.285 mm _____	30. 213.9 mm _____
27. 14.073 mm _____	31. 43.707 mm _____
28. 34.356 mm _____	32. 9.999 mm _____
29. 156.09 mm _____	33. 76.46 mm _____

34. 157.08 mm _____

35. 13.86 mm _____

36. 28.727 mm _____

37. 6.071 mm _____

38. 85.111 mm _____

39. 39.099 mm _____

40. 134.44 mm _____

41. 67.005 mm _____

42. 41.87 mm _____

43. 2.007 mm _____

44. 107.23 mm _____

45. 193.03 mm _____

46. 73.061 mm _____

47. 10.804 mm _____

48. 149.007 mm _____

49. 55.555 mm _____

UNIT **30**   Achievement Review— Section Two

Objective

You should be able to solve the exercises and problems in this Achievement Review by applying the principles and methods covered in units 21–29.

1. Express each of the following lengths as indicated.

   a. 81 inches as feet _____

   b. $6\frac{1}{4}$ feet as inches _____

   c. 9.6 yards as feet _____

   d. 2.7 centimeters as millimeters _____

   e. 0.8 meter as millimeters _____

   f. 218 millimeters as centimeters _____

2. Holes are to be drilled in the length of angle iron as shown. What is the distance between 2 consecutive holes? _____

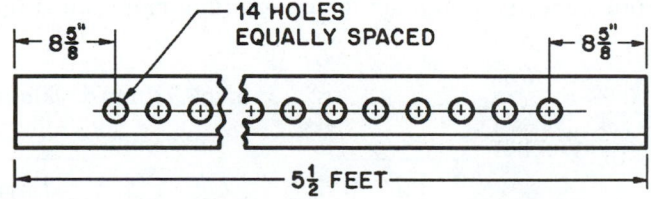

3. How many complete 3-meter lengths of tubing are required to make 250 pieces each 54 millimeters long? Allow a total one-half length of tubing for cutoff and scrap. _____

4. Express each of the following lengths as indicated. When necessary, round the answer to 3 decimal places.

   a. 47 millimeters as inches _____

   b. 5.5 meters as feet _____

   c. 16.8 centimeters as inches _____

   d. 4.75 inches as millimeters _____

   e. 31 inches as centimeters _____

   f. 4.5 feet as meters _____

5. For each of the exercises in the following table, the measurement made and the smallest graduation of the measuring instrument is given. Determine the greatest possible error and the smallest and largest possible actual length measure for each.

	Measurement Made	Smallest Graduation of Measuring Instrument Used	Greatest Possible Error	ACTUAL LENGTH	
				Smallest Possible	Largest Possible
a.	4.28″	0.02″ (steel rule)			
b.	0.8367″	0.0001″ (vernier micrometer)			
c.	46.16 mm	0.02 mm (vernier caliper)			
d.	16.45 mm	0.01 mm (micrometer)			

6. Compute the Absolute Error and Relative Error of each of the values in the following table. Where necessary, round the answers to 3 decimal places.

	True Value	Measured Value
a.	5.963 in	5.960 in
b.	0.392 mm	0.388 mm
c.	7.123°	7.200°

	True Value	Measured Value
d.	0.1070 in	0.0990 in
e.	0.8639 in	0.8634 in
f.	0.713°	0.706°

a. Absolute Error _____

   Relative Error _____

b. Absolute Error _____

   Relative Error _____

c. Absolute Error _____

   Relative Error _____

d. Absolute Error _____

   Relative Error _____

e. Absolute Error _____

   Relative Error _____

f. Absolute Error _____

   Relative Error _____

7. The following dimensions with tolerances are given. Determine the maximum dimension (maximum limit) and the minimum dimension (minimum limit) for each.

a. 1.714″ ± 0.005″

   maximum _____ minimum _____

b. 4.0688″ $^{+0.0000″}_{-0.0012″}$

   maximum _____ minimum _____

c. 5.9047″ $^{+0.0008″}_{-0.0000″}$

   maximum _____ minimum _____

d. 64.91 mm ± 0.08 mm

   maximum _____ minimum _____

e. 173.003 mm $^{+0.000\ mm}_{-0.013\ mm}$

   maximum _____ minimum _____

8. Express each of the following unilateral tolerances as bilateral tolerances having equal plus and minus values.

   a. $0.876'' \begin{smallmatrix} +0.006'' \\ -0.000'' \end{smallmatrix}$

   c. $37.53 \text{ mm} \begin{smallmatrix} +0.00 \text{ mm} \\ -0.03 \text{ mm} \end{smallmatrix}$

   b. $5.2619'' \begin{smallmatrix} +0.0000'' \\ -0.0012'' \end{smallmatrix}$

   d. $78.909 \text{ mm} \begin{smallmatrix} +0.009 \text{ mm} \\ -0.000 \text{ mm} \end{smallmatrix}$

9. The following problems require computations with both clearance fits and interference fits between mating parts. Determine the clearance or interference values as indicated. All dimensions are given in inches.

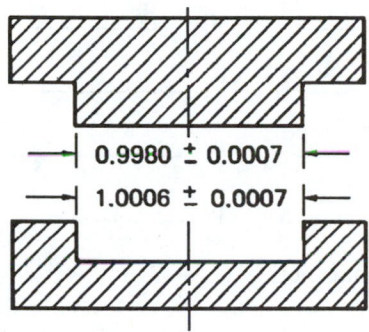
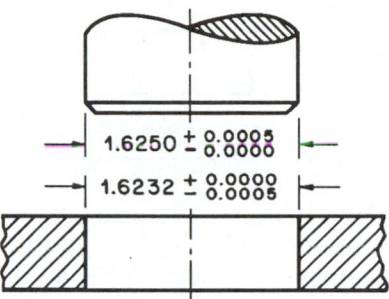

   a. Find the maximum clearance. _____

   b. Find the minimum clearance. _____

   e. Find the maximum interference (allowance). _____

   f. Find the minimum interference. _____

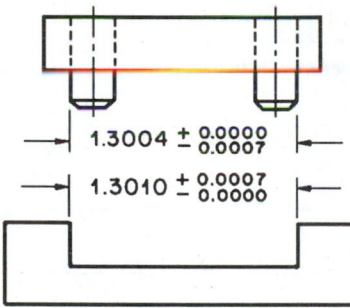
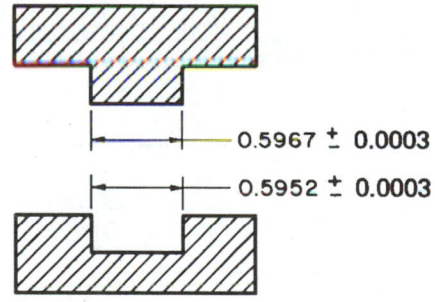

   c. Find the maximum clearance. _____

   d. Find the minimum clearance. _____

   g. Find the maximum interference (allowance). _____

   h. Find the minimum interference. _____

10. Determine the minimum permissible length of distance A of the part shown. All dimensions are in millimeters. _____

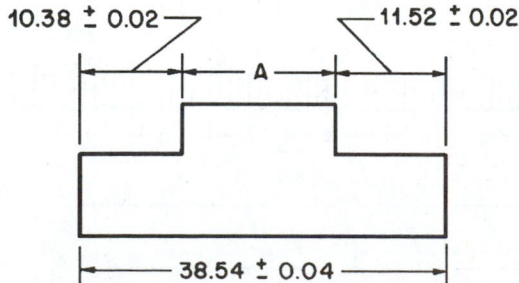

11. Read measurements a–p on the enlarged 32nds and 64ths graduated fractional rule shown.

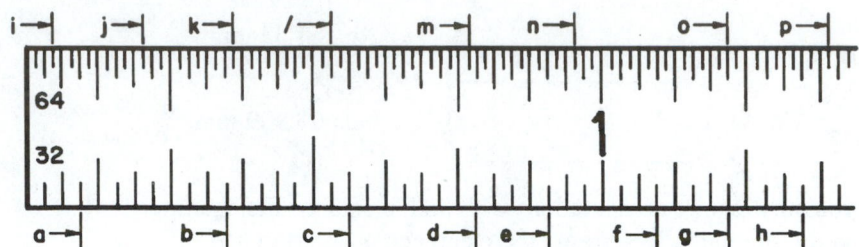

a. _____   e. _____   i. _____   m. _____

b. _____   f. _____   j. _____   n. _____

c. _____   g. _____   k. _____   o. _____

d. _____   h. _____   l. _____   p. _____

12. Read measurements a–p on the enlarged 50th and 100ths graduated decimal-inch rule shown.

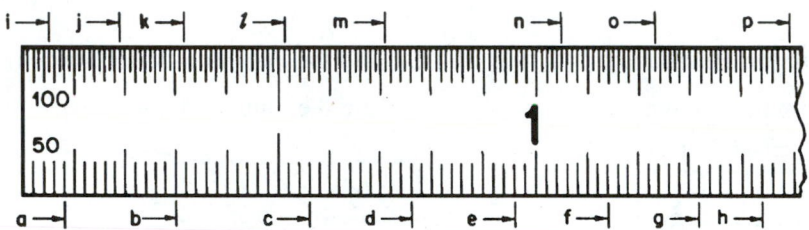

a. _____   e. _____   i. _____   m. _____

b. _____   f. _____   j. _____   n. _____

c. _____   g. _____   k. _____   o. _____

d. _____   h. _____   l. _____   p. _____

13. Read measurements a–p on the enlarged 1 millimeter and 0.5 millimeter graduated metric rule shown.

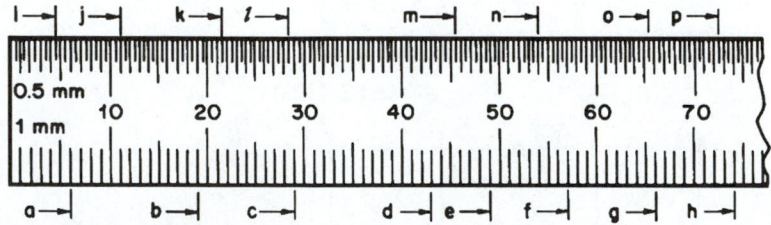

a. _____   e. _____   i. _____   m. _____

b. _____   f. _____   j. _____   n. _____

c. _____   g. _____   k. _____   o. _____

d. _____   h. _____   l. _____   p. _____

14. Read the vernier caliper and height gage measurements for the following settings.

ENGLISH MEASUREMENTS	**METRIC MEASUREMENTS**

a. _____

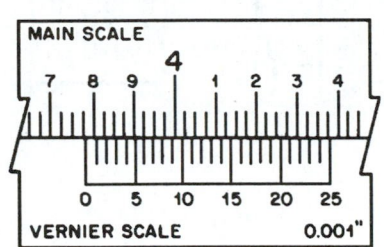

d. _____

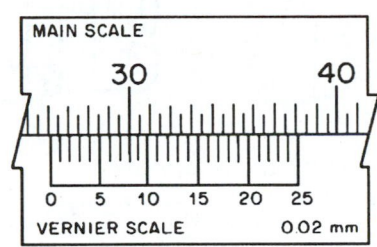

b. _____

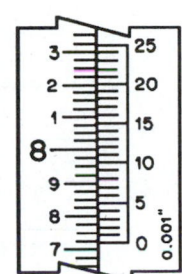

e. _____

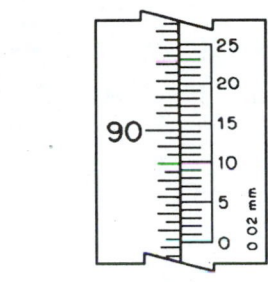

c. _____

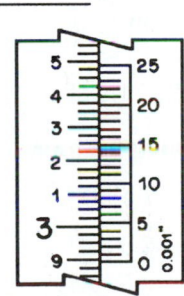

f. _____

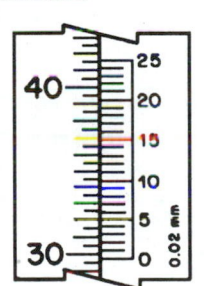

15. Read the settings on the following micrometer scales.

   a. 0.001 Decimal-Inch Micrometer

   (1) _____   (2) _____   (3) _____   (4) _____

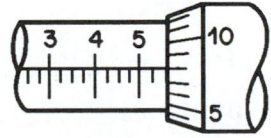 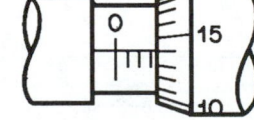 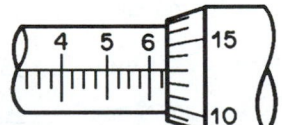

   b. 0.0001 Decimal-Inch Vernier Micrometer

   (1) _____   (2) _____   (3) _____   (4) _____

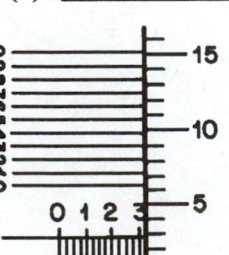 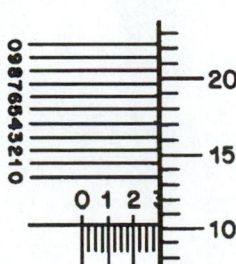  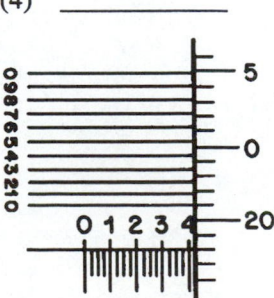

c. 0.01-Millimeter Metric Micrometer

(1) _____ (2) _____ (3) _____ (4) _____

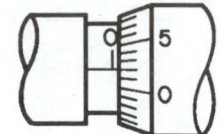 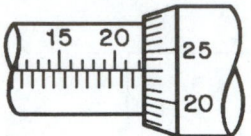 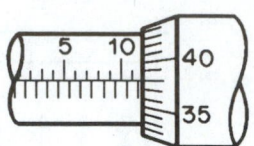

d. 0.002-Millimeter Metric Micrometer

(1) _____ (2) _____ (3) _____

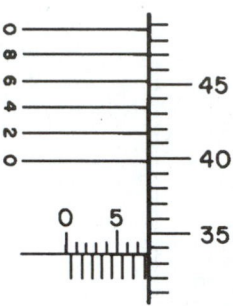 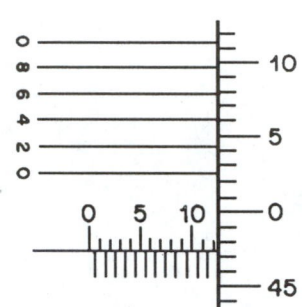 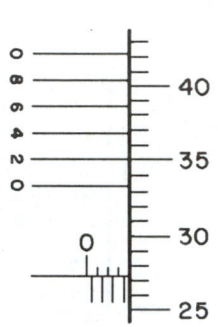

16. Using the Table of Block Thicknesses for an English Gage Block Set found in Unit 29, determine a combination of gage blocks for each of the following dimensions.

   ➤ **Note:** Usually more than one combination of blocks will give the desired dimension.

   a. 0.3784″ _____   d. 5.6467″ _____   g. 7.8895″ _____

   b. 2.5486″ _____   e. 3.0901″ _____   h. 8.0014″ _____

   c. 1.7062″ _____   f. 0.2009″ _____

17. Using the Table of Block Thicknesses for a Metric Gage Block Set found in Unit 29, determine a combination of gage blocks for each of the following dimensions.

   ➤ **Note:** Usually more than one combination of blocks will give the desired dimension.

   a. 67.53 mm _____   d. 13.274 mm _____   g. 99.998 mm _____

   b. 125.22 mm _____   e. 66.066 mm _____   h. 107.071 mm _____

   c. 85.092 mm _____   f. 43.304 mm _____

Section Three
Fundamentals of Algebra

UNIT 31 Symbolism

Objectives After studying this unit you should be able to

- **Express word statements as algebraic expressions.**
- **Express diagram dimensions as algebraic expressions.**
- **Evaluate algebraic expressions by substituting numbers for symbols.**

Algebra is a branch of mathematics in which letters are used to represent numbers. By the use of letters, general rules called *formulas* can be stated mathematically. Algebra is an extension of arithmetic; therefore, the rules and procedures which apply to arithmetic also apply to algebra. Many problems which are difficult or impossible to solve by arithmetic can be solved by algebra.

The basic principles of algebra discussed in this text are intended to provide a practical background for machine shop applications. A knowledge of algebraic fundamentals is essential in the use of trade handbooks and for the solutions of many geometric and trigonometric problems.

Symbolism

Symbols are the language of algebra. Both arithmetic numbers and literal numbers are used in algebra. *Arithmetic numbers* are numbers which have definite numerical values, such as 4, 5.17, and $\frac{7}{8}$. *Literal numbers* are letters which represent arithmetic numbers, such as $a, x, V,$ and P. Depending on how it is used, a literal number can represent one particular arithmetic number, a wide range of numerical values, or all numerical values.

Customarily the multiplication sign ($\times$) is not used in algebra because it can be misinterpreted as the letter x. When a literal number is multiplied by a numerical value, or when two or more literal numbers are multiplied, no sign of operation is required.

Examples
1. 5 times a is written $5a$
2. 17 times c is written $17c$
3. V times P is written VP
4. 6 times a times b times c is written $6abc$

Parentheses ( ) are often used in place of the multiplication sign ($\times$) when numerical values are multiplied; 3×4 is written $3(4)$; $18 \times 3.4 \times 5^2$ is written $18(3.4)(5^2)$.

An *algebraic expression* is a word statement put into mathematical form by using literal numbers, arithmetic numbers, and signs of operation. The following are examples of algebraic expressions.

Example 1 A dimension is increased by 0.5 inch. How long is the increased dimension? All dimensions are in inches.

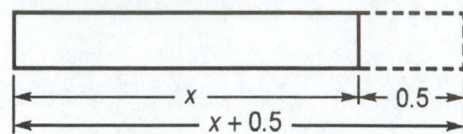

If x is the original dimension, the increased dimension is $x + 0.5''$.   Ans

Example 2 The production rate of a new machine is 4 times as great as an old machine. Write an algebraic expression for the production rate of the new machine.

If the old machine produced y parts per hour, the new machine produces $4y$ parts per hour.   Ans

Example 3 A drill rod is cut in 3 equal pieces. How long is each piece? (Disregard waste)

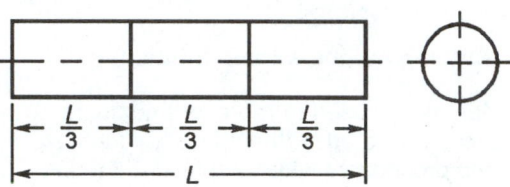

If L is the length of the drill rod, the length of each piece is $\frac{L}{3}$.   Ans

Example 4 In the step block shown, dimension B equals $\frac{3}{4}$ of dimension A and dimension C is twice dimension A. Find the total height of the block.

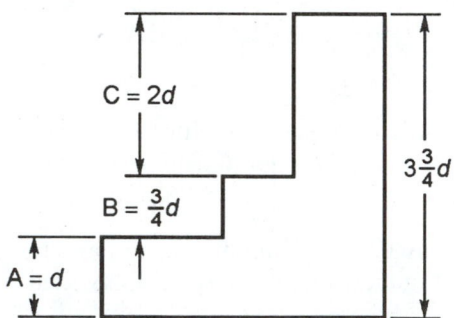

If d is the length of dimension A, dimension B is $\frac{3}{4} d$ and dimension C is $2d$. The total height is $d + \frac{3}{4} d + 2d$ or $3\frac{3}{4} d$.   Ans

> **Note:** If no arithmetic number appears before a literal number, it is assumed that the value is the same as if a one (1) appeared before the letter, $d = 1d$.

Example 5 A plate with 8 drilled holes is shown. The distance from the left edge of the plate to hole 1 and the distance from the right edge of the plate to hole 8 are each represented by a. The distances between holes 1 and 2, holes 2 and 3, and holes 3 and 4 are each represented by b. The distances between holes 4 and 5, holes 5 and 6, holes 6 and 7, and holes 7 and 8 are each represented by c. Find the total length of the plate. All dimensions are in millimeters.

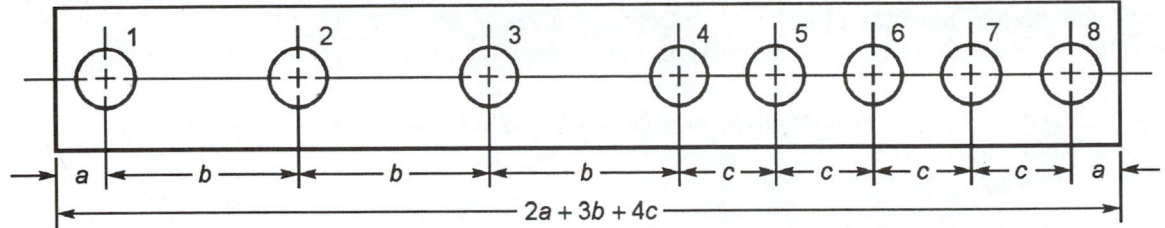

The total length of the plate is $a + b + b + b + c + c + c + c + a$, or $2a + 3b + 4c$.   Ans

➤ **Note:** Only like literal numbers may be arithmetically added.

Evaluation of Algebraic Expressions

The value of an algebraic expression is found by substituting given numerical values for literal values and solving the expression by following the order of operations as in arithmetic.

The order of operations follows:

- Do all operations within the grouping symbol first. Parentheses, the fraction bar, and the radical symbol are used to group numbers. If an expression contains parentheses within parentheses or brackets, do the work within the innermost parentheses first.

- Do powers and roots next. The operations are performed in the order in which they occur. If a root consists of two or more operations within the radical symbol, perform all the operations within the radical symbol, then extract the root.

- Do multiplication and division next in the order in which they occur.

- Do addition and subtraction last in the order in which they occur.

Example 1 The formula for finding the perimeter of a rectangle is given. Find the perimeter of the rectangle shown. All dimensions are in millimeters.

$P = 2L + 2W$
$P = 2(50 \text{ mm}) + 2(30 \text{ mm})$
$P = 100 \text{ mm} + 60 \text{ mm}$
$P = 160 \text{ mm}$ Ans

$P = 2L + 2W$ where
$P =$ perimeter
$L =$ length
$W =$ width

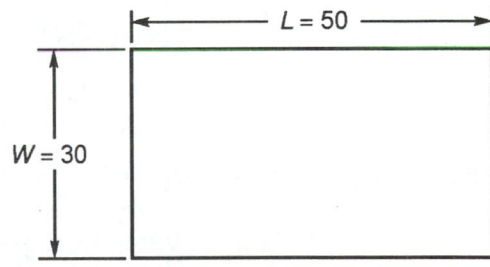

Example 2 The formula for finding the area of a ring is given. Find the area of the ring shown. All dimensions are in inches. Round the answer to 2 decimal places.

$A = \pi R^2 - \pi r^2$ where $A =$ area
$R =$ outside radius
$r =$ inside radius

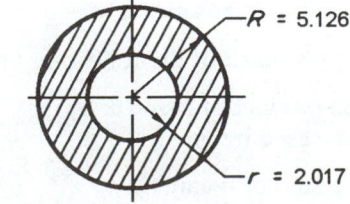

The symbol π (pi) represents a constant value used in mathematical relationships involving circles. It is described in Unit 49 on page 302.

Scientific calculators have the pi key, $\boxed{\pi}$. Depressing the pi key, $\boxed{\pi}$, enters the value of pi to 10 digits (3.141592654) on most calculators. On many calculators, π is the second or third function. Depending on the calculator used, press the $\boxed{\text{SHIFT}}$, $\boxed{\text{2nd}}$, or $\boxed{\text{3rd}}$ key; then press the appropriate key for which π is the second or third function.

$$A = \pi R^2 - \pi r^2$$

$$A = \pi(5.126 \text{ in})^2 - \pi(2.017 \text{ in})^2$$

$\boxed{\pi}\ \boxed{\times}\ 5.126\ \boxed{X^2}\ \boxed{-}\ \boxed{\pi}\ \boxed{\times}\ 2.017\ \boxed{X^2}\ \boxed{=}\ 69.76719217$

$A = 69.77$ sq in   Ans

Example 3   The formula for the approximate perimeter of an ellipse is given. Find the perimeter of the ellipse shown. All dimensions are in inches. Round the answer to 2 decimal places.

$$P = \pi\ \sqrt{2(a^2 + b^2)}\qquad \text{where } P = \text{perimeter}$$
$$a = 0.5 \text{ (major axis)}$$
$$b = 0.5 \text{ (minor axis)}$$

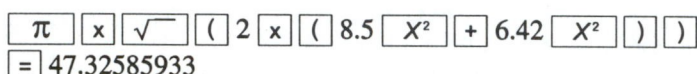

$$P = \pi\ \sqrt{2(8.50^2 + 6.42^2)}$$

$\boxed{\pi}\ \boxed{\times}\ \boxed{\sqrt{}}\ \boxed{(}\ 2\ \boxed{\times}\ \boxed{(}\ 8.5\ \boxed{X^2}\ \boxed{+}\ 6.42\ \boxed{X^2}\ \boxed{)}\ \boxed{)}$
$\boxed{=}\ 47.32585933$

$P = 47.33$ in   Ans

or $\boxed{\pi}\ \boxed{\times}\ \boxed{(}\ 2\ \boxed{\times}\ \boxed{(}\ 8.5\ \boxed{X^2}\ \boxed{+}\ 6.42\ \boxed{X^2}\ \boxed{)}\ \boxed{)}\ \boxed{\sqrt{x}}$
$\boxed{=}\ 47.3258533$

$P = 47.33$ in   Ans

Example 4   Find the value of $\dfrac{3(2b + 3dy)}{4(7d - bd)}$ when $b = 6$, $d = 4$ and $y = 2$.

$$\frac{3[2(6) + 3(4)(2)]}{4[7(4) - 6(4)]} = \frac{3(12 + 24)}{4(28 - 24)} = \frac{3(36)}{4(4)} = \frac{108}{16} = 6.75 \qquad \text{Ans}$$

Example 5   Find the value of $3m[4p + 5(x - m) + p]^2$ when $m = 2$, $p = 3$, $x = 8$.

$$3(2)[4(3) + 5(8 - 2) + 3]^2 = 6[12 + 5(6) + 3]^2 = 6(45)^2$$
$$= 6(2025) = 12{,}150 \qquad \text{Ans}$$

Example 6   Find the value of $\dfrac{6a}{b} + \dfrac{abc}{20}(a^3 - 12b)$ when $a = 5$, $b = 10$, and $c = 8$.

$$\frac{6(5)}{10} + \frac{5(10)(8)}{20}[5^3 - 12(10)] = \frac{30}{10} + \frac{400}{20}(125 - 120)$$
$$= 3 + 20(5) = 3 + 100 = 103 \qquad \text{Ans}$$

APPLICATION

Algebraic Expressions

Express each of the following problems as an algebraic expression.

1. The product of 6 and x increased by y. _____

2. The sum of a and 12. _____

3. Subtract b from 21. _____

4. Subtract 21 from b. _____

5. Divide r by s. _____

6. Twice L minus one-half P. _____

7. The product of x and y divided by the square of m. _____

8. In the part shown, all dimensions are in inches.

   a. What is the total length of this part? _____

   b. What is the length from point A
      to point B? _____

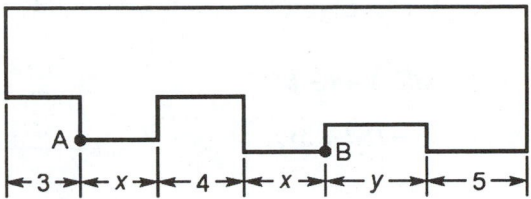

9. Find the distance between the indicated points.

   a. Point A to point B _____

   b. Point F to point C _____

   c. Point B to point C _____

   d. Point D to point E _____

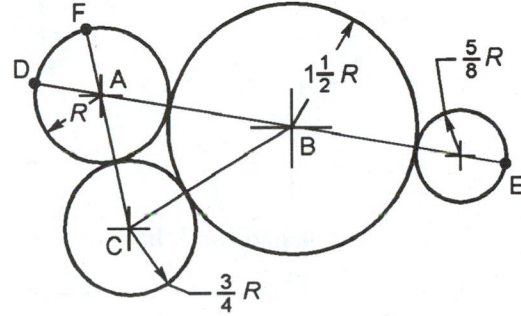

10. What are the lengths of the following dimensions? All dimensions
    are in millimeters.

    a. Dimension A _____

    b. Dimension B _____

    c. Dimension C _____

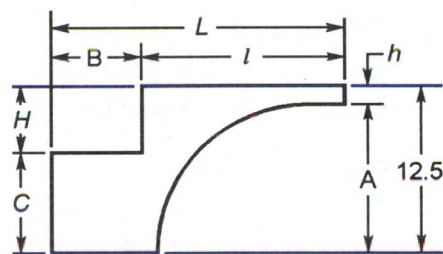

11. Stock is removed from a block in two operations. The original thickness of the
    block is represented by n. The thickness removed by the milling operation is repre-
    sented by p and the thickness removed by the grinding operation is represented by
    t. What is the final thickness of the block? _____

12. Given: s as the length of a side of a hexagon, r as the radius of the inside circle,
    and R as the radius of the outside circle.

    a. What is the length of r if r equals the product
       of 0.866 and the length of a side of the hexagon? _____

    b. What is the length of R if R equals the product
       of 1.155 and the radius of the inside circle? _____

    c. What is the area of the hexagon if the area
       equals the product of 2.598 and the square
       of the radius of the outside circle? _____

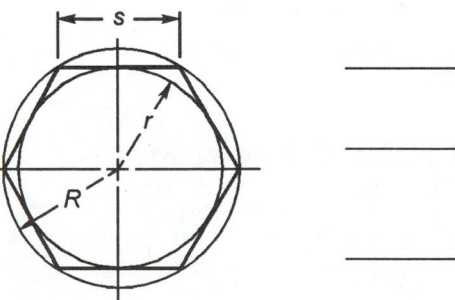

Evaluation of Algebraic Expressions

Substitute the given numbers for letters and find the values of the following expressions.

13. If $a = 5$ and $c = 3$, find

    a. $5a + 3c^2$ _____

    b. $5c + a$ _____

    c. $\dfrac{10c}{a}$ _____

    d. $\dfrac{a + c}{a - c}$ _____

    e. $\dfrac{a + 5c}{ac + a}$ _____

14. If $b = 8$, $d = 4$, and $e = 2$, find

   a. $\dfrac{b}{d} + e - 3$ _____

   b. $bd(3 + 4d - b)$ _____

   c. $5b - (bd + 3)$ _____

   d. $3e(b - e) - d\left(\dfrac{b}{2}\right)$ _____

   e. $\dfrac{12d}{e} - [3b - (d + e) + 4]$ _____

15. If $x = 12$ and $y = 6$, find

   a. $2xy + 7$ _____

   b. $3x - 2y + xy$ _____

   c. $\dfrac{5xy - 2y}{8x - xy}$ _____

   d. $\dfrac{4x - 4y}{3}$ _____

   e. $6x - 3y + xy$ _____

16. If $m = 5$, $p = 4$, and $r = 3$, find

   a. $m + mp^2 - r^3$ _____

   b. $(p + 2)^2(m - r)^2$ _____

   c. $\dfrac{(pr)^2}{2} - pr + m^3$ _____

   d. $\dfrac{p^3 + 3p - 12}{m^2 + 15}$ _____

   e. $\dfrac{r^3}{3p - 9} + m^2(mp - 6r)^2$ _____

For problems 17–28, round the answers to 1 decimal place.

17. All dimensions are in inches.

   a. Find the area (A) of this square.
   $A = \dfrac{1}{2}d^2$ _____

   b. Find the side (S) of this square.
   $S = 0.7071d$ _____

18. All dimensions are in millimeters.

   a. Find the length of this arc (l).
   $l = \dfrac{\pi R\alpha}{180°}$ _____

   b. Find the area of this sector (A).
   $A = \dfrac{1}{2}Rl$ _____

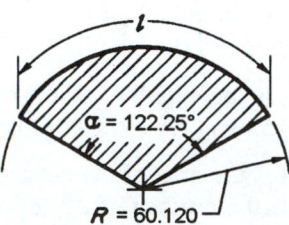

19. All dimensions are in inches. Refer to the triangle shown.

   a. Find S when $S = \dfrac{1}{2}(a + b + c)$. _____

   b. Find the area (A) when
   $A = \sqrt{S(S - a)(S - b)(S - c)}$ _____

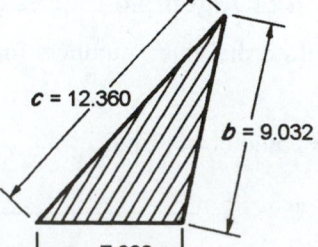

20. All dimensions are in millimeters.

    a. Find the radius of this circle.

$$r = \frac{c^2 + 4h^2}{8h}$$

    ————————

    b. Find the length of the arc (l).

      $l = 0.0175r\alpha$

    ————————

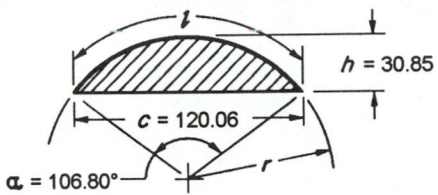

21. All dimensions are in inches. Find the shaded area.

    $$\text{Area} = \frac{(H + h)\,b + ch + aH}{2}$$

    ————————

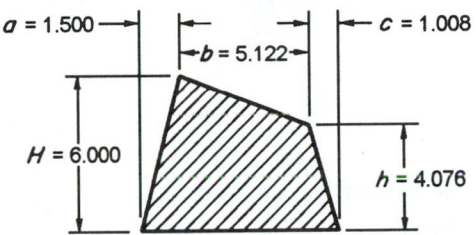

22. All dimensions are in inches. Find the length of belt on the pulleys.

    Length of belt =

    $$2C + \frac{11D + 11d}{7} + \frac{(D - d)^2}{4C}$$

    ————————

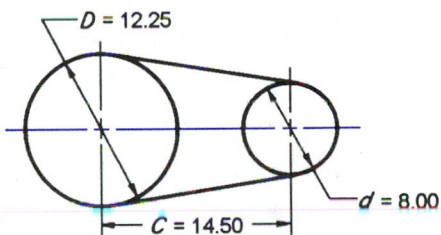

23. All dimensions are in millimeters. Find the shaded area.

    Area = $dt + 2(s + n)$

    ————————

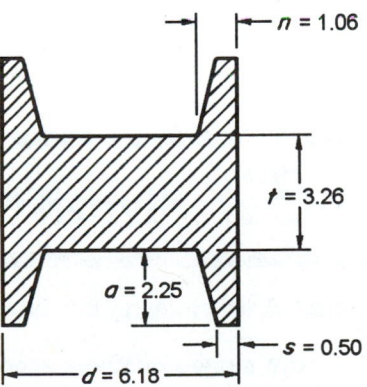

24. All dimensions are in inches. Find the shaded area.

    Area = $\pi(ab - cd)$

    ————————

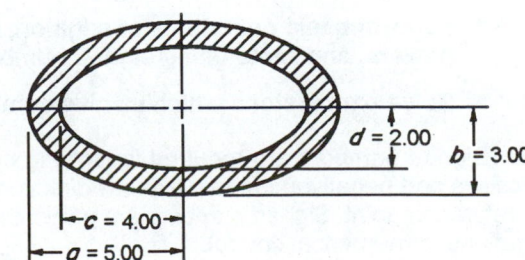

25. All dimensions are in inches. Find the shaded area.

$$\text{Area} = \frac{\pi(R^2 - r^2)}{2}$$ _____

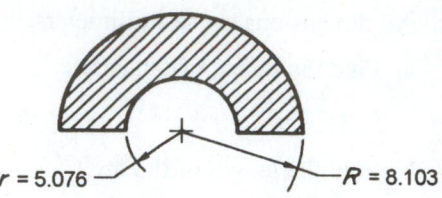

26. All dimensions are in centimeters. Find the shaded area.

$$\text{Area} = t[b + 2(a - t)]$$ _____

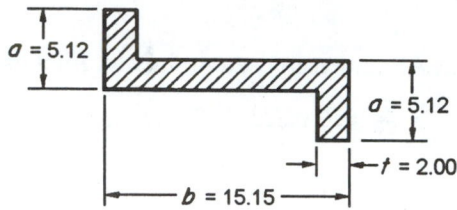

27. All dimensions are in inches.
   a. Find the slant height (S).

   $$S = \sqrt{(R - r)^2 + h^2}$$ _____

   b. Find the volume.

   $$\text{Volume} = 1.05h(R^2 + Rr + r^2)$$ _____

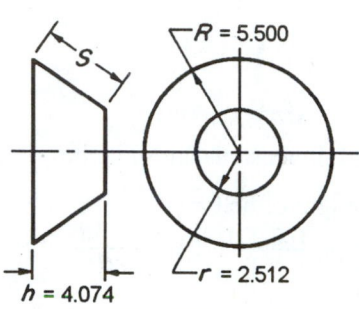

28. All dimensions are in inches. Find the volume.

$$\text{Volume} = \frac{(2a + c)bh}{6}$$ _____

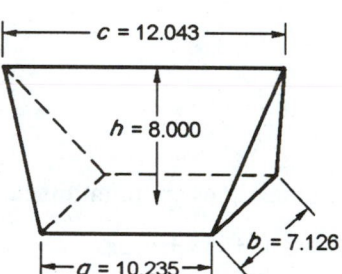

UNIT 32 Signed Numbers

Objectives After studying this unit you should be able to

- **Compare signed numbers according to size and direction using the number scale.**

- **Determine absolute values of signed numbers.**

- **Perform basic operations of addition, subtraction, multiplication, division, powers, and roots using signed numbers.**

- **Solve expressions which involve combined operations of signed numbers.**

Signed numbers are required for solving problems in mechanics and trigonometry. Positive and negative numbers express direction, such as machine table movement from a reference point. Signed numbers are particularly useful in programming machining operations for numerical control.

Meaning of Signed Numbers

Plus and minus signs which you have worked with so far in this book have been *signs of operation*. These are signs used in arithmetic, with the plus sign (+) indicating the operation of addition and the minus sign (−) indicating the operation of subtraction.

In algebra, plus and minus signs are used to indicate both operation and direction from a reference point or zero. A *positive number* is indicated either with no sign or with a plus sign (+) preceding the number. For example, +7 or 7 is a positive number which is 7 units greater than zero. A *negative number* is indicated with a minus sign (−) preceding the number. For example, −7 is a negative number which is 7 units less than zero. Positive and negative numbers are called *signed numbers* or directed numbers.

The Number Scale

The number scale shows the relationship of positive and negative numbers. It shows both distance and direction between numbers. Considering a number as a starting point and counting to a number to the right represents positive (+) direction with numbers increasing in value. Counting to the left represents negative (−) direction with numbers decreasing in value.

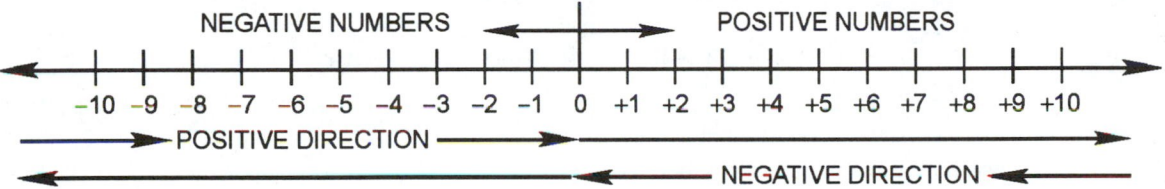

Examples

1. Starting at 0 and counting to the right to +5 represents 5 units in a positive (+) direction; +5 is 5 units greater than 0.

2. Starting at 0 and counting to the left to −5 represents 5 units in a negative (−) direction; −5 is 5 units less than 0.

3. Starting at −2 and counting to the right to +6 represents 8 units in a positive (+) direction; +6 is 8 units greater than −2.

4. Starting at +6 and counting to the left to −2 represents 8 units in a negative (−) direction; −2 is 8 units less than +6.

5. Starting at −3 and counting to the left to −10 represents 7 units in a (−) direction; −10 is 7 units less than −3.

6. Starting at −9 and counting to the right to 0 represents 9 units in a (+) direction; 0 is 9 units greater than −9.

Operations Using Signed Numbers

In order to solve problems in algebra, you must be able to perform the basic operations using signed numbers. The following procedures and examples show how to perform operations of addition, subtraction, multiplication, division, powers, and roots with signed numbers. The procedures for performing certain operations of signed numbers are based on an understanding of absolute value.

The *absolute value* of a number is the number without regard to its sign. For example, the absolute value of +4 is 4, the absolute value of −4 is also 4. Therefore, the absolute value of +4 and −4 is the same value, 4.

The absolute value of −20 is 15 greater than the absolute value of +5; 20 is 15 greater than 5.

Addition of Signed Numbers

Procedure To add two or more positive numbers

- Add the numbers as in arithmetic.

Examples Add the following numbers.

1. +3
   +5
   ‾‾‾
   +8 Ans

2. 15
    7
   ‾‾‾
   22 Ans

3. 2 + 9 + 13 = 24   Ans

4. +12 + (+15) = +27   Ans

Procedure To add two or more negative numbers

- Add the absolute values of the numbers.
- Prefix a minus sign to the sum.

Examples Add the following numbers.

1. −5
   −2
   ‾‾‾
   −7 Ans

2. −13
    − 4
   ‾‾‾
   −15
   −32 Ans

3. −6 + (−5) = −11   Ans

4. −8 + (−10) + (−4) + (−3) = −25   Ans

Procedure To add a positive and a negative number

- Subtract the smaller absolute value from the larger absolute value.
- Prefix the sign of the number having the larger absolute value to the difference.

Examples Add the following numbers.

1. +5
   −3
   ‾‾‾
   +2 Ans

2. −5
   +3
   ‾‾‾
   −2 Ans

3. −17
   +17
   ‾‾‾
    0 Ans

4. +12 + (−8) = +4   Ans

5. −12 + (+8) = −4   Ans

Procedure To add more than two positive and negative numbers

- Add all the positive numbers.
- Add all the negative numbers.
- Add their sums following the procedure for adding signed numbers.

Examples Add the following numbers.

1. −2 + 4 + (−10) + 5 = 9 + (−12) = −3   Ans

2. 8 + 7 + (−6) + 4 + (−3) + (−5) + 10 = 29 −14 = 15   Ans

3. 4 + (−6) + 12 + 3 + (−7) + 1 + (−5) + (−2) = 20 − 20 = 0   Ans

Pressing the change sign key, | +/− |, instructs the calculator to change the sign of the displayed value. Calculations involving negative numbers can be made by using the change sign key. To enter a negative number, enter the absolute value of the number, then press the change sign key.

Examples

1. Add. −25.873 + (−138.029)
   25.873 $\boxed{+/-}$ $\boxed{+}$ 138.029 $\boxed{+/-}$ $\boxed{=}$ −163.902   Ans

2. Add. −6.053 + (−0.072) + (−15.763) + (−0.009)
   6.053 $\boxed{+/-}$ $\boxed{+}$ .072 $\boxed{+/-}$ $\boxed{+}$ 15.763 $\boxed{+/-}$ $\boxed{+}$ .009 $\boxed{+/-}$
   $\boxed{=}$ −21.897   Ans

Certain more advanced calculators permit direct entry of negative values. These calculators do *not* have the change sign key, $\boxed{+/-}$. The subtraction key, $\boxed{-}$ or negative key, $\boxed{(-)}$, is used to enter negative values. The negative sign is entered before the number is entered. A negative value is displayed. To determine if your calculator has this capability, press $\boxed{-}$ or $\boxed{(-)}$, enter a number. The display will show a negative value. For example, −125.87 is entered directly as $\boxed{-}$ or $\boxed{(-)}$ 125.87. The value displayed is −125.87.

Examples

1. Add. −25.873 + (−138.029)
   $\boxed{-}$ or $\boxed{(-)}$ 25.873 $\boxed{+}$ $\boxed{-}$ or $\boxed{(-)}$ 138.029 $\boxed{EXE}$ −163.902   Ans

2. Add. −6.053 + (−0.072) + (−15.763) + (−0.009)
   $\boxed{-}$ or $\boxed{(-)}$ 6.053 $\boxed{+}$ $\boxed{-}$ or $\boxed{(-)}$ .072 $\boxed{+}$ $\boxed{-}$ or $\boxed{(-)}$ 15.763 $\boxed{+}$ $\boxed{-}$ or $\boxed{(-)}$ .009
   $\boxed{EXE}$ −21.897   Ans

Subtraction of Signed Numbers

Procedure   To subtract signed numbers

- Change the sign of the number subtracted (subtrahend) to the opposite sign.
- Follow the procedure for addition of signed numbers.

➤ **Note:** When the sign of the subtrahend is changed, the problem becomes one in addition. Therefore, subtracting a negative number is the same as adding a positive number. Subtracting a positive number is the same as adding a negative number.

Examples

1. Subtract 5 from 8.   8 − (+5) = 8 + (−5) = 3   Ans

2. Subtract 8 from 5.   5 − (+8) = 5 + (−8) = −3   Ans

3. Subtract −5 from 8.   8 − (−5) = 8 + (+5) = 13   Ans

4. Subtract −5 from −8.   −8 − (−5) = −8 + (+5) = −3   Ans

5. −3 − (+7) = −3 + (−7) = −10   Ans

6. 0 − (−14) = 0 + (+14) = 14   Ans

7. 0 − (+14) = 0 + (−14) = −14   Ans

8. −14 − (−14) = −14 + (+14) = 0   Ans

Examples

1. Subtract. −163.94 − (−150.65)
   163.94 $\boxed{+/-}$ $\boxed{-}$ 150.65 $\boxed{+/-}$ $\boxed{=}$ −13.29   Ans
   or $\boxed{-}$ or $\boxed{(-)}$ 163.94 $\boxed{-}$ $\boxed{-}$ or $\boxed{(-)}$ 150.65 $\boxed{EXE}$ −13.29   Ans

2. Subtract. $-27.55 - (-8.64 + 0.74) - (-53.41)$

    27.55 $\boxed{+/-}$ $\boxed{-}$ $\boxed{(}$ 8.64 $\boxed{+/-}$ $\boxed{+}$ .74 $\boxed{)}$ $\boxed{-}$ 53.41 $\boxed{+/-}$ $\boxed{=}$ 33.76   Ans

    or $\boxed{-}$ or $\boxed{(-)}$ 27.55 $\boxed{-}$ $\boxed{(}$ $\boxed{-}$ or $\boxed{(-)}$ 8.64 $\boxed{+}$ .74 $\boxed{)}$ $\boxed{-}$ $\boxed{-}$ or $\boxed{(-)}$ 53.41

    $\boxed{EXE}$ 33.76   Ans

Multiplication of Signed Numbers

Procedure   To multiply two or more signed numbers

- Multiply the absolute values of the numbers.
- Count the number of negative signs.

    If there is an odd number of negative signs, the product is negative.

    If there is an even number of negative signs, the product is positive.

    If all numbers are positive, the product is positive.

It is not necessary to count the number of positive values in an expression consisting of both positive and negative numbers. Count only the number of negative values to determine the sign of the product.

Examples

    1. $4(-3) = -12$ Ans   (There is one negative sign. Since one is an odd number, the product is negative.)

    2. $-4(-3) = +12$ Ans   (There are two negative signs. Since two is an even number, the product is positive.)

    3. $(-2)(-4)(-3)(-1)(-2)(-1) = +48$ Ans   (6 negatives, even number, positive product)

    4. $(-2)(-4)(-3)(-1)(-2) = -48$ Ans   (5 negatives, odd number, negative product)

    5. $(2)(4)(3)(1)(2) = +48$ Ans   (all positives, positive product)

    6. $(2)(-4)(-3)(1)(-2) = -48$ Ans   (3 negatives, odd number, negative product)

    7. $(-2)(4)(-3)(-1)(-2) = +48$ Ans   (4 negatives, even number, positive product)

➤ **Note:** The product of any number or numbers and $0 = 0$; for example, $0(9) = 0$; $0(-9) = 0$; $8(-6)(0)(6) = 0$.

Example   Multiply. $(-8.61)(3.04)(-1.85)(-4.03)(0.162)$. Round the answer to 1 decimal place.

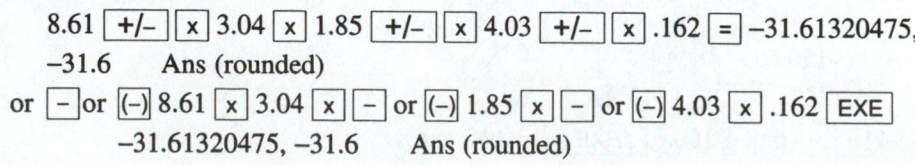

    8.61 $\boxed{+/-}$ $\boxed{x}$ 3.04 $\boxed{x}$ 1.85 $\boxed{+/-}$ $\boxed{x}$ 4.03 $\boxed{+/-}$ $\boxed{x}$ .162 $\boxed{=}$ -31.61320475,

    -31.6 Ans (rounded)

    or $\boxed{-}$ or $\boxed{(-)}$ 8.61 $\boxed{x}$ 3.04 $\boxed{x}$ $\boxed{-}$ or $\boxed{(-)}$ 1.85 $\boxed{x}$ $\boxed{-}$ or $\boxed{(-)}$ 4.03 $\boxed{x}$ .162 $\boxed{EXE}$

    $-31.61320475, -31.6$ Ans (rounded)

Division of Signed Numbers

Procedure   To divide two signed numbers

- Divide the absolute values of the numbers.
- Determine the sign of the quotient.

    If both numbers have the same sign (both negative or both positive) the quotient is positive.

    If the two numbers have unlike signs (one positive and one negative) the quotient is negative.

Examples

1. $\frac{-8}{-2} = +4$ Ans

2. $\frac{8}{2} = +4$ Ans

3. $15 \div 3 = +5$ Ans

4. $-3\overline{)-15} = +5$ Ans

5. $\frac{-30}{5} = -6$ Ans

6. $\frac{30}{-5} = -6$ Ans

7. $-21 \div 3 = -7$ Ans

8. $-3\overline{)21} = -7$ Ans

 Divide. $31.875 \div (-56.625)$. Round the answer to 3 decimal places.

31.875 $\boxed{+}$ 56.625 $\boxed{+/-}$ $\boxed{=}$ $-0.562913907, -0.563$ Ans (rounded)

or 31.875 $\boxed{+}$ $\boxed{-}$ or $\boxed{(-)}$ 56.625 $\boxed{\text{EXE}}$ $-0.562913907, -0.563$ Ans (rounded)

➤ **Note:**   Zero divided by any number equals zero. For example, $0 \div (+3) = 0$, $0 \div (-3) = 0$. Dividing by zero is *not* possible. For example, $+3 \div 0$ and $-3 \div 0$ are not possible.

Powers of Signed Numbers

Procedure   To raise numbers with positive exponents to a power

- Apply the procedure for multiplying signed numbers to raising signed numbers to powers.

Examples

1. $3^2 = +9$ Ans

2. $3^3 = +27$ Ans

3. $2^4 = +16$ Ans

4. $2^5 = +32$ Ans

5. $-3^2 = (-3)(-3) = +9$ Ans

6. $-3^3 = (-3)(-3)(-3) = -27$ Ans

7. $-2^4 = (-2)(-2)(-2)(-2) = +16$ Ans

8. $-2^5 = (-2)(-2)(-2)(-2)(-2) = -32$ Ans

➤ **Note:**
- A positive number raised to any power is positive.
- A negative number raised to an even power is positive.
- A negative number raised to an odd power is negative.

 As presented in Unit 3, the universal power key, $\boxed{y^x}$, $\boxed{x^y}$, or $\boxed{\wedge}$ raises any *positive* number to a power.

Solve. 2.073^5. Round the answer to 2 decimal places.

2.073 $\boxed{y^x}$ 5 $\boxed{=}$ 38.28216674, 38.28   Ans (rounded)

The universal power key can also be used to raise a *negative* number to a power. Most calculators are capable of directly raising a negative number to a power. Use the change sign key $\boxed{+/-}$ or the negative key $\boxed{(-)}$ or $\boxed{-}$ and the universal power key $\boxed{x^y}$, $\boxed{y^x}$, or $\boxed{\wedge}$.

Examples   Round the answers to 1 decimal place.

   1. Solve. $(-3.874)^4$

       3.874 $\boxed{+/-}$ $\boxed{x^y}$ 4 $\boxed{=}$ 225.236342, 225.2   Ans

       or $\boxed{(}$ $\boxed{-}$ or $\boxed{(-)}$ 3.874 $\boxed{)}$ $\boxed{x^y}$ or $\boxed{\wedge}$ 4 $\boxed{\text{EXE}}$ 225.236342, 225.2   Ans

      ➤ **Note:**   –3.874 must be enclosed within parentheses.

   2. Solve. $(-3.874)^5$

       3.874 $\boxed{+/-}$ $\boxed{x^y}$ 5 $\boxed{=}$ –872.565589, –872.6   Ans

       or $\boxed{(}$ $\boxed{-}$ or $\boxed{(-)}$ 3.874 $\boxed{)}$ $\boxed{x^y}$ or $\boxed{\wedge}$ 5 $\boxed{\text{EXE}}$ –872.565589, –872.6   Ans

With calculators that are *not* capable of directly raising a *negative* number to a power, enter the absolute value of the negative number and calculate as a positive number. Assign a plus sign or a minus sign to the displayed calculator answer, following the procedure for signs of powers of negative numbers.

Procedure   To raise numbers with negative exponents to a power

- Invert the number.
- Change the negative exponent to a positive exponent.

Examples

   1. $3^{-2} = \frac{3^{-2}}{1} = \frac{1}{3^2} = \frac{1}{9}$ or 0.111   Ans (rounded)

   2. $2^{-3} = \frac{2^{-3}}{1} = \frac{1}{2^3} = \frac{1}{8}$ or 0.125   Ans

   3. $-4^{-3} = \frac{-4^{-3}}{1} = \frac{1}{-4^3} = \frac{1}{-64}$ or –0.016   Ans (rounded)

 Depending on the calculator used, a *negative* exponent is entered with the change sign key $\boxed{+/-}$ or the negative key $\boxed{(-)}$ or $\boxed{-}$. The rest of the procedure is the same as used with positive exponents.

Examples   Round the answers to 3 decimal places.

   1. Calculate. 3.162^{-3}

       3.162 $\boxed{y^x}$ 3 $\boxed{+/-}$ $\boxed{=}$ 0.031631108, 0.032   Ans

       or 3.162 $\boxed{x^y}$ or $\boxed{\wedge}$ $\boxed{-}$ 3 $\boxed{\text{EXE}}$ 0.0316311078, 0.032   Ans

   2. Calculate. $(-3.162)^{-3}$

       The solutions shown are with calculators capable of directly raising a negative number to a power.

       3.162 $\boxed{+/-}$ $\boxed{y^x}$ 3 $\boxed{+/-}$ $\boxed{=}$ –0.031631108, –0.032   Ans

       or $\boxed{(}$ $\boxed{-}$ 3.162 $\boxed{)}$ $\boxed{x^y}$ or $\boxed{\wedge}$ $\boxed{-}$ 3 $\boxed{\text{EXE}}$ –0.031631108, –0.032   Ans

Roots of Signed Numbers

When either a positive number or a negative number are squared, a positive number results. For example, $3^2 = 9$ and $(-3)^2 = 9$. Therefore, every positive number has two square roots, one positive root and one negative root. The square roots of 9 are +3 and −3. The expression $\sqrt{9}$ is used to indicate the positive or *principal root*, +3 or 3. The expression $-\sqrt{9}$ is used to indicate the negative root, −3. The expression $\pm\sqrt{9}$ indicates both the positive and negative square roots, ±3. The principal cube root of 8 is 2, $\sqrt[3]{8} = 2$. The principal cube root of −8 is −2, $\sqrt[3]{-8} = -2$. In this book, only principal roots are to be determined or used in problem solving.

Examples

1. $\sqrt{36} = \sqrt{(6)(6)} = 6$ Ans

2. $\sqrt[4]{16} = \sqrt[4]{(2)(2)(2)(2)} = 2$ Ans

3. $\sqrt[3]{-27} = \sqrt{(-3)(-3)(-3)} = -3$ Ans

4. $\sqrt[5]{32} = \sqrt[5]{(2)(2)(2)(2)(2)} = 2$ Ans

5. $\sqrt[3]{\dfrac{-8}{27}} = \sqrt[3]{\dfrac{(-2)(-2)(-2)}{(3)(3)(3)}} = \dfrac{-2}{3}$ Ans

As presented in Unit 16, not all calculators have the root key $\boxed{\sqrt[x]{\ }}$, $\boxed{\sqrt[x]{y}}$, or $\boxed{x^{\frac{1}{y}}}$. The following examples show two basic methods of calculating roots of *positive* numbers.

1. Solve. $\sqrt[4]{562.824}$. The procedure shown is used with calculators that have the root key $\boxed{\sqrt[x]{\ }}$.

   ➤ **Note:** On certain calculators roots are a second function.

   4 $\boxed{\sqrt[x]{\ }}$ 562.824 $\boxed{\text{EXE}}$ 4.870719863   Ans

2. Solve. $\sqrt[4]{562.824}$. The procedures shown are used with calculators that have root key $\boxed{\sqrt[x]{y}}$ or $\boxed{x^{\frac{1}{y}}}$ and where roots are second functions. The procedures vary somewhat depending on the calculator used.

   562.824 $\boxed{\text{2nd}}$ $\boxed{x\sqrt{y}}$ 4 $\boxed{=}$ 4.870719863   Ans

   or 562.824 $\boxed{\text{SHIFT}}$ $\boxed{x^{\frac{1}{y}}}$ 4 $\boxed{=}$ 4.870719863   Ans

Most calculators are capable of directly computing roots of *negative* numbers. The following examples show the procedures for calculating roots of *negative* numbers depending on the make and model calculator.

1. Solve. $\sqrt[5]{-85.376}$. The procedure shown is used with a calculator that has the root key $\boxed{\sqrt[x]{\ }}$.

   ➤ **Note:** Roots may be a second function on certain calculators.

   5 $\boxed{\sqrt[x]{\ }}$ $\boxed{-}$ 85.376 $\boxed{\text{EXE}}$ −2.433700665   Ans

   or 5 $\boxed{\sqrt[x]{\ }}$ 85.376 $\boxed{+/-}$ $\boxed{=}$ −2.433700665   Ans

2. Solve. $\sqrt[5]{-85.376}$. The procedures shown are used with calculators that have the root key $\boxed{\sqrt[x]{y}}$ or $\boxed{x^{\frac{1}{y}}}$ and where roots are second functions.

   85.376 $\boxed{+/-}$ $\boxed{\text{2nd}}$ $\boxed{x\sqrt{y}}$ 5 $\boxed{=}$ −2.433700665   Ans

   or 85.376 $\boxed{+/-}$ $\boxed{\text{SHIFT}}$ $\boxed{x^{\frac{1}{y}}}$ 5 $\boxed{=}$ −2.433700665   Ans

With a calculator that is *not* capable of directly computing roots of negative numbers, enter the absolute value of the negative number and calculate as a positive number. Assign a plus or a minus sign to the displayed calculator answer following the procedure for signs of roots of negative numbers.

The square root of a negative number has no solution in the real number system. For example, $\sqrt{-4}$ has no solution; $\sqrt{-4}$ is not equal to $\sqrt{(-2)(-2)}$ or is not equal to $\sqrt{(+2)(+2)}$. Any even root (even index) of a negative number has no solution in the real number system. For example, $\sqrt[4]{-16}$ and $\sqrt[6]{-64}$ have no solution.

Expressing Numbers with Fractional Exponents as Radicals

Procedure   To simplify numbers with fractional exponents

- Write the numerator of the fractional exponent as the power of the radicand.
- Write the denominator of the fractional exponent as the root index of the radicand.
- Simplify.

Examples

1. $25^{1/2} = \sqrt[2]{25^1} = \sqrt{25} = \sqrt{(5)(5)} = 5$ Ans

2. $8^{1/3} = \sqrt[3]{8^1} = \sqrt{(2)(2)(2)} = 2$ Ans

3. $8^{2/3} = \sqrt[3]{8^2} = \sqrt[3]{64} = \sqrt[3]{(4)(4)(4)} = 4$ Ans

4. $36^{-1/2} = \dfrac{1}{36^{1/2}} = \dfrac{1}{\sqrt{36}} = \dfrac{1}{\sqrt{(6)(6)}} = \dfrac{1}{6}$ Ans

Use the universal power key, $\boxed{y^x}$, $\boxed{x^y}$, or $\boxed{\wedge}$ with fractional exponents.

Example   Solve. $8.732^{-2/3}$

8.732 $\boxed{y^x}$ or $\boxed{x^y}$ $\boxed{(}$ 2 $\boxed{+}$ 3 $\boxed{)}$ $\boxed{+/-}$ $\boxed{=}$ 0.235825546   Ans

or 8.732 $\boxed{\wedge}$ $\boxed{(}$ $\boxed{-}$ or $\boxed{(-)}$ 2 $\boxed{+}$ 3 $\boxed{)}$ $\boxed{\text{EXE}}$ 0.235825546   Ans

Combined Operations of Signed Numbers

Expressions consisting of two or more operations of signed numbers are solved using the same order of operations as in arithmetic.

Example   Compute the value of $50 + (-2)[6 + (-2)^3(4)]$.

$50 + (-2)[6 + (-2)^3(4)] = 50 + (-2)[6 + (-8)(4)] =$

$50 + (-2)[6 + (-32)] = 50 + (-2)(-26) = 50 + 52 = 102$ Ans

Examples

➤ **Note:** The universal power key is a second function on certain calculators.

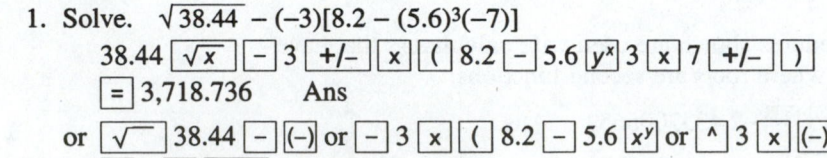

1. Solve. $\sqrt{38.44} - (-3)[8.2 - (5.6)^3(-7)]$
   38.44 $\boxed{\sqrt{x}}$ $\boxed{-}$ 3 $\boxed{+/-}$ $\boxed{\times}$ $\boxed{(}$ 8.2 $\boxed{-}$ 5.6 $\boxed{y^x}$ 3 $\boxed{\times}$ 7 $\boxed{+/-}$ $\boxed{)}$
   $\boxed{=}$ 3,718.736   Ans

   or $\boxed{\sqrt{\ }}$ 38.44 $\boxed{-}$ $\boxed{(-)}$ or $\boxed{-}$ 3 $\boxed{\times}$ $\boxed{(}$ 8.2 $\boxed{-}$ 5.6 $\boxed{x^y}$ or $\boxed{\wedge}$ 3 $\boxed{\times}$ $\boxed{(-)}$
   or $\boxed{-}$ 7 $\boxed{)}$ $\boxed{\text{EXE}}$ 3,718.376   Ans

2. Solve. $18.32 - (-4.52) + \dfrac{\sqrt[4]{93.724 - 6.023}}{-1.236^3}$. Round the answer to 2 decimal places. The solutions shown are with calculators capable of directly computing powers of negative numbers.

18.32 ⎡−⎤ 4.52 ⎡+/−⎤ ⎡+⎤ ⎡(⎤ 93.724 ⎡−⎤ 6.023 ⎡)⎤ ⎡2nd⎤ ⎡x√y⎤ 4 ⎡+⎤ 1.236
⎡+/−⎤ ⎡yˣ⎤ 3 ⎡=⎤ 21.21932578, 21.22   Ans

or 18.32 ⎡−⎤ 4.52 ⎡+/−⎤ ⎡+⎤ ⎡(⎤ 93.724 ⎡−⎤ 6.023 ⎡)⎤ ⎡SHIFT⎤ ⎡x^(1/y)⎤ 4 ⎡+⎤ 1.236
⎡+/−⎤ ⎡xʸ⎤ 3 ⎡=⎤ 21.21932578, 21.22   Ans

or 18.32 ⎡−⎤ ⎡(−)⎤ or ⎡−⎤ 4.52 ⎡+⎤ 4 ⎡ʸ√⎤ ⎡(⎤ 93.724 ⎡−⎤ 6.023 ⎡)⎤ ⎡+⎤ ⎡(⎤ ⎡(−)⎤
or ⎡−⎤ 1.236 ⎡)⎤ ⎡xʸ⎤ or ⎡^⎤ 3 ⎡EXE⎤ 21.21932578, 21.22   Ans

APPLICATION

The Number Scale

1. Refer to the number scale and give the direction (+ or −) and the number of units counted going from the first to the second number.

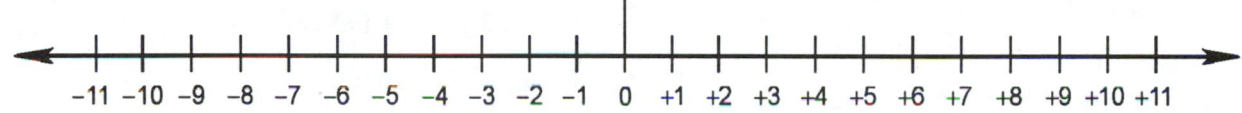

-11 -10 -9 -8 -7 -6 -5 -4 -3 -2 -1 0 +1 +2 +3 +4 +5 +6 +7 +8 +9 +10 +11

a. −11 to −2 _____

b. −8 to −3 _____

c. −6 to 0 _____

d. −2 to −8 _____

e. +2 to −8 _____

f. +3 to +10 _____

g. +10 to −10 _____

h. +10 to 0 _____

i. +4 to +7 _____

j. +9 to +1 _____

k. +11 to 0 _____

l. 0 to −6 _____

m. −7.5 to +10 _____

n. +10 to −7.5 _____

o. −10.8 to −4.3 _____

p. −2.3 to −0.8 _____

q. $+7\frac{1}{2}$ to $2\frac{1}{4}$ _____

r. $+5\frac{3}{4}$ to 0 _____

Comparing Signed Numbers

2. Select the greater of the two signed values and indicate the number of units by which it is greater.

a. +5, −14 _____

b. +7, −3 _____

c. −8, −1 _____

d. +8, +13 _____

e. +20, −22 _____

f. −16, −4 _____

g. +14.3, +23 _____

h. −1.8, +1.8 _____

i. +17.6, −21.9 _____

3. List the following signed numbers in order of increasing value starting with the smallest number.

a. +17, −1, +2, 0, −18, +4, −25 _____

b. −5, +5, 0, +13, +27, −21, −2, −19 _____

c. +10, −10, −7, +7, 0, +25, −25, +14 _____

d. 0, 15, −3.6, −2.5, −14.9, +17, +0.3 _____

e. $-16, +14\frac{1}{8}, -13\frac{7}{8}, +6, -3\frac{5}{8}$ _____

Absolute Value

4. Express each of the following pairs of signed numbers as absolute values and subtract the smaller absolute value from the larger absolute value.

   a. +23, −14 _____   c. −6, +6 _____   e. −16, +16 _____

   b. −17, +9 _____   d. +25, +13 _____   f. −33.7, −29.7 _____

➤ **Note:**   For problems 5–35 which follow, round the answers to 3 decimal places wherever necessary.

Addition of Signed Numbers

5. Add the following signed numbers as indicated.

   a. +15 + (+8) _____   l. $18\frac{5}{8} + \left(-21\frac{3}{4}\right)$ _____

   b. 7 + (+18) + 5 _____

   c. 0 + (+25) _____   m. $-13 + \left(-\frac{5}{16}\right)$ _____

   d. −8 + (−15) _____   n. −4.25 + (−7) + (−3.22) _____

   e. −18 + (−4) + (−11) _____   o. 18.07 + (−17.64) _____

   f. +12 + (−5) _____   p. 16 + (−4) + (−11) _____

   g. +18 + (−26) _____   q. −53.07 + (−6.37) + 19.82 _____

   h. −20 + (+19) _____   r. 30.88 + (−0.95) +1.32 _____

   i. −23 + 17 _____   s. −12.77 + (−9) + (−7.61) + 0.48 _____

   j. −25 + 3 _____   t. 2.53 + 16.09 + (−54.05) + 21.37 _____

   k. $-9\frac{1}{4} + \left(-3\frac{3}{4}\right)$ _____

Subtraction of Signed Numbers

6. Subtract the following signed numbers as indicated.

   a. −10 − (−4) _____   l. −50.2 − (+51) _____

   b. +5 − (−13) _____   m. +50.2 − (−51) _____

   c. −22 − (−14) _____   n. 0.03 − (+0.06) _____

   d. +17 − (+6) _____   o. $-10\frac{1}{2} - \left(-7\frac{1}{4}\right)$ _____

   e. +40 − (+40) _____

   f. −40 − (−40) _____   p. $5\frac{7}{8} - \left(-4\frac{1}{8}\right)$ _____

   g. −40 − (+40) _____   q. (6 + 10) − (−7 + 9) _____

   h. 0 − (−12) _____   r. (−14 + 5) − (2 − 10) _____

   i. −52 − (−8) _____   s. (7.23 − 6.81) − (−10.73) _____

   j. 16.5 − (+14.3) _____   t. [−8.76 + (−5.83)] − [12.06 − (−0.97)] _____

   k. −18.4 − (−14.3) _____

Multiplication of Signed Numbers

7. Multiply the following signed numbers as indicated.

a. $(-4)(6)$ _____

b. $(-4)(-6)$ _____

c. $(+10)(-3)$ _____

d. $(-10)(-3)$ _____

e. $(-5)(7)$ _____

f. $(-2)(-14)$ _____

g. $0(-16)$ _____

h. $(6.5)(-5)$ _____

i. $(-3.2)(-0.1)$ _____

j. $(-0.06)(-0.60)$ _____

k. $\left(1\frac{1}{2}\right)\left(-\frac{3}{4}\right)$ _____

l. $\frac{1}{4}(0)$ _____

m. $(-2)(-2)(-2)$ _____

n. $(-2)(+2)(+2)$ _____

o. $(8)(-4)(3)(0)(-1)$ _____

p. $(-3.86)(-2.1)(27.85)(-32.56)$ _____

q. $(8)(-2.65)(0.5)(-1)$ _____

r. $(-6.3)(-0.35)(2)(-1)(0.05)$ _____

s. $(-4.03)(-0.25)(-3)(-0.127)$ _____

t. $(-0.03)(-100)(-0.10)$ _____

Division of Signed Numbers

8. Divide the following signed numbers as indicated.

a. $-10 \div (-5)$ _____

b. $-10 \div (+2.5)$ _____

c. $+18 \div (+9)$ _____

d. $-21 \div 3$ _____

e. $-30 \div (-6)$ _____

f. $+48 \div (-6)$ _____

g. $-35 \div 7$ _____

h. $\frac{-16}{-4}$ _____

i. $\frac{0}{-10}$ _____

j. $\frac{-48}{-8}$ _____

k. $-\frac{1}{2} \div \left(-\frac{1}{2}\right)$ _____

l. $\frac{-60}{-0.5} - 6 \div \frac{3}{4}$ _____

m. $\frac{-10}{-2.5} + \frac{1}{3} \div \left(-\frac{2}{3}\right)$ _____

n. $\frac{-17.92}{3.28}$ _____

o. $0.562 \div (-0.821)$ _____

p. $-29.96 \div 5.35$ _____

q. $-4.125 \div (-0.75)$ _____

r. $-41.87 \div 7.9$ _____

s. $-20.47 \div 0.537$ _____

t. $-44.876 \div (-7.836)$ _____

Powers of Signed Numbers

9. Raise the following signed numbers to the indicated powers.

a. $(-2)^2$ _____

b. 2^3 _____

c. $(-2)^3$ _____

d. $(-4)^3$ _____

e. $(-2)^4$ _____

f. $(-2)^5$ _____

g. $(-6)^2$ _____

h. $(-5)^3$ _____

i. $(-2)^6$ _____

j. $(-1.6)^2$ _____

k. $(-0.4)^3$ _____

l. 0.93^6 _____

m. $(-1.58)^2$ _____

n. $(-0.85)^3$ _____

o. 0.73^3 _____

p. $\left(-\frac{2}{3}\right)^3$ _____

q. $(-1.038)^{-5}$ _____

r. 17.66^{-2} _____

s. $(-0.83)^{-3}$ _____

t. $(-6.087)^{-4}$ _____

Roots of Signed Numbers

10. Determine the indicated root of the following signed numbers.

a. $\sqrt[3]{64}$ _____

b. $\sqrt[3]{-64}$ _____

c. $\sqrt[3]{-27}$ _____

d. $\sqrt[3]{-1000}$ _____

e. $\sqrt[3]{1000}$ _____

f. $\sqrt[5]{-32}$ _____

g. $\sqrt[3]{125}$ _____

h. $\sqrt[5]{+32}$ _____

i. $\sqrt[3]{+1}$ _____

j. $\sqrt[3]{-1}$ _____

k. $\sqrt[7]{-1}$ _____

l. $\sqrt[3]{216}$ _____

m. $\sqrt[3]{\frac{-8}{-64}}$ _____

n. $\sqrt[3]{\frac{+8}{-27}}$ _____

o. $\sqrt[4]{\frac{+1}{+16}}$ _____

p. $\sqrt[3]{\frac{+27}{-125}}$ _____

q. $\sqrt[3]{-236.539}$ _____

r. $\sqrt[5]{-86.009}$ _____

s. $\sqrt[3]{\frac{-97.326}{123.592}}$ _____

t. $\frac{\sqrt[3]{-89.096}}{-17.323}$ _____

Expressing Numbers with Fractional Exponents as Radicals

11. Determine the value of the following.

a. $9^{½}$ _____

b. $81^{½}$ _____

c. $8^{⅓}$ _____

d. $64^{⅓}$ _____

e. $-8^{⅓}$ _____

f. $16^{¼}$ _____

g. $-125^{⅓}$ _____

h. $125^{⅓}$ _____

i. $273.19^{⅔}$ _____

j $41.673^{-½}$ _____

k. $8.007^{⅔}$ _____

l. $67.725^{-⅔}$ _____

Combined Operations of Signed Numbers

Solve each of the following problems using the proper order of operations.

12. $19 - (3)(-2) + (-5)^2$ _____

13. $4 - 5(8 - 10)$ _____

14. $-2(4 + 2) + 3(5 - 7)$ _____

15. $5 - 3(8 - 6) - [1 + (-6)]$ _____

16. $\dfrac{2(-1)(-3) - (6)(5)}{3(7) - 9}$ _____

17. $(-3)^3 + 3^3 - (-6)(3) - \left(\dfrac{-6}{2}\right)$ _____

18. $5^2 + \sqrt[3]{-8} + (-4)(0)(-3)$ _____

19. $[4^2 + (2)(5)(-3)]^2 + 2(-3)^3$ _____

20. $(-2)^3 + \sqrt{16} - (5)(3)(8)$ _____

21. $\dfrac{2(-5)^2}{2(5)} - \dfrac{(-4)^3}{18 + (-2)}$ _____

22. $(-2.87)^3 + \sqrt{15.93} - (5.63)(4)(-5.26)^3$ _____

23. $\dfrac{2(-5.16)^2}{3.07(4.98)} - \dfrac{(-4.66)^3}{18.37 + (-2.02)}$ _____

24. $(-2.46)^3 + \sqrt[3]{(-3.86)(-10.41) - (-6.16)}$ _____

25. $10.78^{-2} + [43.28 + (9)(-0.563)]^{-3}$ _____

Substitute the given numbers for letters in the following expressions and solve.

26. Find $6xy + 15 - xy$ when $x = -2$ and $y = 7$.   _____

27. Find $\dfrac{-3ab - 2bc}{abc - 35}$ when $a = -3$, $b = 10$, and $c = -4$.   _____

28. Find $(x - y)(3x - 2y)$ when $x = -5$ and $y = -7$.   _____

29. Find $\dfrac{d^3 + 4f - fh}{h^2 - (2 + d)}$ when $d = -2$, $f = -4$, and $h = 4$.   _____

30. Find $\dfrac{x^2}{n} - \dfrac{21 + y^3}{xy}$ when $n = 5$, $x = -5$, and $y = -1$.   _____

31. Find $\sqrt{6(ab - 6)} - (b)^3$ when $a = -6$ and $b = -2$.   _____

32. $\dfrac{x^2}{n} - \dfrac{21 + y^3}{xy}$; $n = 5.31$, $x = -5.67$, $y = -1.87$ _____

33. $\sqrt{6(ab - 6)} - (c)^3$; $a = -6.07$, $b = -2.91$, $c = 1.56$ _____

34. $5\sqrt[3]{e} + (ef - d) - (d)^3$; $d = -10.55$, $e = 8.26$, $f = -7.09$ _____

35. $\dfrac{\sqrt[4]{(mpt + pt + 19)}}{t^2 + 2p - 7}$; $m = 2$, $p = -2.93$, $t = -5.86$ _____

UNIT 33 Algebraic Operations of Addition, Subtraction, and Multiplication

Objectives   After studying this unit you should be able to

- Perform the basic algebraic operations of addition, subtraction, and multiplication.
- Express decimal numbers in scientific notation form and multiply and divide using scientific notation.

A knowledge of basic algebraic operations is essential in order to solve equations. For certain applications, formulas given in machine trade handbooks cannot be used directly as given, but must be rearranged. Formulas are rearranged by using the principles of algebraic operations.

Definitions

It is important to understand the following definitions in order to apply procedures which are required for solving problems involving basic operations.

A *term* of an algebraic expression is that part of the expression which is separated from the rest by a plus or a minus sign. For example, $4x + \frac{ab}{2x} - 12 + 3ab^2x - 8a\sqrt{b}$ is an expression that consists of five terms: $4x$, $\frac{ab}{2x}$, 12, $3ab^2x$, and $8a\sqrt{b}$.

A *factor* is one of two or more literal and/or numerical values of a term that are multiplied. For example, 4 and x are each factors of $4x$; 3, a, b^2 and x are each factors of $3ab^2x$; 8, a, and $\sqrt{b}$ are each factors of $8a\sqrt{b}$.

➤ **Note:** *It is absolutely necessary that you distinguish between factors and terms.*

A *numerical coefficient* is the number factor of a term. The letter factors of a term are the *literal factors*. For example, in the term $5x$, 5 is the numerical coefficient; x is the literal factor. In the term $\frac{1}{3}ab^2c^3$, $\frac{1}{3}$ is the numerical coefficient; a, b^2, and c^3 are the literal factors.

Like terms are terms that have identical literal factors including exponents. The numerical coefficients do not have to be the same. For example, $6x$ and $13x$ are like terms; $15ab^2c^3$, $3.2ab^2c^3$, and $\frac{1}{8}ab^2c^3$ are like terms.

Unlike terms are terms which have different literal factors or exponents. For example, $12x$ and $12y$ are unlike terms. The terms $15xy$, $3x^2y$, and $4x^2y^2$ are unlike terms. Although the literal factors are x and y in each of the terms, these literal factors are raised to different powers.

Addition

Terms must be like terms to be added. The addition of unlike terms can only be indicated. As in arithmetic, like things can be added, but unlike things cannot be added. For example, 4 inches + 5 inches = 9 inches. Both values are inches; therefore, they can be added. But 4 inches + 5 pounds cannot be added because they are unlike things.

Procedure To add like terms

- Add the numerical coefficients applying the procedure for addition of signed numbers. If a term does not have a numerical coefficient, the coefficient 1 is understood: $x = 1x$, $abc = 1abc$, $n^2rs^3 = 1n^2rs^3$.

- Leave the literal factors unchanged.

Examples Add the following like terms.

1. $\begin{array}{l} 3x \\ \underline{12x} \\ 15x \end{array}$ Ans

2. $\begin{array}{l} x \\ \underline{-14x} \\ -13x \end{array}$ Ans

3. $\begin{array}{l} -5xy^2 \\ \underline{+5xy^2} \\ 0 \end{array}$ Ans

4. $\begin{array}{l} 6x^2y^3 \\ \underline{-13x^2y^3} \\ -7x^2y^3 \end{array}$ Ans

5. $\begin{array}{l} 2(a+b) \\ -3(a+b) \\ \underline{7(a+b)} \\ 6(a+b) \end{array}$ Ans

Procedure   To add unlike terms

- The addition of unlike terms can only be indicated.

Examples   Add the following unlike terms.

1.   15
$$\frac{x}{15 + x} \quad \text{Ans}$$

2. $7x$
$$\frac{8y}{7x + 8y} \quad \text{Ans}$$

3. $3x$
$$\frac{-7x^2}{3x + (-7x^2)} \quad \text{Ans}$$

4. $8a$
$$\frac{\begin{array}{l}-6b\\2c\end{array}}{8a + (-6b) + 2c} \quad \text{Ans}$$

Procedure   To add two or more expressions that consist of two or more terms

- Group the like terms in the same column.

- Add like terms and indicate the addition of unlike terms.

Examples   Add the following expressions.

1. $12x - 2xy + 6x^2y^3$ and $-4x - 7xy + 5x^2y^3$

Group like terms in the same column.

Add like terms.

$$\begin{array}{l}12x - 2xy + 6x^2y^3\\-4x - 7xy + 5x^2y^3\\\hline 8x - 9xy + 11x^2y^3 \quad \text{Ans}\end{array}$$

2. $6a - 7b$ and $18b - 3ab + a$ and $-14a + ab^2 - 5ab$

Group like terms.

Add like terms and indicate the addition of unlike terms.

$$\begin{array}{l}6a - 7b\\a + 18b - 3ab\\-14a - 5ab + ab^2\\\hline -7a + 11b - 8ab + ab^2 \quad \text{Ans}\end{array}$$

Subtraction

As in addition, terms must be like terms to be subtracted. The subtraction of unlike terms can only be indicated. The same principles apply in arithmetic. For example, 8 feet − 3 feet = 5 feet, but 8 feet − 3 ounces cannot be subtracted because they are unlike things.

Procedure   To subtract like terms

- Subtract the numerical coefficients applying the procedure for subtraction of signed numbers.

- Leave the literal factors unchanged.

Examples   Subtract the following like terms as indicated.

1. $18ab - 7ab = 11ab$ Ans

2. $bx^2y^3 - 13bx^2y^3 = -12bx^2y^3$ Ans

3. $-5x^2y - 8x^2y = -13x^2y$ Ans

4. $-24dmr - (-24dmr) = 0$ Ans

Procedure   To subtract unlike terms

- The subtraction of unlike terms can only be indicated.

Examples   Subtract the following unlike terms as indicated.

1. $3x^2 - (+2x) = 3x^2 - 2x$ Ans

2. $-13abc - (+8abc^2) = -13abc - 8abc^2$ Ans

3. $-2xy - (-7y) = -2xy + 7y$ Ans

Procedure   To subtract expressions that consist of two or more terms

- Group like terms in the same column.

- Subtract like terms and indicate the subtractions of the unlike terms.

➤ **Note:** Each term of the subtrahend is subtracted following the procedure for subtraction of signed numbers.

Examples   Subtract the following expressions as indicated.

1. Subtract. $7a + 3b - 3d$ from $8a - 7b + 5d$

   Group like terms in the same column.

   Change the sign of each term in the subtrahend and follow the procedure for addition of signed numbers.

$$
\begin{aligned}
8a - 7b + 5d &= \quad 8a - 7b + 5d \\
-(7a + 3b - 3d) &= + (-7a - 3b + 3d) \\
\hline
 & \qquad a - 10b + 8d \quad \text{Ans}
\end{aligned}
$$

2. Subtract as indicated: $(3x^2 + 5x - 12xy) - (7x^2 - x - 3x^3 + 6y)$

$$
\begin{aligned}
3x^2 + 5x - 12xy &= \quad 3x^2 + 5x - 12xy \\
-(7x^2 - x - 3x^3 + 6y) &= + (-7x^2 + x + 3x^3 - 6y) \\
\hline
 & \quad -4x^2 + 6x - 12xy + 3x^3 - 6y \quad \text{Ans}
\end{aligned}
$$

Multiplication

It was shown that unlike terms could not be added or subtracted. In multiplication, the exponents of the literal factors do not have to be the same to multiply the values. For example, x^2 can be multiplied by x^4. The term x^2 means $(x)(x)$. The term x^4 means $(x)(x)(x)(x)$.

$$(x^2)(x^4) = (x)(x)(x)(x)(x)(x) = x^{2+4} = x^6$$

Procedure   To multiply two or more terms

- Multiply the numerical coefficients following the procedure for multiplication of signed numbers.

- Add the exponents of the same literal factors.

- Show the product as a combination of all numerical and literal factors.

Examples   Multiply as indicated.

1. Multiply. $(-3x^2)(6x^4)$

   Multiply numerical coefficients $(-3)(6) = -18$

   Add exponents of like literal factors $(x^2)(x^4) = x^{2+4} = x^6$

   Show product as combination of all numerical and literal factors.
   $(-3x^2)(6x^4) = -18x^6$ Ans

2. $(3a^2b^3)(7ab^3) = (3)(7)(a^{2+1})(b^{3+3}) = 21a^3b^6$ Ans

3. $(-4a)(-7b^2c^2)(-2ac^3d^3) = (-4)(-7)(-2)(a^{1+1})(b^2)(c^{2+3})d^3 = -56a^2b^2c^5d^3$ Ans

Procedure   To multiply expressions that consist of more than one term within an expression

- Multiply each term of one expression by each term or the other expression.
- Combine like terms.

Before applying the procedure to algebraic expressions, two examples are given to show that the procedure is consistent with arithmetic.

Examples in Arithmetic:

1. Multiply. $3(4 + 2)$

   From arithmetic: $3(4 + 2) = 3(6) = 18$ Ans

   From algebra:
   Multiply each term of $3(4 + 2) = 3(4) + 3(2) = 12 + 6 = 18$ Ans
   one expression by each
   term of the other expression.

   Combine like terms.

2. Multiply. $(5 + 3)(2 + 4)$

   From arithmetic: $(5 + 3)(2 + 4) = (8)(6) = 48$ Ans

   From algebra: Multiply each term of one expression by each term of the other expression.

	Step 1	Step 2	Step 3	Step 4
$(5 + 3)\ (2 + 4) =$	$5(2)$ +	$5(4)$ +	$3(2)$ +	$3(4)$ =
	10 +	20 +	6 +	12 = 48 Ans

   Combine like terms.

Examples in Algebra:

1. $3a(6 + 2a^2) = (3a)(6) + 3a(2a^2) = 18a + 6a^3$ Ans

2. $-5x^2y(3xy - 4x^3y^2 + 5y) = -5x^2y(3xy) - 5x^2y(-4x^3y^2) - 5x^2y(5y)$
   $= -15x^3y^2 + 20x^5y^3 - 25x^2y^2$

3. $(3c + 5d^{\,2})(4d^{\,2} - 2c)$

   Multiply each term of one expression by each term of the other expression.

	Step 1	Step 2	Step 3	Step 4
$(3c + 5d^2)\ (4d^2 + 2c) =$	$3c(4d^2)$ +	$3c(-2c)$ +	$5d^2(4d^2)$ +	$5d^2(-2c)$ =
	$12cd^2$ +	$(-6c^2)$ +	$20d^4$ +	$(-10cd^2)$

   Combine like terms.

   Combine

   $12cd^2 + (-6c^2) + 20d^4 + (-10cd^2) = 2cd^2 + (-6c^2) + 20d^4$ or $2cd^2 - 6c^2$
   $+ 20d^4$ Ans

APPLICATION

Addition of Single Terms

Add the terms in the following expressions.

1. $18y + y$ _____

2. $15xy + 7xy$ _____

3. $-15xy + (-7xy)$ _____

4. $22m^2 + (-m^2)$ _____

5. $-5x^2y + 5x^2y$ _____

6. $4c^3 + 0$ _____

7. $-9pt + (-pt)$ _____

8. $0.4x + (-0.8x)$ _____

9. $8.3a^2b + 6.9a^2b$ _____

10. $-0.04y + 0.07y$ _____

11. $\frac{1}{2}xy + \frac{3}{4}xy$ _____

12. $2\frac{3}{4}c^2d + \left(-3\frac{1}{8}c^2d\right)$ _____

13. $-2.06gh^3 + (-0.85gh^3)$ _____

14. $-50.6abc + 50.5abc$ _____

15. $4P + (-6P) + P + 12P$ _____

16. $-0.3dt^2 + (-1.7dt^2) + (-dt^2)$ _____

17. $\frac{1}{4}xy + \frac{7}{8}xy + xy + (-4xy)$ _____

18. $20.06D + (-19.97D) + (-0.7D)$ _____

19. $6M + 0.6M + 0.06M + 0.006M$ _____

20. $-3xy^2 + 8xy^2 + 7.8xy^2$ _____

21. The machined plate distances shown are dimensioned, in millimeters, in terms of x. Determine dimensions A–G.

a. _____
b. _____
c. _____
d. _____
e. _____
f. _____
g. _____

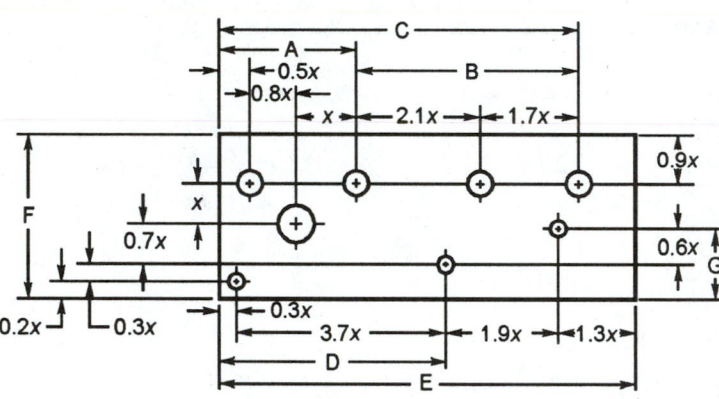

Addition of Expressions with Two or More Terms

Add the following expressions.

22. $-5x + 7xy - 8y$
$\underline{-9x - 12xy + 13y}$

23. $3a - 11d - 8m$
$\underline{- a + 11d - 3m}$

24. $-6ab - 5a^2b^2 - 3a^3b$
$-5ab + 14a^2b^2 - 12a^3b$
$-9ab - 7a^2b^2 + a^3b$
$\underline{ab \qquad\quad - 2a^3b}$

25. $(3xy^2 + x^2y - x^2y^2), (2x^2y + x^2y^2)$ _____

26. $(10a - 5b), (-12a - 7b), (11a + b)$ _____

27. $(x^3 + 5), (3x - 7x^2 + 7), (x - 3x^3)$ _____

28. $(b^4 + 4b^3c - 2b^2c), (4b^3c - 7bc)$ _____

29. $(x^2 - 4xy), (4xy - y^2), (-x^2 + y^2)$ _____

30. $(1.3M - 3N), (-8M + 0.5N), (20M + 0.4N)$ _____

31. $(c + 3.6cd - 4.9d), (-1.4c + 8.6d)$ _____

Subtraction of Single Terms

Subtract the following terms as indicated.

32. $7xy^2 - (-18xy^2)$ _____

33. $3xy - xy$ _____

34. $-3xy - xy$ _____

35. $-3xy - (-xy)$ _____

36. $9ab - (-9ab)$ _____

37. $-5a^2 - (5a^2)$ _____

38. $0.7a^2b^2 - 2.3a^2b^2$ _____

39. $0 - (-12mn^3)$ _____

40. $-8mn^3 - 0$ _____

41. $\frac{7}{8}x^2 - \left(-\frac{3}{8}x^2\right)$ _____

42. $13a - 9a^2$ _____

43. $-13a - (-7a^2)$ _____

44. $0.2xy - 0.9xy^2$ _____

45. $-ax^2 - ax^2$ _____

46. $\frac{1}{2}dt - \left(-\frac{3}{8}dt\right)$ _____

47. $\frac{1}{2}d^2t^2 - \left(-\frac{1}{2}d^2t^2\right)$ _____

48. $21 - 3x$ _____

49. $3x - 21$ _____

50. $-3.2d - 6.4d$ _____

51. $-1.4xy - (-1.4xy)$ _____

Subtraction of Expressions with Two or More Terms

Subtract the following expressions as indicated.

52. $(2a^2 - 3a) - (7a^2 - 10a)$ _____

53. $(4x^2 + 8xy) - (3x^2 + 5xy)$ _____

54. $(9b^2 + 1) - (9b^2 - 1)$ _____

55. $(9b^2 - 1) - (9b^2 - 1)$ _____

56. $(xy^2 - x^2y^2 + x^3y^2) - 0$ _____

57. $(2a^3 - 0.3a^2) - (-a^3 + a^2 - a)$ _____

58. $(5x + 3xy - 7y) - (3y^2 - x^2y)$ _____

59. $(-d^2 - dt + dt^2) - (-4 + dt)$ _____

60. $(15L - 12H) - (-12L + 6H - 4)$ _____

61. $(9.08e + 14.76f) - (e - f - 10.03)$ _____

Multiplication of Single Terms

Multiply the following terms as indicated.

62. $(-5b^2c)(3b^3)$ _____

63. $(x)(x^2)$ _____

64. $(-3a^2)(-5a^4)$ _____

65. $(8ab^2c)(7a^3bc^2)$ _____

66. $(-x^3y^3)(5a^3b)$ _____

67. $(-3xy)(0)$ _____

68. $(7ab^4)(3a^4b)$ _____

69. $(-3d^5r^4)(-d^3)$ _____

70. $(-3d^5r^4)(-d^3)(-1)$ _____

71. $(0.3x^2y^4)(0.7x^5)$ _____

72. $\left(\frac{1}{4}a^3\right)\left(\frac{3}{8}a^2\right)$ _____

73. $(-5x)(0)(-5x)$ _____

74. $(m^2t)(st^2)$ _____

75. $(-1.6bc)(2.1)$ _____

76. $(abc^3)(c^3d)$ _____

77. $(2x^6y^6)(-x^2)$ _____

78. $\left(-\frac{2}{3}mt\right)(t^4)$ _____

79. $(7ab^3)(-7a^3b)$ _____

80. $(-0.3a^3b^2)(-4b^3)$ _____

81. $(-x^2y)(-xy)(-x)$ _____

82. $(d^4m^2)(-1)(-m^3)$ _____

Multiplication of Expressions with Two or More Terms

Multiply the following expressions as indicated and combine like terms where possible.

83. $-5xy(2xy^2 - 3x^4)$ _____

84. $3a^2(-a^2 + a^3b)$ _____

85. $-2a^3b^2(4ab^3 - b^2 - 2)$ _____

86. $xy^2(x^2 + y^3 + xy)$ _____

87. $-4(dt + t^2 - 1)$ _____

88. $(m^2t^3s^4)(-m^4s^2 + m - s^5)$ _____

89. $(3x + 7)(x^2 + 9)$ _____

90. $(7x^2 - y^3)(-2x^3 + y^2)$ _____

91. $(5ax^3 + bx)(2a^2x^3 + b^2x)$ _____

92. $(-3a^2b^3 + 5xy^2)(4a^2b^3 - 5xy)$ _____

UNIT 34 Algebraic Operations of Division, Powers, and Roots

Objectives After studying this unit you should be able to

- Perform the basic algebraic operations of division, powers, and roots.
- Remove parentheses which are preceded by a plus or minus sign.
- Simplify algebraic expressions which involve combined operations.
- Write decimal numbers as scientific notation.
- Compute expressions using scientific notation.

Division

As with multiplication, the exponents of the literal factors do not have to be the same to divide the values. For example, x^4 can be divided by x.

$$\frac{x^4}{x} = \frac{(x)(x)(x)(x)}{x} = x^{4-1} = x^3 \qquad \text{Ans}$$

Procedure To divide two terms

- Divide the numerical coefficients following the procedure for division of signed numbers.
- Subtract the exponents of the literal factors of the divisor from the exponents of the same letter factors of the dividend.
- Combine numerical and literal factors.

This division procedure is consistent with arithmetic.

Example in Arithmetic:

Divide. $\frac{2^5}{2^2}$

From arithmetic: $\frac{2^5}{2^2} = \frac{(2)(2)(2)(2)(2)}{(2)(2)} = (2)(2)(2) = 8 \qquad \text{Ans}$

From algebra: $\frac{2^5}{2^2} = 2^{5-2} = 2^3 = 8 \qquad \text{Ans}$

Examples in Algebra:

1. Divide $-16x^3$ by $8x$.

   Divide the numerical coefficients following the procedure for signed numbers.

   $-16 \div 8 = -2$

   Subtract the exponents of the literal factors in the divisor from the exponents of the same letter factors in the dividend.

   $x^3 \div x = x^{3-1} = x^2$

   Combine the numerical and literal factors.

   $\dfrac{-16x^3}{8x} = -2x^2$ Ans

2. $\dfrac{-30a^3b^5c^2}{-5a^2b^3} = \left(\dfrac{-30}{-5}\right)(a^{3-2})(b^{5-3})(c^2) = 6ab^2c^2$ Ans

In arithmetic, any number except 0 divided by itself equals 1. For example, $4 \div 4 = 1$. Applying the division procedure $4 \div 4 = 4^{1-1} = 4^0$. Therefore, $4^0 = 1$. Any number except 0 raised to the zero power equals 1.

Example 1 $\dfrac{5^3}{5^3} = 5^{3-3} = 5^0 = 1$ Ans

Example 2 $\dfrac{a^3b^2c}{a^3b^2c} = (a^{3-3})(b^{2-2})(c^{1-1}) = a^0b^0c^0 = (1)(1)(1) = 1$ Ans

Procedure   To divide when the divisor consists of one term and the dividend consists of more than one term

- Divide each term of the dividend by the divisor following the procedure for division of signed numbers.

- Combine terms.

This division procedure is consistent with arithmetic.

Example in Arithmetic:

Divide. $\dfrac{6 + 8}{2}$

From arithmetic: $\dfrac{6 + 8}{2} = \dfrac{14}{2} = 7$ Ans

From algebra: $\dfrac{6 + 8}{2} = \dfrac{6}{2} + \dfrac{8}{2} = 3 + 4 = 7$ Ans

Example in Algebra:

Divide. $\dfrac{-20xy^2 + 15x^2y^3 + 35x^3y}{-5xy}$

$\dfrac{-20xy^2 + 15x^2y^3 + 35x^3y}{-5xy} = \dfrac{-20xy^2}{-5xy} + \dfrac{15x^2y^3}{-5xy} + \dfrac{35x^3y}{-5xy} = 4y - 3xy^2 - 7x^2$ Ans

Powers

Procedure   To raise a single term to a power

- Raise the numerical coefficients to the indicated power following the procedure for powers of signed numbers.

- Multiply each of the literal factor exponents by the exponent of the power to which it is raised.

- Combine numerical and literal factors.

This power procedure is consistent with arithmetic.

Example in Arithmetic:

Raise to the indicated power. $(2^2)^3$

From arithmetic: $(2^2)^3 = (4)^3 = (4)(4)(4) = 64$ Ans

From algebra: $(2^2)^3 = 2^{2(3)} = 2^6 = (2)(2)(2)(2)(2)(2) = 64$ Ans

Examples in Algebra:

1. Raise to the indicated power. $(5x^3)^2$

   Raise the numerical coefficient to the indicated power following the procedure for powers of signed numbers.

   $5^2 = 25$

   Multiply each literal factor exponent by the exponent of the power to which it is to be raised.

   $(x^3)^2 = x^{3(2)} = x^6$

   Combine numerical and literal factors.

   $(5x^3)^2 = 25x^6$ Ans

➤ **Note:** $(x^3)^2$ is not the same as $x^3 x^2$.
$(x^3)^2 = (x^3)(x^3) = (x)(x)(x)(x)(x)(x) = x^6$
$x^3 x^2 = (x)(x)(x)(x)(x) = x^5$

2. $(-3a^2b^4c)^3 = (-3)^3 a^{2(3)} b^{4(3)} c^{1(3)} = -27a^6 b^{12} c^3$ Ans

3. $\left[-\dfrac{1}{2} x^3 (yd^2)^3 r^4\right]^2 = \left[-\dfrac{1}{2} x^3 y^3 d^6 r^4\right]^2 = \dfrac{1}{4} x^6 y^6 d^{12} r^8$ Ans

Procedure To raise two or more terms to a power

- Apply the procedure for multiplying expressions that consist of more than one term.

Example Solve. $(2x + 4y^3)^2$

$$(2x + 4y^3)^2 = (2x + 4y^3)(2x + 4y^3)$$

$$\quad\quad\quad\quad\quad Step\ 1 \quad Step\ 2 \quad Step\ 3 \quad Step\ 4$$
$$\quad\quad\quad\quad\quad \downarrow \quad\quad \downarrow \quad\quad \downarrow \quad\quad \downarrow$$
$$= 2x(2x) + 2x(4y^3) + 4y^3(2x) + 4y^3(4y^3) =$$

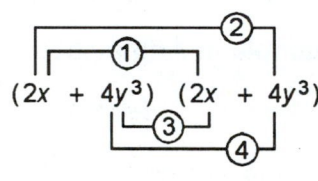

$$4x^2 + 8xy^3 + 8xy^3 + 16y^6 = 4x^2 + 16xy^3 + 16y^6 \quad \text{Ans}$$
$$\underset{Combine}{\underline{\quad\quad\quad\quad\quad}}$$

Roots

Procedure To extract the root of a term

- Determine the root of the numerical coefficient following the procedure for roots of signed numbers.

- The roots of the literal factors are determined by dividing the exponent of each literal factor by the index of the root.

- Combine the numerical and literal factors.

This procedure for extracting roots is consistent with arithmetic.

Example in Arithmetic:

Find the indicated root. $\sqrt{2^6}$

From arithmetic: $\sqrt{2^6} = \sqrt{(2)(2)(2)(2)(2)(2)} = \sqrt{64} = 8$ Ans

From algebra: $\sqrt{2^6} = 2^{6 \div 2} = 2^3 = (2)(2)(2) = 8$ Ans

Examples in Algebra:

1. $\sqrt{25a^6b^4c^8} = \sqrt{25}(a^{6\div2})(b^{4\div2})(c^{8\div2}) = 5a^3b^2c^4$ Ans

2. $\sqrt[3]{-27d^3x^9y^2} = \sqrt[3]{-27}(d^{3\div3})(x^{9\div3})\sqrt[3]{y^2} = -3dx^3\sqrt[3]{y^2}$ Ans

3. $\sqrt[4]{\dfrac{16}{81}d^8t^{12}y^2} = \sqrt[4]{\dfrac{16}{81}}(d^{8\div4})(t^{12\div4})(y^{2\div4}) = \dfrac{2}{3}d^2t^3y^{1/2} = \dfrac{2}{3}d^2t^3\sqrt{y}$ Ans

➤ **Note:** Roots of expressions that consist of two or more terms *cannot* be extracted by this procedure. This fact is consistent with arithmetic.

$\sqrt{3^2 + 4^2} = \sqrt{9 + 16} = \sqrt{25} = 5$ but $\sqrt{3^2} + \sqrt{4^2} = 3 + 4 = 7$.
5 does *not* equal 7 therefore $\sqrt{3^2 + 4^2} \neq \sqrt{3^2} + \sqrt{4^2}$

Removal of Parentheses

In certain expressions terms are enclosed within parentheses which are preceded by a plus or minus sign. In order to combine like terms, it is necessary to first remove parentheses.

Procedure   To remove parentheses preceded by a plus sign

- Remove the parentheses without changing the signs of any terms within the parentheses.
- Combine like terms.

Example $5a + (4b + 7a - 3d) = 5a + 4b + 7a - 3d = 12a + 4b - 3d$ Ans

Procedure   To remove parentheses preceded by a minus sign

- Remove the parentheses and change the sign of each term within the parentheses.
- Combine like terms.

Example $-(7a^2 + b - 3) + 12 - (-b + 5) = -7a^2 - b + 3 + 12 + b - 5$
$= -7a^2 + 10$ Ans

Combined Operations

Procedure   To solve expressions consisting of two or more different operations

- Apply the proper order of operations.

Order of Operations

- First, do all operations within grouping symbols. Grouping symbols are parentheses ( ), brackets [ ], and braces { }.
- Second, do powers and roots.
- Next, do multiplication and division operations in order from left to right.
- Last, do addition and subtraction operations in order from left to right.

Examples

1. $10x - 3x(2 + x - 4x^2) = 10x - 6x - 3x^2 + 12x^3$

$$= 4x - 3x^2 + 12x^3 \quad \text{Ans}$$

2. $15a^6b^3 + (2a^2b)^3 - \dfrac{a^7(b^3)^2}{ab^3} = 15a^6b^3 + 8a^6b^3 - \dfrac{a^7b^6}{ab^3}$

$$= 15a^6b^3 + 8a^6b^3 - a^6b^3 = 22a^6b^3 \quad \text{Ans}$$

3. $-4a[15 - 3(2a + ab) + a] -2a^2b = -4a(15 - 6a - 3ab + a) -2a^2b$

$$= -60a + 24a^2 + 12a^2b - 4a^2 - 2a^2b$$

$$= -60a + 20a^2 + 10a^2b \quad \text{Ans}$$

Scientific Notation

In scientific applications and certain technical fields, computations with very large and very small numbers are required. The numbers in their regular or standard form are inconvenient to read, write, and to use in computations. For example, copper expands 0.00000900 per unit of length per degree Fahrenheit. Scientific notation simplifies reading, writing, and computing with large and small numbers.

In scientific notation, a number is written as a whole number or decimal between 1 and 10 multiplied by 10 with a suitable exponent. For example, a value of 325,000 is written in scientific notation as 3.25×10^5.

The effect of multiplying a number by 10 is to shift the position of the decimal point. Changing a number from the standard decimal form to scientific notation involves counting the number of decimal places the decimal point must be shifted.

Examples   Express the following values using scientific notation.

1. 146,000

    a. Write the number as a value between 1 and 10: 1.46

    b. Count the number of places the decimal point is shifted to determine the exponent of 10: 1 46000. The decimal point is shifted 5 places. The exponent of 10 is 5: 10^5

    c. Multiply 1.46×10^5

       $146,000 = 1.46 \times 10^5$ Ans

2. 6 3 150 000. $= 6.315 \times 10^7$ Ans

    └ *Shift 7 places*

3. 9 7.856 $= 9.7856 \times 10^1$ Ans

    └ *Shift 1 place*

In scientific notation, positive decimal numbers less than 1 have negative exponents of 10.

Examples   Express the following values using scientific notation.

1. 0.02 89 $= 2.89 \times 10^{-2}$ Ans

    └ *Shift 2 places.* Observe that the decimal point is shifted to the right, resulting in a negative exponent.

2. 0.00003 18 $= 3.18 \times 10^{-5}$ Ans

    └ *Shift 5 places*

3. 0.8 59 $= 8.59 \times 10^{-1}$ Ans

    └ *Shift 1 place*

To express a number given in scientific notation as a decimal number, shift the decimal point in the reverse direction and attach required zeros.

Examples   Express the following values in decimal form.

1. $4.3 \times 10^3 = 4{,}300.$ Ans

   Shift 3 places. Attach required zeros.

2. $8.907 \times 10^5 = 8{,}90{,}700.$ Ans

   Shift 5 places. Attach required zeros.

3. $3.8 \times 10^{-4} = 0.0003\ 8$ Ans

   Shift 4 places. Attach required zeros.

Scientific notation is used primarily for multiplication and division operations. The procedures presented in Units 29 and 30 for the algebraic operations of multiplication and division are applied to operations involving scientific notation.

Examples   Compute the following expressions.

1. $(2.8 \times 10^3) \times (3.5 \times 10^5)$

   a. Multiply the decimals: $2.8 \times 3.5 = 9.8$

   b. The product of the 10's equals 10 raised to a power which is the sum of the exponents:
   $$10^3 \times 10^5 = 10^{3+5} = 10^8$$

   c. Combine both parts (9.8 and 10^8) as a product:
   $$(2.8 \times 10^3) \times (3.5 \times 10^5) = 9.8 \times 10^8 \quad \text{Ans}$$

2. $340{,}000 \times 7{,}040{,}000$

   Rewrite the numbers in scientific notation and solve:

   $$340{,}000 \times 7{,}040{,}000 \quad = (3.4 \times 10^5) \times (7.04 \times 10^6)$$
   $$= (3.4 \times 10^5) \times (7.04 \times 10^6) = 23.936 \times 10^{11}.$$

   Notice that the decimal part is greater than 10. Rewrite the decimal part and solve: $23.936 = 2.3936 \times 10^1$.

   $$(2.3936 \times 10^1) \times 10^{11} = 2.3936 \times 10^{12} \quad \text{Ans}$$

3. $840{,}000 + 0.0006$
   $$840{,}000 + 0.0006 = (8.4 \times 10^5) + (6 \times 10^{-4})$$
   $$= (8.4 + 6) \times (10^5 + 10^{-4})$$
   $$= 1.4 \times 10^{5-(-4)}$$
   $$= 1.4 \times 10^9 \quad \text{Ans}$$

 With 10-digit calculators, the number shown in the calculator display is limited to 10 digits. Calculations with answers that are greater than 9,999,999,999 or less than 0.000000001 are automatically expressed in scientific notation.

Example

1. $80000000\ \boxed{\times}\ 400000\ \boxed{=}\ 3.2\ ^{13}$ (Answer displayed as 3.2 ^{13})

   There is some variation among calculators as to how the answer is displayed. Many calculators display the answer as shown, with a space between the 3.2 and the 13 and the 13 smaller in size than the 3.2.

   The display shows the number (mantissa) and the exponent of 10; it does *not* show the 10. The displayed answer of 3.2 ^{13} does *not* mean that 3.2 is raised to the thirteenth power. The display 3.2 ^{13} means 3.2×10^{13}; $80{,}000{,}000 \times 400{,}000 = 3.2 \times 10^{13}$.

2. .0000007 $\boxed{\times}$ .000002 $\boxed{=}$ 1.4 $^{-12}$ (Answer is displayed as 1.4 $^{-12}$)

The display 1.4 $^{-12}$ means 1.4×10^{-12}; $0.0000007 \times 0.000002 = 1.4 \times 10^{-12}$.

Numbers in scientific notation can be directly entered in a calculator. For calculations whose answer does *not exceed* the number of digits in the calculator display, the answer is displayed in standard decimal form.

The answer is displayed in decimal (standard) form with certain calculators with the exponent entry key, $\boxed{\text{EE}}$, or exponent key, $\boxed{\text{EXP}}$.

Example   Solve. $(3.86 \times 10^3) \times (4.53 \times 10^4)$

$3.86 \boxed{\text{EE}} 3 \boxed{\times} 4.53 \boxed{\text{EE}} 4 \boxed{=} 174858000$ Ans

or $3.86 \boxed{\text{EXP}} 3 \boxed{\times} 4.53 \boxed{\text{EXP}} 4 \boxed{\text{EXE}}$ or $\boxed{=} 174858000$ Ans

The answer is displayed in standard form.

For calculations with answers that *exceed* the number of digits in the calculator display, the answer is displayed in scientific notation. Both calculators with the $\boxed{\text{EE}}$ key or $\boxed{\text{EXP}}$ key display the answer in scientific notation.

Example   Solve. $\dfrac{(-1.96 \times 10^7) \times (2.73 \times 10^5)}{8.09 \times 10^{-4}}$

1. Using the $\boxed{\text{EE}}$ key:

$1.96 \boxed{+/-}\boxed{\text{EE}} 7 \boxed{\times} 2.73 \boxed{\text{EE}} 5 \boxed{\div} 8.09 \boxed{\text{EE}} 4 \boxed{+/-} \boxed{=}$
-6.614091471 ^{15},
$-6.614091471 \times 10^{15}$ Ans

2. Using the $\boxed{\text{EXP}}$ key:

$1.96 \boxed{+/-}\boxed{\text{EXP}} 7 \boxed{\times} 2.73 \boxed{\text{EXP}} 5 \boxed{\div} 8.09 \boxed{\text{EXP}} 4 \boxed{+/-} \boxed{=}$
-6.614091471 ^{15},
$-6.614091471 \times 10^{15}$ Ans

or $\boxed{(-)}$ or $\boxed{-}$ $1.96 \boxed{\text{EXP}} 7 \boxed{\times} 2.73 \boxed{\text{EXP}} 5 \boxed{\div} 8.09 \boxed{\text{EXP}} \boxed{(-)}$ or $\boxed{-} 4$
$\boxed{\text{EXE}} -6.614091471$ ^{15},
$-6.614091471 \times 10^{15}$ Ans

APPLICATION

Division of Single Terms

Divide the following terms as indicated.

1. $\dfrac{4x^2}{2x}$ _____

2. $\dfrac{-16a^4b^5}{4ab^3}$ _____

3. $\dfrac{FS^2}{-FS^2}$ _____

4. $\dfrac{-FS^2}{-FS^2}$ _____

5. $0 \div 14mn$ _____

6. $(-42a^5d^2) \div (-6a^2d^2)$ _____

7. $(-3.6H^2P) \div (0.6HP)$ _____

8. $DM^2 \div (-1)$ _____

9. $3.7ab \div ab$ _____

10. $0.8PV^2 \div (-0.2V)$ _____

11. $1\frac{1}{4}c^2d^3 \div \frac{1}{4}cd^2$ _____

12. $\left(-\frac{1}{3}x^3y^3\right) \div \frac{1}{9}x^3$ _____

13. $-6g^3h^2 \div \left(-\frac{3}{4}gh\right)$ _____

14. $-24x^2y^5 \div (-0.5x^2y^4)$ _____

15. $x^2y^3z^4 \div xy^3z^2$ _____

16. $18a^2bc^2y \div (-a^2)$ _____

17. $0.25P^2V \div 0.0625$ _____

18. $-0.08xy \div 0.02y$ _____

19. $-\frac{3}{4}FS^3 \div (-3S)$ _____

20. $-9.6x^2yz \div (-1.2x)$ _____

Division of Expressions with Two or More Terms in the Dividend

Divide the following expressions as indicated.

21. $(8x^3 + 12x^2) \div x$ _____

22. $(12x^3y^3 - 8x^2y^2) \div 4xy$ _____

23. $(9x^6y^3 - 6x^2y^5) \div (-3xy^2)$ _____

24. $(2x - 4y) \div 4$ _____

25. $(15a^2 + 25a^5) \div (-a)$ _____

26. $(-18a^2b^7 - 12a^5b^5) \div (-6a^2b^5)$ _____

27. $(14cd - 35c^2d - 7) \div (-7)$ _____

28. $(0.8x^5y^6 + 0.2x^4y^7) \div (2x^2y^4)$ _____

29. $(-0.9a^2x - 0.3ax^2 + 0.6) \div (-0.3)$ _____

30. $(5y^2 - 25xy^2 - 10y^4) \div 5y^2$ _____

31. $\left(\frac{1}{2}a^2c - \frac{3}{4}a^3c^2 - ac^3\right) \div \frac{1}{8}ac$ _____

32. $(-2.5e^2\ f - 0.5ef^2 + e^2f^2) \div 0.5f$ _____

Powers of Single Terms

Raise the following terms to indicated powers.

33. $(3ab)^2$ _____

34. $(-4xy)^3$ _____

35. $(2x^2y)^3$ _____

36. $(4a^4b^3)^2$ _____

37. $(-3c^3d^2e^4)^3$ _____

38. $(2MS^2)^2$ _____

39. $(-7x^4y^5)^2$ _____

40. $(-3N^2P^2T^3)^4$ _____

41. $(a^3bc^2)^3$ _____

42. $(-2a^2bc^3)^3$ _____

43. $(-x^4y^5z)^3$ _____

44. $(8C^3FH^2)^2$ _____

45. $(0.4x^3y)^3$ _____

46. $(-0.5c^2d^3e)^3$ _____

47. $(4.3M^2N^2P)^2$ _____

48. $\left(\frac{3}{4}abc^3\right)^3$ _____

49. $[-8(a^2b^3)^2c]^3$ _____

50. $[-3x^2(y^2)^2z^3]^3$ _____

51. $[0.6d^3(ef^2)^3]^2$ _____

52. $[(-2x^2y)^2(xy^2)^2]^3$ _____

Powers of Expressions of Two or More Terms

Raise the following terms to the indicated powers and combine like terms where possible.

53. $(3x^2 - 5y^3)^2$ _____

54. $(a^4 + b^3)^2$ _____

55. $(5t^2 - 6x)^2$ _____

56. $(a^2b^3 + ab^3)^2$ _____

57. $(0.4d^2t^3 - 0.2t)^2$ _____

58. $(-0.2x^2y - y^4)^2$ _____

59. $\left(\frac{2}{3}c^2d + \frac{3}{4}cd^2\right)^2$ _____

60. $[(x^2)^3 - (y^3)^2]^2$ _____

61. $[(-a^4b)^2 + (x^2y)^3]^2$ _____

Roots

Determine the roots of the following terms.

62. $\sqrt{16c^2d^6}$ _____

63. $\sqrt{m^6n^4s^2}$ _____

64. $\sqrt[3]{64x^3y^9}$ _____

65. $\sqrt{81x^8y^6}$ _____

66. $\sqrt[3]{p^9t^6w^3}$ _____

67. $\sqrt[3]{-27x^6y^{12}}$ _____

68. $\sqrt{0.25h^4y^2}$ _____

69. $\sqrt{0.16a^8c^2f^6}$ _____

70. $\sqrt{\dfrac{4}{9}a^2b^4c^6}$ _____

71. $\sqrt{\dfrac{1}{16}x^2y^2}$ _____

72. $\sqrt[3]{\dfrac{8}{27}m^6n^3}$ _____

73. $\sqrt[3]{-64d^6t^9}$ _____

74. $\sqrt[4]{16x^4y^8}$ _____

75. $\sqrt[5]{32h^{10}}$ _____

76. $\sqrt{25ab^2}$ _____

77. $\sqrt[3]{64a^3c}$ _____

78. $\sqrt[3]{-\dfrac{1}{64}x^3y^6z^2}$ _____

79. $\sqrt{\dfrac{9}{16}a^2bc^2}$ _____

80. $\sqrt[3]{27d^3e^6f^2}$ _____

81. $\sqrt[5]{-32a^5b^3}$ _____

Removal of Parentheses

Remove parentheses and combine like terms where possible.

82. $6a + (3a - 2a^2 + a^3)$ _____

83. $9b - (15b^2 - c + d)$ _____

84. $15 + (x^2 - 10)$ _____

85. $-(ab + a^2b - a)$ _____

86. $-10c^3 - (-8c^3 - d + 12)$ _____

87. $-(16 + xy - x) + (-x)$ _____

88. $-25a^2b - (-2a^2b - a + b^2)$ _____

89. $15 - (r^2 + r) + (r^2 - 14)$ _____

90. $-(a^2 + b^2) + (a^2 + b^2)$ _____

91. $-(3x + xy - 6) + 18 + (x + xy)$ _____

92. $20 + (cd - c^2d + d) + 14 - (cd + d)$ _____

93. $20 - (cd - c^2d + d) - 14 + (cd + d)$ _____

Combined Operations

Simplify the following expressions.

94. $15 - 2(3xy)^2 + x^2y^2 - 8$ _____

95. $5(a^2 - b) + a^2 - b$ _____

96. $(2 - c^2)(2 + c^2) + 2c$ _____

97. $\dfrac{ab}{a} - \left(\dfrac{-a^2b}{a^2} - \dfrac{a^3b}{a^3}\right)$ _____

98. $\dfrac{4 - 8x + 16x^2}{2} + \dfrac{3x^4}{x^2}$ _____

99. $\dfrac{16xy^8}{2xy^2} - (y^2)^3 + 15$ _____

100. $\dfrac{\sqrt{25x^2}}{-5}(3xy^3) - (-10)$ _____

101. $\sqrt{\dfrac{64d^6}{9}} + d^2$ _____

102. $\dfrac{12x^6 + 16x^4y}{(2x)^2} - (16x^4y^2)^{\frac{1}{2}}$ _____

103. $-5a(-8 + (ab^2)^3 - 12)$ _____

104. $5a[-6 + (ab^2)^3 - 10]$ _____

105. $(10f^6 + 12f^4h) \div \sqrt{4f^4}$ _____

Rewriting Numbers in Scientific Notation

Rewrite the following standard form numbers in scientific notation.

106. 625 _____

107. 80,000 _____

108. 1,320,000 _____

109. 976,000 _____

110. 0.0073 _____

111. 0.015 _____

112. 0.00004 _____

113. 0.2 _____

114. 39 _____

115. 0.00039 _____

116. 175,000 _____

117. 0.00175 _____

Rewriting Scientific Notation Values

Rewrite the following scientific notation values in standard decimal form.

118. 3×10^3 _____

119. 1.6×10^5 _____

120. 8.5×10^2 _____

121. 5.09×10^6 _____

122. 4.7×10^{-1} _____

123. 6.32×10^{-5} _____

124. 1.05×10^{-3} _____

125. 3.123×10^{-6} _____

126. 7.312×10^4 _____

127. 7.321×10^{-4} _____

128. 2.09×10^6 _____

129. 2.09×10^{-2} _____

Multiplying and Dividing in Scientific Notation

The following problems are given in scientific notation. Solve and leave answers in scientific notation. Round the answers (mantissas) to 2 decimal places.

130. $(2.50 \times 10^3) \times (5.10 \times 10^5)$ _____

131. $(3.10 \times 10^{-3}) \times (5.20 \times 10^{-4})$ _____

132. $(-7.60 \times 10^4) \times (1.90 \times 10^5)$ _____

133. $(2.43 \times 10^{-6}) \div (7.60 \times 10^3)$ _____

134. $(8.51 \times 10^7) \div (6.30 \times 10^{-5})$ _____

135. $\dfrac{(1.25 \times 10^4) \times (6.30 \times 10^5)}{(7.83 \times 10^3)}$ _____

136. $\dfrac{(8.76 \times 10^{-5}) \times (1.05 \times 10^9)}{(6.37 \times 10^3)}$ _____

137. $\dfrac{(5.50 \times 10^4) \times (-6.00 \times 10^6)}{(6.92 \times 10^{-3})}$ _____

138. $\dfrac{(8.46 \times 10^{-5})}{(3.90 \times 10^7) \times (6.77 \times 10^{-3})}$ _____

The following problems are given in decimal (standard) form. Calculate and give answers in scientific notation. Round the answers (mantissas) to 2 decimal places.

139. $1510 \times 30,500$ _____

140. 0.000300×0.00210 _____

141. $-56,100 \times 781,000$ _____

142. $61,770 \times 53,100$ _____

143. $0.0000821 \div -315$ _____

144. $\dfrac{-0.00623 \times 742,000}{651,000}$ _____

145. $\dfrac{65,300 \times 517,000}{0.00786}$ _____

146. $\dfrac{-0.000829}{405,000 \times 0.00312}$ _____

147. $\dfrac{518,000 \times 0.00612}{37,400 \times 0.0000830}$ _____

148. The amount of expansion of metal when heated is computed as follows:

$$\text{Expansion} = \text{original length} \times \text{linear expansion per unit of length}$$
$$\text{per degree Fahrenheit} \times \text{temperature change.}$$

Calculate the amount of expansion for the metals shown in the table. Give the answers in decimal (standard) form to 3 decimal places.

	Metal	Original Length of Metal	Linear Expansion Per Unit of Length Per Degree Fahrenheit	Original Temperature	Temperature to Which Heated
a.	Aluminum	6.7520 in	1.244×10^{-5}	68.0°F	225.0°F
b.	Copper	35.750 ft	9.000×10^{-6}	35.0°F	97.0°F
c.	Carbon Steel	3.0950 in	6.330×10^{-6}	84.0°F	743.0°F

a. _____

b. _____

c. _____

UNIT 35 Introduction to Equations

Objectives After studying this unit you should be able to

- **Express word problems as equations.**
- **Express problems given in graphic form as equations.**
- **Solve simple equations using logical reasoning.**

It is essential that the skilled machine technician understand equations and their applications. The solution of equations is required to compute problems using trade handbook formulas. Often machine shop problems are solved using a combination of equations, with elements of geometry and trigonometry.

Expression of Equality

An *equation* is a mathematical statement of equality between two or more quantities and always contains the equal sign (=). The value of all quantities on the left side of the equal sign equals the value of all quantities on the right side of the equal sign. A *formula* is a particular type of an equation which states a mathematical rule.

The following are examples of simple equations:

$$7 + 2 = 5 + 4$$
$$3 \times 5\frac{1}{2}'' = 16\frac{1}{2}''$$
$$a + b = c - d$$

$$\frac{12}{3} + 2 \times 5 = 18 - 4$$
$$360° = 5 \times 80° - 40°$$
$$\frac{xy}{2} = x + y$$

Because it expresses the equality of the quantities on the left and on the right of the equal sign, an equation is a balanced mathematical statement. An equation may be considered similar to a balanced scale as illustrated in A. The total weight on the left side of the scale equals the total weight on the right side; therefore, the scale balances.

$$3 \text{ pounds} + 5 \text{ pounds} + 2 \text{ pounds} = 4 \text{ pounds} + 6 \text{ pounds}$$
$$10 \text{ pounds} = 10 \text{ pounds}$$

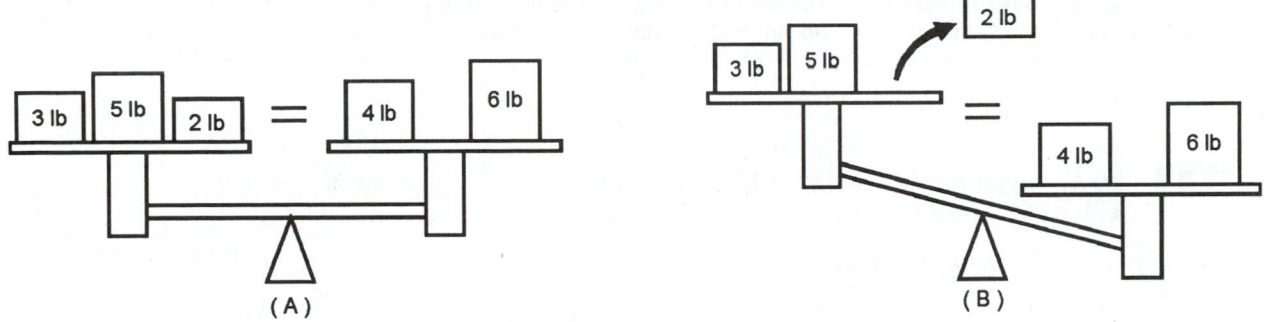

When the 2-pound weight is removed from the scale, the scale is no longer in balance as illustrated in B.

$$3 \text{ pounds } + 5 \text{ pounds } \neq 4 \text{ pounds } + 6 \text{ pounds}$$
$$8 \text{ pounds } \neq 10 \text{ pounds}$$

The Unknown Quantity

In general, an equation is used to determine the numerical value of an unknown quantity. Although any letter or symbol can be used to represent the unknown quantity, the letter x is commonly used.

The first letter of the unknown quantity is often used to represent a quantity. Some common letter designations are

L to represent length P to represent pressure
A to represent area F to represent feed of cutter
t to represent time W to represent weight
D to represent diameter h to represent height

Writing Equations from Word Statements

An equation asks a question. It asks for the value of the unknown which makes the left side of the equation equal to the right side. The question asked may not be in equation form; instead it may be expressed in words.

It is important to develop the ability to express word statements as mathematical symbols, or equations. A problem must be fully understood before it can be written as an equation.

Whether the word problem is simple or complex, a definite logical procedure should be followed to analyze the problem. A few or all of the following steps may be required, depending on the complexity of the particular problem.

- Carefully read the entire problem, several times if necessary.
- Break the problem down into simpler parts.
- It is sometimes helpful to draw a simple picture as an aid in visualizing the various parts of the problem.
- Identify and list the unknowns. Give each unknown a letter name, such as x.
- Decide where the equal sign should be, and group the parts of the problem on the proper side of the equal sign.
- Check. Are the statements on the left equal to the statements on the right of the equal sign?
- After writing the equation, check it against the original problem, step-by-step. Does the equation state mathematically what the problem states in words?

The following examples illustrate the method of writing equations from given word statements. After each equation is written the value of the unknown quantity is obtained. No specific procedures are given at this time in solving for the unknowns. The unknown quantity values are determined by logical reasoning.

Example 1 What weight must be added to 15 pounds so that it will be in balance with a 22-pound weight?

Ask the question: 15 pounds + what weight = 22 pounds?

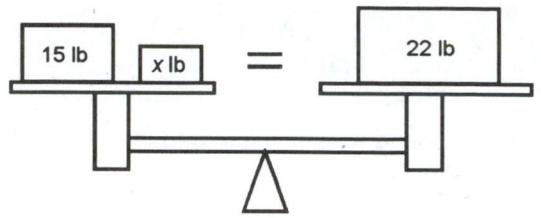

To help visualize the problem, a picture is shown.

Let x represent the unknown weight.

Write an equation. 15 lb + x = 22 lb

Ask the question: What number added to 15 pounds equals 22 pounds? Since 7 pounds added to 15 pounds equals 22 pounds, x = 7 lb   Ans

Check the answer by substituting 7 pounds for x in the original equation.

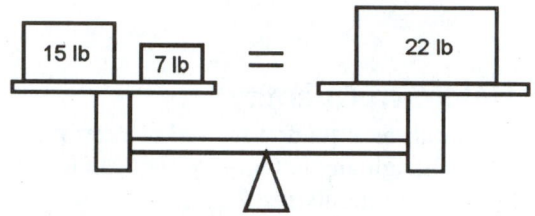

$$15 \text{ lb} + 7 \text{ lb} = 22 \text{ lb}$$
$$22 \text{ lb} = 22 \text{ lb} \quad \text{Ck}$$

The equation is balanced.

Example 2 A $9\frac{1}{2}$-inch piece is cut from a 12-inch length of bar stock. Find the length of the unused piece. Make no allowance for thickness of the cut.

Ask the question: What number subtracted from 12 inches = $9\frac{1}{2}$ inches?

A picture of the problem is shown. All dimensions are in inches. Let x represent the number of inches cut off.

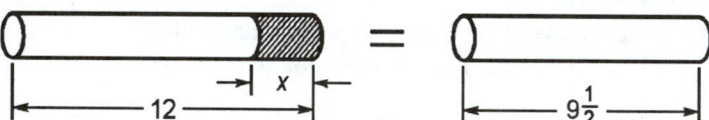

Express the problem as an equation. $12'' - x = 9\frac{1}{2}''$

Since $2\frac{1}{2}$ inches subtracted from 12 inches is equal to $9\frac{1}{2}$ inches,

$x = 2\frac{1}{2}''$ Ans

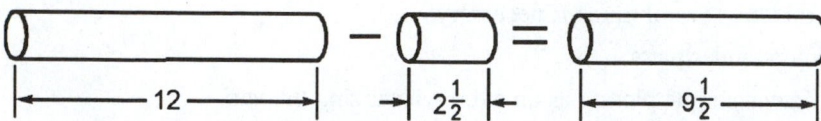

Check the answer by substituting $2\frac{1}{2}$ inches for x in the original equation.

$$12'' - 2\frac{1}{2}'' = 9\frac{1}{2}''$$
$$9\frac{1}{2}'' = 9\frac{1}{2}'' \quad \text{Ck}$$

The equation is balanced.

Example 3 The sum of two angles equals 90°. One angle is twice as large as the other. What is the size of the smaller angle?

An angle + an angle twice as large = 90°.

A picture of the problem is shown.

Let x represent the smaller angle. Let $2x$ represent the larger angle.

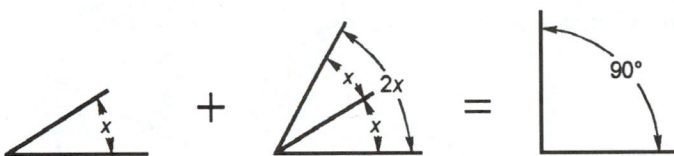

Express the problem as an equation. $x + 2x = 90°$ or $3x = 90°$

Ask the question: What number multiplied by 3 = 90°?

Since 3 multiplied by 30° = 90°, $x = 30°$

The smaller angle $x = 30°$ and the larger angle = $2x$ or 60° as shown.   Ans

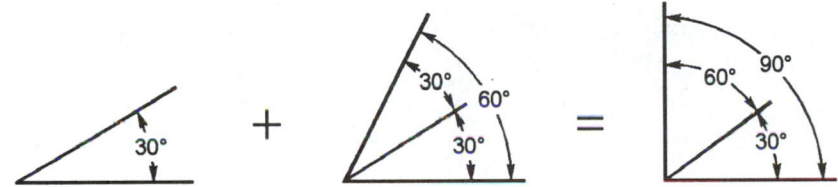

Check the answer by substituting 30° for x in the original equation.

$$30° + 2(30°) = 90°$$
$$90° = 90°   \text{Ck}$$

The equation is balanced.

Example 4 Three gage blocks are used to tilt a sine plate. The total height of the three blocks is 2.75 inches. The bottom block is 4 times as thick as the middle block. The middle block is twice as thick as the top block. How thick is each block?

Convert the problem from word form to equation form.

Let x represent the thickness of the thinnest block, the top block.

The middle block is twice as thick as the top block, or $2x$.

The bottom block is four times as thick as the middle block, or $(4)(2x) = 8x$.

The sum of the three blocks = 2.75″.

Therefore, $x + 2x + 8x = 2.75″$, or $11x = 2.75″$.

A picture of the problem is shown. All dimensions are in inches.

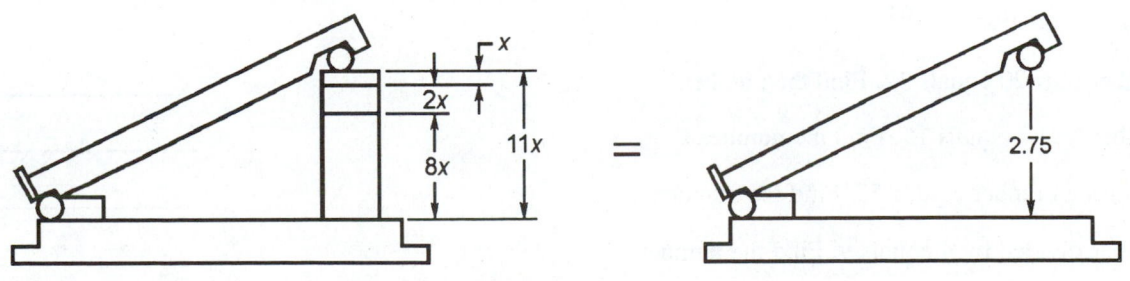

Ask the question: What number multiplied by 11 = 2.75″? Since $11 \times 0.25″$ = 2.75″, $x = 0.25″$.

The top block is x or 0.25″.   Ans

The middle block is $2x$ or $2(0.25″) = 0.50″$.   Ans

The bottom block is $8x$ or $8(0.25″) = 2.00″$.   Ans

The thickness of each block is shown. All dimensions are in inches.

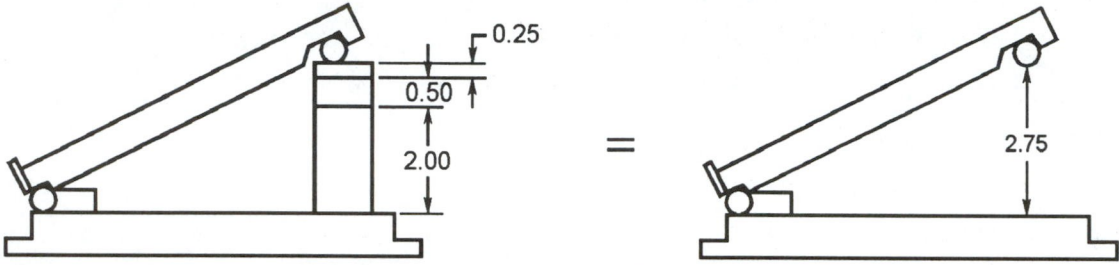

Check the answer by substituting 0.25″ for x in the original equation:

$$x + 2x + 4(2x) = 2.75″$$

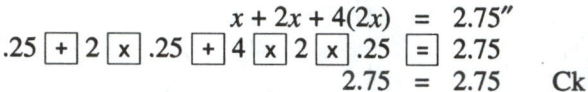

$$2.75 = 2.75 \quad \text{Ck}$$

The equation is balanced.

In many cases the problems to be solved in actual machine shop applications will be more difficult than the preceding examples. It is essential, therefore, to be able to use the procedure shown to analyze the problem, determine the unknowns, and set up the equation.

Checking the Equation

In the final step in each of the preceding examples, the value found for the unknown was substituted in the original equation to prove that it was the correct value. If an equation is properly written and if both sides of the equation are equal, the equation is balanced and the solution is correct.

All work in a machine shop should be checked and rechecked to prevent errors. It is important that you check your computations. When working with equations on the job, checking your work is essential. Errors in computation can often be costly in terms of time, labor, and materials.

APPLICATION

Express each of the following word problems as equations. Let the unknown number equal x and find the value of the unknown. Check the equation by comparing it to the word problem. Does the equation state mathematically what the problem states in words? Check whether the equation is balanced by substituting the value of the unknown in the equation.

1. A number plus 20 equals 32. Find the number.   _____

2. A number less 7 equals 15. Find the number.   _____

3. Five times a number equals 55. Find the number.   _____

4. A number divided by 4 equals 9. Find the number.   _____

5. Thirty-two divided by a number equals 8. Find the number.   _____

6. A number plus twice the number equals 36. Find the number.   _____

7. Five times a number minus the number equals 48. Find the number.   _____

8. Seven times a number plus eight times the number equals 60. Find the number.   _____

9. Sixty divided by 3 times a number equals 4. Find the number.   _____

10. A piece of bar stock 32 inches long is cut into two unequal lengths. One piece is 3 times as long as the other. How long is each piece?   _____

11. Three blocks are used to tilt a sine plate. The total height of the three blocks is 4.5 inches. The first block is 3 times as thick as the second block. The second block is twice as thick as the third block. How thick is each block?   _____

12. Five holes are drilled in a steel plate on a bolt circle. There are 300° between hole 1 and hole 5. The number of degrees between any two consecutive holes doubles in going from hole 1 to hole 5. Find the number of degrees between the indicated holes.

   a. 1 and 2   _____

   b. 2 and 3   _____

   c. 3 and 4   _____

   d. 4 and 5   _____

13. The total amount of stock milled off an aluminum casting in two cuts is 8.58 millimeters. The roughing cut is 6.35 millimeters greater than the finish cut. What is the depth of the finish cut?   _____

In each of the following problems, refer to the corresponding figure. Write an equation, solve for x, and check.

14. All dimensions are in inches.

   _____ x = _____

16. All dimensions are in inches.

   _____ x = _____

15. All dimensions are in millimeters.

   _____ x = _____

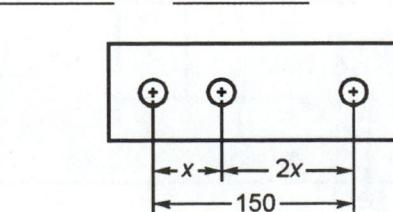

17. _____ x = _____

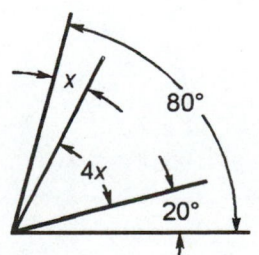

18. All dimensions are in inches.

_____ x = _____

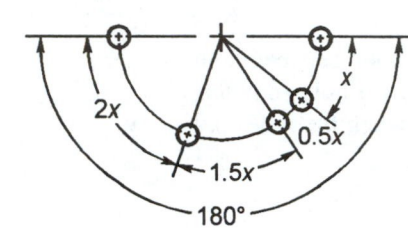

19. _____ x = _____

20. All dimensions are in millimeters.

_____ x = _____

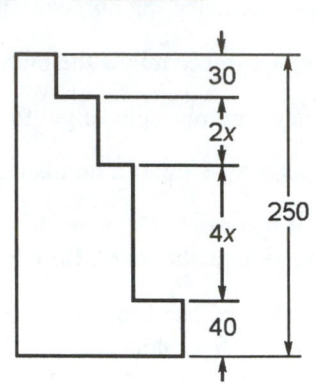

21. All dimensions are in inches.

_____ x = _____

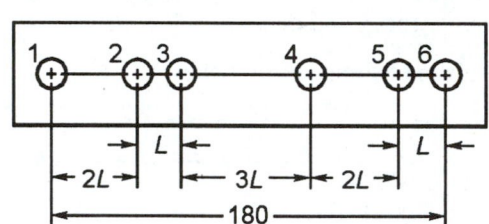

For each of the following problems, refer to the given figure, solve for the unknowns, and check.

22. Find the distances between the indicated holes. All dimensions are in millimeters.

   a. Hole 1 to Hole 2 _____
   b. Hole 2 to Hole 3 _____
   c. Hole 3 to Hole 4 _____
   d. Hole 4 to Hole 5 _____
   e. Hole 5 to Hole 6 _____
   f. Hole 2 to Hole 4 _____
   g. Hole 3 to Hole 6 _____

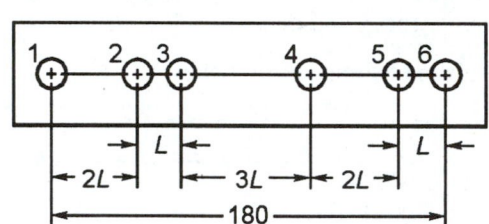

23. Find the distances between the indicated points. All dimensions are in inches.

   a. A and B _____
   b. B and D _____
   c. E and F _____

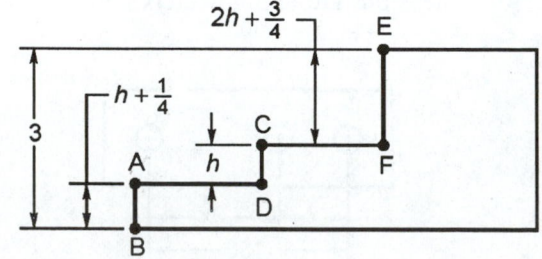

24. Find the value of each of the four angles.

   a. $\angle 1$ _____

   b. $\angle 2$ _____

   c. $\angle 3$ _____

   d. $\angle 4$ _____

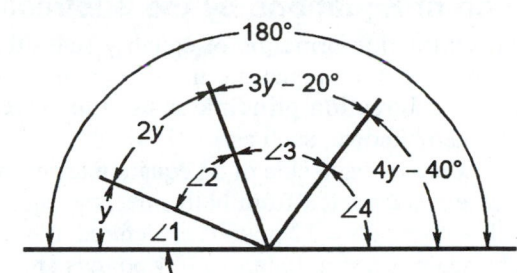

Solve for the unknown values in the following equations.

25. $x + 9x = 30$ _____

26. $x + 3 = 12$ _____

27. $2y + 5y + 3y = 70$ _____

28. $32 = 17 + y$ _____

29. $18 - a = 12$ _____

30. $b - 13 = 80$ _____

31. $3b + 5b - 2b = 96$ _____

32. $3(5a) = \frac{6(30)}{2}$ _____

33. $\frac{1}{2}x = 42$ _____

34. $\frac{x}{4} = 15$ _____

35. $\frac{27}{x} = 9$ _____

36. $\frac{d}{6} + 4 = 9$ _____

37. $0.75x - 0.5x = \frac{18 + 30}{4}$ _____

38. $6(2.5x) + 5x = 80$ _____

39. $\frac{2y + 4y + 6y}{3} = 80$ _____

40. $27 - (3)(6) = b + 3$ _____

UNIT 36 Solution of Equations by the Subtraction, Addition, and Division Principles of Equality

Objectives After studying this unit you should be able to

- Solve equations using the subtraction principle of equality.
- Solve equations using the addition principle of equality.
- Solve equations using the division principle of equality.
- Solve equations using transposition.

Principles of Equality

In actual practice, equations cannot always be solved by inspection or common sense. There are specific procedures for solving equations using the fundamental principles of equality. The principles of equality that will be presented are those of subtraction, addition, and division in this unit. Multiplication, root, and power principles are presented in Unit 37.

Solution of Equations by the Subtraction Principle of Equality

The subtraction principle of equality states that if the same number is subtracted from both sides of an equation, the sides remain equal, and the equation remains balanced. The subtraction principle is used to solve an equation in which a number is added to the unknown, such as $x + 15 = 20$.

The values on each side of an equation are equal and an equation is balanced. If the same value is subtracted from both sides, the equation remains balanced. The equation 8 pounds + 4 pounds = 12 pounds is pictured. If 4 pounds are removed from the left side only, the scale is not in balance. If 4 pounds are removed from both the left and right sides, the scale remains in balance.

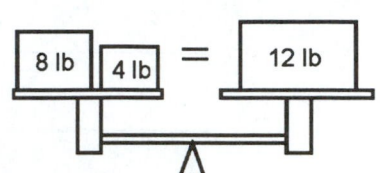 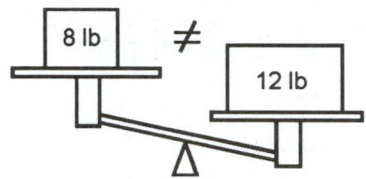 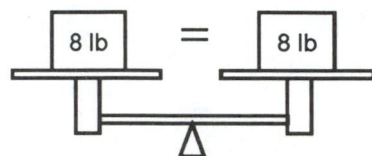

8 lb + 4 lb = 12 lb   8 lb + 4 lb − 4 lb ≠ 12 lb   8 lb + 4 lb − 4 lb = 12 lb − 4 lb
12 lb = 12 lb   8 lb ≠ 12 lb   8 lb = 8 lb

Procedure   To solve an equation in which a number is added to the unknown

- Subtract the number which is added to the unknown from both sides of the equation.
- Check.

Examples

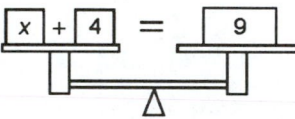

1. $x + 4 = 9$. Solve for x.

Subtract 4 from both sides of the equation.

$$\begin{array}{rcl} x + 4 &=& 9 \\ - 4 &=& -4 \\ \hline x &=& 5 \quad \text{Ans} \end{array}$$

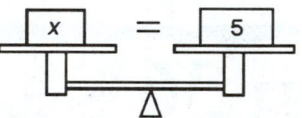

Check.

$$\begin{array}{rcl} x + 4 &=& 9 \\ 5 + 4 &=& 9 \\ \hline 9 &=& 9 \quad \text{Ck} \end{array}$$

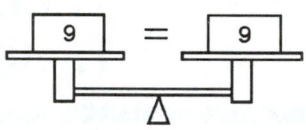

2. In the part shown, determine dimension y. All dimensions are in inches. Write an equation.

Subtract 5.5″ from both sides.

$$\begin{array}{rcl} 5.5'' + y &=& 17'' \\ - 5.5'' &=& -5.5'' \\ \hline y &=& 11.5'' \quad \text{Ans} \end{array}$$

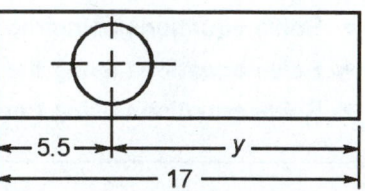

Check.

$$\begin{array}{rcl} 5.5'' + y &=& 17'' \\ 5.5'' + 11.5'' &=& 17'' \\ \hline 17'' &=& 17'' \quad \text{Ck} \end{array}$$

3. $-39 = P + 18$. Solve for P.

$$\begin{array}{rcl} -39 &=& P + 18 \\ -18 &=& -18 \\ \hline -57 &=& P \quad \text{Ans} \end{array}$$

Check. $\begin{array}{rcl} -39 &=& P + 18 \\ -39 &=& -57 + 18 \\ -39 &=& -39 \quad \text{Ck} \end{array}$

4. $W + 4\frac{3}{4} = 12$. Solve for W.

$$-4\frac{3}{4} = -4\frac{3}{4}$$

$$W = 7\frac{1}{4} \quad \text{Ans}$$

Check. $W + 4\frac{3}{4} = 12$

$$7\frac{1}{4} + 4\frac{3}{4} = 12$$

$$12 = 12 \quad \text{Ck}$$

Transposition

With your instructor's permission, an alternate method of solving certain equations may be used. The alternate method is called transposition. *Transposition* or transposing a term means that a term is moved from one side of an equation to the opposite side with the sign changed.

Transposition is not a mathematical process although it is based on the addition and subtraction principles of equality. Transposition should only be used after the principles of equality are fully understood and applied.

Transposition is a quick and convenient means of solving equations in which a term is added to or subtracted from the unknown. The purpose of using transposition is the same as that of using the addition and subtraction principles of equality. Both methods involve getting the unknown term to stand alone on one side of the equation in order to determine the value of the unknown.

The following example is solved by applying the subtraction principle of equality and transposition. Notice that when applying the subtraction principle of equality, a term is eliminated on one side of the equation and appears on the other side with the sign changed.

Example 1   Solve for x.

$$x + 15 = 25$$

Method 1:   The Subtraction Principle of Equality

$$x + 15 = 25$$
$$-15 = -15$$
$$x = 25 - 15 \qquad \leftarrow \text{Observe that } +15 \text{ is eliminated from the left side}$$
$$x = 10 \quad \text{Ans} \qquad \text{of the equation and appears as } -15 \text{ on the right side.}$$

Method 2:   Transposition

$$x + 15 = 25$$
$$x \;(+15) = 25 - 15$$

$$x = 25 - 15 \qquad \leftarrow \text{Observe that this expression is identical to the}$$
$$x = 10 \quad \text{Ans} \qquad \text{expression obtained when applying the subtraction}$$
$$\text{principle of equality.}$$

The following examples are solved by transposition.

Examples

1. $y + 10.7 = 18$.   Solve for y.

Move $+10.7$ from the left side of the equation to the right side and change to -10.7.

$$y + 10.7 = 18$$
$$y = 18 - 10.7$$
$$y = 7.3 \quad \text{Ans}$$

2. $T + 6\frac{1}{8} = -19$.   Solve for T. $T + 6\frac{1}{8} = -19$

Move $+6\frac{1}{8}$ from the left side $T = -19 - 6\frac{1}{8}$

of the equation to the right side $T = -25\frac{1}{8}$ Ans

and change to $-6\frac{1}{8}$.

Solution of Equations by the Addition Principle of Equality

The addition principle of equality states that if the same number is added to both sides of an equation, the sides remain equal and the equation remains balanced. The addition principle is used to solve an equation in which a number is subtracted from the unknown, such as $x - 17 = 30$.

Procedure   To solve an equation in which a number is subtracted from the unknown

- Add the number, which is subtracted from the unknown, to both sides of the equation.

- Check.

Examples

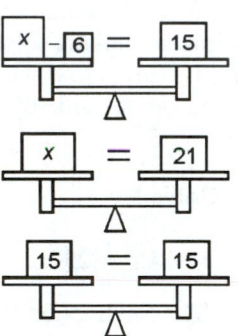

1. $x - 6 = 15$.   Solve for x.

   Add 6 to both sides of the equation
   $$x - 6 = 15$$
   $$\underline{+ 6 = +6}$$
   $$x = 21 \qquad \text{Ans}$$

   Check.
   $$x - 6 = 15$$
   $$21 - 6 = 15$$
   $$15 = 15 \qquad \text{Ck}$$

2. A 7-inch piece is cut from the height of a block as shown. The remaining block is 10 inches high. What is the height of the original block? All dimensions are in inches. Make no allowance for thickness of cut.

   Let y = the height of the original block

   Write an equation.
   $$y - 7'' = 10''$$

   Add 7" to both sides
   of the equation.
   $$\underline{+ 7'' = +7''}$$
   $$y = 17'' \qquad \text{Ans}$$

   Check.
   $$y - 7'' = 10''$$
   $$17'' - 7'' = 10''$$
   $$10'' = 10'' \qquad \text{Ck}$$

3. $-35 = P - 20.4$.   Solve for P.
   $$-35 = P - 20.4$$
   $$\underline{+20.4 = \quad + 20.4}$$
   $$- 14.6 = P \qquad \text{Ans}$$

   Check.
   $$-35 = P - 20.4$$
   $$-35 = -14.6 - 20.4$$
   $$-35 = -35 \qquad \text{Ck}$$

The following examples are solved by transposition.

Examples

1. $x - 4 = 19$.   Solve for x.

   Move -4 from the left side of the equation to the right and change to $+4$.

   $$x - 4 = 19$$
   $$x = 19 + 4$$
   $$x = 23 \qquad \text{Ans}$$

2. $y - 16.9 = 30$.   Solve for y.

   Move -16.9 from the left side of the equation to the right and change to $+16.9$.

   $$y - 16.9 = 30$$
   $$y = 30 + 16.9$$
   $$y = 46.9 \qquad \text{Ans}$$

Solution of Equations by the Division Principle of Equality

The division principle of equality states that if both sides of an equation are divided by the same number, the sides remain equal and the equation remains balanced. The division principle is used to solve an equation in which a number is multiplied by the unknown, such as $3x = 18$.

Procedure   To solve an equation in which a number is multiplied by the unknown

- Divide both sides of the equation by the number which multiplies the unknown.
- Check.

Examples

1. $6x = 24$.   Solve for x.

   Divide both sides of the equation by 6.

   $$6x = 24$$
   $$\frac{6x}{6} = \frac{24}{6}$$
   $$x = 4 \qquad \text{Ans}$$

   Check.

   $$6x = 24$$
   $$6(4) = 24$$
   $$24 = 24 \qquad \text{Ck}$$

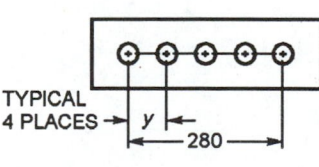

2. A part is shown.   Solve for y.

   All dimensions are in millimeters.

   Write an equation.

   $$4y = 280 \text{ mm}$$

   Divide both sides of the equation by 4.

   $$\frac{4y}{4} = \frac{280}{4} \text{ mm}$$
   $$y = 70 \text{ mm} \qquad \text{Ans}$$

   Check.

   $$4y = 280$$
   $$4(70 \text{ mm}) = 280 \text{ mm}$$
   $$280 \text{ mm} = 280 \text{ mm} \qquad \text{Ck}$$

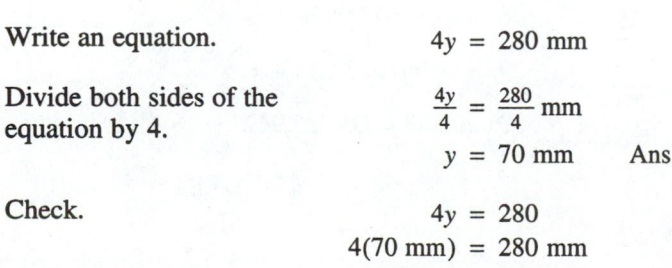

TYPICAL 4 PLACES

3. $-14.4 = 3.2F.$ Solve for F.

$$-14.4 = 3.2F$$
$$\frac{-14.4}{3.2} = \frac{3.2F}{3.2}$$
$$-4.5 = F \quad \text{Ans}$$

Check. $\quad -14.4 = 3.2F$
$$-14.4 = 3.2(-4.5)$$
$$-14.4 = -14.4 \quad \text{Ck}$$

4. $7\frac{1}{4}A = 21\frac{3}{4}.$ Solve for A.

$$7\frac{1}{4}A = 21\frac{3}{4}$$
$$\frac{7\frac{1}{4}A}{7\frac{1}{4}} = \frac{21\frac{3}{4}}{7\frac{1}{4}}$$
$$A = 3 \quad \text{Ans}$$

Check. $\quad 7\frac{1}{4}A = 21\frac{3}{4}$
$$7\frac{1}{4}(3) = 21\frac{3}{4}$$
$$21\frac{3}{4} = 21\frac{3}{4} \quad \text{Ck}$$

APPLICATION

Solution by the Subtraction Principle of Equality

Solve each of the following equations using the subtraction principle of equality. Check each answer.

1. $P + 15 = 22$ _____

2. $x + 18 = 27$ _____

3. $M + 24 = 43$ _____

4. $y + 48 = 82$ _____

5. $13 = T + 9$ _____

6. $37 = D + 2$ _____

7. $62 = a + 19$ _____

8. $y + 16 = 15$ _____

9. $C + 34 = 12$ _____

10. $x + 6 = -13$ _____

11. $y + 30 = -23$ _____

12. $x + 63 = 17$ _____

13. $10 + R = 53$ _____

14. $51 = 48 + E$ _____

15. $-36 = 14 + x$ _____

16. $H + 7.6 = 14.7$ _____

17. $22.5 = L + 3.7$ _____

18. $-36.2 = y + 6.2$ _____

19. $78.09 = x + 61.95$ _____

20. $F + 0.007 = 1.006$ _____

21. $T + 9.07 = 9.07$ _____

22. $H + 3\frac{1}{4} = 6\frac{1}{2}$ _____

23. $-\frac{7}{8} = x + \frac{3}{4}$ _____

24. $20\frac{3}{16} = A + 17\frac{1}{8}$ _____

25. $39\frac{5}{8} = y + 40\frac{7}{8}$ _____

26. $1\frac{7}{16} = W + \frac{9}{16}$ _____

27. $x + 13\frac{1}{8} = -10$ _____

28. $0.023 = 1.009 + H$ _____

29. $-14.067 = 3.034 + x$ _____

30. $20.863 = D + 25.942$ _____

Write an equation for each of the following problems, solve for the unknown, and check.

31. All dimensions are in inches.
Find x.   _____ $x =$ _____

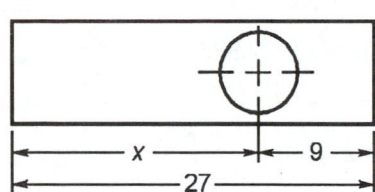

34. All dimensions are in inches.
Find T.   _____ $T =$ _____

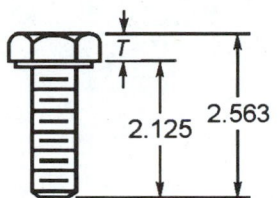

32. All dimensions are in millimeters.
Find y.   _____ $y =$ _____

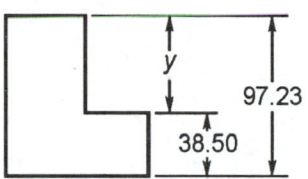

35. All dimensions are in millimeters.
Find x.   _____ $x =$ _____

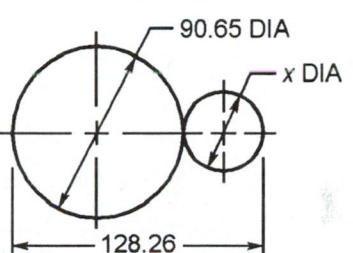

33. All dimensions are in inches.
Find r.   _____ $r =$ _____

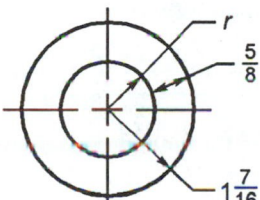

36. All dimensions are in inches.
Find H.   _____ $H =$ _____

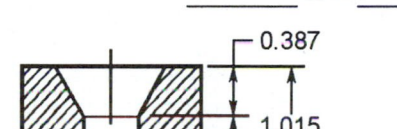

37. The height of 2 gage blocks is 0.8508 inch. One block is 0.750 inch thick. What is the thickness of the other block?   _____

38. Three holes are drilled on a horizontal line in a housing. The center distance between the first hole and the second hole is 193.75 millimeters and the center distance between the first hole and the third hole is 278.12 millimeters. What is the distance between the second hole and the third hole?   _____

39. A milling cut of $\frac{9}{32}$ inch is required to provide a reference surface on a rough casting which is $7\frac{5}{8}$ inches high. What is the height of the casting after the milling cut?   _____

40. A shaft rotates in a bearing which is 0.3968 inch in diameter. The total clearance between the shaft and bearing is 0.0008 inch. What is the diameter of the shaft?   _____

For each of the following problems, substitute the given values in the formula and solve for the unknown. Check.

41. One of the formulas used in computing spur gear dimensions is $D_o = D + 2a$. Determine D when $a = 0.1429$ inch and $D_o = 4.7144$ inches.   _____

42. A formula used to compute the dimensions of a ring is $D = d + 2T$. Determine d when $D = 52.0$ millimeters and $T = 9.40$ millimeters.   _____

43. A formula used in relation to the depth of a gear tooth is $WD = a + d$. Determine d when $WD = 0.3082$ inch and $a = 0.1429$ inch.   _____

Solution by the Addition Principle of Equality

Solve each of the following equations using the addition principle of equality. Check each answer.

44. $T - 12 = 34$ _____

45. $x - 9 = -19$ _____

46. $B - 4 = 9$ _____

47. $P - 48 = 87$ _____

48. $y - 23 = -20$ _____

49. $16 = M - 12$ _____

50. $-40 = E - 21$ _____

51. $47 = R - 36$ _____

52. $h - 8 = 12$ _____

53. $T - 19 = -5$ _____

54. $-22 = x - 31$ _____

55. $39 = F - 39$ _____

56. $W - 18 = 33$ _____

57. $N - 2.4 = 6.9$ _____

58. $A - 0.8 = 0.3$ _____

59. $x - 10.09 = -13.78$ _____

60. $5.07 = r - 3.07$ _____

61. $-30.003 = x - 29.998$ _____

62. $91.96 = L - 13.74$ _____

63. $P - 0.02 = 0.07$ _____

64. $G - 59.875 = 48.873$ _____

65. $x - 8.12 = -13.01$ _____

66. $D - \frac{1}{2} = \frac{1}{2}$ _____

67. $y - \frac{7}{8} = -\frac{3}{8}$ _____

68. $15\frac{5}{8} = H - 2\frac{7}{8}$ _____

69. $-46\frac{3}{32} = x - 29\frac{15}{16}$ _____

70. $C - 5\frac{7}{16} = -5\frac{7}{16}$ _____

71. $W - 10.0039 = 8.0481$ _____

72. $-14\frac{15}{32} = y - 14\frac{7}{16}$ _____

73. $E - 29.8936 = 18.3059$ _____

Write an equation for each of the following problems, solve for the unknown, and check.

74. The bushing shown has a body diameter of 44.45 millimeters, which is 14.29 millimeters less than the head diameter. What is the size of the head diameter? All dimensions are in millimeters. _____

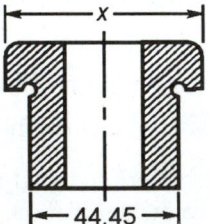

75. The flute length of the reamer shown is $1\frac{1}{8}$ inches, which is $3\frac{3}{8}$ inches less than the shank length. How long is the shank? All dimensions are in inches. _____

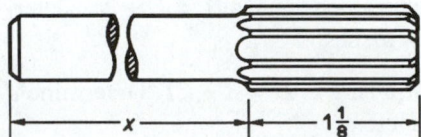

76. A hole is countersunk as shown to a depth of 0.250 inch. The depth of the counter-sink is 1.650 inches less than the depth of the 0.625-inch hole. Find the depth of the countersink. All dimensions are in inches.

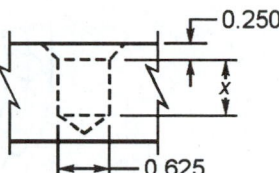

For each of the following problems, substitute the given values in the formula and solve for the unknown. Check each answer.

77. The total taper of a shaft equals the diameter of the large end minus the diameter of the small end, $T = D - d$. Determine D when $T = 22.5$ millimeters and $d = 30.8$ millimeters.

78. Using the spur gear formula, $D_R = D - 2d$, compute the pitch diameter (D) when the root diameter $(D_R) = 3.0118$ inches and the dedendum $(d) = 0.1608$ inch.

79. Using a sheet metal formula, $W = L.S. - 4S$, determine the length size $(L.S.)$ when $W = 382$ millimeters and $S = 112$ millimeters.

Solution by the Division Principle of Equality

Solve each of the following equations using the division principle of equality. Check each answer.

80. $4D = 32$

81. $7x = -21$

82. $15M = 75$

83. $54 = 9P$

84. $-27 = 3y$

85. $54 = 6x$

86. $10y = 0.80$

87. $18T = 41.4$

88. $12x = -54$

89. $-x = 19$

90. $0 = 7H$

91. $-5C = 0$

92. $7.1E = 21.3$

93. $0.6L = 12$

94. $-2.7x = 23.76$

95. $0.1y = -0.18$

96. $13.2W = 0$

97. $-x = -19.75$

98. $0.125P = 1.500$

99. $9.37R = 103.07$

100. $-0.66x = 4.752$

101. $\frac{1}{4}D = 8$

102. $24 = \frac{3}{8}B$

103. $-\frac{1}{2}y = 36$

104. $1\frac{5}{8}L = 9\frac{3}{4}$

105. $-48\frac{3}{8} = 10\frac{3}{4}x$

106. $-\frac{7}{16} = -\frac{7}{16}y$

107. $50.98W = 10.196$

108. $-0.006x = 4.938$

109. $-\frac{3}{16} = -1\frac{1}{16}y$

Write an equation for each of the following problems, solve for the unknown.

110.   All dimensions are in millimeters.
Find x.   _____ $x =$ _____

111.   Find x.   _____ $x =$ _____

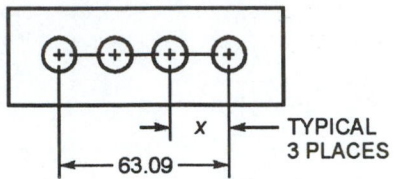

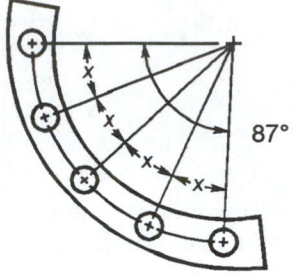

87°

112.   The feed of a drill is the depth of material that the drill penetrates in one revolution. The total depth of penetration equals the product of the number of revolutions and the feed. Compute the feed of a drill which cuts to a depth of 3.300 inches while turning 500.0 revolutions.

_____

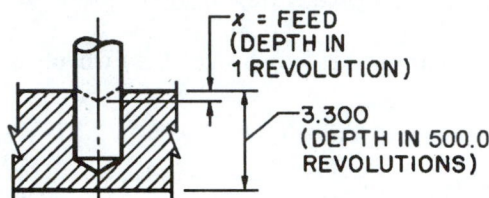

$x =$ FEED
(DEPTH IN
1 REVOLUTION)

3.300
(DEPTH IN 500.0
REVOLUTIONS)

For each of the following problems, substitute the given values in the formula and solve for the unknown. Check each answer. Round the answers to 2 decimal places.

113.   The circumference of a circle (C) equals π (approximately 3.1416) times the diameter (d) of the circle, $C = \pi d$. Determine d when $C = 392.50$ millimeters.

_____

114.   The depth (d) of a sharp V-thread is equal to 0.866 times the pitch (p). $d = 0.866p$. Determine p when $d = 0.125$ inch.

_____

115.   The length of cut (L) in inches of a workpiece in a lathe is equal to the product of the cutting time (T) in minutes, the tool feed (F) in inches per revolution, and the number of revolutions per minute (N) of the workpiece, $L = TFN$. Determine N when $L = 9.50$ inches, $T = 3.00$ minutes, and $F = 0.050$ inch per revolution.

_____

UNIT 37   Solution of Equations by the Multiplication, Root, and Power Principles of Equality

Objectives   **After studying this unit you should be able to**

- **Solve equations using the multiplication principle of equality.**
- **Solve equations using the root principle of equality.**
- **Solve equations using the power principle of equality.**

Solution of Equations by the Multiplication Principle of Equality

The multiplication principle of equality states that if both sides of an equation are multiplied by the same number, the sides remain equal, and the equation remains balanced.

The multiplication principle is used to solve an equation in which the unknown is divided by a number, such as $\frac{x}{4} = 10$.

Procedure To solve an equation in which the unknown is divided by a number

- Multiply both sides of the equation by the number which divides the unknown.
- Check.

Examples

1. $\frac{x}{3} = 7$. Solve for x.

$$\frac{x}{3} = 7$$

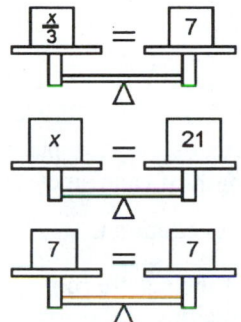

Multiply both sides of the equation by 3.

$$3\left(\frac{x}{3}\right) = 3(7)$$
$$x = 21 \quad \text{Ans}$$

Check.

$$\frac{x}{3} = 7$$
$$\frac{21}{3} = 7$$
$$7 = 7 \quad \text{Ck}$$

2. The length of bar stock shown is cut into 5 equal pieces. Each piece is 4.5 inches long. Find y, the length of the bar before it was cut. All dimensions are in inches. Make no allowance for thickness of cuts.

Write the equation.

$$\frac{y}{5} = 4.5''$$

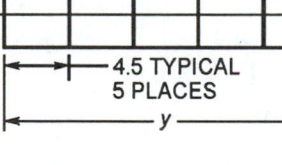

Multiply both sides of the equation by 5.

$$5\left(\frac{y}{5}\right) = 5(4.5'')$$
$$y = 22.5'' \quad \text{Ans}$$

4.5 TYPICAL
5 PLACES

Check.

$$\frac{y}{5} = 4.5''$$
$$\frac{22.5''}{5} = 4.5''$$
$$4.5'' = 4.5'' \quad \text{Ck}$$

3. $6\frac{1}{8} = \frac{F}{-5}$. Solve for F.

$$6\frac{1}{8} = \frac{F}{-5}$$

$$-5\left(6\frac{1}{8}\right) = -5\left(\frac{F}{-5}\right)$$

Check. $\quad 6\frac{1}{8} = \frac{F}{-5}$

$$6\frac{1}{8} = \frac{-30\frac{5}{8}}{-5}$$

$$6\frac{1}{8} = 6\frac{1}{8} \quad \text{Ck}$$

$$-30\frac{5}{8} = F \quad \text{Ans}$$

Solution of Equations by the Root Principle of Equality

The root principle of equality states that if the same root of both sides of an equation is taken, the sides remain equal, and the equation remains balanced.

The root principle is used to solve an equation that contains an unknown which is raised to a power, such as $x^2 = 36$.

Procedure To solve an equation in which an unknown is raised to a power

- Extract the root of both sides of the equation which leaves the unknown with an exponent of one.
- Check.

Examples

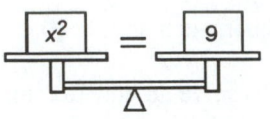

1. $x^2 = 9$. Solve for x.

$$x^2 = 9$$

Extract the square root of both sides of the equation.

$$\sqrt{x^2} = \sqrt{9}$$
$$x = 3 \quad \text{Ans}$$

Check.

$$x^2 = 9$$
$$3^2 = 9$$
$$9 = 9 \quad \text{Ck}$$

2. The area of a square piece of sheet steel shown equals 16 square feet. What is the length of each side(s)?

Write an equation.

$$s^2 = 16 \text{ sq ft}$$

Extract the square root of both sides of the equation.

$$\sqrt{s^2} = \sqrt{16 \text{ sq ft}}$$
$$s = 4 \text{ ft} \quad \text{Ans}$$

Check.

$$s^2 = 16 \text{ sq ft}$$
$$(4 \text{ ft})^2 = 16 \text{ sq ft}$$
$$16 \text{ sq ft} = 16 \text{ sq ft} \quad \text{Ck}$$

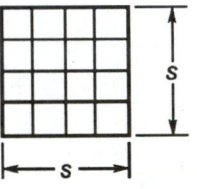

3. Solve for T.

$$T^3 = -64$$

$$\sqrt[3]{T^3} = \sqrt[3]{-64}$$
$$T = -4 \quad \text{Ans}$$

Check.

$$T^3 = -64$$
$$(-4)^3 = -64$$
$$-64 = -64 \quad \text{Ck}$$

4. Solve for V.

$$V^2 = \frac{9}{64}$$

$$\sqrt{V^2} = \sqrt{\frac{9}{64}}$$

$$V = \frac{3}{8} \quad \text{Ans}$$

Check.

$$V^2 = \frac{9}{64}$$
$$\left(\frac{3}{8}\right)^2 = \frac{9}{64}$$
$$\frac{9}{64} = \frac{9}{64} \quad \text{Ck}$$

Solution of Equations by the Power Principle of Equality

The power principle of equality states that if both sides of an equation are raised to the same power, the sides remain equal and the equation remains balanced. The power principle is used to solve an equation that contains a root of the unknown, such as $\sqrt{x} = 8$.

Procedure To solve an equation which contains a root of the unknown

- Raise both sides of the equation to the power which leaves the unknown with an exponent of one.
- Check.

Examples

1. $\sqrt{x} = 8$. Solve for x.

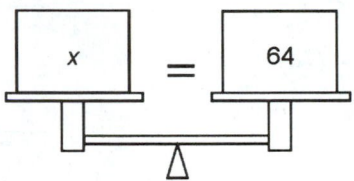

$$\sqrt{x} = 8$$

Square both sides of
the equation.

$$\left(\sqrt{x}\right)^2 = 8^2$$
$$x = 64 \quad \text{Ans}$$

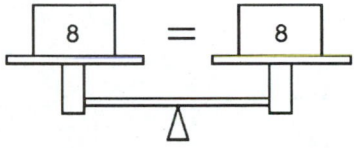

Check.

$$\sqrt{x} = 8$$
$$\sqrt{64} = 8$$
$$8 = 8 \quad \text{Ck}$$

2. The length of a side of the cube shown equals 2.8620 inches. The cube root of the volume equals the length of a side. Find the volume of the cube. Round the answer to 3 decimal places.

Let $V =$ the volume of the cube.

Write the equation.

$$\sqrt[3]{V} = 2.8620 \text{ in}$$

Cube both sides of the equation.

$$\left(\sqrt[3]{V}\right)^3 = (2.8620 \text{ in})^3$$
$$V = 23.443 \text{ cu in} \quad \text{Ans}$$

Check.

$$\sqrt[3]{V} = 2.8620 \text{ in}$$
$$\sqrt[3]{23.443 \text{ cu in}} = 2.8620 \text{ in}$$
$$2.8620 \text{ in} = 2.8620 \text{ in} \quad \text{Ck}$$

2.862 $\boxed{x^y}$ or $\boxed{y^x}$ 3 $\boxed{=}$ 23.44276793, 23.443 cu in   Ans
or 2.862 $\boxed{\wedge}$ 3 $\boxed{\text{EXE}}$ 23.44276793, 23.443 cu in   Ans

➤ **Note:** $\boxed{\sqrt[x]{y}}$ and $\boxed{\sqrt[x]{}}$ are second functions on certain calculators.
   Check.   23.443 $\boxed{\sqrt[x]{y}}$ 3 $\boxed{=}$ 2.862
   or 3 $\boxed{\sqrt[x]{}}$ 23.443 $\boxed{=}$ or $\boxed{\text{EXE}}$ 2.862
   2.862 in = 2.862 in

APPLICATION

Solution by the Multiplication Principle of Equality

Solve each of the following equations using the multiplication principle of equality. Check each answer.

1. $\dfrac{P}{5} = 6$ _____

2. $\dfrac{M}{12} = 5$ _____

3. $D + 9 = 7$ _____

4. $3 = L + 8$ _____

5. $3 = W + 9$ _____

6. $\dfrac{N}{12} = -2$ _____

7. $\frac{C}{14} = 0$ _____

8. $\frac{x}{-10} = 9$ _____

9. $\frac{E}{-2} = -18$ _____

10. $13 = y + (-4)$ _____

11. $\frac{F}{4.3} = 5$ _____

12. $\frac{A}{-0.5} = 24$ _____

13. $S \div (7.8) = 3$ _____

14. $x + (-0.3) = 16$ _____

15. $-20 = \frac{y}{0.3}$ _____

16. $\frac{T}{-1.8} = 2.4$ _____

17. $0 = H + (-3.8)$ _____

18. $M + 9.5 = -12$ _____

19. $\frac{y}{-0.1} = -0.01$ _____

20. $\frac{R}{12.6} = 0.002$ _____

21. $1.04 = \frac{H}{0.06}$ _____

22. $\frac{B}{\frac{1}{2}} = 7$ _____

23. $V + 1\frac{1}{4} = 3$ _____

24. $\frac{x}{\frac{3}{8}} = -\frac{1}{4}$ _____

25. $D + \left(-\frac{1}{16}\right) = -32$ _____

26. $4 = y + \left(-\frac{7}{8}\right)$ _____

27. $\frac{1}{2} = \frac{T}{1\frac{1}{2}}$ _____

28. $H + (-2) = 7\frac{9}{16}$ _____

29. $\frac{M}{0.009} = 100$ _____

30. $x \div (6.004) = -0.2125$ _____

Write an equation for each of the following problems, solve for the unknown, and check.

31. All dimensions are in millimeters.
Find x. _____ $x =$ _____

108.78 TYPICAL
4 PLACES

32. Find x. _____ $x =$ _____

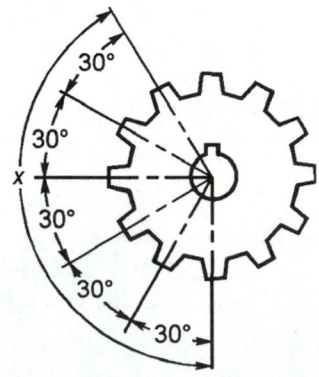

33. A 10-inch sine plate is tilted at an angle of 45° as shown. The gage block height divided by 10 equals 0.70711 inch. Compute the height of the gage blocks. All dimensions are in inches.

_____

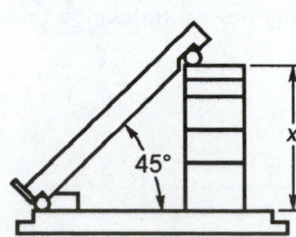

34. The width of a rectangular sheet of metal shown is equal to the area of the sheet divided by its length. Compute the area of a sheet which is $3\frac{1}{4}$ feet wide and $5\frac{1}{2}$ feet long.

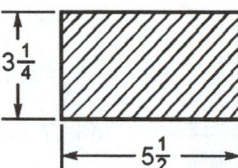

_____

35. The depth of an American Standard thread shown divided by 0.6495 is equal to the pitch. Compute the depth of a thread with a 0.0500-inch pitch. All dimensions are in inches. Round the answer to 3 decimal places.

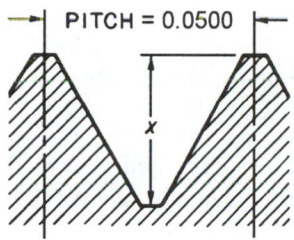

_____

    For each of the following problems, substitute the given values in the formula and solve for the unknown. Check each answer.

36. In mechanical energy applications, force (F) in pounds equals work (W) in foot-pounds divided by distance (D) in feet, $F = \frac{W}{D}$. Determine W when $F = 150.0$ pounds and $D = 7.500$ feet.

_____

37. The diameter (D) of a circle equals the circle circumference (C) divided by 3.1416, $D = \frac{C}{3.1416}$. Determine C when $D = 52.14$ millimeters. Round the answer to 1 decimal place.

_____

38. The pitch (P) of a spur gear equals the number of gear teeth (N) divided by the pitch diameter (D), $P = \frac{N}{D}$. Determine N when $P = 5$ teeth per inch and $D = 5.6000$ inches.

_____

Solution by the Root Principle of Equality

    Solve each of the following equations using the root principle of equality. Round the answers to 3 decimal places where necessary.

39. $S^2 = 16$ _____   47. $L^3 = -125$ _____

40. $P^2 = 81$ _____   48. $T^3 = 0$ _____

41. $81 = M^2$ _____   49. $10000 = L^2$ _____

42. $49 = B^2$ _____   50. $-125 = x^3$ _____

43. $D^3 = 64$ _____   51. $\frac{9}{25} = W^2$ _____

44. $x^3 = -64$ _____

45. $144 = F^2$ _____   52. $C^2 = \frac{1}{16}$ _____

46. $-64 = y^3$ _____   53. $P^2 = \frac{9}{25}$ _____

54. $M^3 = \frac{1}{64}$ _____

55. $-\frac{1}{8} = y^3$ _____

56. $D^3 = \frac{64}{27}$ _____

57. $G^3 = \frac{64}{125}$ _____

58. $x^3 = \frac{-64}{125}$ _____

59. $E^2 = 0.04$ _____

60. $0.64 = H^2$ _____

61. $W^2 = 2.753$ _____

62. $0.0017 = R^2$ _____

63. $N^3 = 0.123$ _____

64. $-0.123 = x^3$ _____

65. $7.843 = F^4$ _____

66. $T^2 = 7.056$ _____

67. $y^3 = 0.0393$ _____

68. $-2.127 = y^5$ _____

Write an equation for each of the following problems, solve for the unknown, and check.

69. The area of a square equals the length of a side squared, $A = s^2$. For each area of a square given, compute the length of a side. Round the answers to 3 decimal places where necessary.

a. 36 square inches _____ $s =$ _____

b. $\frac{25}{64}$ square foot _____ $s =$ _____

c. 1.44 square meters _____ $s =$ _____

d. 64.700 square meters _____ $s =$ _____

e. 0.049 square foot _____ $s =$ _____

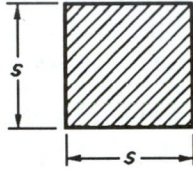

70. The volume of a cube equals the length of a side cubed, $V = s^3$. For each volume of a cube given, compute the length of a side. Round the answers to 3 decimal places where necessary.

a. 125 cubic inches _____ $s =$ _____

b. $\frac{27}{216}$ cubic foot _____ $s =$ _____

c. 0.642 cubic meter _____ $s =$ _____

d. 92.76 cubic millimeters _____ $s =$ _____

e. 0.026 cubic foot _____ $s =$ _____

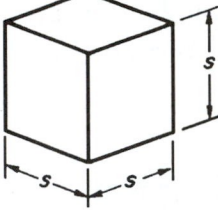

Solution by the Power Principle of Equality

Solve each of the following equations using the power principle of equality. Check all answers. Round the answers to 3 decimal places where necessary.

71. $\sqrt{C} = 6$ _____

72. $\sqrt{T} = 12$ _____

73. $\sqrt{P} = 1.2$ _____

74. $0.8 = \sqrt{M}$ _____

75. $0.82 = \sqrt{F}$ _____

76. $\sqrt[3]{V} = 3$ _____

77. $\sqrt[3]{H} = 1.7$ _____

78. $\sqrt[3]{x} = -4$ _____

79. $-0.1 = \sqrt[3]{y}$ _____

80. $\sqrt[4]{M} = 3$ _____

81. $\sqrt{A} = 0$ _____

82. $\sqrt[5]{N} = 1$ _____

83. $-2 = \sqrt[5]{y}$ _____

84. $0.3 = \sqrt[4]{D}$ _____

85. $\sqrt[3]{x} = -0.6$ _____

86. $\sqrt[4]{P} = 0.1$ _____

87. $0.1 = \sqrt[3]{B}$ _____

88. $\frac{1}{4} = \sqrt{A}$ _____

89. $\sqrt{R} = \frac{3}{8}$ _____

90. $\sqrt[3]{V} = \frac{2}{3}$ _____

91. $\sqrt[4]{F} = \frac{1}{4}$ _____

92. $-\frac{3}{5} = \sqrt[3]{y}$ _____

93. $\frac{5}{8} = \sqrt{H}$ _____

94. $\sqrt{P} = 1.256$ _____

95. $\sqrt[3]{B} = 2.868$ _____

96. $\sqrt[5]{x} = -1.090$ _____

97. $0.7832 = \sqrt[3]{y}$ _____

98. $0.364 = \sqrt[3]{y}$ _____

99. $\sqrt[3]{x} = -2.9631$ _____

100. $\sqrt[5]{x} = 0.797$ _____

Write an equation for each of the following problems, solve for the unknown, and check. Round the answers to 3 decimal places where necessary.

101. The length of a side of a square equals the square root of the area, $s = \sqrt{A}$. For each side of a square given, compute the area.

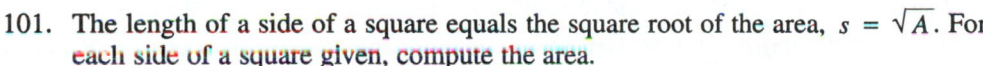

a. 3.4″ _____ $A =$ _____

b. 0.75′ _____ $A =$ _____

c. 0.652 m _____ $A =$ _____

d. 2.162 mm _____ $A =$ _____

e. 1.290″ _____ $A =$ _____

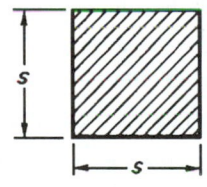

102. The length of a side of a cube equals the cube root of the volume, $s = \sqrt[3]{V}$. For each side of a cube given, compute the volume. Round the answers to 2 decimal places where necessary.

a. 3.300″ _____ $V =$ _____

b. 0.900′ _____ $V =$ _____

c. 0.62 m _____ $V =$ _____

d. 4.073 mm _____ $V =$ _____

e. 1.281″ _____ $V =$ _____

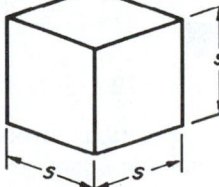

UNIT 38 Solution of Equations Consisting of Combined Operations and Rearrangement of Formulas

Objectives After studying this unit you should be able to

- **Solve equations involving several operations.**
- **Rearrange formulas in terms of any letter value.**
- **Substitute values in formulas and solve for unknowns.**

Often in actual occupational applications, the formulas used result in complex equations. These equations require the use of two or more principles of equality for their solutions. For example,

$$0.13x - 4.73(x + 6.35) = 5.06x - 2.87$$

requires a definite procedure in determining the value of x. Use of proper procedure results in the unknown standing alone on one side of the equation with its value on the other.

Procedure for Solving Equations Consisting of Combined Operations

It is essential that the steps used in solving an equation be taken in the following order. Some or all of these steps may be used depending upon the particular equation.

- Remove parentheses.

- Combine like terms on each side of the equation.

- Apply the addition and subtraction principles of equality to get all unknown terms on one side of the equation and all known terms on the other side.

- Combine like terms.

- Apply the multiplication and division principles of equality.

- Apply the power and root principles of equality.

- ➤ **Note:** Always solve for a positive unknown. A positive unknown may equal a negative value, but a negative unknown is not a solution. For example, $x = -10$ is correct, but $-x = 10$ is incorrect. When solving equations where the unknown remains a negative value, multiply both sides of the equation by -1. Multiplying a negative unknown by -1 results in a positive unknown. For example, multiply both sides of $-x = 10$ by -1: $(-1)(-x) = (-1)(10)$, $x = -10$.

Examples

1. $5x + 7 = 22$.   Solve for x.

   The operations involved are multiplication and addition. Follow the procedure for solving equations consisting of combined operations.

   Apply the subtraction principle. Subtract 7 from both sides of the equation.

$$\begin{aligned} 5x + 7 &= 22 \\ -7 &= -7 \\ \hline 5x &= 15 \end{aligned}$$

   Apply the division principle. Divide both sides of the equation by 5.

$$\frac{5x}{5} = \frac{15}{5}$$
$$x = 3 \quad \text{Ans}$$

Check.

$$5x + 7 = 22$$
$$5(3) + 7 = 22$$
$$15 + 7 = 22$$
$$22 = 22 \qquad Ck$$

2. $6x + 4x = 3x - 5x + 19 + 5.$ Solve for x.

$$6x + 4x = 3x - 5x + 19 + 5$$

Combine like terms on each side of the equation.

$$10x = -2x + 24$$

Apply the addition principle. Add $2x$ to both sides of the equation.

$$\underline{+2x = +2x \qquad\quad}$$
$$12x = 24$$

Apply the division principle. Divide both sides of the equation by 12.

$$\frac{12x}{12} = \frac{24}{12}$$
$$x = 2 \qquad Ans$$

Check.

$$6x + 4x = 3x - 5x + 19 + 5$$
$$6(2) + 4(2) = 3(2) - 5(2) + 19 + 5$$
$$12 + 8 = 6 - 10 + 19 + 5$$
$$20 = 20 \qquad Ck$$

3. $9x + 7(x + 3) = 25.$ Solve for x.

$$9x + 7(x + 3) = 25$$

Remove parentheses.

$$9x + 7x + 21 = 25$$

Combine like terms.

$$16x + 21 = 25$$

Apply the subtraction principle. Subtract 21 from both sides of the equation.

$$\underline{\quad - 21 = -21}$$
$$16x = \quad 4$$

Apply the division principle. Divide both sides of the equation by 16.

$$\frac{16x}{16} = \frac{4}{16}$$
$$x = \frac{1}{4} \qquad Ans$$

Check.

$$9x + 7(x + 3) = 25$$
$$9\left(\frac{1}{4}\right) + 7\left(\frac{1}{4} + 3\right) = 25$$
$$2\frac{1}{4} + 22\frac{3}{4} = 25$$
$$25 = 25 \qquad Ck$$

4. $-x = 14.$ Solve for x.

$$-x = 14$$

Apply the multiplication principle. Multiply both sides of the equation by -1.

$$(-1)(-x) = (-1)(14)$$
$$x = -14 \qquad Ans$$

Check.

$$-x = 14$$
$$-(-14) = 14$$
$$14 = 14 \qquad Ck$$

5. $\frac{x^2}{4} - 32 = -23$.   Solve for x.

$$\frac{x^2}{4} - 32 = -23$$

Apply the addition principle.
Add 32 to both sides of the equation.

$$\frac{+\ 32\ =\ +32}{\frac{x^2}{4}\ =\quad 9}$$

Apply the multiplication principle.
Multiply both sides of the equation
by 4.

$$4\left(\frac{x^2}{4}\right) = 4(9)$$

$$x^2 = 36$$

Apply the root principle. Extract the
square root of both sides of the
equation.

$$\sqrt{x^2} = \sqrt{36}$$
$$x = 6 \qquad \text{Ans}$$

Check.

$$\frac{x^2}{4} - 32 = -23$$

$$\frac{6^2}{4} - 32 = -23$$

$$\frac{36}{4} - 32 = -23$$

$$9 - 32 = -23$$
$$-23 = -23 \qquad \text{Ck}$$

6. $6\sqrt[3]{x} = 4(\sqrt[3]{x} + 1.5)$.   Solve for x.

$$6\sqrt[3]{x} = 4(\sqrt[3]{x} + 1.5)$$

Remove parentheses.

$$6\sqrt[3]{x} = 4\sqrt[3]{x} + 6$$

Apply the subtraction principle.
Subtract $4\sqrt[3]{x}$ from both sides of
the equation.

$$\frac{-4\sqrt[3]{x} = -4\sqrt[3]{x}}{2\sqrt[3]{x} = 6}$$

Apply the division principle. Divide
both sides of the equation by 2.

$$\frac{2\sqrt[3]{x}}{2} = \frac{6}{2}$$

$$\sqrt[3]{x} = 3$$

Apply the power principle. Raise
both sides of the equation to the
third power.

$$(\sqrt[3]{x})^3 = 3^3$$
$$x = 27 \qquad \text{Ans}$$

Check.

$$6\sqrt[3]{27} = 4(\sqrt[3]{27} + 1.5)$$
$$6(3) = 4(3 + 1.5)$$
$$18 = 4(4.5)$$
$$18 = 18 \qquad \text{Ck}$$

Substituting Values and Solving Formulas

Manufacturing applications often require solving formulas in which all but one numerical value for letter values is known. The unknown letter value can appear anywhere within the formula. To determine the numerical value of the unknown, write the original formula, substitute the known number values for their respective letter values, and simplify. Then follow the procedure given for solving equations consisting of combined operations.

Example   An open belt pulley system is shown. The larger pulley diameter is 6.25 inches and the smaller pulley diameter is 4.25 inches. The belt length is 56.0 inches. Find the distance between pulley centers using this formula found in a trade handbook. Round the answer to 1 decimal place.

$$L = 3.14(0.5D + 0.5d) + 2x$$

where L = belt length
D = the diameter of the larger pulley
d = the diameter of the smaller pulley
x = the distance between pulley centers

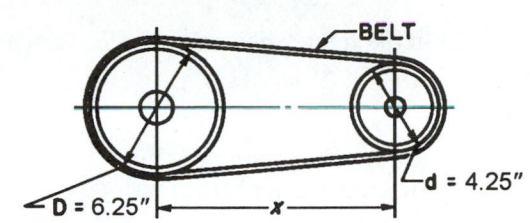

Write the formula.	$L = 3.14(0.5D + 0.5d) + 2x$

Substitute the known numerical values for their respective letter values and simplify.

$$56.0 \text{ in} = 3.14[0.5(6.25 \text{ in}) + 0.5(4.25 \text{ in})] + 2x$$

$$3.14 \boxed{\times} \boxed{(} .5 \boxed{\times} 6.25 \boxed{+} .5 \boxed{\times} 4.25 \boxed{)}$$
$$\boxed{=} 16.485$$

$$56.0 \text{ in} = 3.14(5.25 \text{ in}) + 2x$$
$$56.0 \text{ in} = 16.485 \text{ in} + 2x$$

Apply the subtraction principle. Subtract 16.485 inches from both sides.

$$\begin{array}{r} 56.0 \text{ in} = 16.485 \text{ in} + 2x \\ -16.485 \text{ in} = -16.485 \text{ in} \\ \hline \end{array}$$

Apply the division principle. Divide both sides by 2.

$$\frac{39.515 \text{ in}}{2} = \frac{2x}{2}$$
$$19.7575 \text{ in} = x, x = 19.8 \text{ in} \quad \text{Ans (rounded)}$$

Check.

$$L = 3.14(0.5D + 0.5d) + 2x$$
$$56 = 3.14 \boxed{\times} \boxed{(} .5 \boxed{\times} 6.25 \boxed{+} .5 \boxed{\times} 4.25 \boxed{)}$$
$$\boxed{+} 2 \boxed{\times} 19.7575 \boxed{=} 56$$
$$56 \text{ in} = 56 \text{ in}$$

Rearranging Formulas

A formula that is used to find a particular value must sometimes be rearranged to solve for another value. Consider the letter to be solved for as the unknown term and the other letters in the formula as the known values. The formula must be rearranged so that the unknown term is on one side of the equation and all other values are on the other side. A formula is rearranged by using the same procedure that is used for solving equations consisting of combined operations.

Problems are often solved more efficiently by first rearranging formulas than by directly substituting values in the original formula and solving for the unknown. This is particularly true in solving more complex formulas which involve many operations. Also, it is sometimes necessary to solve for the same unknown after a formula has been rearranged using different known values. Since the formula has been rearranged in terms of the specific unknown, solutions are more readily computed.

First rearranging formulas and then substituting known values enables you to solve for the unknown using a calculator for continuous operations. This is illustrated in Example 4.

Examples   Given the following formulas, rearrange and solve for the designated letter.

1. $A = bh$.   Solve for h.

$$A = bh$$

Apply the division principle. Divide both sides of the equation by b.

$$\frac{A}{b} = \frac{bh}{b}$$
$$\frac{A}{b} = h \quad \text{Ans}$$

2. In the figure shown $L = a + b$. Solve for a.

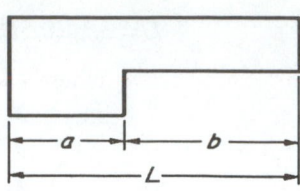

$$L = a + b$$

Apply subtraction principle.
Subtract b from both sides of
the equation.

$$\begin{array}{rcl} -b &=& -b \\ \hline L - b &=& a \quad \text{Ans} \end{array}$$

3. A screw thread is checked using a micrometer and 3 wires as shown. The measurement is checked using the following formula. Solve the formula for W.

THREAD CHECKING

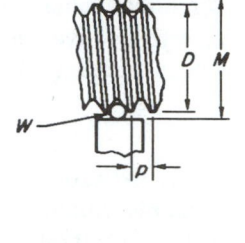

$M = D - 1.5155P + 3W$ where M = measurement over the wires
                                                   D = major diameter
                                                   P = pitch
                                                   W = wire size

Apply subtraction principle.
Subtract D from both sides
of the equation.

$$\begin{array}{rcl} M &=& D - 1.5155P + 3W \\ -D &=& -D \\ \hline M - D &=& -1.5155P + 3W \end{array}$$

Apply addition principle.
Add $1.5155P$ to both sides
of the equation.

$$\begin{array}{rcl} +1.5155P &=& +1.5155P \\ \hline M - D + 1.5155P &=& 3W \end{array}$$

Apply division principle.
Divide both sides of the
equation by 3.

$$\frac{M - D + 1.5155P}{3} = \frac{3W}{3}$$
$$\frac{M - D + 1.5155P}{3} = W \quad \text{Ans}$$

4. A slot is cut in the circular piece shown. The piece has a radius (R) of 97.60 millimeters. The number of millimeters in the width is represented by W. Dimension A is 20.20 millimeters. This formula is found in a machine trade handbook.

$$A = R - \sqrt{R^2 - 0.2500\ W^2}$$

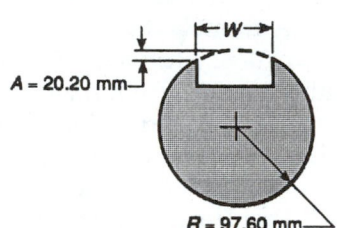

Solve for W.

Apply the subtraction principle.
Subtract R from both sides of
the equation.

$$\begin{array}{rcl} A &=& R - \sqrt{R^2 - 0.2500\ W^2} \\ -R &-& R \\ \hline A - R &=& -\sqrt{R^2 - 0.2500\ W^2} \end{array}$$

Apply the power principle.
Square both sides of the
equation.

$$(A - R)^2 = (-\sqrt{R^2 - 0.2500\ W^2})^2$$
$$(A - R)^2 = R^2 - 0.2500\ W^2$$

Apply the subtraction
principle. Subtract R^2 from
both sides of the equation.

$$\begin{array}{rcl} (A - R)^2 &=& R^2 - 0.2500\ W^2 \\ -R^2 && -R^2 \\ \hline \end{array}$$

Apply the division principle.
Divide both sides of the
equation by −0.2500.

$$\frac{(A - R)^2 - R^2}{-0.2500} = \frac{-0.2500\ W^2}{-0.2500}$$

Apply the root principle.
Take the square root of
both sides.

$$\sqrt{\frac{(A - R)^2 - R^2}{-0.2500}} = \sqrt{W^2}$$
$$\sqrt{\frac{(A - R)^2 - R^2}{-0.2500}} = W$$

Substitute the given numerical values for their respective letter values and find dimension W.

$$W = \sqrt{\frac{(20.20 - 97.60)^2 - 97.60^2}{-0.2500}}$$

$W = $ $\boxed{(}$ $\boxed{(}$ 20.2 $\boxed{-}$ 97.6 $\boxed{)}$ $\boxed{X^2}$ $\boxed{-}$ 97.6 $\boxed{X^2}$ $\boxed{=}$ $\boxed{+}$ .25 $\boxed{+/-}$ $\boxed{)}$ $\boxed{\sqrt{x}}$
$\rightarrow$ 118.911732

or $W = $ $\boxed{\sqrt{\ }}$ $\boxed{(}$ $\boxed{(}$ $\boxed{(}$ 20.2 $\boxed{-}$ 97.6 $\boxed{)}$ $\boxed{X^2}$ $\boxed{-}$ 97.6 $\boxed{X^2}$ $\boxed{)}$ $\boxed{+}$ $\boxed{(-)}$ .25 $\boxed{)}$
$\boxed{EXE}$ 118.911732   or .25 $\boxed{+/-}$ $\boxed{)}$ $\boxed{=}$ $\hookleftarrow$

$W = $ 118.9 mm   Ans (rounded)

Check.   Substitute numerical values in the *original* formula.

20.20 = 97.6 $\boxed{-}$ $\boxed{(}$ 97.6 $\boxed{X^2}$ $\boxed{-}$ .25 $\boxed{x}$ 118.911732 $\boxed{X^2}$ $\boxed{)}$ $\boxed{\sqrt{x}}$
$\boxed{=}$ 20.20000001

or 20.20 = 97.6 $\boxed{-}$ $\boxed{\sqrt{\ }}$ $\boxed{(}$ 97.6 $\boxed{X^2}$ $\boxed{-}$ .25 $\boxed{x}$ 118.911732 $\boxed{X^2}$ $\boxed{)}$ $\boxed{EXE}$
or $\boxed{=}$ 20.20000001

20.20 = 20.20

APPLICATION

Equations Consisting of Combined Operations

Solve for the unknown and check each of the following combined operations equations.

1. $5x - 33 = 12$ _____

2. $10M + 5 + 4M = 89$ _____

3. $8E - 14 = 2E + 28$ _____

4. $4B - 7 = B + 21$ _____

5. $7T - 14 = 0$ _____

6. $6N + 4 = 84 + N$ _____

7. $2.5A + 8 = 15 - 4.5$ _____

8. $12 - (-x + 8) = 18$ _____

9. $3H + (2 - H) = 20$ _____

10. $12 = -(2 + C) - (4 + 2C)$ _____

11. $-5(R + 6) = 10(R - 2)$ _____

12. $0.29E = 9.39 - 0.01E$ _____

13. $7.2F + 5(F - 8.1) = 0.6F + 15.18$ _____

14. $\frac{P}{7} + 8 = 6.3$ _____

15. $\frac{1}{4}W + (W - 8) = \frac{3}{4}$ _____

16. $\frac{1}{8}D - 3(D - 7) = 5\frac{1}{8}D - 3$ _____

17. $0.58y = 18.3 - 0.02y$ _____

18. $2H^2 - 20 = (H + 4)(H - 4)$ _____

19. $4A^2 + 3A + 36 = 8A^2 + 3A$ _____

20. $x(4 + x) + 20 = x^2 - (x - 5)$ _____

21. $\left(\frac{b}{2}\right)^3 + 34 = 42$ _____

22. $3F^3 + F(F + 8) = 8F + F^2 + 81$ _____

23. $9 + y^2 = (y - 4)(y - 1)$ _____

24. $\frac{1}{4}(2B - 12) + B^2 = \frac{1}{2}B + 22$ _____

25. $-4(y - 1.5) = 2\sqrt{y} - 4y$ _____

26. $14\sqrt{x} = 6(\sqrt{x} + 8) + 16$ _____

27. $8.12P^2 + 6.83P + 5.05 = 16.7P^2 + 6.83P$ _____

28. $7.3\sqrt{x} = 3(\sqrt{x} + 8.06) - 4.59$ _____

29. $\sqrt{B^2} - 2.53B = -2.53(B - 3.95)$ _____

30. $(2y)^3 - 2.80(5.89 + 3y) = -23.87 - 8.40y$ _____

Substituting Values and Solving Formulas

The following formulas are used in the machine trades. Substitute the given values in each formula and solve for the unknown. Round the answers to 3 decimal places where necessary.

31. $F = 2.380P + 0.250$
    Given: $F = 2.125$.
    Solve for P. _____

32. $a = 3H \div 8$
    Given: $a = 0.1760$.
    Solve for H. _____

33. $H.P. = 0.000016MN$
    Given: $H.P. = 22, N = 50.8$.
    Solve for M. _____

34. $N = 0.707DP_n$
    Given: $N = 24, P_n = 8$.
    Solve for D. _____

35. $S = T - \frac{1.732}{N}$
    Given: $S = 0.4134, N = 20$.
    Solve for T. _____

36. $a = \frac{D_2 - D_1}{2}$
    Given: $a = 0.250, D_1 = 0.875$.
    Solve for D_2. _____

37. $S = \frac{0.290W}{t^2}$
    Given: $S = 1000, t = 0.750$.
    Solve for W. _____

38. $W = St(0.55d^2 - 0.25d)$
    Given: $W = 1150, d = 0.750$.
    Solve for St. _____

39. $M = E - 0.866P + 3W$
    Given: $M = 3.3700, E = 3.2000, P = 0.125$.
    Solve for W. _____

40. $S = \frac{L_1}{L_2}\left[\frac{1}{2}(D_1 - D_2)\right]$
    Given: $S = \frac{1}{4}, L_1 = 16, L_2 = 4, D_2 = 2\frac{1}{2}$.
    Solve for D_1. _____

41. $C = \frac{\pi DN}{12}$
    Given: $C = 210, D = 6, \pi = 3.1416$.
    Solve for N. _____

42. $D_o = \frac{P_c(N + 2)}{\pi}$
    Given: $D_o = 4.3750, \pi = 3.1416, P_c = 0.3927$.
    Solve for N. _____

43. $S = \sqrt{\frac{d^2}{4} + h^2}$
    Given: $S = 12.700, d = 6$.
    Solve for h. _____

44. $C = 2\sqrt{h(2r - h)}$
    Given: $C = 7.600, h = 3.750$.
    Solve for r. _____

Rearranging Formulas

The following formulas are used in machine trade calculations. Rearrange the formulas in terms of the designated values.

45. The dimensions shown can be found using these two formulas.

    (1) $A = ab$ (2) $d = \sqrt{a^2 + b^2}$

    a. Solve formula (1) for a. _____
    b. Solve formula (1) for b. _____
    c. Solve formula (2) for a. _____
    d. Solve formula (2) for b. _____

46. The radii shown in this figure can be found using these two formulas.

    (1) $R = 1.155r$ (2) $A = 2.598R^2$

    a. Solve formula (1) for r. _____
    b. Solve formula (2) for R. _____

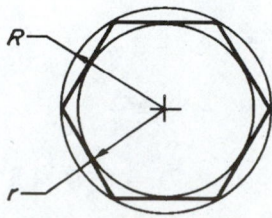

47. The dimensions shown can be found using these two formulas.

(1) $FW = \sqrt{D_o^2 - D^2}$ (2) $D_o = 2C - d + 2a$

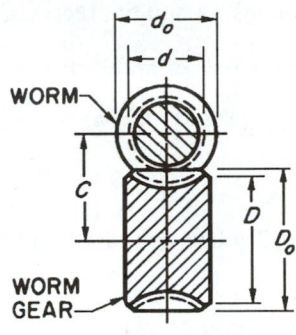

a. Solve formula (1) for D_o. _____

b. Solve formula (1) for D. _____

c. Solve formula (2) for d. _____

d. Solve formula (2) for a. _____

48. $A = \pi(R^2 - r^2)$

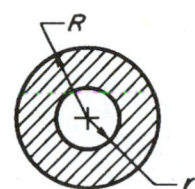

a. Solve for R. _____

b. Solve for r. _____

49. $M = D - 1.5155P + 3W$

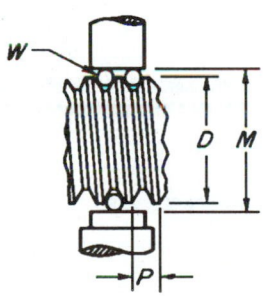

a. Solve for D. _____

b. Solve for P. _____

c. Solve for W. _____

50. $\angle A + \angle B + \angle C = 180°$

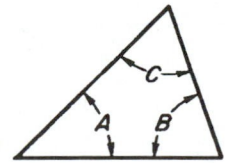

a. Solve for $\angle A$. _____

b. Solve for $\angle B$. _____

c. Solve for $\angle C$. _____

51. $L = 3.14(0.5D + 0.5d) + 2x$

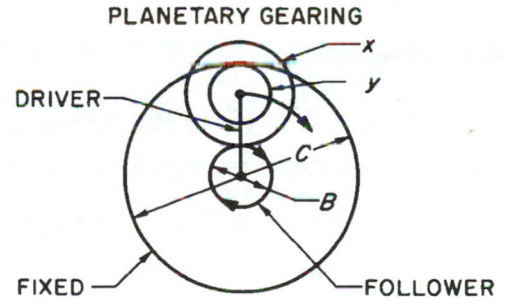

PULLEYS-OPEN BELT

a. Solve for D. _____

b. Solve for d. _____

c. Solve for x. _____

52. $By(F - 1) = Cx$

PLANETARY GEARING

a. Solve for x. _____

b. Solve for B. _____

c. Solve for C. _____

53. $Ca = S(C - F)$

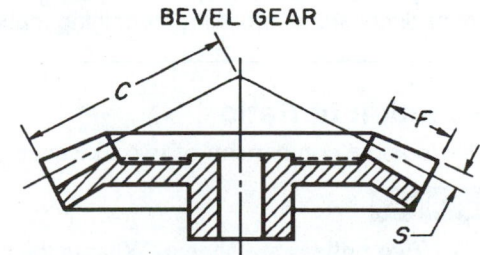

BEVEL GEAR

a. Solve for S. _____

b. Solve for C. _____

For problems 54 and 55, rearrange each formula for the designated letter and solve.

54. The horsepower of an electric motor is found with this formula.

$$hp = \frac{6.2832\ T\ (rpm)}{33,000}$$ where hp = horsepower

T = torque in pound feet (lb ft)

rpm = revolutions per minute

Solve for T when $hp = 1.50$ and rpm $= 2,250$. Round the answer to 2 decimal places.

_____

55. A tapered pin is shown.

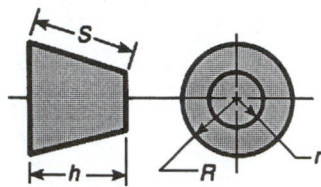

a. Solve for h when $R = 2.38$ cm, $r = 1.46$ cm, and $V = 69.5$ cm^3. Round the answer to 2 decimal places.

$$V = 1.05(R^2 + Rr + r^2)h$$

_____

b. Solve for h when $S = 0.875$ in, $R = 0.420$ in, and $r = 0.200$ in. Round the answer to 3 decimal places.

$$S = \sqrt{(R - r)^2 + h^2}$$

_____

UNIT **39** Ratio and Proportion

Objectives   After studying this unit you should be able to

- Write comparisons as ratios.
- Express ratios in lowest terms.
- Solve for the unknown term of a proportion.
- Substitute given numerical values for symbols in a proportion and solve for the unknown term.

The ability to solve practical machine shop problems using ratio and proportion is a requirement for the skilled machinist. Ratio and proportion are used for calculating gear and pulley speeds and sizes, for computing thread cutting values on a lathe, for computing taper dimensions, and for determining machine cutting times.

Description of Ratio

Ratio is the comparison of two like quantities.

Examples

1. Two pulleys are shown. What is the ratio of the diameter of the small pulley to the diameter of the larger pulley? All dimensions are in inches.

The ratio is 3 to 5.   Ans

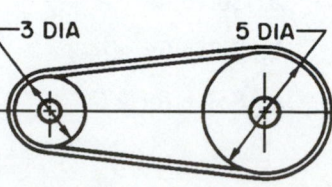

2. A triangle with given lengths of 3 meters, 4 meters, and 5 meters for sides a, b, and c is shown.

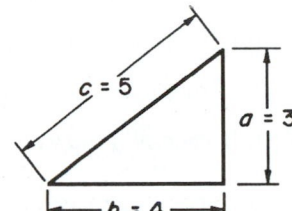

a. What is the ratio of side a to side b?

The ratio is 3 to 4.   Ans

b. What is the ratio of side a to side c?

The ratio is 3 to 5.   Ans

c. What is the ratio of side b to side c?

The ratio is 4 to 5.   Ans

The terms of a ratio are the two numbers that are compared. *Both terms of a ratio must be expressed in the same units.*

Example   Two pieces of bar stock are shown. What is the ratio of the short piece to the long piece?

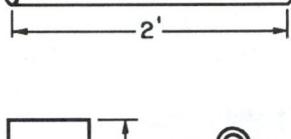

The terms cannot be compared to a ratio until the 2-foot length is expressed as 24 inches.

The ratio is 11 to 24.   Ans

It is impossible to express two quantities as ratios if the terms have unlike units that cannot be expressed as like units. Inches and pounds as shown cannot be compared as ratios.

Expressing Ratios.   Ratios are expressed in the following ways.

* With a colon between the two terms, such as 4:7. The ratio 4:7 is read as 4 to 7.

* With a division sign separating the two numbers, such as $4 \div 7$ or as a fraction, $\frac{4}{7}$.

Order of Terms

The terms of a ratio must be compared in the order in which they are given. The first term is the numerator of a fraction and the second is the denominator.

Examples

1. $1 \text{ to } 3 = 1 \div 3 = \frac{1}{3}$ Ans
     3. $x{:}y = x \div y = \frac{x}{y}$ Ans

2. $3 \text{ to } 1 = 3 \div 1 = \frac{3}{1}$ Ans
     4. $y{:}x = y \div x = \frac{y}{x}$ Ans

Expressing Ratios in Lowest Terms

Generally, a ratio should be expressed in lowest fractional terms.

Examples

1. $3{:}9 = \frac{3}{9} = \frac{1}{3}$ Ans

2. $40{:}15 = \frac{40}{15} = \frac{8}{3}$ Ans

3. $\frac{3}{8} : \frac{9}{16} = \frac{3}{8} \div \frac{9}{16} = \frac{3}{8} \times \frac{16}{9} = \frac{2}{3}$ Ans

4. $10{:}\frac{5}{6} = 10 \div \frac{5}{6} = \frac{10}{1} \times \frac{6}{5} = \frac{12}{1}$ Ans

5. $x^3{:}x^2 = \frac{x^3}{x^2} = \frac{x}{1}$ Ans

6. $4ab{:}6a = \frac{4ab}{6a} = \frac{2b}{3}$ Ans

Description of Proportions

A *proportion* is an expression that states the equality of two ratios.

Expressing Proportions. Proportions are expressed in the following two ways.

- 3:4::6:8, which is read as 3 is to 4 as 6 is to 8.

- $\frac{3}{4} = \frac{6}{8}$. This equation form is generally the way that proportions are used.

A proportion consists of four terms. The first and the fourth term are called *extremes* and the second and third terms are called *means*.

Examples

1. 2:3::4:6   2 and 6 are the extremes; 3 and 4 are the means.   Ans

2. $\frac{5}{6} = \frac{10}{12}$ 5 and 12 are the extremes; 6 and 10 are the means.   Ans

In a proportion the product of the means equals the product of the extremes. If the terms are cross multiplied, their products are equal.

Examples

1. $\frac{3}{4} = \frac{6}{8}$

  Cross multiply, $\frac{3}{4} \diagdown\!\!\!\!\!\diagup \frac{6}{8}$

$$3 \times 8 = 4 \times 6$$
$$24 = 24$$

2. $\frac{a}{b} = \frac{c}{d}$

  Cross multiply, $\frac{a}{b} \diagdown\!\!\!\!\!\diagup \frac{c}{d}$

$$a \times d = b \times c$$
$$ad = bc$$

The method of cross multiplying is used in solving proportions which have an unknown term. Since a proportion is an equation, the principles used for solving equations are applied in determining the value of the unknown after the terms have been cross multiplied.

Examples   Solve for the value of x.

1. $\frac{3}{4} = \frac{x}{16}$

  Cross multiply.

  Apply the division principle of equality. Divide both sides of the equation by 4.

  Check.

$\frac{3}{4} = \frac{x}{16}$

$4x = 3(16)$
$4x = 48$
$\frac{4x}{4} = \frac{48}{4}$
$x = 12$ Ans

$\frac{3}{4} = \frac{x}{16}$
$\frac{3}{4} = \frac{12}{16}$
$\frac{3}{4} = \frac{3}{4}$ Ck

2. $\frac{7}{x} = \frac{8}{15}$

$8x = 7(15)$
$8x = 105$
$\frac{8x}{8} = \frac{105}{8}$
$x = 13\frac{1}{8}$ Ans

Check. $\frac{7}{x} = \frac{8}{15}$

$\frac{7}{13\frac{1}{8}} = \frac{8}{15}$

$\frac{8}{15} = \frac{8}{15}$ Ck

3. $\dfrac{x}{7.5} = \dfrac{23.4}{20}$

$20x = 7.5(23.4)$

$20x = 175.5$

$\dfrac{20x}{20} = \dfrac{175.5}{20}$

$x = 8.775$ Ans

Solving by calculator: $x = 7.5\ \boxed{\times}\ 23.4\ \boxed{\div}\ 20$

$\boxed{=}\ 8.775$ Ans

Check. $\dfrac{x}{7.5} = \dfrac{23.4}{20}$

$\dfrac{8.775}{7.5} = \dfrac{23.4}{20}$

$1.17 = 1.17$ Ck

4. $\dfrac{a}{b} = \dfrac{c}{x}$

$ax = bc$

$\dfrac{ax}{a} = \dfrac{bc}{a}$

$x = \dfrac{bc}{a}$ Ans

Check. $\dfrac{a}{b} = \dfrac{c}{x}$

$\dfrac{a}{b} = \dfrac{c}{\frac{bc}{a}}$

$\dfrac{a}{b} = \dfrac{a}{b}$ Ck

APPLICATION

Ratios

Express the following ratios in lowest fractional form.

1. 6:21   _____

2. 21:6   _____

3. 2:11   _____

4. 7:21   _____

5. 12″:46″   _____

6. 3 lb:21 lb   _____

7. 13 mi:9 mi   _____

8. 156 mm:200 mm   _____

9. $3a^2b:9ab$ _____

10. $xy:x^2y$ _____

11. $\dfrac{2}{3} : \dfrac{1}{2}$ _____

12. $\dfrac{1}{2} : \dfrac{2}{3}$ _____

Related Ratio Problems

13. Length A in this figure is 3 inches and length B is 2.5 feet. Determine the ratio of length A to length B in lowest fractional form.   _____

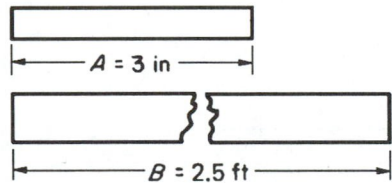

14. The diameters of pulleys E, F, G, and H are given in the table. Determine the ratios in lowest fractional form.   _____

	DIAMETERS (Inches)				RATIOS							
	E	F	G	H	$\frac{E}{F}$	$\frac{E}{G}$	$\frac{E}{H}$	$\frac{F}{G}$	$\frac{F}{H}$	$\frac{G}{H}$	$\frac{G}{E}$	$\frac{H}{F}$
a.	8	6	4	3								
b.	10	8	5	4								
c.	12	9	6	3								
d.	15	12	10	6								

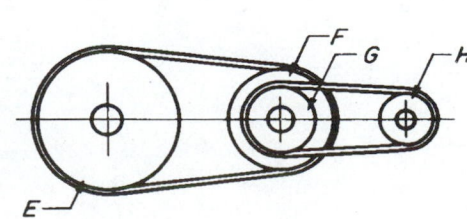

15. Refer to the hole locations given for the plate. Determine the ratios in lowest fractional form. All dimensions are in millimeters.

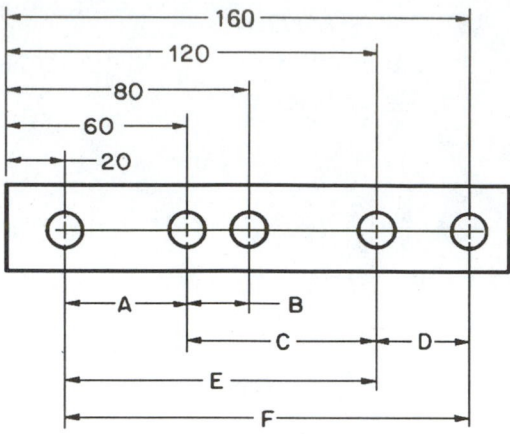

   a. Dimension A to dimension B.   _____

   b. Dimension A to dimension C.   _____

   c. Dimension C to dimension D.   _____

   d. Dimension C to dimension E.   _____

   e. Dimension D to dimension F.   _____

   f. Dimension F to dimension B.   _____

   g. Dimension F to dimension C.   _____

   h. Dimension E to dimension A.   _____

   i. Dimension D to dimension B.   _____

   j. Dimension C to dimension F.   _____

16. Gear A is turning at 120 revolutions per minute and gear B is turning at 18 revolutions per second. Determine the ratio of the speed of gear A to the speed of gear B. _____

GEAR A
120 r/min

GEAR B
18 r/s

Proportions

Solve for the unknown value in each of the following proportions. Check each answer. Round the answers to 3 decimal places where necessary.

17. $\dfrac{x}{2} = \dfrac{6}{24}$ _____

18. $\dfrac{3}{A} = \dfrac{15}{30}$ _____

19. $\dfrac{7}{9} = \dfrac{E}{45}$ _____

20. $\dfrac{3}{13} = \dfrac{24}{y}$ _____

21. $\dfrac{15}{c} = \dfrac{5}{4}$ _____

22. $\dfrac{P}{27} = \dfrac{1}{3}$ _____

23. $\dfrac{6}{7} = \dfrac{15}{F}$ _____

24. $\dfrac{12}{H} = \dfrac{4}{25}$ _____

25. $\dfrac{T}{6.6} = \dfrac{7.5}{22.0}$ _____

26. $\dfrac{2.4}{3} = \dfrac{M}{0.8}$ _____

27. $\dfrac{4}{4.1} = \dfrac{8}{L}$ _____

28. $\dfrac{3.4}{y} = \dfrac{1}{-9}$ _____

29. $\dfrac{A}{5} = \dfrac{3.2}{A}$ _____

30. $\dfrac{\frac{3}{8}}{N} = \dfrac{\frac{1}{2}}{4}$ _____

31. $\dfrac{3}{\frac{1}{4}} = \dfrac{5}{F}$ _____

32. $\dfrac{G}{\frac{1}{4}} = \dfrac{\frac{7}{8}}{\frac{3}{8}}$ _____

33. $\dfrac{7}{\frac{-1}{8}} = \dfrac{x}{\frac{9}{16}}$ _____

34. $\dfrac{4}{R} = \dfrac{2R}{12.5}$ _____

35. $\dfrac{11}{8} = \dfrac{E + 3}{12}$ _____

38. $\dfrac{E - 15}{E + 7.53} = \dfrac{0.36}{1.86}$ _____

36. $\dfrac{M - 5}{12} = \dfrac{15}{9}$ _____

39. $\dfrac{8.62M + 23.30}{12.36} = \dfrac{7.62M + 0.05}{0.86}$ _____

37. $\dfrac{6.08}{3H^2 - 12} = \dfrac{5.87}{12.53}$ _____

40. $\dfrac{P^2 - 186.73}{5.65P} = \dfrac{-23.30P}{3.04}$ _____

Related Proportion Problems

41. The proportion $\dfrac{A}{B} = \dfrac{C}{D}$ compares the sides of the two illustrated similar triangles. Determine the missing values in the table.

	A	B	C	D
a.	18″	4.5″		3″
b.	$6\frac{1}{2}$″	$1\frac{5}{8}$″	$4\frac{1}{2}$″	
c.	87.5 mm		75 mm	62.5 mm
d.		25.8 mm	20.6 mm	16.4 mm

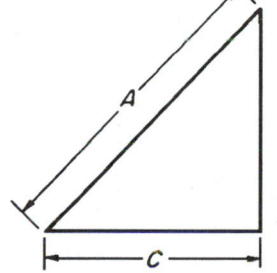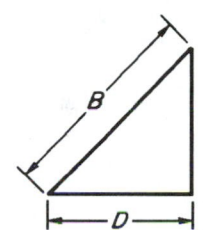

42. Where machine parts are doweled in position, it is good practice to extend the pin 1 to $1\frac{1}{2}$ times its diameter into the mating part. Use the following proportion to determine the value of each unknown in the table. Round the answers to 3 decimal places where necessary.

$$\dfrac{N}{1} + \dfrac{L}{D}$$

where
N = the number of times the pin extension is greater than the pin diameter
L = the length of the pin extension
D = the pin diameter

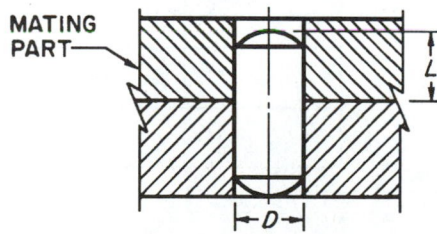

MATING PART

	N	D	L
a.	1.250	7.940 mm	
b.	$1\frac{1}{4}$	$\frac{1}{2}$″	
c.	$1\frac{1}{2}$		$\frac{3}{4}$″
d.	1.375		8.730 mm
e.	1.250	16.120 mm	

	N	D	L
f.	1.375		1.032″
g.	1.250	0.875″	
h.	1.500	3.680 mm	
i.	1.125		0.281″
j.	1.000	7.500 mm	

43. It is sometimes impractical to make engineering drawings full size. If the part to be drawn is very large or small, a scale drawing is generally made. The scale which is shown on the drawing compares the lengths of the lines on the drawing to the dimensions on the part. A scale on a drawing which states $\frac{1''}{4} = 1''$ means the drawing is one-quarter the size of the part. It is expressed as a ratio of 1:4 or $\frac{1}{4}$. A scale drawing which states $2'' = 1''$ means that the drawing is double the size of the part. It is expressed as a ratio of 2:1 or $\frac{2}{1}$. The actual dimensions of a steel support are given in the figure. All dimensions are in inches. Using the formula given, compute the lengths on a drawing for each unknown in the table. Round the answers to 3 decimal places where necessary.

$$\frac{\text{numerator of scale ratio}}{\text{denominator of scale ratio}} = \frac{\text{drawing length}}{\text{part dimension}}$$

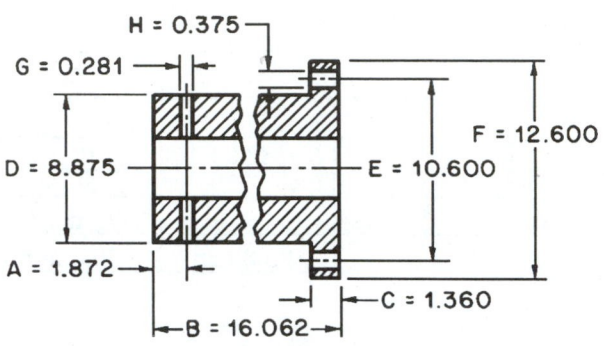

	Scale	Drawing Length
a.	$\frac{1''}{2} = 1''$	$B =$
b.	$4'' = 1''$	$G =$
c.	$\frac{1''}{4} = 1''$	$B =$
d.	$2'' = 1''$	$C =$
e.	$1\frac{1''}{2} = 1''$	$A =$
f.	$\frac{3''}{4} = 1''$	$E =$
g.	$3'' = 1''$	$H =$
h.	$\frac{1''}{8} = 1''$	$F =$

	Scale	Drawing Length
i.	$\frac{1''}{2} = 1''$	$E =$
j.	$6'' = 1''$	$G =$
k.	$\frac{3''}{4} = 1''$	$F =$
l.	$1\frac{1''}{2} = 1''$	$C =$
m.	$\frac{1''}{2} = 1''$	$F =$
n.	$3'' = 1''$	$G =$
o.	$\frac{1''}{4} = 1''$	$B =$
p.	$2'' = 1''$	$A =$

44. This figure shows the relationship of gears in a lathe using a simple gear train. The proportion given is used for lathe thread cutting computations using simple gearing. The fixed stud gear and the spindle gear have the same number of teeth. Determine the missing values for each of the following problems.

$$\frac{N_L}{N_C} = \frac{T_S}{T_L}$$

where $N_L =$ number of threads per inch on the lead screw

         $N_C =$ number of threads per inch to be cut

         $T_S =$ number of teeth on stud gear

         $T_L =$ number of teeth on lead screw gear

➤ **Note:**   Intermediate gears only change direction.

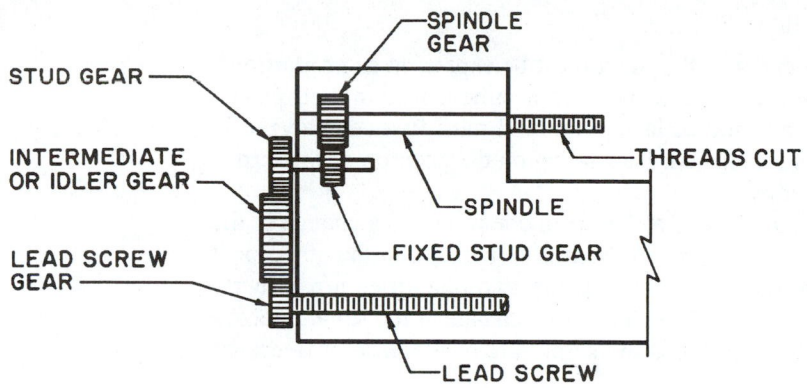

a. If $N_L = 4$, $N_C = 8$, and $T_S = 32$, find T_L. _____

b. If $N_L = 7$, $T_S = 35$, and $N_C = 15$, find T_L. _____

c. If $N_C = 10$, $N_L = 6$, and $T_L = 40$, find T_S. _____

d. If $N_L = 8$, $T_L = 42$, and $T_S = 28$, find N_C. _____

45. A template is shown on the left. A drafter makes an enlarged drawing of the template as shown on the right. The original length of 1.80 inches on the enlarged drawing is 3.06 inches as shown. Determine the lengths of A, B, C, and D.

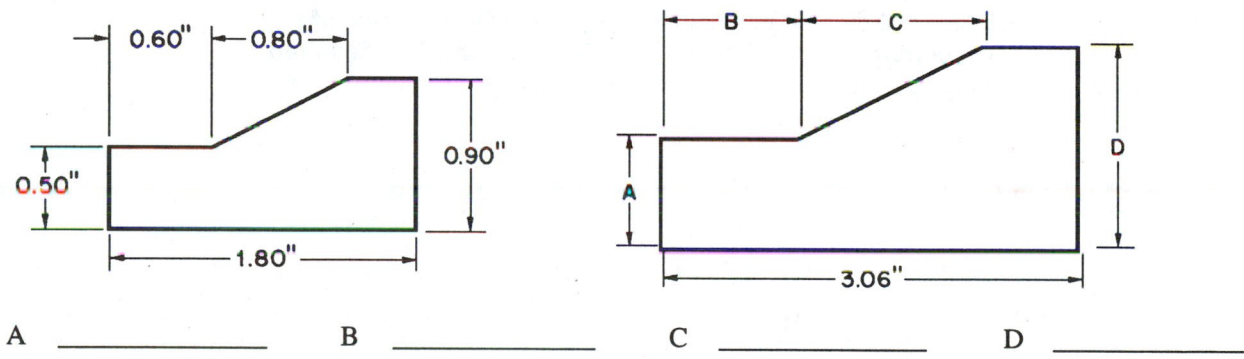

A _____ B _____ C _____ D _____

UNIT **40** Direct and Inverse Proportions

Objectives   After studying this unit you should be able to

- **Analyze problems to determine whether quantities are directly or inversely proportional.**

- **Set up and solve direct and inverse proportions.**

Many shop problems are solved by the use of proportions. A machinist may be required to express word statements or other given data as proportions. Generally, three of the four terms of a proportion must be known in order to solve the proportion. When setting up a proportion it is important that the terms be placed in their proper positions.

Direct Proportions

In actual practice, word statements or other data must be expressed as proportions. When a proportion is set up, the terms of the proportion must be placed in their proper positions. A problem that is set up and solved as a proportion must first be analyzed in order to determine where the terms are placed. Depending on the position of the terms, proportions are either direct or inverse.

Two quantities are *directly proportional* if a change in one produces a change in the other in the same direction. If an increase in one produces an increase in the other, or if a decrease in one produces a decrease in the other, the two quantities are directly proportional. The proportions discussed will be those that change at the same rate. An increase or decrease in one quantity produces the same rate of increase or decrease in the other quantity.

When setting up a direct proportion in fractional form, the numerator of the first ratio must correspond to the numerator of the second ratio. The denominator of the first ratio must correspond to the denominator of the second ratio.

Example 1   If 120 parts are produced in 2 hours, how many parts are produced in 3 hours?

Analyze the problem.   An increase in time (from 2 hours to 3 hours) will produce an increase in the number of pieces produced. Production increases as time increases. The proportion is direct.

Set up the direct proportion.   Let x represent the number of parts that are produced in 3 hours. The numerator of the first ratio must correspond to the numerator of the second ratio; 2 hours corresponds to 120 parts. The denominator of the first ratio must correspond to the denominator of the second ratio; 3 hours corresponds to x.

Solve for x.

$$\frac{2 \text{ hours}}{3 \text{ hours}} = \frac{120 \text{ parts}}{x}$$
$$2x = 3(120 \text{ parts})$$
$$2x = 360 \text{ parts}$$
$$x = 180 \text{ parts} \qquad \text{Ans}$$

Check.

$$\frac{2 \text{ hours}}{3 \text{ hours}} = \frac{120 \text{ parts}}{x}$$

$$\frac{2 \text{ hours}}{3 \text{ hours}} = \frac{120 \text{ parts}}{180 \text{ parts}}$$

$$\frac{2}{3} = \frac{2}{3} \qquad \text{Ck}$$

Example 2   A tapered shaft is one that varies uniformly in diameter along its length. The shaft shown is 15.000 inches long with a 1.200-inch diameter on the large end. A 9.000-inch piece is cut from the shaft. Determine the diameter at the large end of the 9.000-inch piece.

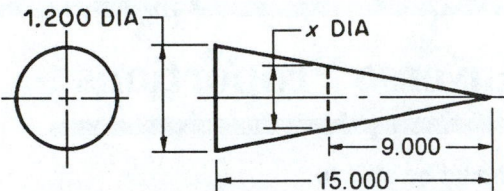

Analyze the problem.   As the length decreases from 15.000 inches to 9.000 inches, the diameter also decreases at the same rate. The proportion is direct.

Set up the proportion.   Let x represent the diameter at the large end of the 9.000-inch piece. The numerator of the first ratio must correspond to the numerator of the second ratio; the 15.000-inch piece has a 1.200-inch diameter at the large end. The denominator of the first ratio must correspond to the denominator of the second ratio; the 9.000-inch piece has a diameter of x at the large end.

Solve for x.

$$\frac{\overset{5}{\cancel{15.000 \text{ inches}}}}{\underset{3}{\cancel{9.000 \text{ inches}}}} = \frac{1.200\text{-inch DIA}}{x \text{ DIA}}$$

$$5x = 3(1.200 \text{ inches})$$
$$5x = 3.600 \text{ inches}$$
$$x = 0.720 \text{ inch} \qquad \text{Ans}$$

Check.

$$\frac{15.000 \text{ inches}}{9.000 \text{ inches}} = \frac{1.200\text{-inch DIA}}{x \text{ DIA}}$$

$$\frac{15.000 \text{ inches}}{9.000 \text{ inches}} = \frac{1.200\text{-inch DIA}}{0.720\text{-inch DIA}}$$

$$1.67 = 1.67 \qquad \text{Ck}$$

Inverse Proportions

Two quantities are *inversely or indirectly proportional* if a change in one produces a change in the other in the opposite direction. If an increase in one produces a decrease in the other, or if a decrease in one produces an increase in the other, the two quantities are inversely proportional. For example, if one quantity increases by 4 times its original value, the other quantity decreases by 4 times its value or is $\frac{1}{4}$ of its original value. Notice 4 or $\frac{4}{1}$ inverted is $\frac{1}{4}$.

When setting up an inverse proportion in fractional form, the numerator of the first ratio must correspond to the denominator of the second ratio. The denominator of the first ratio must correspond to the numerator of the second ratio.

Example 1   Two gears in mesh are shown. The driver gear has 40 teeth and revolves at 360 revolutions per minute. Determine the number of revolutions per minute of a driven gear with 16 teeth.

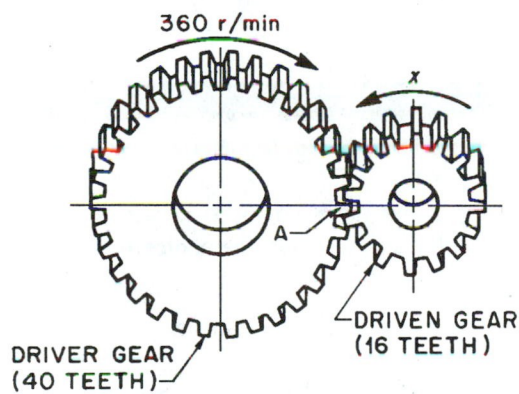

360 r/min

A

DRIVER GEAR
(40 TEETH)

DRIVEN GEAR
(16 TEETH)

Analyze the problem.   When the driver turns one revolution, 40 teeth pass point A. The same number of teeth on the driven gear must pass point A. Therefore, the driven gear turns more than one revolution for each revolution of the driver gear. The gear with 16 teeth (driven gear) revolves at greater revolutions per minute than the gear with 40 teeth (driver gear). A decrease in the number of teeth produces an increase in revolutions per minute. The proportion is inverse.

Set up the proportion.   Let x represent the revolutions per minute of the gear with 16 teeth. The numerator of the first ratio must correspond to the denominator of the second ratio; the gear with 40 teeth revolves at 360 r/min. The denominator of the first ratio must correspond to the numerator of the second ratio; the gear with 16 teeth revolves at x.

Solve for x.

$$\frac{\overset{5}{\cancel{40 \text{ teeth}}}}{\underset{2}{\cancel{16 \text{ teeth}}}} = \frac{x}{360 \text{ r/min}}$$

$$2x = 1800 \text{ r/min}$$
$$x = 900 \text{ r/min} \qquad \text{Ans}$$

Check.

$$\frac{40 \text{ teeth}}{16 \text{ teeth}} = \frac{x}{360 \text{ r/min}}$$

$$\frac{40 \text{ teeth}}{16 \text{ teeth}} = \frac{900 \text{ r/min}}{360 \text{ r/min}}$$

$$2.5 = 2.5 \qquad \text{Ck}$$

Example 2 Five identical machines produce the same parts at the same rate. The 5 machines complete the required number of parts in 1.8 hours. How many hours does it take 3 machines to produce the same number of parts?

Analyze the problem. A decrease in the number of machines (from 5 to 3) requires an increase in time. Time increases as the number of machines decreases; therefore, the proportion is inverse.

Set up the proportion. Let x represent the time required by 3 machines to produce the parts.

$$\frac{5 \text{ machines}}{3 \text{ machines}} = \frac{x}{1.8 \text{ hours}}$$

Notice that the numerator of the first ratio corresponds to the denominator of the second ratio; 5 machines corresponds to 1.8 hours. The denominator of the first ratio corresponds to the numerator of the second ratio; 3 machines correspond to x.

Solve for x.

$$\frac{5}{3} = \frac{x}{1.8 \text{ hours}}$$
$$3x = 5(1.8 \text{ hours})$$
$$\frac{3x}{3} = \frac{9 \text{ hours}}{3}$$
$$x = 3 \text{ hours} \qquad \text{Ans}$$

Check.

$$\frac{5}{3} = \frac{x}{1.8 \text{ hours}}$$
$$\frac{5}{3} = \frac{3 \text{ hours}}{1.8 \text{ hours}}$$
$$1.\overline{6} = 1.\overline{6} \qquad \text{Ck}$$

APPLICATION

Tapers

Taper is the difference between the diameters at each end of a part. Tapers are expressed as the difference in diameters for a particular length along the centerline of a part.

Note: All dimensions are in millimeters.   *Note:* All dimensions are in inches.

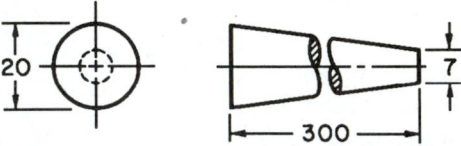   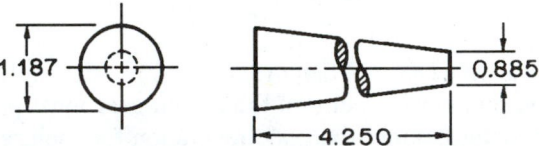

20 mm − 7 mm = 13 mm taper per 300 mm   1.187″ − 0.885″ = 0.302″ mm taper per 4.250″

1. A plug gage tapers 3.10 mm along a 38.00 mm length. Set up a proportion and determine the amount of taper in the workpiece for each of the following problems. Express the answers to 2 decimal places.

	WORKPIECE THICKNESS	PROPORTION	TAPER IN WORKPIECE
a.	18.40 mm		
b.	31.75 mm		
c.	14.28 mm		
d.	28.58 mm		
e.	28.60 mm		

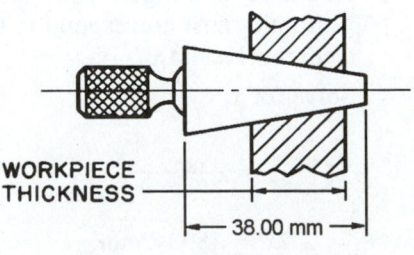

WORKPIECE THICKNESS

38.00 mm

2. A reamer tapers 0.130″ along a 4.250″ length. Set up a proportion and determine length A for each of the following problems. Express the answers to 3 decimal places.

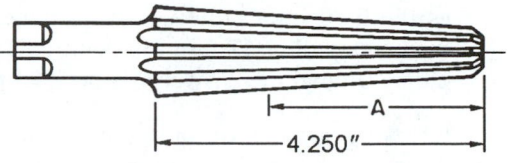

	TAPER IN LENGTH A	PROPORTION	LENGTH A
a.	0.030″		
b.	0.108″		
c.	0.068″		
d.	0.008″		
e.	0.093″		

3. A micrometer reading is made at dimension D on a tapered shaft. For each of the problems use the dimensions given in the table, compute the taper, set up a proportion, and determine diameter C to 3 decimal places.

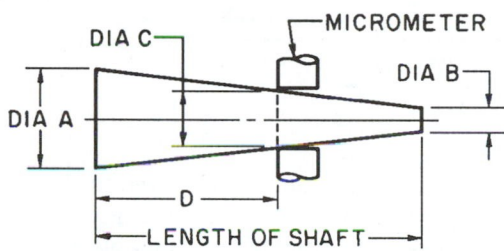

	LENGTH OF SHAFT	DIAMETER A	DIAMETER B	DIMENSION D	DIAMETER C
a.	10.200″	1.500″	0.700″	6.500″	
b.	8.750″	1.250″	0.375″	4.875″	
c.	550.000 mm	106.250 mm	62.500 mm	337.500 mm	
d.	147.500 mm	22.500 mm	10.000 mm	112.500 mm	
e.	8.800″	1.325″	0.410″	8.620″	

Proportions

Analyze each of the following problems to determine whether the problem is a direct or inverse proportion. Set up the proportion and solve.

4. A sheet of steel $8\frac{1}{4}$ feet long weights 325 pounds. A piece $2\frac{1}{2}$ feet long is sheared from the sheet. Determine the weight of the $2\frac{1}{2}$-foot piece to the nearest whole pound.

5. If 1350 parts are produced in 6.75 hours, find the number of parts produced in 8.25 hours.

6. The production rate for each of 3 machines is the same. Using these 3 machines, 720 parts are produced in 1.6 hours. How many hours will it take 2 of these machines to produce 720 parts?

7. Two forgings are made of the same stainless steel alloy. A forging which weighs 76.00 kilograms contains 0.38 kilogram of chromium. How many kilograms of chromium does the second forging contain if it weighs 96.00 kilograms? Round the answer to 2 decimal places.

Gears and Pulleys

8. A belt connects a 10.00-inch diameter pulley which rotates at 160.0 rpm with a 6.50-inch diameter pulley. An 8.00-inch diameter pulley is fixed to the same shaft as the 6.50-inch pulley. A belt connects the 8.00-inch pulley with a 3.50-inch diameter pulley. Determine the revolutions per minute of the 3.50-inch diameter pulley. Round the answer to 1 decimal place. _____

9. Of two gears that mesh, the one which has the greater number of teeth is called the gear, and the one which has the fewer teeth is called the pinion. For each of the problems, set up a proportion, and determine the unknown value, x. Round the answers to 1 decimal place where necessary.

	NUMBER OF TEETH ON GEAR	NUMBER OF TEETH ON PINION	SPEED OF GEAR (rpm)	SPEED OF PINION (rpm)
a.	48	20	100.0	$x =$
b.	32	24	$x =$	210.0
c.	35	$x =$	160.0	200.0
d.	$x =$	15	150.0	250.0
e.	54	26	80.0	$x =$

10. The figure shows a compound gear train. Gears B and C are keyed to the same shaft; therefore, they turn at the same speed. Gear A and gear C are driving gears. Gear B and gear D are driven gears. Set up a proportion for each problem and determine the unknown values, x, y, and z in the table. Round the answers to 1 decimal place where necessary.

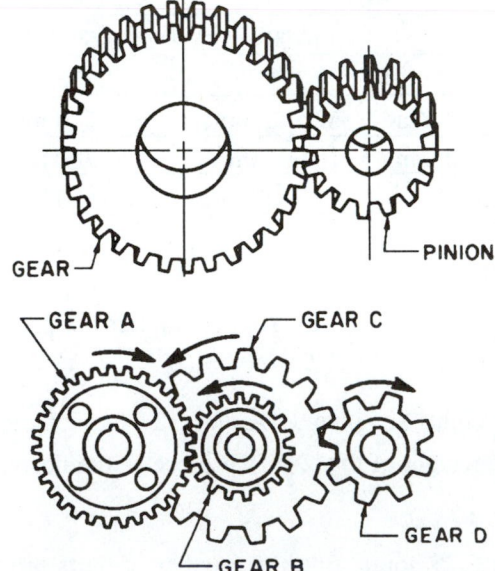

	NUMBER OF TEETH				SPEED (rpm)			
	Gear A	Gear B	Gear C	Gear D	Gear A	Gear B	Gear C	Gear D
a.	80	30	50	20	120.0	$x =$	$y =$	$z =$
b.	60	$x =$	45	$y =$	100.0	300.0	$z =$	450.0
c.	$x =$	24	60	36	144.0	$y =$	$z =$	280.0
d.	55	25	$x =$	15	$y =$	$z =$	175.0	350.0

UNIT **41** Applications of Formulas to Cutting Speed, Revolutions Per Minute, and Cutting Time

Objectives   After studying this unit you should be able to

- Solve cutting speed, revolutions per minute, and cutting time problems by substitution in given formulas.

- Solve production time and cutting feed problems by rearranging and combining formulas.

In order to perform cutting operations efficiently, a machine must be run at the proper cutting speed. Proper cutting speed is largely determined by the type of material that is being cut, the feed and depth of cut, the cutting tool, and the machine characteristics. The machinist must be able to determine proper cutting speeds by using trade handbook data and formulas.

Cutting Speed Using English Units of Measure

Cutting speeds or surface speeds for lathes, drills, milling cutters, and grinding wheels are computed using the same formula. On the lathe, the workpiece revolves. On drill presses, milling machines, and grinders the tool revolves. Speeds are computed in reference to the tool rather than the workpiece. The speed of a revolving object equals the product of the circumference times the number of revolutions per minute made by the object. Generally, diameters are expressed in inches. In order to express inches per minute as feet per minute, it is necessary to divide by 12.

➤ **Note:** $\pi = 3.1416$ rounded to 4 decimal places. When solving problems with a calculator use the $\boxed{\pi}$ key.

$$C = \frac{3.1416\,DN}{12} \qquad \text{where} \qquad C = \text{cutting speed in feet per minute (fpm)}$$
$$\text{or } C = \frac{\pi DN}{12} \qquad\qquad\qquad D = \text{diameter in inches}$$
$$N = \text{revolutions per minute (rpm)}$$

Lathe

The *cutting speed* of a lathe is the number of feet that the revolving workpiece travels past the cutting edge of the tool in one minute.

Example   A steel shaft 2.500 inches in diameter is turned in a lathe at 184.0 rpm. Determine the cutting speed to 1 decimal place.

$$C = \frac{\pi DN}{12} = \frac{\pi(2.500)(184.0)}{12}$$

 $\boxed{\pi}$ $\boxed{\times}$ 2.5 $\boxed{\times}$ 184 $\boxed{\div}$ 12 $\boxed{=}$ 120.428, 120.4 fpm   Ans

Milling Machine, Drill Press, and Grinder

The *cutting speed* or surface speed of a drill press, milling machine, and grinder is the number of feet that a point on the circumference of the tool travels in 1 minute.

Example 1 A 10-inch diameter grinding wheel runs at 1910 rpm. Determine the surface speed to the nearer whole number.

$$C = \frac{3.1416DN}{12} = \frac{3.1416(10)(1910)}{12} = 5000.37, 5000 \text{ fpm} \qquad \text{Ans}$$

Example 2 Determine the cutting speed to the nearer whole number of a $3\frac{1}{2}$-inch diameter milling cutter revolving at 120 rpm.

$$C = \frac{\pi DN}{12} = \frac{\pi(3.5)(120)}{12}$$

$\boxed{\pi} \; \boxed{x} \; 3.5 \; \boxed{x} \; 120 \; \boxed{+} \; 12 \; \boxed{=} \; 109.9557, 110 \text{ fpm} \qquad \text{Ans}$

Revolutions Per Minute Using English Units of Measure

The cutting speed formula is rearranged in terms of N in order to determine the revolutions per minute of a workpiece or tool.

$$C = \frac{3.1416DN}{12}$$
$$12C = 3.1416DN$$
$$\frac{12C}{3.1416D} = N \text{ or } N = \frac{12C}{3.1416D}$$

Lathe

Example An aluminum cylinder with a 6.000-inch outside diameter is turned in a lathe at a cutting speed of 225 feet per minute. Determine the revolutions per minute to the nearest whole revolution.

$$N = \frac{12C}{3.1416D} = \frac{12(225)}{3.1416(6)} = 143.239, 143 \text{ rpm} \qquad \text{Ans}$$

Milling Machine, Drill Press, and Grinder

Example 1 A $\frac{1}{2}$-inch diameter twist drill has a cutting speed of 60.0 feet per minute. Determine the revolutions per minute to the nearest whole revolution.

$$N = \frac{12C}{3.1416D} = \frac{12(60)}{3.1416(0.5)} = 458.366, 458 \text{ rpm} \qquad \text{Ans}$$

Example 2 A 6.00-inch diameter grinding wheel operates at a cutting speed of 6000 feet per minute. Determine the revolutions per minute to the nearest whole revolution.

$$N = \frac{12C}{\pi D} = \frac{12(6000)}{\pi(6.00)}$$

$12 \; \boxed{x} \; 6000 \; \boxed{+} \; \boxed{(} \; \boxed{\pi} \; \boxed{x} \; 6 \; \boxed{)} \; \boxed{=} \; 3819.7186, 3820 \text{ rpm} \qquad \text{Ans}$

Cutting Time Using English Units of Measure

The same formula is used to compute cutting times for machines which have a revolving workpiece, such as the lathe, as is used for machines which have a revolving tool, such as the milling machine, drill press and grinder. Cutting time is determined by the length or depth to be cut in inches, the revolutions per minute of the revolving workpiece or revolving tool, and the tool feed in inches for each revolution of the workpiece or tool.

$$T = \frac{L}{FN}$$ where T = cutting time per cut in minutes

L = length of cut in inches
F = tool feed in inches per revolution
N = speed of revolving workpiece or tool in revolutions per minute

Lathe

Example 1   How many minutes are required to take one cut 22.00 inches in length on a steel shaft when the lathe feed is 0.050 inch per revolution and the shaft turns 152 rpm? Round the answer to 1 decimal place.

$$T = \frac{L}{FN} = \frac{22.00}{0.050(152)} = 2.895, 2.9 \text{ min} \quad \text{Ans}$$

Example 2   A 3.250-inch diameter cast iron sleeve which is 20.00 inches long is turned in a lathe to a 2.450-inch diameter. Roughing cuts are each made to a 0.125-inch depth of cut. One finish cut using a 0.025-inch depth of cut is made. The feed is 0.100 inch per revolution for roughing and 0.030 inch for finishing. Roughing cuts are made at 150 rpm and the finish cut at 200 rpm. What is the total cutting time required?

Compute the total depth of cut.

$$\frac{3.250'' - 2.450''}{2} = \frac{0.800''}{2} = 0.400''$$

Compute the number of roughing cuts required.

$$\frac{0.400'' - 0.025''}{0.125''} = 3$$

Compute the time required for one roughing cut.

$$T = \frac{20.00}{(0.100)(150)} = 1.33 \text{ min}$$

Compute the total time for roughing.

$$3 \times 1.33 \text{ min} = 4.0 \text{ min}$$

Compute the time required for finishing.

$$T = \frac{20.00}{(0.030)(200)} = 3.3 \text{ min}$$

Compute the total cutting time.

$$4.0 \text{ min} + 3.3 \text{ min} = 7.3 \text{ min} \quad \text{Ans}$$

Milling Machine, Drill Press, and Grinder

Example 1   Determine the cutting time required to drill through a workpiece which is 3.600 inches thick with a drill revolving 300 rpm and a feed of 0.025 inch per revolution.

$$T = \frac{L}{FN} = \frac{3.600}{0.025(300)} = 0.48 \text{ min} \quad \text{Ans}$$

Example 2   A milling machine cutter makes 460 rpm with a table feed of 0.020 inch per revolution. Four cuts are required to mill a slot in an aluminum plate 28.68 inches long. Compute the total cutting time. Round the answer to 1 decimal place.

$$\text{Total Cutting Time} = \frac{L}{FN} \times 4 = \frac{28.68}{0.020(460)} \times 4$$

28.68 $\boxed{\div}$ $\boxed{(}$ .02 $\boxed{\times}$ 460 $\boxed{)}$ $\boxed{\times}$ 4 $\boxed{=}$ 12.46956, 12.5 min   Ans

Cutting Speed Using Metric Units of Measure

Diameters are expressed in millimeters. Cutting speeds are expressed in meters per minute. The symbol for meters per minute is m/min. In order to express speed in millimeters per minute as meters per minute, it is necessary to divide by 1000 or to move the decimal point 3 places to the left.

$$C = \frac{3.1416\,DN}{1000}$$ where C = cutting speed in meters per minute
D = diameter in millimeters
N = revolutions per minute

Example   A medium-steel shaft is cut in a lathe using a high-speed tool. The shaft has a diameter of 55 millimeters and is turning at 260 revolutions per minute. Determine the cutting speed to the nearest whole number.

$$C = \frac{3.1416\,DN}{1000} = \frac{3.1416\,(55)(260)}{1000} = 45 \text{ m/min}   \text{Ans}$$

Revolutions Per Minute Using Metric Units of Measure

In the metric system, the symbol for revolutions per minute is r/min. The cutting speed formula is rearranged in terms of N in order to determine the revolutions per minute of a workpiece or tool.

$$C = \frac{3.1416\,DN}{1000}$$

$$1000\,C = 3.1416\,DN$$

$$\frac{1000\,C}{3.1416\,D} = N \text{ or } N = \frac{1000\,C}{3.1416\,D}$$

Example   A high-speed steel milling cutter with a 45 millimeter diameter and a cutting speed of 12 meters per minute is used for a roughing operation on an annealed chromium-nickel steel workpiece. Determine the revolutions per minute to the nearest whole number.

$$N = \frac{1000\,C}{3.1416\,D} = \frac{1000\,(12)}{3.1416\,(45)}$$

$1000 \;\boxed{\times}\; 12 \;\boxed{+}\; \boxed{(}\; \boxed{\pi} \;\boxed{\times}\; 45 \;\boxed{)}\; \boxed{=}\; 84.88264$, 85 r/min   Ans

Cutting Time Using Metric Units of Measure

Cutting time is determined by the length or depth to be cut in millimeters, the revolutions per minute of the revolving workpiece or revolving tool, and the tool feed in millimeters for each revolution of the workpiece or tool.

$$T = \frac{L}{FN}$$ where T = cutting time per cut in minutes
L = length of cut in millimeters
F = tool feed in millimeters per revolution
N = r/min of revolving workpiece or tool

Example   An 88-millimeter diameter cast iron cylinder is turned in a lathe at 260 revolutions per minute. Each length of cut is 700 millimeters and 5 cuts are required. A carbide tool is fed into the workpiece at 0.40 millimeter per revolution. What is the total cutting time? Round the answer to the nearest minute.

Calculate the time required for one cut.

$$T = \frac{L}{FN} = \frac{700}{0.40\,(260)} = 6.73 \text{ min}$$

Calculate the total cutting time.

$$5(6.73 \text{ min}) = 33.65 \text{ min, } 34 \text{ min}   \text{Ans}$$

Using Data from a Cutting Speed Table

Tables of cutting speeds have been developed which are used in determining machine spindle speed (revolutions per minute) settings. The tables take into consideration the material to be cut and the tool material.

In addition to the material being cut and the type of tool used, other factors must be taken into consideration. Variables are considered, such as the depth and width of cut, the design of the cutting tool, the rate of feed, the coolant used, and the finish required.

Because cutting speed depends upon many factors, data given in cutting speed tables should be considered as recommended values. Generally it is not possible to set machines to an exact calculated spindle speed. Therefore, a simplified spindle speed formula is used in computing revolutions per minute. In the simplified formula, 3.1416 is rounded to 3.

$$N = \frac{12C}{3D} = \frac{4C}{D}$$

Comprehensive detailed cutting speed tables are available which list cutting speeds for specific materials based on material alloy composition, hardness, and condition. Some tables list cutting speeds separately for rough and finish cuts.

Selected materials are listed in the following table with their respective cutting speeds using high-speed steel and carbide tools.

| | CUTTING SPEEDS: FEET PER MINUTE (fpm) | | | | | | |
| | Turning | | Milling | | Drilling | Reaming | |
Material	High-Speed Steel Tool	Carbide Tool	High-Speed Steel Tool	Carbide Tool	High-Speed Steel Tool	High-Speed Steel Tool	Carbide Tool
Carbon Steel (1020), BHN 175–225	100	350	70–130	200–400	70	40	175
Alloy Steel (4320), BHN 220–275	70	300	50–100	225–450	60	40	150
Malleable Cast Iron (32510), BHN 110–160	200	600	130–225	400–800	130	90	240
Stainless Steel (305), BHN 225–275	60	200	50–80	175–275	40	25	100
Aluminum (5052)	600	1200	500–800	1000–1800	250	250	700
Brass, annealed	300	650	250–450	500–900	160	160	320
Manganese Bronze, cold drawn	250	550	200–350	450–650	140	120	275
Beryllium Copper, annealed	100	200	80–140	180–275	60	50	180

Revolutions per minute are generally computed using table cutting speeds with the simplified revolutions per minute formula. Where a range of cutting speed table values is listed, use the average of the low and high speeds given. For example, the cutting speed for milling the alloy steel shown in the table with a high-speed steel cutter is listed as 50–100 feet per minute. Use the average cutting speed of 75 feet per minute $\left(\frac{50 + 100}{2} = 75\right)$.

After revolutions per minute are calculated, generally, the machine spindle speed is set to the closest spindle speed below the calculated revolutions per minute. The spindle speed may then be increased or decreased depending on the performance of the operation.

Example 1 Calculate the revolutions per minute required to turn a 3.500-inch diameter piece of stainless steel using a carbide toolbit. Express the answer to the nearest revolution per minute.

Refer to the table of cutting speeds. The recommended cutting speed is 200 feet per minute.

$$N = \frac{4C}{D} = \frac{4(200)}{3.500} = 228.571, \; 229 \text{ rpm} \qquad \text{Ans}$$

Example 2 A carbon steel plate is milled using a 2.75-inch diameter high-speed steel cutter. Compute, to the nearest whole number, the revolutions per minute.

Refer to the table of cutting speeds. The recommended cutting speed is 100 feet per minute.

$$\left(\frac{70 + 130}{2} = 100\right) \quad N = \frac{4C}{D} = \frac{4(100)}{2.75} = 145.455, \; 145 \text{ rpm} \qquad \text{Ans}$$

APPLICATION

Cutting Speeds

Given the workpiece or tool diameters and the revolutions per minute, determine the cutting speeds in the following tables to the nearest whole number. Use $C = \frac{3.1416DN}{12}$ or $\frac{\pi DN}{12}$ for English units and $C = \frac{3.1416DN}{1000}$ or $\frac{\pi DN}{1000}$ for metric units.

	Workpiece or Tool Diameter	Revolutions per Minute	Cutting Speed (fpm)
1.	0.475"	460	
2.	2.750"	50	
3.	4.000"	86	
4.	0.850"	175	
5.	1.750"	218	

	Workpiece or Tool Diameter	Revolutions per Minute	Cutting Speeds (m/min)
6.	190.00 mm	59	
7.	53.98 mm	764	
8.	3.25 mm	1525	
9.	133.35 mm	254	
10.	6.35 mm	4584	

Revolutions Per Minute

Given the cutting speed and the tool or workpiece diameter, determine the revolutions per minute in the following tables to the nearest whole number. Use $N = \frac{12C}{3.1416D}$ or $\frac{12C}{\pi D}$ for English units and $N = \frac{1000C}{3.1416D}$ or $\frac{1000C}{\pi D}$ for metric units.

	Cutting Speed (fpm)	Workpiece or Tool Diameter	Revolutions per Minute
11.	70	2.400"	
12.	120	0.750"	
13.	90	8.000"	
14.	180	8.000"	
15.	200	0.375"	

	Cutting Speed (m/min)	Workpiece or Tool Diameter	Revolutions per Minute
16.	130	25.50 mm	
17.	100	66.70 mm	
18.	30	6.35 mm	
19.	180	15.80 mm	
20.	150	114.30 mm	

Cutting Time

Given the number of cuts, the length of cut, the revolutions per minute of the workpiece or tool, and the tool feed, determine the total cutting time in the table to 1 decimal place. Use $T = \frac{L}{FN}$.

	Number of Cuts	Feed (per revolution)	Length of Cut	Revolutions per Minute	Tool Cutting Time (Minutes)
21.	1	0.002″	20″	2100	
22.	1	0.12 mm	925 mm	610	
23.	4	0.008″	8″	350	

Cutting Speed and Surface Speed Problems

Compute the following problems. Express the answers to the nearer whole number. Use $C = \frac{3.1416DN}{12}$ or $\frac{\pi DN}{12}$ for English units and $C = \frac{3.1416DN}{1000}$ or $\frac{\pi DN}{1000}$ for metric units.

24. A $3\frac{1}{2}$-inch diameter high-speed steel cutter, running at 55 rpm, is used to rough mill a steel casting. What is the cutting speed?   _____

25. A 50-millimeter diameter carbon steel drill running at 286 r/min is used to drill an aluminum plate. Find the cutting speed.   _____

26. What is the surface speed of a 16-inch diameter surface grinder wheel running at 1194 rpm?   _____

27. A medium-steel shaft is cut in a lathe using a high-speed steel tool. Determine the cutting speed if the shaft is 2.125 inches in diameter, and is turning at 275 rpm.   _____

28. A finishing cut is taken on a brass workpiece using a 100-millimeter diameter carbon steel milling cutter. What is the cutting speed when the cutter is run at 86 r/min?   _____

Revolutions Per Minute Problems

Compute the following problems. Express the answers to the nearest whole number. Use $N = \frac{12C}{3.1416D}$ or $\frac{12C}{\pi D}$ for English units and $N = \frac{1000C}{3.1416D}$ or $\frac{1000C}{\pi D}$ for metric units.

29. Grooves are cut in a stainless steel plate using a 3.750-inch diameter carbide milling cutter with a cutting speed of 180 feet per minute. Determine the revolutions per minute.   _____

30. An annealed cast iron housing is drilled with a cutting speed of 20 meters per minute using a 22-millimeter diameter carbon steel drill. Find the revolutions per minute.   _____

31. A grinding operation is performed using a 150-millimeter diameter wheel with a cutting speed of 1800 meters per minute. Determine the revolutions per minute.   _____

32. Determine the revolutions per minute of an aluminum alloy rod 1.250 inches in diameter with a cutting speed of 550 feet per minute.   _____

33. A high-speed steel milling cutter with a 1.750-inch diameter and a cutting speed of 40 feet per minute is used for a roughing operation on an annealed chromium-nickel steel workpiece. Find the revolutions per minute.   _____

Cutting Time Problems

Compute the following problems. Express the answers to 1 decimal place. Use

$$T = \frac{L}{FN}.$$

34. Cast iron, $3\frac{1}{4}$ inches in diameter, is turned in a lathe at 270 rpm. Each length of cut is 27.00 inches and five cuts are required. A carbide tool is fed into the work at 0.015 inch per revolution. What is the total cutting time?   _____

35. A slot 812.00 millimeters long is cut into a carbon steel baseplate with a feed of 0.80 millimeter per revolution. Find the cutting time using a 75-millimeter diameter carbide milling cutter running at 640 r/min.   _____

36. Fifteen 3.20 millimeter diameter holes each 57.15 millimeters deep are drilled in an aluminum workpiece. The high-speed steel drill runs at 9200 r/min with a feed of 0.05 millimeter per revolution. Determine the total cutting time.   _____

37. Thirty 2-inch diameter stainless steel shafts are turned in a lathe at 250 rpm. Two cuts each 14.5 inches long are required using a feed of 0.020 inch per revolution. Setup and handling time averages 3 minutes per piece. Calculate the total production time.   _____

38. Seven brass pates 9.00 inches wide and 21.00 inches long are machined with a milling cutter along the length of the plates. The entire top face of each plate is milled. The width of each cut allowing for overlap is $2\frac{1}{4}$ inches. Using a feed of 0.020 inch per revolution and 525 rpm, determine the total cutting time.   _____

Complex Problems

The solution of the following problems requires more than one formula and the rearrangement of formulas.

Use $C = \frac{3.1416\,DN}{12}$ or $\frac{\pi DN}{12}$ for English units and

$C = \frac{3.1416\,DN}{1000}$ or $\frac{\pi DN}{1000}$ for metric units.

$N = \frac{12C}{3.1416\,D}$ or $\frac{12C}{\pi D}$ for English units and

$N = \frac{1000\,C}{3.1416\,D}$ or $\frac{1000\,C}{\pi D}$ for metric units.

$T = \frac{L}{FN}$

39. A 3.000-inch diameter cylinder is turned for an 11.300-inch length of cut. The cutting speed is 300 feet per minute and the cutting time is 1.02 minutes. Calculate the tool feed in inches per revolution. Round the answer to 3 decimal places.   _____

40. A combination drilling and countersinking operation on bronze round stock is performed on an automatic screw machine. The length of cut per piece is $1\frac{3}{4}$ inches. The total cutting time for 2300 pieces is $6\frac{1}{2}$ hours running at 1600 rpm. What is the tool feed in inches per revolution? Round the answer to 3 decimal places.   _____

41. Steel shafts, $1\frac{1}{4}$ inches in diameter, are turned on an automatic machine. One finishing operation is required for a 16.5-inch length of cut. The tool feed is 0.015 inch per revolution using a cutting speed of 200 feet per minute. Determine the number of hours of cutting time required for 1500 shafts. Round the answer to the nearest hour.

_____

42. A carbide milling cutter is used for machining a 560.00-millimeter length of stainless steel. The cutting time is 11.95 minutes, the cutting speed is 60.000 meters per minute, and the feed is 0.250 millimeter per revolution. What is the diameter of the carbide milling cutter? Round the answer to 1 decimal place.

_____

43. Aluminum baseplates are produced that are $1\frac{5}{8}$-inches thick. Six $\frac{1}{4}$-inch diameter holes are drilled in each plate using a feed of 0.004 inch per revolution and a cutting speed of 300 feet per minute. Setup and handling time is estimated at 0.5 minute per piece. What is the total number of hours required to produce 850 aluminum baseplates? Round the answer to 1 decimal place.

_____

Cutting Speed Table

Refer to cutting speed table on page 241. Use the table values and the simplified revolutions per minute formula, $N = \frac{4C}{D}$. Compute the revolutions per minute to the nearer revolution for each problem in the following table.

	Material Machined	Cutting Operation	Tool Material	Tool or Workpiece Diameter (inches)	Speed (rpm)
44.	Aluminum (5052)	Milling	High-Speed Steel	3.500	
45.	Stainless Steel (305), BHN 225–275	Turning	Carbide	5.200	
46.	Alloy Steel (4320), BHN 220–275	Reaming	High-Speed Steel	0.480	
47.	Manganese Bronze, cold drawn	Drilling	High-Speed Steel	0.375	
48.	Brass, annealed	Milling	High-Speed Steel	4.000	
49.	Carbon Steel (1020), BHN 175–225	Turning	Carbide	6.100	
50.	Beryllium Copper, annealed	Drilling	High-Speed Steel	1.100	
51.	Malleable Cast Iron (32510), BHN 110–160	Milling	Carbide	3.000	
52.	Alloy Steel (4320), BHN 220–275	Milling	Carbide	2.500	
53.	Aluminum (5052)	Turning	High-Speed Steel	5.800	
54.	Carbon Steel (1020), BHN 175–225	Milling	Carbide	4.500	
55.	Brass, annealed	Turning	High-Speed Steel	2.750	
56.	Stainless Steel (305), BHN 225–275	Reaming	Carbide	0.620	
57.	Malleable Cast Iron (32510), BHN 110–160	Turning	Carbide	7.000	
58.	Carbon Steel (1020), BHN 175–225	Drilling	High-Speed Steel	0.375	

UNIT 42 Applications of Formulas to Spur Gears

Objectives After studying this unit you should be able to

- Identify the proper gear formula to use depending on the unknown and the given data.
- Compute gear part dimensions by substituting known values directly into formulas.
- Compute gear part dimensions by rearranging given formulas in terms of the unknowns.
- Compute gear part dimensions by the application of two or more formulas in order to determine an unknown.

Gears have wide application in machine technology. They are basic to the design and operation of machinery. Most machine shops are equipped to cut gears, and some shops specialize in gear design and manufacture. It is essential that the machinist and drafter have an understanding of gear parts and the ability to determine gear dimensions by the use of trade handbook formulas.

Description of Gears

Gears are used for transmitting power by rotary motion between shafts. Gears are designed to prevent slippage and to insure positive motion while maintaining a high degree of accuracy of the speed ratios between driving and driven gears. The shape of the gear tooth is of primary importance in providing a smooth transmission of motion. The shape of most gear teeth is an *involute curve*. This curve is formed by the path of a point on a straight line as it rolls along a circle. *Spur gears* are gears that are in mesh between parallel shafts. Of two gears in mesh, the smaller gear is called the *pinion* and the larger gear is called the *gear*.

Spur Gear Definitions

Spur gears and the terms that are applied to these gears are shown. It is essential to study the figures and gear terms before computing gear problems by the use of formulas.

Pitch Circles are the imaginary circles of two meshing gears that make contact with each other. The circles are the basis of gear design and gear calculations.

Pitch Diameter is the diameter of the pitch circle.

Root Circle is a circle which coincides with the bottoms of the tooth spaces.

Root Diameter is the diameter of the root circle.

Outside Diameter is the diameter measured to the tops of the gear teeth.

Addendum is the height of the tooth above the pitch circle.

Dedendum is the depth of the tooth space below the pitch circle.

Whole Depth is the total depth of the tooth space. It is equal to the addendum plus the dedendum.

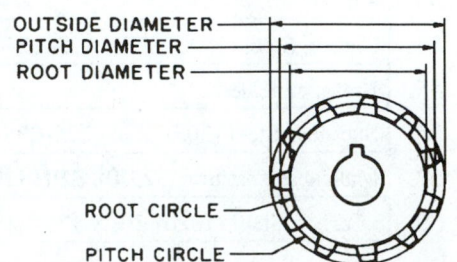

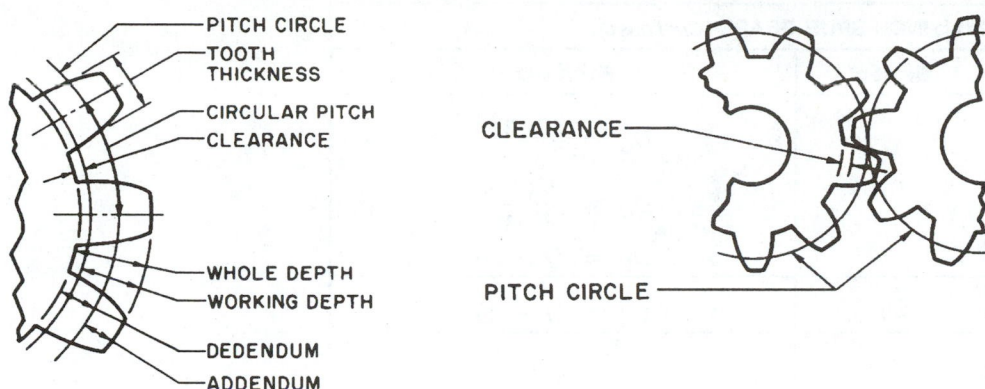

Working Depth is the total depth of mating teeth when two gears are in mesh. It is equal to twice the addendum.

Clearance is the distance between the top of a tooth and the bottom of the mating tooth space of two gears in mesh. It is equal to the whole depth minus the working depth.

Tooth Thickness (Circular) is the length of the arc, on the pitch circle, between the two sides of a tooth.

Circular Pitch is the length of the arc measured on the pitch circle between the centers of two adjacent teeth. It is equal to the circumference of the pitch circle divided by the number of teeth on the gear.

Diametral Pitch (Pitch) is the ratio of the number of gear teeth to the number of inches of pitch diameter. It is equal to the number of gear teeth for each inch of pitch diameter.

When the pitch of a gear is mentioned, the reference is to diametral pitch, rather than circular pitch. For example, if a gear has 28 teeth and a pitch diameter of 4 inches, it has a pitch (diametral pitch) of $\frac{28}{4}$ or 7. It has 7 teeth per inch of pitch diameter, and it is called a 7-pitch gear. It will only mesh with other 7-pitch gears. Gears must have the same pitch in order to mesh.

Gearing–Diametral Pitch System

The diametral pitch system is the system of gear design which is generally applied to decimal-inch dimensional gears. The following table lists the symbols and formulas used in the diametral pitch system.

DECIMAL-INCH SPUR GEARS (American National Standard)		
Term	**Symbol**	**Formulas**
Pitch (diametral pitch)	P	$P = \dfrac{N}{D}$ $P = \dfrac{3.1416}{P_C}$
Circular Pitch	P_C	$P_C = \dfrac{3.1416D}{N}$ $P_C = \dfrac{3.1416}{P}$
Pitch Diameter	D	$D = \dfrac{N}{P}$ $D = \dfrac{NP_C}{3.1416}$

DECIMAL-INCH SPUR GEARS (*Continued*)		
Term	*Symbol*	*Formulas*
Outside Diameter	D_O	$D_O = \dfrac{N + 2}{P}$ $D_O = \dfrac{P_C(N + 2)}{3.1416}$ $D_O = D + 2a$
Root Diameter	D_R	$D_R = D - 2d$
Addendum	a	$a = \dfrac{1}{P}$ * $a = 0.3183P_C$ *
Dedendum	d	$d = \dfrac{1.157}{P}$ * $d = 0.3683P_C$ *
Whole Depth	WD	$WD = \dfrac{2.157}{P}$ * $WD = 0.6866P_C$ * $WD = a + d$ *
Working Depth	W_D	$W_D = \dfrac{2.000}{P}$ * $W_D = 0.6366P_C$ *
Clearance	c	$c = \dfrac{0.157}{P}$ * $c = 0.050P_C$ *
Tooth Thickness	T	$T = \dfrac{1.5708}{P}$
Number of Teeth	N	$N = PD$ $N = \dfrac{3.1416D}{P_C}$

Note: Formulas for $14\frac{1}{2}$-degree Involute and Composite Full-Depth Teeth and 20-degree Involute Full-Depth Teeth.

Gear Calculations

Most gear calculations are made by identifying the proper formula which is given in terms of the unknown, and substituting the known dimensions. It is sometimes necessary to rearrange a formula in terms of a particular unknown. The solution of a problem may require the substitution of values in two or more formulas.

Examples Refer to the Decimal-Inch Spur Gears Table on page 247.

1. Determine the pitch diameter of a 5-pitch gear which has 28 teeth.

Identify the formula whose parts consist of pitch diameter, pitch, and number of teeth.

$$D = \frac{N}{P}$$

Solve.

$$D = \frac{28}{5}$$
$$D = 5.6000 \text{ inches} \qquad \text{Ans}$$

2. Determine the outside diameter of a gear which has 16 teeth and a circular pitch of 0.7854 inch.

Identify the formula whose parts consist of outside diameter, number of teeth, and circular pitch.

$$D_O = \frac{P_C(N + 2)}{3.1416}$$

Solve.

$$D_O = \frac{0.7854(16 + 2)}{3.1416}$$

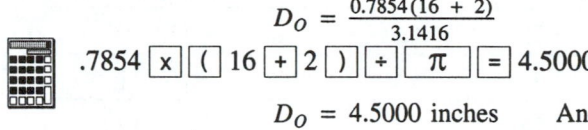 .7854 $\boxed{\times}$ $\boxed{(}$ 16 $\boxed{+}$ 2 $\boxed{)}$ $\boxed{\div}$ $\boxed{\pi}$ $\boxed{=}$ 4.5000

$$D_O = 4.5000 \text{ inches} \qquad \text{Ans}$$

3. Determine the circular pitch of a gear with a whole depth dimension of 0.3081 inch.

Identify the formula whose parts consist of circular pitch and whole depth.

$$WD = 0.6866 P_C$$

The formula must be rearranged in terms of circular pitch.

$$P_C = \frac{WD}{0.6866}$$

Solve.

$$P_C = \frac{0.3081}{0.6866}$$

$$P_C = 0.4487 \text{ inch} \qquad \text{Ans}$$

4. Determine the addendum of a gear which has an outside diameter of 3.0000 inches and a pitch diameter of 2.7500 inches.

Identify the formula whose parts consist of addendum, outside diameter, and pitch diameter.

$$D_O = D + 2a$$

The formula must be rearranged in terms of the addendum.

$$D_O - D = 2a$$

$$a = \frac{D_O - D}{2}$$

Solve.

$$a = \frac{3.0000 - 2.7500}{2}$$

$$a = 0.1250 \text{ inch} \qquad \text{Ans}$$

5. Determine the working depth of a gear which has 46 teeth and a pitch diameter of 11.5000 inches.

There is no single formula in the table that consists of working depth, number of teeth, and pitch diameter. Therefore, it is necessary to substitute in two formulas in order to solve the problem.

Observe $W_D = \frac{2.0000}{P}$. The pitch must be found first.

$$P = \frac{N}{D}$$

$$P = \frac{46}{11.5000}$$

$$P = 4$$

Solve for W_D.

$$W_D = \frac{2.000}{P}$$

$$W_D = \frac{2.000}{4}$$

$$W_D = 0.5000 \text{ inch} \qquad \text{Ans}$$

Gearing–Metric Module System

The *module system* of gear design is generally the system which is used with metric system units of measure. The *module* of a gear equals the pitch diameter divided by the number of teeth. In the metric system, the module of a gear means the pitch diameter in millimeters is divided by the number of teeth. Module is an actual dimension in millimeters, not a ratio as with diametral pitch. For example, if a gear has 20 teeth and a 50 millimeter pitch diameter, the module is 2.5 millimeters (50 mm ÷ 20). A module of 2.5 millimeters means that there are 2.5 millimeters of pitch diameter per tooth.

A partial list of a standard series of modules (in millimeters) is listed as follows:

1	2	3	4	6	9
1.25	2.25	3.25	4.5	6.5	10
1.5	2.5	3.5	5	7	11
1.75	2.75	3.75	5.5	8	12

The relation between module and various gear parts using a metric module system is shown in the following table.

METRIC SPUR GEARS		
Term	**Symbol**	**Formulas**
Module	m	$m = \dfrac{D}{N}$
Circular Pitch	P_C	$P_C = \dfrac{m}{0.3183}$
Pitch Diameter	D	$D = mN$
Outside Diameter	D_O	$D_O = m(N + 2)$
Addendum	a	$a = m$
Dedendum	d	$d = 1.157m$ * $d = 1.167m$ **
Whole Depth	WD	$WD = 2.157m$ * $WD = 2.167m$ **
Working Depth	W_D	$W_D = 2m$
Clearance	c	$c = 0.157m$ * $c = 0.1667m$ **
Tooth Thickness	T	$T = 1.5708m$

* When clearance = 0.157 × module
** When clearance = 0.1667 × module

Examples   Refer to the Metric Spur Gears Table.

1. Determine the circular pitch of a 6 millimeter module gear.

    Circular Pitch = Module ÷ 0.3183
    Circular Pitch = 6 mm ÷ 0.3183 = 18.850 mm   Ans

2. Determine the outside diameter of a 3.5 millimeter module gear with 20 teeth.

    Outside Diameter = (Number of teeth + 2) × Module
    Outside Diameter = (20 + 2) × 3.5 mm = 22 × 3.5 mm = 77.000 mm   Ans

3. Compute the dedendum of a 4.5 millimeter module gear designed with a clearance of 0.157 × module.

    When *Clearance = 0.157 × Module, the Dedendum = 1.157 × Module*
    Dedendum = 1.157 × 4.5 mm = 5.207 mm   Ans

4. Compute the whole depth of 7 millimeter module gear designed with a clearance of $0.1667 \times$ module.

When *Clearance* $= 0.1667 \times Module$, *the Whole Depth* $= 2.167 \times Module$
Whole Depth $= 2.167 \times 7$ mm $= 15.169$ mm   Ans

APPLICATION

Gearing–Diametral Pitch System

Refer to the Decimal-Inch Spur Gears Table on page 247 for each of the following gearing problems.

	GIVEN VALUES	FIND	ANSWER
1.	Circular Pitch = 1.5708″	Pitch	
2.	Pitch = 10	Circular Pitch	
3.	Pitch Diameter = 5.2000″ Number of Teeth = 26	Circular Pitch	
4.	Pitch Diameter = 12.5714″ Number of Teeth = 44	Pitch	
5.	Pitch = 7 Number of Teeth = 26	Pitch Diameter	
6.	Circular Pitch = 0.3142″ Number of Teeth = 12	Pitch Diameter	
7.	Pitch = 18	Circular Pitch	
8.	Pitch diameter = 0.7273″ Number of Teeth = 16	Pitch	
9.	Pitch = 12 Pitch Diameter = 1.1667″	Number of Teeth	
10.	Circular Pitch = 0.6283″ Pitch Diameter = 8.4000″	Number of Teeth	
11.	Number of Teeth = 56 Pitch = 8	Outside Diameter	
12.	Pitch = 14	Addendum	
13.	Pitch Diameter = 1.3333″ Dedendum = 0.0643″	Root Diameter	
14.	Pitch = 3.4	Whole Depth	
15.	Circular Pitch = 0.2856″	Working Depth	
16.	Circular Pitch = 1.4650″	Clearance	
17.	Pitch = 20	Tooth Thickness	
18.	Pitch Diameter = 3.5000″ Addendum = 0.1818″	Outside Diameter	
19.	Circular Pitch = 0.0954″	Dedendum	
20.	Circular Pitch = 0.8976″	Addendum	

	GIVEN VALUES	FIND	ANSWER
21.	Addendum = 0.1429″ Dedendum = 0.1653″	Whole Depth	
22.	Pitch = 4	Dedendum	
23.	Circular Pitch = 0.3076″	Whole Depth	
24.	Pitch = 17	Clearance	
25.	Pitch = 9	Working Depth	

Refer to the Decimal-Inch Spur Gears Table on page 247. The formula in terms of the unknown is not given. Choose the formula that consists of the given parts, rearrange in terms of the unknown, and solve.

	GIVEN VALUES	FIND	ANSWER
26.	Addendum = 0.0857″	Circular Pitch	
27.	Addendum = 0.0666″	Pitch	
28.	Addendum = 0.2000″ Outside Diameter = 4.8000″	Pitch Diameter	
29.	Outside Diameter = 2.7144″ Number of Teeth = 17	Pitch	
30.	Outside Diameter = 4.3750″ Circular Pitch = 0.3927″	Number of Teeth	
31.	Working Depth = 0.0769″	Pitch	
32.	Working Depth = 0.4500″	Circular Pitch	
33.	Outside Diameter = 4.7144″ Pitch Diameter = 4.4286″	Addendum	

Refer to the Decimal-Inch Spur Gears Table on page 247. No single formula is given which consists of the given parts and the unknown. Two or more formulas, some in rearranged form, must be used in solving these problems.

	GIVEN VALUES	FIND	ANSWER
34.	Number of Teeth = 72 Pitch Diameter = 6.0000″	Addendum	
35.	Number of Teeth = 44 Pitch Diameter = 3.6667″	Dedendum	
36.	Number of Teeth = 10 Pitch Diameter = 2.5000″	Whole Depth	
37.	Number of Teeth = 90 Pitch Diameter = 12.8571″	Working Depth	
38.	Pitch Diameter = 1.0625″ Pitch = 16	Outside Diameter	
39.	Pitch Diameter = 2.9167″ Pitch = 12	Root Diameter	

	GIVEN VALUES	FIND	ANSWER
40.	Number of Teeth = 29 Pitch Diameter = 2.0714″	Root Diameter	
41.	Number of Teeth = 75 Pitch Diameter = 6.8182″	Clearance	
42.	Addendum = 0.1429″	Tooth Thickness	
43.	Pitch Diameter = 1.0455″ Addendum = 0.0455″	Number of Teeth	

Backlash is the amount that a tooth space is greater than the engaging tooth on the pitch circles of two gears. Determine the average backlash of each of the following using this formula.

$$\text{Average backlash} = \frac{0.030}{P}$$

44. A 7-pitch gear _____

45. A 20-pitch gear _____

46. A 3.5-pitch gear _____

47. A gear with a whole depth of 0.2606″ _____

48. A gear with a working depth of 0.1176″ _____

49. A gear with a pitch diameter of 4.800″ and 24 teeth _____

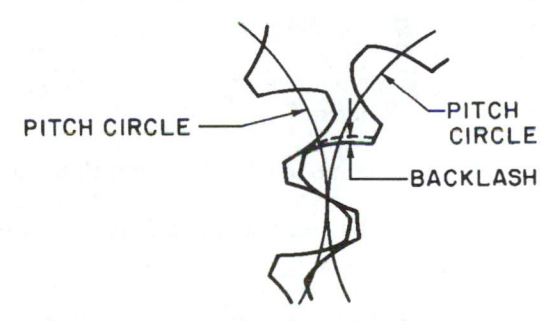

The center distance of a pinion and a gear is the distance between the centers of the pitch circles. Determine the center distance of each of the following using this formula.

$$\text{Center distance} = \frac{\text{pitch diameter of gear} + \text{pitch diameter of pinion}}{2}$$

50. A pinion with a pitch diameter of 2.8300 inches and a gear with a pitch diameter of 4.1667 inches _____

51. A pinion with a pitch diameter of 4.8889 inches and a gear with a pitch diameter of 8.6752 inches _____

52. A 9-pitch pinion and gear; the pinion has 23 teeth and the gear has 38 teeth _____

53. A 16-pitch pinion and gear; the pinion has 18 teeth and the gear has 44 teeth _____

54. A gear and pinion with a circular pitch of 0.1745 inch; the gear has 55 teeth and the pinion has 37 teeth _____

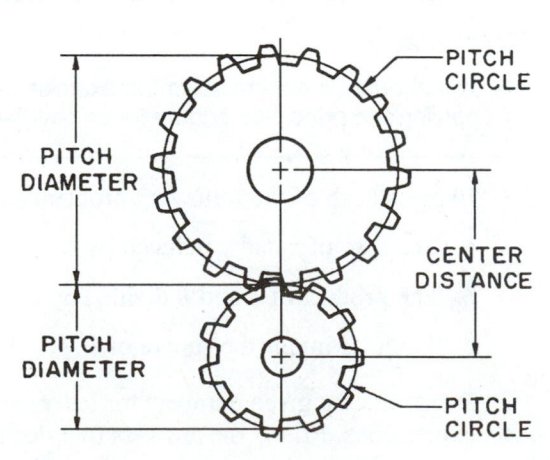

Gearing–Metric Module System

Refer to the Metric Spur Gears Table on page 250 and determine the values in the following table.

	Module	Number of Teeth	a. Pitch Diameter	b. Circular Pitch	c. Outside Diameter	d. Addendum	e. Working Depth	f. Tooth Thickness
55.	6.5 mm	18						
56.	9 mm	24						
57.	2.5 mm	10						
58.	3.75 mm	16						
59.	10 mm	26						

Solve the following metric module system gearing problems. Certain problems require rearranging the data given in Metric Spur Gears Table on page 250.

60. Compute the whole depth of a 5 millimeter module gear designed with a clearance of $0.157 \times$ module.   _____

61. What is the number of teeth on a 4 millimeter module gear with a pitch diameter of 120 millimeters?   _____

62. What is the module of a gear which has a working depth of 13 millimeters?   _____

63. Compute the dedendum of a 7 millimeter module gear designed with a clearance of $0.1667 \times$ module.   _____

64. What is the module of a gear with 38 teeth and an outside diameter of 220 millimeters?   _____

UNIT 43 Achievement Review— Section Three

Objective

You should be able to solve the exercises and problems in this Achievement Review by applying the principles and methods covered in units 31–42.

1. Express each of the following problems as an algebraic expression.

   a. The sum of x and y reduced by c.   _____

   b. The product of a and b divided by d.   _____

   c. Twice M minus the square of P.   _____

2. Substitute the given numbers for letters and find the value for each of the following expressions. Round the answers to 3 decimal places where necessary.

   a. Find $(5a + 6b) \div 4b$ when $a = 4$ and $b = 5$.   _____

   b. Find $3xy - (2x + y)$ when $x = 6$ and $y = 3$.   _____

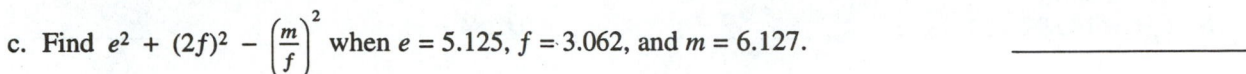

c. Find $e^2 + (2f)^2 - \left(\dfrac{m}{f}\right)^2$ when $e = 5.125$, $f = 3.062$, and $m = 6.127$.   _____

d. Find $\sqrt{hr + 5p}\ \left(\dfrac{5h}{p} + h\right)$ when $h = 10.26$, $p = 8.00$, and $r = 6.59$.   _____

3. Perform the operation or operations as indicated for each of the following exercises. Round the answers to 3 decimal places where necessary.

a. $-25 + (-12)$ _____

b. $24 + (-8)$ _____

c. $-1.8 - (12.6)$ _____

d. $18(-4)$ _____

e. $(-0.3)(-2.6)$ _____

f. $-18 \div 3$ _____

g. $-12.8 \div (-0.4)$ _____

h. $(-6)^2$ _____

i. $(-5)^3$ _____

j. $\sqrt[3]{-27}$ _____

k. $\left(-\dfrac{1}{4}\right)^2$ _____

l. $\sqrt[3]{\dfrac{27.063}{8.920}}$ _____

m. $(-4.02)^2 + \sqrt[3]{8.96} - (3.86)(-5.66)$ _____

n. $[4(-10.66)(0.37)] \div (12 + 18.95)$ _____

4. The following expressions consist of literal terms. Perform the indicated operations.

a. $-8P + 21P$ _____

b. $-0.05H^2 - 1.13H^2$ _____

c. $12d + 8d^2 - 7d + 14d^2$ _____

d. $(5a - 2a^2) - (6a - 4a^2)$ _____

e. $(-10x)(9x^2y)$ _____

f. $(-5.9e^2f^2)(-f^2)$ _____

g. $(16x^2 - 4x^3) \div 0.5x$ _____

h. $(0.6f^2g^3 - fg - 2f^2) \div 0.2f$ _____

i. $(6x^2 - y^3)(-3x^3 + y^2)$ _____

j. $(-4a^3bc^2)^3$ _____

k. $[(xy^2)^3 - (x^2y^2)]^2$ _____

l. $\sqrt[3]{-27a^6b^3c^9}$ _____

m. $\sqrt{\dfrac{25}{64}\ e^4g^2d}$ _____

n. $-6(x^2 + y - 2x)$ _____

o. $36 - (m^3 + m) + (m^3 - 12)$ _____

p. $-3a(2a)^2 + \sqrt{36a^4}$ _____

q. $9y - x[-8 + (xy)^2 - y] + 12x$ _____

r. $b(b + 7m)^2 - b(b - 7m)^2$ _____

5. Solve for the unknown in each of the following equations using one of the six principles of equality. Check each answer. Round the answers to 3 decimal places where necessary.

a. $x + 12 = 33$ _____

b. $y - 15 = 23$ _____

c. $-32 = B - 46$ _____

d. $H + 11.7 = 43.9$ _____

e. $14.3 = x + 53.6$ _____

f. $R - 7.8 = -9.2$ _____

g. $7y = -84$ _____

h. $1.3E = 7.54$ _____

i. $11.22 = \dfrac{L}{6.6}$ _____

j. $\dfrac{x}{-0.8} = 8.48$ _____

k. $6.75 = -20.25x$ _____

l. $\dfrac{3}{4}x = 2\dfrac{1}{4}$ _____

m. $s^2 = 81$ _____

n. $x^3 = \dfrac{-8}{27}$ _____

o. $\sqrt{M} = 12.892$ _____   q. $-0.0284 = y^3$ _____

p. $\sqrt[3]{V} = 5.873$ _____   r. $\sqrt[3]{B} = \dfrac{3.866}{4.023}$ _____

6. Solve for the unknown and check each of the following combined operations equations. Round the answers to 3 decimal places where necessary.

   a. $32 - (-P + 18) = 45$ _____   f. $12x^2 - 53 = (x - 3)(x + 3)$ _____

   b. $10(M - 4) = -5(M - 4)$ _____   g. $-8.53(G - 3.67) = 5.7(-18.36)$ _____

   c. $7.1E + 3(E - 6) = 0.5E + 22.8$ _____   h. $(T - 7.8)(T - 8) = T^2 + 0.3$ _____

   d. $\dfrac{H}{4} + 7.8 = 13.6$ _____   i. $59.66\sqrt{x} = 8.71(1.07\sqrt{x} + 55.32)$ _____

   e. $\dfrac{1}{4}F + 6\left(F - \dfrac{1}{2}\right) = 15\dfrac{3}{4}$ _____   j. $H + 5.023\sqrt{H} = -9.777\sqrt{H} + H$ _____

7. In each of the following formulas, substitute given numerical values for letter values and solve for the unknown and check. Round the answers to 3 decimal places where necessary.

   a. $N = 0.707DP_n$.   Solve for D when $N = 36$ and $P_n = 12$. _____

   b. $I = \dfrac{nE}{R + nr}$.   Solve for E when $I = 0.3$, $n = 5$, $R = 6$, and $r = 4$. _____

   c. $S = \dfrac{0.290W}{t^2}$.   Solve for W when $S = 600$ and $t = 0.375$. _____

   d. $c = \sqrt{a^2 + b^2}$.   Solve for b when $a = 8.053$ and $c = 10.096$. _____

   e. $V = 1.570h(R^2 + r^2)$.   Solve for R when $V = 105.823$, $h = 5.897$, and $r = 2.023$. _____

8. Rearrange each of the following formulas in terms of the designated letter.

   a. $E = I(R + r)$.   Solve for r. _____

   b. $D_O = 2C - d + 2a$.   Solve for a. _____

   c. $M = D - 1.5155P + 3W$.   Solve for W. _____

   d. $r = \sqrt{x^2 + y^2}$.   Solve for y. _____

   e. $L = 3.14(0.5D + 0.5d) + 2x$.   Solve for x. _____

   f. $HP = \dfrac{D^2N}{2.5}$.   Solve for D. _____

9. Solve for the unknown value in each of the following proportions and check. Round the answers to 3 decimal places where necessary.

   a. $\dfrac{P}{12.8} = \dfrac{3}{2}$ _____   e. $\dfrac{6.5}{M} = \dfrac{8.2}{41}$ _____

   b. $\dfrac{3.6}{0.9} = \dfrac{E}{2.7}$ _____   f. $\dfrac{22.517}{13.503} = \dfrac{1.297}{x}$ _____

   g. $\dfrac{20.021}{5.773T} = \dfrac{1.892}{4.518}$ _____

   c. $\dfrac{H}{\frac{1}{4}} = \dfrac{\frac{1}{2}}{\frac{3}{8}}$ _____   h. $\dfrac{10.360}{7.890} = \dfrac{2.015}{N - 2.515}$ _____

   d. $\dfrac{7}{6} = \dfrac{C + 7}{12}$ _____

10. Analyze each of the following problems to determine whether the problem is a direct or inverse proportion and solve. Round the answers to 3 decimal places where necessary.

    a. A reamer tapers 0.0975 inch along a 3.2625-inch length. What is the amount of taper along a 2.1250-inch length?

    b. A machine produces 2550 parts in 8.5 hours. How many parts are produced by the machine in 10 hours?

    c. Of two gears that mesh, one gear with 12 teeth revolves at 420 rpm. What is the revolutions per minute of the other gear which has 16 teeth?

11. Solve the following cutting speed and gear problems.

    a. A steel shaft 2.300 inches in diameter is turned in a lathe at 250 rpm. Determine the cutting speed to the nearer whole number.

    $$C = \frac{3.1416\,DN}{12}$$

    b. Determine the revolutions per minute to the nearer whole number of an aluminum cylinder 40.00 millimeters in diameter with a cutting speed of 160 meters per minute.

    $$N = \frac{1000\,C}{3.1416\,D}$$

    c. Twenty 5.50-millimeter diameter holes each 58.00 millimeters deep are drilled in a workpiece. The drill turns at 6500 r/min with a feed of 0.05 millimeter per revolution. Determine the total cutting time to the nearest hundredth minute.

    $$T = \frac{L}{FN}$$

    d. What is the pitch of a gear with 44 teeth and a pitch diameter of 12.5714 inches?

    $$P = \frac{N}{D}$$

    e. Determine the whole depth to 4 decimal places of a gear with 20 teeth and a pitch diameter of 5.0000 inches.

    $$P = \frac{N}{D} \text{ and } WD = \frac{2.157}{P}$$

12. The following problems are given in scientific notation. Solve and leave answers in scientific notation. Round the answers (mantissas) to 2 decimal places.

    a. $(3.76 \times 10^4) \times (2.87 \times 10^3)$ _____

    b. $(8.63 \times 10^7) \div (5.77 \times 10^{-5})$ _____

    c. $\dfrac{(9.76 \times 10^{-5}) \times (1.77 \times 10^9)}{(5.87 \times 10^3)}$ _____

    d. $\dfrac{(9.09 \times 10^{-5})}{(4.72 \times 10^6) \times (6.15 \times 10^{-3})}$ _____

13. The following problems are given in decimal (standard) form. Calculate and give answers in scientific notation. Round the answers (mantissas) to 2 decimal places.

    a. 0.00021×0.00039 _____

    b. $1{,}476{,}000 \times -0.0000373$ _____

    c. $\dfrac{0.0000287 \times 216{,}000{,}000}{0.00981}$ _____

    d. $\dfrac{0.00503 \times 0.000406}{416{,}000 \times 0.00392}$ _____

Section Four
Fundamentals of Plane Geometry

UNIT 44 Introduction to Geometric Figures

Objectives After studying this unit you should be able to

- Add, subtract, multiply, and divide angles in terms of degrees, minutes, and seconds.
- Express decimal degrees as degrees, minutes, and seconds.
- Express degrees, minutes, and seconds as decimal degrees.

The fundamental principles of geometry generally applied to machine shop problems are those used to make the calculations required for machining parts from engineering drawings. An engineering drawing is an example of applied geometry.

Plane Geometry

Plane geometry is the branch of mathematics that deals with points, lines, and various figures that are made of combinations of points and lines. The figures lie on a flat surface, or *plane*. Examples of plane geometry are the views of a part as shown on an engineering drawing.

Since geometry is fundamental to machine technology, it is essential to understand the definitions and terms of geometry. It is equally important to be able to apply the geometric principles in problem solving. The methods and procedures used in problem solving are the same as those required for the planning, making, and checking of machined parts.

Procedure To solve a geometry problem

- Study the figure.
- Relate it to the principle or principles that are needed for the solution.
- Base all conclusions on fact: given information and geometric principles.
- Do not assume that something is true because of its appearance or because of the way it is drawn.

➤ **Note:** The same requirements are applied in reading engineering drawings.

Axioms and Postulates

In the study of geometry certain basic statements called *axioms* are assumed to be true without requiring proof. Axioms or postulates may be compared to the rules of a game. Some axioms or postulates are listed. Others will be given as they are required for problem solving.

Things equal to the same thing, or to equal things, are equal to each other. Equals may be substituted for equals.

If equals are added to or subtracted from equals, the sums or remainders are equal.

If equals are multiplied or divided by equals, the products or quotients are equal.

The whole is equal to the sum of its parts.

Only one straight line can be drawn between two given points.

Through a given point, only one line can be drawn parallel to a given line.

Two straight lines can intersect at one point only.

Points and Lines

A *point* has no size or form; it has only location. A point is shown as a dot. Each point is usually named by a capital letter as shown.

A •   B •

C •

D •

A *line,* as it is used in this book, always means a straight line. A line other than a straight line, such as a curved line, is identified. In this book, no distinction between a line and a line segment is made.

Parallel lines do not meet regardless of how far they are extended. They are the same distance apart (*equidistant*) at all points. The symbol ‖ means parallel. In the figure, line AB is parallel to line CD; therefore, AB and CD are equidistant (distance *x*) at all points.

Perpendicular lines meet or intersect at a right, or 90°, angle. The symbol ⊥ means perpendicular. These figures are examples of perpendicular lines.

Oblique lines meet or intersect at an angle other than 90°. They are neither parallel nor perpendicular. These figures are examples of oblique lines.

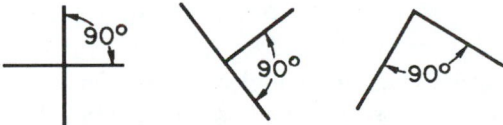

Angles

An *angle* is a figure which consists of two lines that meet at a point called the vertex. The symbol ∠ means angle. The size of an angle is determined by the number of degrees one side of the angle is rotated from the other. The length of the side does not determine the size of the angle. For example, ∠1 is equal to ∠2. The rotation of side AC from side AB is equal to the rotation of side DF from side DE although the lengths of the sides are not equal.

Units of Angular Measure

The degree is the basic unit of angular measure. The symbol (°) means degree. A radius which is rotated one revolution makes a complete circle or 360°. In the English system computations and measurements are in degrees and minutes. Degrees, minutes, and seconds are used for applications which require precise angular measure. In the metric system computations and measurements are in decimal degrees; however, degrees, minutes, and seconds are sometimes used.

A circle may be thought of as a ray with a fixed end point. The ray is rotated. One rotation makes a complete circle of 360° as shown.

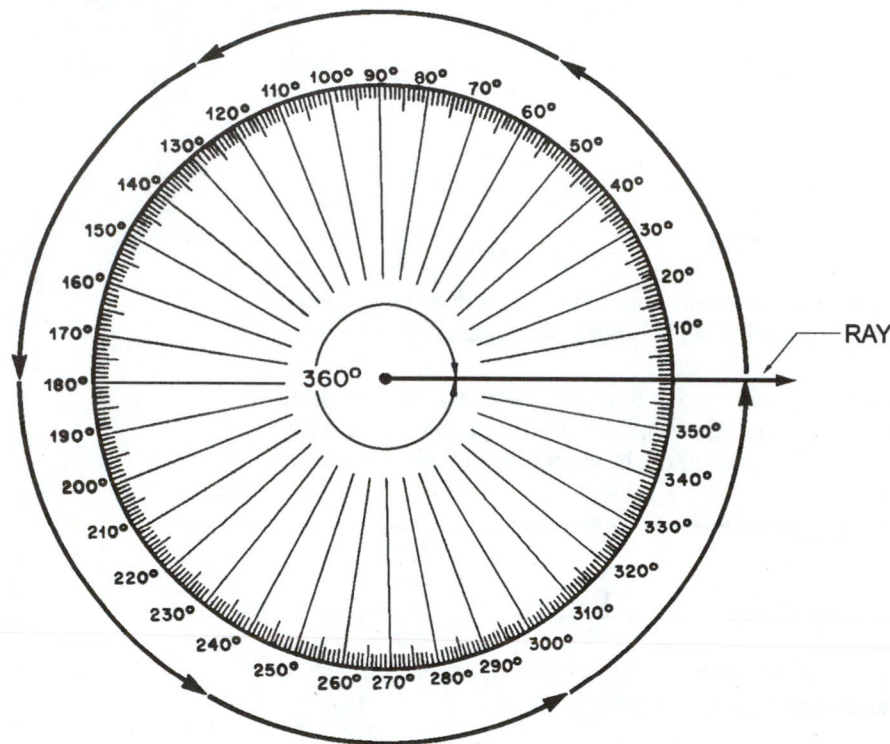

The degree of precision required in computing and measuring angles depends on how the angle is used. Some manufactured parts are designed and processed to a very high degree of precision.

In metric calculations, the decimal degree is generally the preferred unit of measurement. In the English system, angular measure is expressed in these ways.

- As decimal degrees, such as 6.5 degrees and 108.274 degrees.

- As fractional degrees, such as $12\frac{1}{4}$ degrees and $53\frac{1}{10}$ degrees.

- As degrees, minutes, and seconds, such as 37 degrees, 18 minutes and 123 degrees, 46 minutes, 53 seconds.

Decimal and fractional degrees are added, subtracted, multiplied, and divided the same as any other numbers.

Units of Angular Measure in Degrees, Minutes, and Seconds

A degree is divided in 60 equal parts called *minutes*. The symbol for minute is '. A minute is divided in 60 equal parts called *seconds*. The symbol for second is ". The relationship between degrees, minutes, and seconds is shown in the following chart.

1 Circle = 360 Degrees (°)	1 Degree (°) = $\frac{1}{360}$ of a Circle
1 Degree (°) = 60 Minutes (′)	1 Minute (′) = $\frac{1}{60}$ Degree (°)
1 Minute (′) = 60 Seconds (″)	1 Second (″) = $\frac{1}{60}$ Minute (′)

Expressing Decimal Degrees as Degrees, Minutes, and Seconds

The measure of an angle given in the form of decimal degrees, such as 47.1938°, must often be expressed as degrees, minutes, and seconds.

Procedure   To express decimal degrees as degrees, minutes, and seconds

- Multiply the decimal part of the degrees by 60′ in order to obtain minutes.
- If the number of minutes obtained is not a whole number, multiply the decimal part of the minutes by 60″ in order to obtain seconds. Round to the nearer whole second if necessary.
- Combine degrees, minutes, and seconds.

Example   Express 47.1938° as degrees, minutes and seconds.

Multiply 0.1938 by 60′ to obtain minutes.   60′(0.1938) = 11.6280′

Multiply 0.6280 by 60″ to obtain seconds.   60″(0.6280) = 38″
Round to the nearer whole second.

Combine degrees, minutes, and seconds.   47° + 11′ + 38″ = 47°11′38″   Ans

Expressing Degrees, Minutes, and Seconds as Decimal Degrees

Often an angle given in degrees and minutes is to be expressed as decimal degrees. This is often the case when computations involve metric system units of measure.

Procedure   To express degrees and minutes as decimal degrees

- Divide the minutes by 60 to obtain the decimal degree.
- Combine whole degrees and the decimal degree. Round the answer to 2 decimal places.

Example   Express 76°29′ as decimal degrees.

Divide 29′ by 60 to obtain decimal degree.   29′ ÷ 60 = 0.48°
Combine whole degrees with the decimal   76° + 0.48° = 76.48°   Ans
degree.

When working with English and metric units of measure, it may be necessary to express angles given in degrees, minutes, and seconds as angles in decimal degrees.

Procedure   To express degrees, minutes, and seconds as decimal degrees

- Divide the seconds by 60 in order to obtain the decimal minute.
- Combine whole minutes and the decimal minute.
- Divide the total minutes by 60 in order to obtain the decimal degree.
- Combine whole degrees and the decimal degree. Round the answer to 4 decimal places.

Example Express 23°18'44" as decimal degrees.

Divide 44" by 60 to obtain the decimal minute.	$44'' \div 60 = 0.7333'$
Combine whole minutes and the decimal minute.	$18' + 0.7333' = 18.7333'$
Divide 18.7333' by 60 to obtain the decimal degree.	$18.7333' \div 60 = 0.3122°$
Combine whole degrees with the decimal degree.	$23° + 0.3122° = 23.3122°$ Ans

There are two basic formats used in degrees, minutes, seconds and decimal degrees conversions. Depending on the make and model of your calculator, one of the two formats should apply.

1. Calculators with a $\boxed{\circ\,'\,''}$ key (degrees, minutes, seconds)

    To convert degrees, minutes, seconds to decimal degrees:

    Enter degrees, press $\boxed{\circ\,'\,''}$, enter minutes, press $\boxed{\circ\,'\,''}$, enter seconds, press $\boxed{\circ\,'\,''}$.

    The angle is directly displayed as decimal degrees.

Example Convert 53°47'25" to decimal degrees.

53 $\boxed{\circ\,'\,''}$ 47 $\boxed{\circ\,'\,''}$ 25 $\boxed{\circ\,'\,''}$ → 53.79027778° Ans

On certain calculators the execute key $\boxed{\text{EXE}}$ must be pressed last to convert display to decimal degrees.

To convert decimal degrees to degrees, minutes, seconds:

Enter decimal degrees, press $\boxed{\text{SHIFT}}$, press $\boxed{\leftarrow}$

➤ **Note:** ← is the second function of the primary function key $\boxed{\circ\,'\,''}$.

a. The angle is directly displayed as degrees, minutes, seconds.

Example Convert 53.79027778° to degrees, minutes, seconds.

53.79027778 $\boxed{\text{SHIFT}}$ $\boxed{\leftarrow}$ → 53°47'25" Ans

➤ **Note:** Your calculator may display 50°47°25, which is interpreted as 50°47'25".

or

b. The execute key $\boxed{\text{EXE}}$ must be pressed directly after the angle is entered.

Example Convert 53.79027778° to degrees, minutes, seconds.

53.79027778 $\boxed{\text{EXE}}$ $\boxed{\text{SHIFT}}$ $\boxed{\leftarrow}$ → 53°47'25" Ans

2. Calculators with ▶ DD (decimal degrees) and ▶ DMS (degrees, minutes, seconds) functions with $\boxed{\text{2nd}}$ and $\boxed{\text{3rd}}$ function keys

    ➤ **Note:** ▶ DD is the second function of the $\boxed{\blacktriangleright\text{DD}}$ key and ▶ DMS is the third function of the $\boxed{\blacktriangleright\text{DD}}$ key.

    With certain model calculators without the $\boxed{\text{3rd}}$ function key, the procedures will be somewhat different than the procedures given. If so, refer to your user's guide or manual.

To convert degrees, minutes, seconds to decimal degrees:

Enter degrees, press $\boxed{\cdot}$, enter minutes, enter seconds, press $\boxed{\text{2nd}}$,
press $\boxed{\blacktriangleright \underline{\underline{DD}}}$.

Example   Convert 53°47′25″ to decimal degrees.

53 $\boxed{\cdot}$ 4725 $\boxed{\text{2nd}}$ $\boxed{\blacktriangleright \underline{\underline{DD}}}$ → 53.79027778°   Ans

➤ **Note:**   When minutes and seconds have only one digit, enter zeros to position the digits in their proper place positions.

Example   Convert 7°8′4″ to decimal degrees.

7 $\boxed{\cdot}$ 0804 $\boxed{\text{2nd}}$ $\boxed{\blacktriangleright \underline{\underline{DD}}}$ → 7.134444444°   Ans

To convert decimal degrees to degrees, minutes, seconds:

Enter decimal degrees, press $\boxed{\text{3rd}}$, press $\boxed{\blacktriangleright \text{DMS}}$.

Example   Convert 53.79027778° to degrees, minutes, seconds.

53.79027778 $\boxed{\text{3rd}}$ $\boxed{\blacktriangleright \text{DMS}}$ → 53°47′25″0, 53°47′25″   Ans

Arithmetic Operations on Angular Measure in Degrees, Minutes, and Seconds

The division of minutes and seconds permit very precise computations and measurements. In machining operations, dimensions at times are computed to seconds in order to ensure the proper functioning of parts.

When computing with degrees, minutes, and seconds, it is sometimes necessary to exchange units. When exchanging units, keep in mind that 1 degree equals 60 minutes and 1 minute equals 60 seconds. These examples illustrate adding, subtracting, multiplying, and dividing angles in degrees, minutes, and seconds.

Arithmetic operations with degrees, minutes, and seconds computed with a calculator require entering an angle as degrees, minutes, and seconds, converting to decimal degrees, and converting back to degrees, minutes, and seconds. Calculator applications for each of the arithmetic operations are shown.

Adding Angles Expressed in Degrees, Minutes, and Seconds

Example 1   Determine ∠1.

∠1 = 15°18′ + 63°37′

$$\begin{array}{r} 15°18′ \\ + \ 63°37′ \\ \hline 78°55′ \end{array} \quad \text{Ans}$$

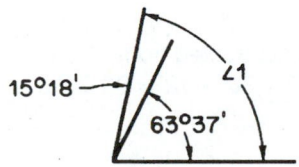

Example 2   Determine ∠2.

∠2 = 43°37′ + 82°54′

$$\begin{array}{r} 43°37′ \\ + \ 82°54′ \\ \hline 125°91′ = 126°31′ \end{array} \quad \text{Ans}$$

➤ **Note:**   91′ = 60′ + 31′ = 1°31′. Add, 125° + 1°31′ = 126°31′.

Example 3   Determine ∠3.

∠3 = 78°43′27″ + 29°38′52″

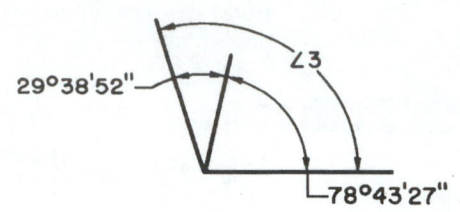

$$\begin{array}{r} 78°43′27″ \\ + \ 29°38′52″ \\ \hline 107°81′79″ = 107°82′19″ = 108°22′19″ \quad \text{Ans} \end{array}$$

➤ **Note:**   79″ = 60″ + 19″ = 1′19″ therefore, 107°81′79″ = 107°82′19″.
   82′ = 60′ + 22′ = 1°22′ therefore, 107°82′19″ = 108°22′19″.

Example 4   Determine ∠3.

∠3 = 78°43′27″ + 29°38′52″

78 [° ′ ″] 43 [° ′ ″] 27 [° ′ ″] [+] 29 [° ′ ″] 38 [° ′ ″] 52 [° ′ ″] [=] [SHIFT] [←] →
108°22′19″   Ans

or   78 [·] 4327 [2nd] [►DD] [+] 29 [·] 3852 [2nd] [►DD] [=] [3rd] [►DMS] →
108°22′19″   Ans

Subtracting Angles Expressed in Degrees, Minutes, and Seconds

Example 1   Determine ∠1.

∠1 = 123°47′32″ − 86°13′07″

$$\begin{array}{r} 123°47′32″ \\ - \ 86°13′07″ \\ \hline 37°34′25″ \quad \text{Ans} \end{array}$$

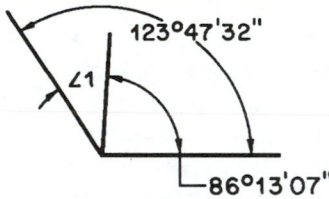

Example 2   Determine ∠2.

∠2 = 97°12′ − 45°26′

$$\begin{array}{r} 97°12′ = 96°72′ \\ - \ 45°26′ = 45°26′ \\ \hline 51°46′ \quad \text{Ans} \end{array}$$

➤ **Note:**   Since 26′ cannot be subtracted from 12′, 1° is exchanged for 60′.
   97°12′ = 96° + 1° + 12′ = 96° + 60′ + 12′ = 96°72′

Example 3   Determine ∠3.

∠3 = 57°13′28″ − 44°19′42″

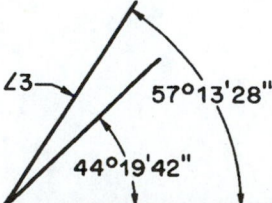

$$\begin{array}{r} 57°13′28″ = 56°73′28″ = \quad 56°72′88″ \\ - \ 44°19′42″ = 44°19′42″ = - \ 44°19′42″ \\ \hline 12°53′46″ \quad \text{Ans} \end{array}$$

➤ **Note:**   Since 19′ cannot be subtracted from 13′, and 42″ cannot be subtracted
   from 28″, 1° is exchanged for 60′ and 1′ is exchanged for 60″.
   57°13′28″ = 56° + 1° + 13′ + 28″ = 56°73′28″ = 56°72′ + 1′ + 28″ = 56°72′88″

Example 4   Determine ∠3.

∠3 = 57°13′28″ − 44°19′42″

57 ⬚ 13 ⬚ 28 ⬚ − 44 ⬚ 19 ⬚ 42 ⬚ = | SHIFT | ← →
12°53′46″   Ans

or   57 ⬚ 1328 | 2nd | ►DD | − 44 ⬚ 1942 | 2nd | ►DD | = | 3rd | ►DMS →
12°53′46″   Ans

Multiplying Angles Expressed in Degrees, Minutes, and Seconds

Example 1   Five holes are drilled on a circle as shown. The angular measure between two consecutive holes is 32°18′. Determine the angular measure, ∠1, between hole 1 and hole 5.

∠1 = 4(32°18′)

$$\begin{array}{r} 32°18′ \\ \times\ 4 \\ \hline 128°72′ = 129°12′ \quad Ans \end{array}$$

➤ **Note:** 72′ = 1°12′

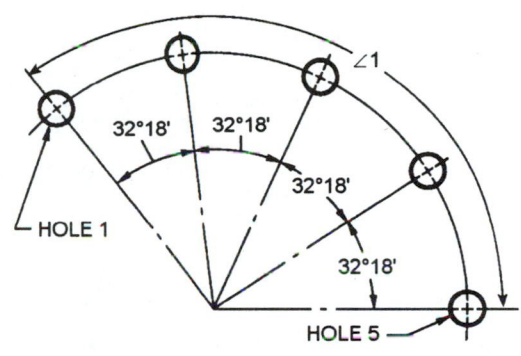

Example 2   Determine ∠2 when x = 41°27′42″.

∠2 = 5x = 5(41°27′42″)

$$\begin{array}{r} 41°\ 27′\ 42″ \\ \times\ 5 \\ \hline 205°135′210″ = 205°138′30″ = 207°18′30″ \quad Ans \end{array}$$

➤ **Note:** 210″ = 3′30″ and 138′ = 2°18′.

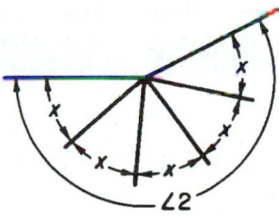

Example 3   Determine ∠2 when x = 41°27′42″.

∠2 = 5x = 5(41°27′42″)

5 | x | 41 ⬚ 27 ⬚ 42 ⬚ = | SHIFT | ← → 207°18′30″   Ans

or   5 | x | 41 ⬚ 2742 | 2nd | ►DD | = | 3rd | ►DMS → 207°18′30″   Ans

Dividing Angles Expressed in Degrees, Minutes, and Seconds

Example 1   Determine ∠1 and ∠2.

∠1 = ∠2 = 104°58′ ÷ 2

$$\begin{array}{r} 52°29′ \\ 2\overline{)104°58′} \quad Ans \end{array}$$

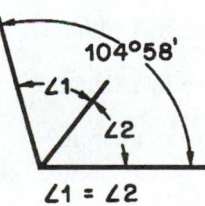

Example 2 Determine ∠1, ∠2, and ∠3.

∠1 = ∠2 = ∠3 = 128°37'21" ÷ 3

Divide 128° by 3.

$$\begin{array}{r} 42° \\ 3\overline{)128°} \\ \underline{126} \\ 2° \end{array}$$

128° ÷ 3 = 42° plus a remainder of 2°.

Add the 2° (120') to the 37'.
120' + 37' = 157'
Divide 157' by 3.
157' ÷ 3 = 52' plus a remainder of 1'.

$$\begin{array}{r} 52' \\ 3\overline{)157'} \\ \underline{156} \\ 1' \end{array}$$

Add the 1' (60") to the 21".
60" + 21" = 81"
Divide 81" by 3.
81" ÷ 3 = 27"

$$\begin{array}{r} 27" \\ 3\overline{)81"} \end{array}$$

Combine.   42° + 52' + 27" = 42°52'27"   Ans

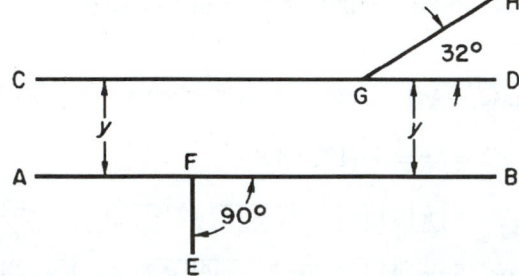

∠1 = ∠2 = ∠3

Example 3 Determine ∠1, ∠2, and ∠3.

∠1 = ∠2 = ∠3 = 128°37'21" ÷ 3

128 [° ' "] 37 [° ' "] 21 [° ' "] [+] 3 [=] [SHIFT] [◄——] → 42°52'27"   Ans

or   128 [·] 3721 [2nd] [►DD] [+] 3 [=] [3rd] [►DMS] → 42°52'27"   Ans

---

APPLICATION

Definitions and Terms

1. Refer to the figure and identify each of the following as parallel, perpendicular, or oblique lines.

   a. Line AB and line CD   _____

   b. Line AB and EF   _____

   c. Line CD and GH   _____

2. a. How many degrees are in a circle?   _____

   b. How many minutes are in 1 degree?   _____

   c. How many seconds are in 1 minute?   _____

   d. How many seconds are in 1 degree?   _____

   e. How many minutes are in a circle?   _____

3. Write the symbols for the following words.

a. parallel _____   d. minute _____

b. perpendicular _____   e. second _____

c. degree _____

Expressing Decimal Degrees As Degrees and Minutes

Express the following decimal degrees as degrees and minutes. When necessary, round the answer to the nearer whole minute.

4. 13.50° _____   9. 93.15° _____

5. 67.85° _____   10. 81.08° _____

6. 48.10° _____   11. 6.47° _____

7. 117.70° _____   12. 125.91° _____

8. 18.60° _____   13. 77.67° _____

Expressing Decimal Degrees As Degrees, Minutes, and Seconds

Express the following decimal degrees as degrees, minutes, and seconds. When necessary, round the answer to the nearer whole second.

14. 52.1380° _____   19. 103.0090° _____

15. 212.0710° _____   20. 37.9365° _____

16. 7.9250° _____   21. 89.9056° _____

17. 44.4440° _____   22. 182.0692° _____

18. 73.9330° _____   23. 19.8973° _____

Expressing Degrees and Minutes As Decimal Degrees

Express the following degrees and minutes as decimal degrees. Round the answer to 2 decimal places.

24. 22°40′ _____   29. 56°48′ _____

25. 107°45′ _____   30. 87°37′ _____

26. 6°10′ _____   31. 2°19′ _____

27. 87°16′ _____   32. 32°08′ _____

28. 122°07′ _____   33. 79°59′ _____

Expressing Degrees, Minutes, and Seconds As Decimal Degrees

Express the following degrees, minutes, and seconds as decimal degrees. Round the answer to 4 decimal places.

34. 28°18′30″ _____   38. 176°27′18″ _____

35. 57°08′45″ _____   39. 2°07′13″ _____

36. 130°50′10″ _____   40. 19°49′59″ _____

37. 98°20′25″ _____   41. 61°12′06″ _____

Adding Angles Expressed in Degrees, Minutes, and Seconds

42. Determine ∠1. _____

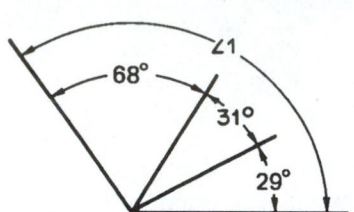

43. Determine ∠2. _____

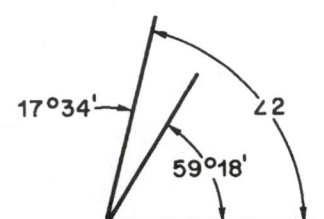

44. Determine ∠3. _____

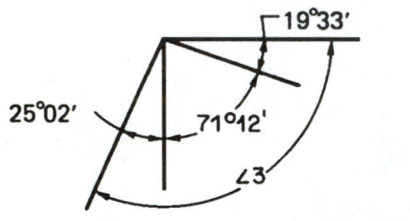

45. Determine ∠1 + ∠2 + ∠3. _____

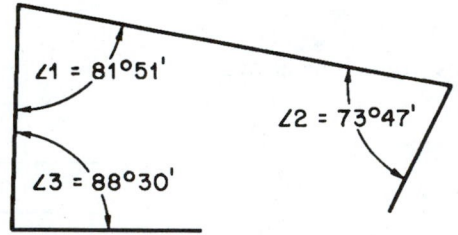

46. Determine ∠5. _____

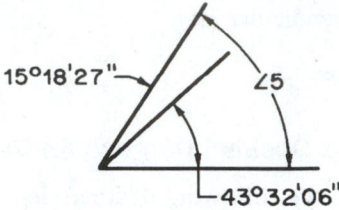

47. Determine ∠6. _____

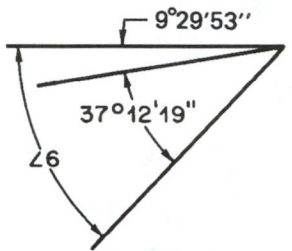

48. Determine ∠7 + ∠8 + ∠9. _____

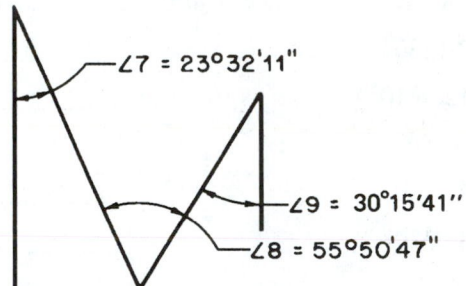

49. Determine ∠1 + ∠2 + ∠3 + ∠4 + ∠5. _____

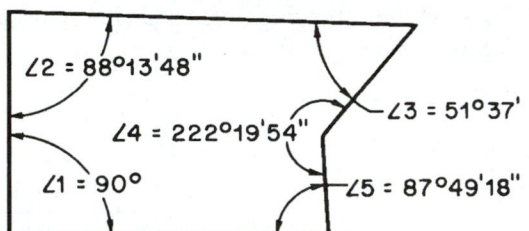

Subtracting Angles Expressed in Degrees, Minutes, and Seconds

Subtract the angles in each of the following exercises.

50. 114° − 89° _____

51. 92°35′ − 76°26′ _____

52. 63°23′ − 32°58′ _____

53. 122°36′17″ − 13°15′08″ _____

54. 49°34′12″ − 19°13′42″ _____

55. Determine ∠1. _____

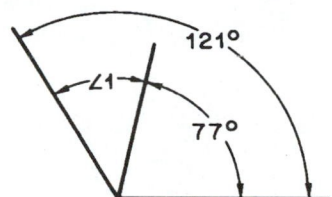

57. Determine ∠2. _____

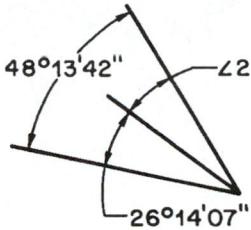

56. Determine ∠3 – ∠2. _____

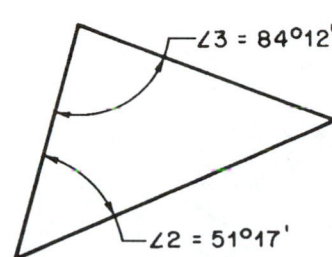

58. Determine ∠1 – ∠2. _____

59. In the figure shown ∠6 = 720° – (∠1 + ∠2 + ∠3 + ∠4 + ∠5).
    Determine ∠6. _____

➤ **Note:** 720° = 719°59′60″

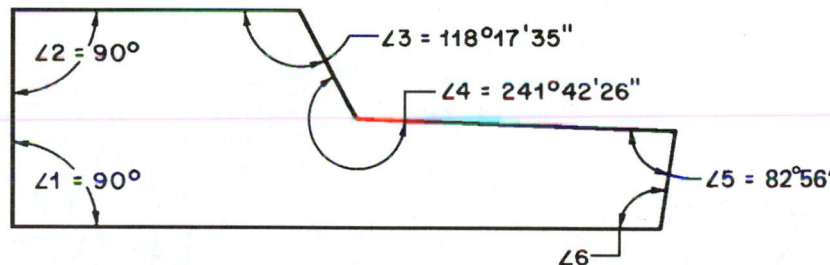

Multiplying Angles Expressed in Degrees, Minutes, and Seconds

Multiply the angles in each of the following exercises.

60. 7(15°) _____

61. 3(29°19′) _____

62. 2(43°43′) _____

63. 5(22°10′13″) _____

64. 8(43°23′28″) _____

65. In the figure shown, ∠1 = ∠2 = 42°.
    Determine ∠3. _____

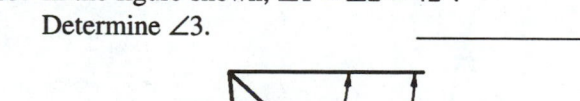

66. If x = 39°14′, find ∠4. _____

67. In the figure shown, $\angle1 = \angle2 = \angle3 = \angle4 = \angle5 = 54°03'$.
    Determine $\angle6$. _____

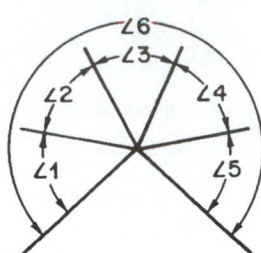

Dividing Angles Expressed in Degrees, Minutes, and Seconds

Divide the angles in each of the following exercises.

68. $94° \div 2$ _____

69. $87° \div 2$ _____

70. $105°20' \div 4$ _____

71. If $\angle1 = \angle2$, find $\angle1$. _____

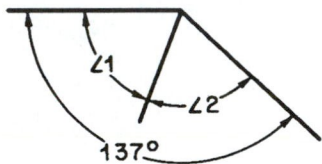

72. Determine x. _____

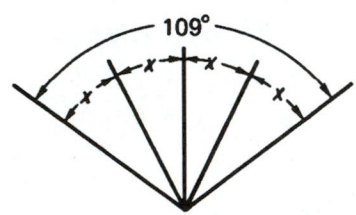

73. Determine y. _____

74. If $\angle1 = \angle2 = \angle3 = \angle4$, find $\angle1$. _____

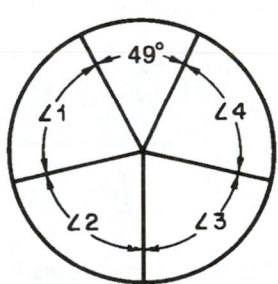

75. The sum of the angles in the figure shown equal $1440°$.

    If $\angle1 = \angle2 = \angle3 = \angle4 = \angle5 = \angle6$ and $\angle7 = \angle8 = \angle9 = \angle10 = 118°14'23''$,
    find $\angle1$. _____

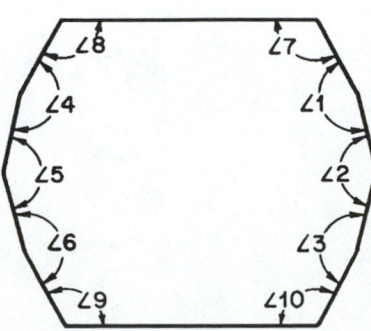

UNIT 45 Protractors—Simple and Vernier

Objectives After studying this unit you should be able to

- Measure angles with a simple protractor.
- Lay out angles with a simple protractor.
- Read settings on a vernier bevel protractor.
- Compute complements and supplements of angles.

Protractors are used for measuring, drawing, and laying out angles. Various types of protractors are available, such as the simple semicircular protractor, swinging blade protractor, and bevel protractor. The type of protractor used depends on its application and the degree of precision required. Protractors have wide occupational use, particularly in the metal and woodworking trades.

Simple Semicircular Protractor

A simple semicircular protractor has two scales, each graduated from 0° to 180° so that it can be read from either the left or right side. The vertex of the angle to be measured or drawn is located at the center of the base of the protractor.

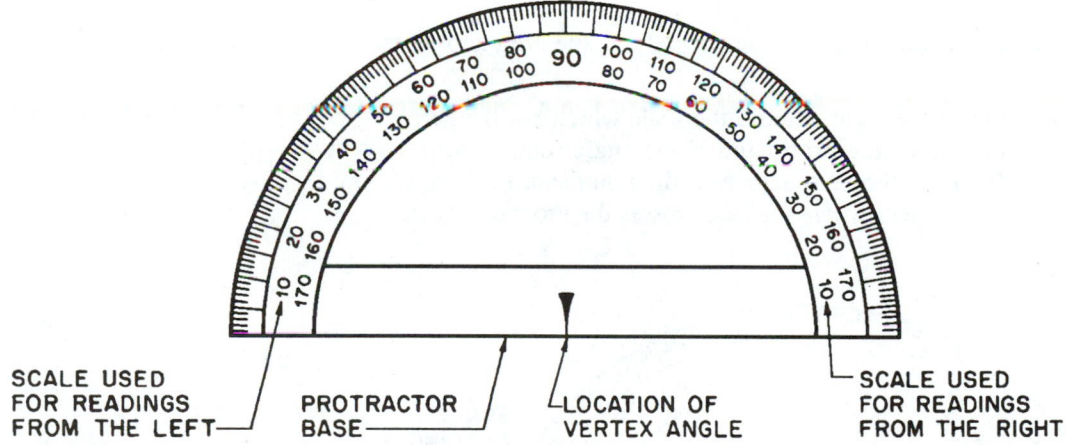

SCALE USED FOR READINGS FROM THE LEFT   PROTRACTOR BASE   LOCATION OF VERTEX ANGLE   SCALE USED FOR READINGS FROM THE RIGHT

Procedure To lay out a given angle

- Draw a baseline.
- On the baseline mark a point as the vertex.
- Place the protractor base on the baseline with the protractor center on the vertex.
- If the angle rotates from the right, choose the scale which has a zero degree reading on the right side of the protractor. If the angle rotates from the left, choose the scale which has a zero degree reading on the left side of the protractor. At the scale reading for the angle being drawn, mark a point.
- Remove the protractor and connect the two points.

Example Lay out an angle of 105°.

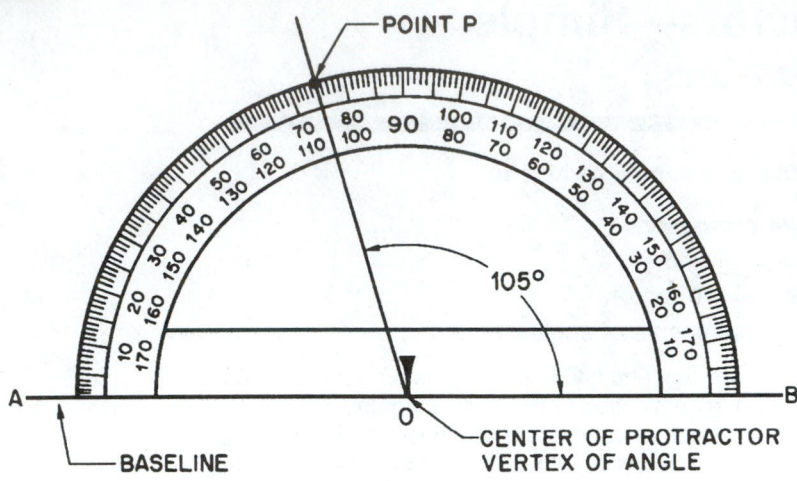

Draw baseline AB.

On AB mark point O as the vertex.

Place the protractor base on AB with the protractor center on point O.

The angle is rotated from the right. The inside scale has a zero degree reading on the right side of the protractor. Use the inside scale and mark a point (point P) at the scale reading of 105°.

Remove the protractor and connect points P and O.

Procedure To measure a given angle

- Place the protractor base on one side of the angle with the protractor center on the angle vertex.

- If the angle rotates from the right, choose the scale which has the zero degree reading on the right side of the protractor. If the angle rotates from the left, choose the scale which has the zero degree reading on the left side of the protractor. Read the measurement where the side crosses the protractor scale.

Example 1 Measure ∠1.

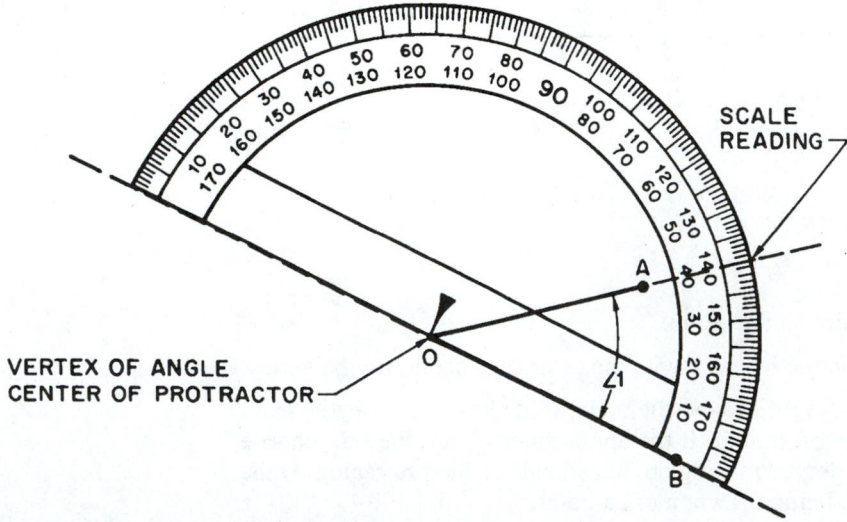

Extend the sides OA and OB of ∠1 as shown.

Place the protractor base on side OB with the protractor center on the angle vertex, point O.

Angle 1 is rotated from the right. The angle measurement is read from the inside scale since the inside scale has a zero degree (0°) reading on the right side of the protractor base. Read the measurement where the extension of side OA crosses the protractor scale. Angle 1 = 40°   Ans

Example 2   Measure ∠2.

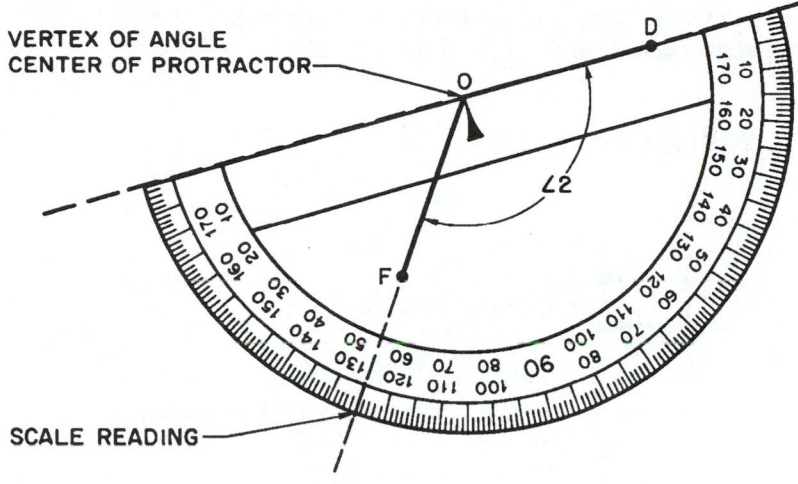

Extend the sides OD and OF of ∠2 as shown.

The protractor is positioned upside down. Place the protractor base on the side OD with the protractor center on the angle vertex point O.

Angle 2 is rotated from the right. The angle measurement is read from the outside scale since the outside scale has a zero degree (0°) reading on the right side of the protractor base. Read the measurement where the extension of side OF crosses the protractor scale. Angle 2 = 125°   Ans

Bevel Protractor with Vernier Scale

The bevel protractor is the most widely used vernier protractor in the machine shop. A vernier bevel protractor is shown.

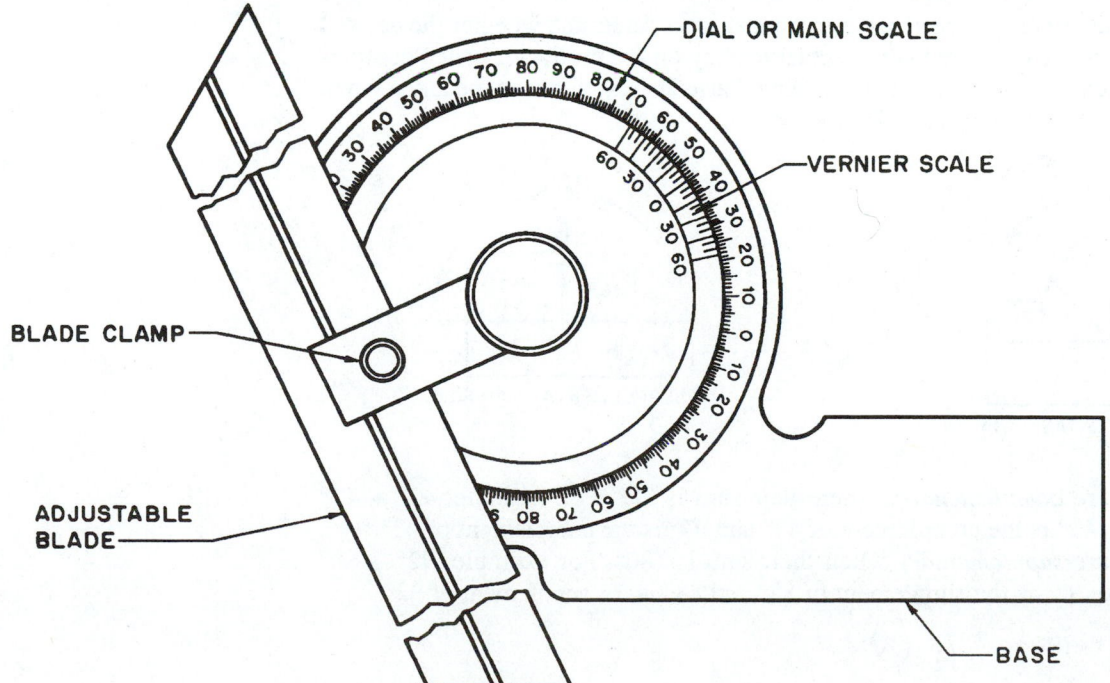

A vernier bevel protractor consists of a fixed dial or main scale. The main scale is divided into four sections, each from 0° to 90°. The vernier scale rotates within the main scale. A blade which can be adjusted to required positions is rotated to a desired angle.

The vernier scale permits accurate readings to $\frac{1}{12}$ degree or 5 minutes. The vernier scale is divided into 24 units, with 12 units on each side of zero. The divisions on the vernier scale are in minutes. Each division is equal to 5 minutes.

The left vernier scale is used when the vernier zero is to the left of the dial zero. The right vernier scale is used when the vernier zero is to the right of the dial zero.

Example   Read the setting on the vernier protractor shown.

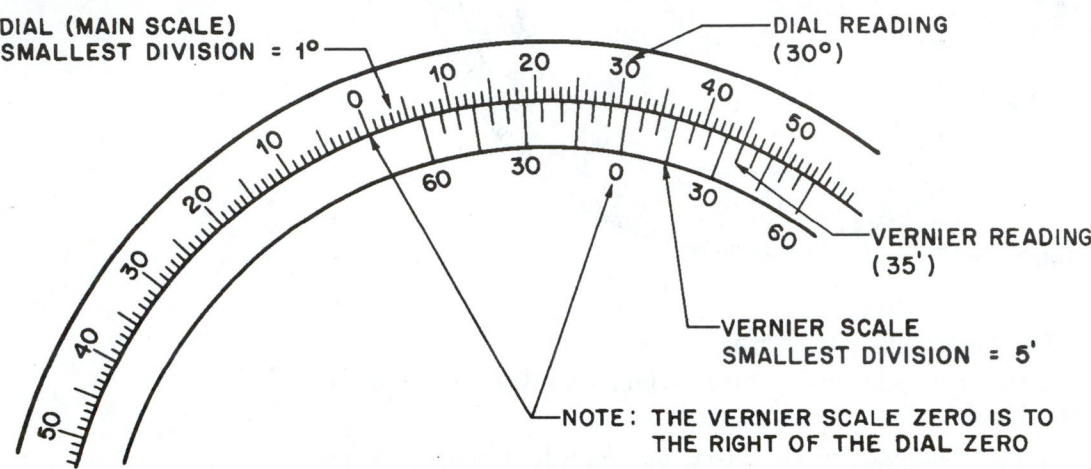

The zero mark on the vernier scale is just to the right of the 30° division of the dial scale. The vernier zero is to the right of the dial zero; therefore, the right vernier scale is read. The 35′ vernier graduation coincides with a dial graduation. The protractor reading is 30°35′.   Ans

Complements and Supplements of Scale Readings

When using the bevel protractor, the machinist must determine whether the desired angle of the part being measured is the actual reading on the protractor or the complement or the supplement of the protractor reading. Particular caution must be taken when measuring angles close to 45° and 90°.

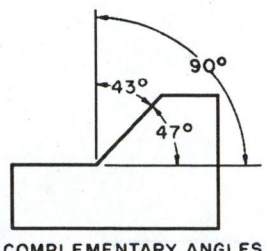

COMPLEMENTARY ANGLES

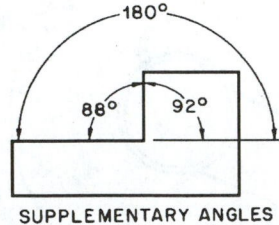

SUPPLEMENTARY ANGLES

Two angles are *complementary* when their sum is 90°. For example, 43° + 47° = 90°. Therefore, 43° is the complement of 47° and 47° is the complement of 43°.

Two angles are *supplementary* when their sum is 180°. For example, 92° + 88° = 180°. Therefore, 92° is the supplement of 88° and 88° is the supplement of 92°.

APPLICATION

Simple Protractor

1. Write the values of angles A–J on the protractor scale shown.

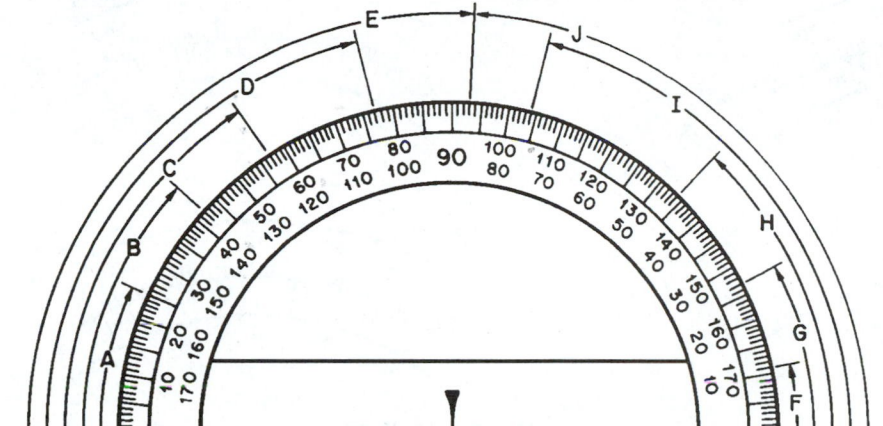

A = _____
B = _____
C = _____
D = _____
E = _____
F = _____
G = _____
H = _____
I = _____
J = _____

2. Using a protractor, lay out the following angles.

   a. 19°   c. 80°   e. 4°   g. 97°   i. 150°

   b. 65°   d. 12°   f. 92°   h. 123°   j. 166°

3. Lay out a 3-sided closed figure (triangle) of any size containing angles of 47° and 105°. Measure the third angle. How many degrees are contained in the third angle?

   _____

4. Lay out a 4-sided figure (quadrilateral) of any size containing angles of 89°, 69°, and 124°. Measure the fourth angle. How many degrees are contained in the fourth angle?

   _____

5. Measure each of the following angles, $\angle 1$–$\angle 14$, to the nearer degree. Extend the sides of the angles if necessary.

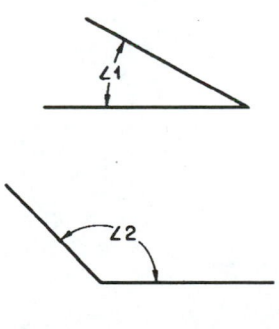
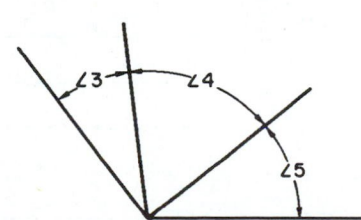
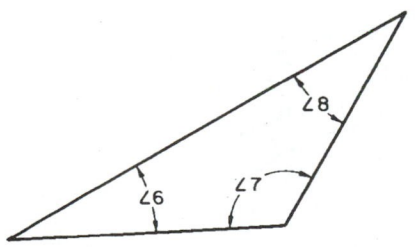
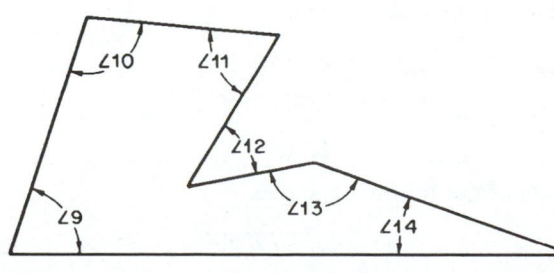

$\angle 1$ = _____
$\angle 2$ = _____
$\angle 3$ = _____
$\angle 4$ = _____
$\angle 5$ = _____
$\angle 6$ = _____
$\angle 7$ = _____
$\angle 8$ = _____
$\angle 9$ = _____
$\angle 10$ = _____
$\angle 11$ = _____
$\angle 12$ = _____
$\angle 13$ = _____
$\angle 14$ = _____

Vernier Protractor

Write the values of the settings on the following vernier protractor scales.

6. _____

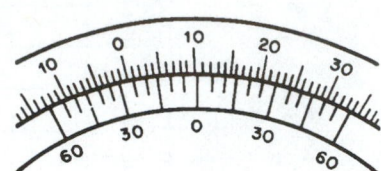

7. _____

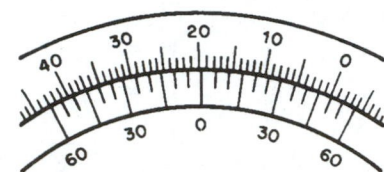

8. _____

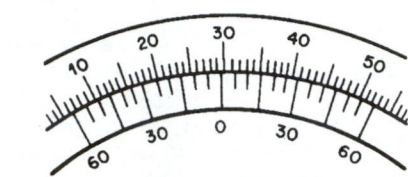

9. _____

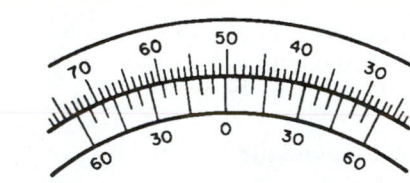

10. _____

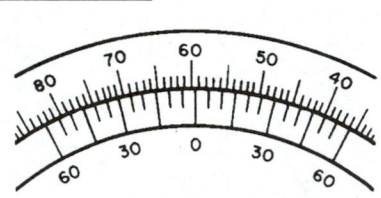

11. _____

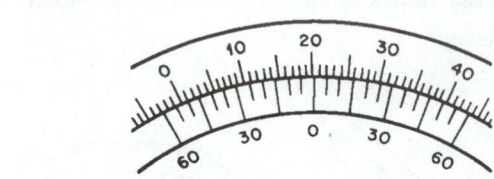

12. _____

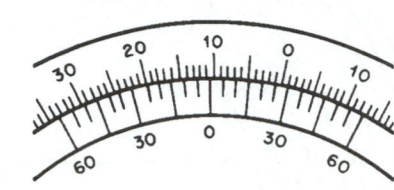

13. _____

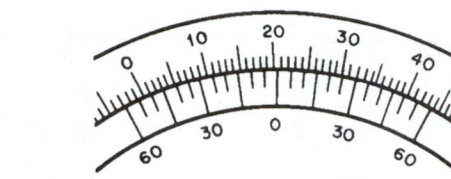

14. _____

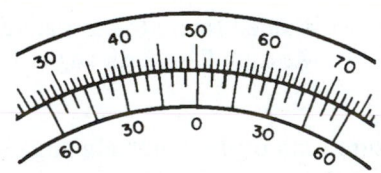

Complementary and Supplementary Angles

15. Write the complements of the following angles.

   a. 43° _____ d. 5° _____ g. 21°43′ _____

   b. 76° _____ e. 67°49′ _____ h. 78°19′27″ _____

   c. 17° _____ f. 45°19′ _____ i. 59°0′59″ _____

16. Write the supplements of the following angles.

   a. 13° _____ d. 179°59′ _____ g. 2°43′20″ _____

   b. 65° _____ e. 0°49′ _____ h. 68°21′29″ _____

   c. 91° _____ f. 89°59′ _____ i. 133°32′08″ _____

UNIT **46** Angles

Objectives   After studying this unit you should be able to

- Identify different types of angles.
- Determine unknown angles in geometric figures using the principles of opposite, alternate interior, corresponding, parallel, and perpendicular angles.

Naming Angles

Angles are named by a number, a letter, or three letters. When an angle is named with three letters, the vertex must be the middle letter. For example, the angle shown can be called ∠1, ∠C, ∠ACB, or ∠BCA.

In cases where a point is the vertex of more than one angle, a single letter cannot be used to name an angle. For example, since point E is the vertex of three different angles, the single letter E cannot be used in naming the angle.

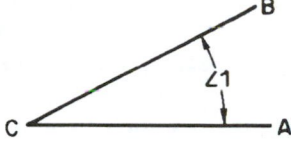

∠1 is called ∠GEH or ∠HEG.
∠2 is called ∠FEG or ∠GEF.
∠3 is called ∠FEH or ∠HEF.

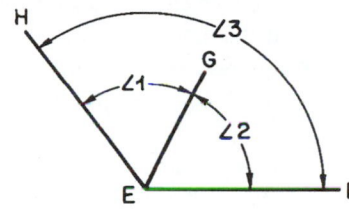

Types of Angles

An *acute angle* is an angle that is less than 90°. Angle 1 is acute.

A *right angle* is an angle of 90°. Angle A is a right angle.

An *obtuse angle* is an angle greater than 90° but less than 180°. Angle ABC is an obtuse angle.

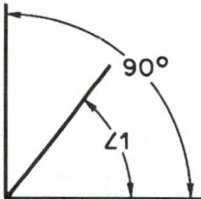
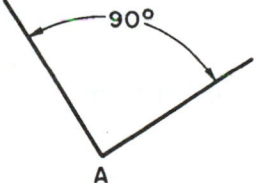
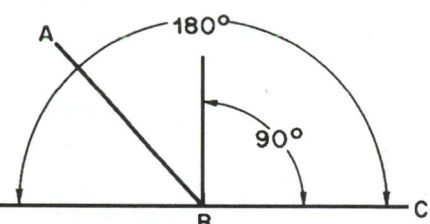

A *straight angle* is an angle of 180°. A straight line is a straight angle. Line EFG is a straight angle.

A *reflex angle* is an angle greater than 180° and less than 360°. Angle 3 is a reflex angle.

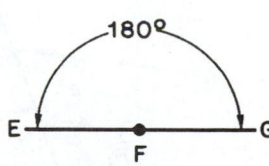
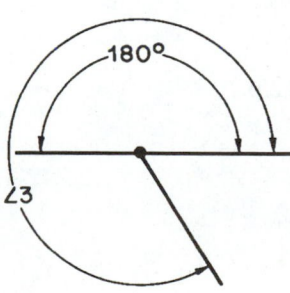

Adjacent Angles

Two angles are *adjacent* if they have a common side and a common vertex. Angle 1 and angle 2 shown are adjacent since they both contain the common side BC and the common vertex B. Angle 4 and angle 5 shown are not adjacent. The angles do not have a common vertex.

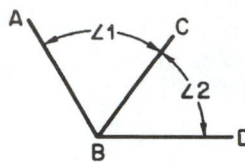 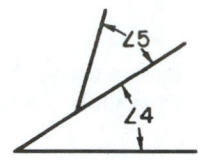

Angles Formed by a Transversal

A *transversal* is a line that intersects (cuts) two or more lines. Line EF is a transversal since it cuts lines AB and CD.

> *Alternate interior angles* are pairs of interior angles on opposite sides of the transversal. The angles have different vertices. For example, angles 3 and 5 and angles 4 and 6 are alternate interior angles.

> *Corresponding angles* are pairs of angles, one interior and one exterior with both angles on the same side of the transversal. The angles have different vertices. For example, angles 1 and 5, 2 and 6, 3 and 7, and 4 and 8 are corresponding angles.

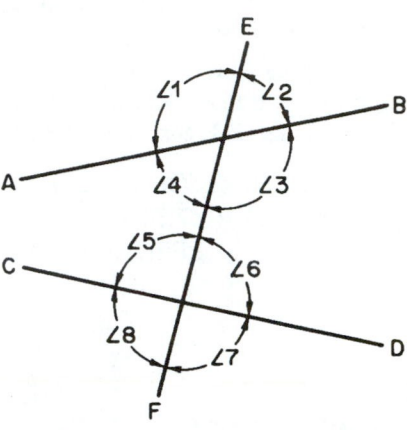

Geometric Principles

In this book, geometric postulates, theorems, and corollaries are grouped together and are called geometric principles. *Geometric principles* are statements of truth which are used as geometric rules. The principles will not be proved, but will be used as the basis for problem solving.

> ➤ **Principle 1**
> **If two lines intersect, the opposite or vertical angles are equal.**

Given: AB intersects CD.

Conclusion: $\angle 1 = \angle 3$ and $\angle 2 = \angle 4$.

> ➤ **Principle 2**
> **If two parallel lines are intersected by a transversal, the alternate interior angles are equal.**

Given: AB ‖ CD.

Conclusion: $\angle 3 = \angle 5$ and $\angle 4 = \angle 6$.

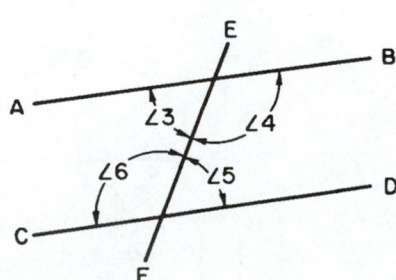

➤ **If two lines are intersected by a transversal and a pair of alternate interior angles are equal, the lines are parallel.**

Given: $\angle 1 = \angle 2$.

Conclusion: AB ‖ CD.

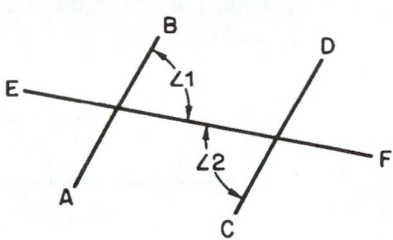

➤ **Principle 3**
If two parallel lines are intersected by a transversal, the corresponding angles are equal.

Given: AB ‖ CD.

Conclusion: $\angle 1 = \angle 5$, $\angle 2 = \angle 6$, $\angle 3 = \angle 7$ and $\angle 4 = \angle 8$.

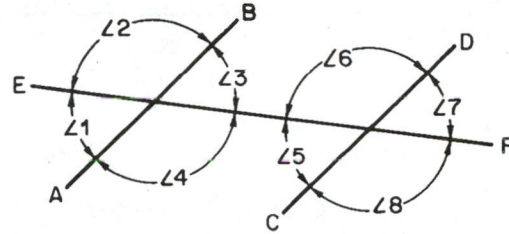

➤ **If two lines are intersected by a transversal and a pair of corresponding angles are equal, the lines are parallel.**

Given: $\angle 1 = \angle 2$.

Conclusion: AB ‖ CD.

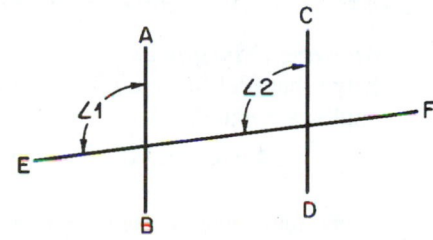

➤ **Principle 4**
Two angles are either equal or supplementary if their corresponding sides are parallel.

Given: AB ‖ FG and BC ‖ DE.

Conclusion: $\angle 1 = \angle 3$ and $\angle 1$ and $\angle 2$ are supplementary.

$(\angle 1 + \angle 2 = 180°)$

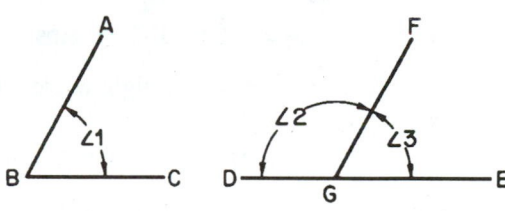

➤ **Principle 5**
Two angles are either equal or supplementary if their corresponding sides are perpendicular.

Given: AB ⊥ DH and BC ⊥ EF.

Conclusion: $\angle 1 = \angle 2$; $\angle 1$ and $\angle 3$ are supplementary.

$(\angle 1 + \angle 3 = 180°)$

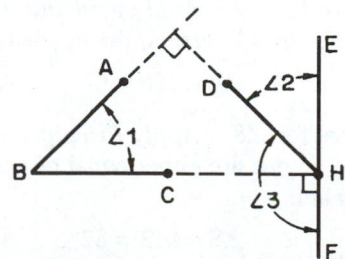

The following example illustrates the method of solving angular measure problems. Values of angles are determined by applying angular geometric principles and the fact that a straight angle (straight line) contains 180°.

Example Given: AB ∥ CD, EF ∥ GH, ∠1 = 115°, and ∠2 = 82°. Determine the values of ∠3 through ∠9.

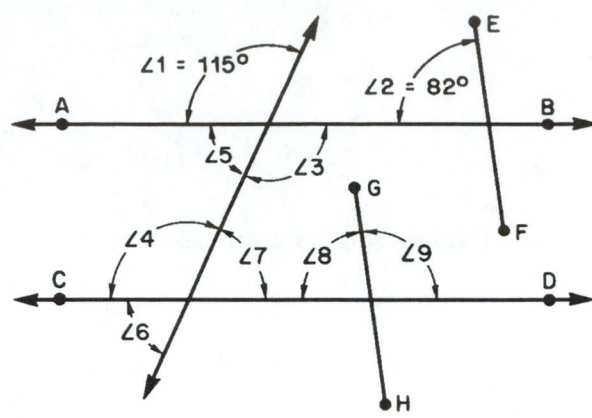

Solve for ∠3. *Apply Principle 1.*
If two lines intersect, the opposite or vertical angles are equal.

$$\angle 3 = \angle 1 = 115° \quad \text{Ans}$$

Solve for ∠4. *Apply either Principle 2 or Principle 3.*

Applying Principle 2:
If two parallel lines are intersected by a transversal, the alternate interior angles are equal.

$$\angle 4 = \angle 3 = 115° \quad \text{Ans}$$

or,

Applying Principle 3:
If two parallel lines are intersected by a transversal, the corresponding angles are equal.

$$\angle 4 = \angle 1 = 115° \quad \text{Ans}$$

Solve for ∠5. Since a straight angle (straight line) contains 180°, ∠5 and ∠1 are supplementary.

$$\angle 5 = 180° - \angle 1 = 180° - 115° = 65° \quad \text{Ans}$$

Solve for ∠6. *Apply Principle 3.*
If two parallel lines are intersected by a transversal, the corresponding angles are equal.

$$\angle 6 = \angle 5 = 65° \quad \text{Ans}$$

Solve for ∠7. *Apply Principle 1.*
If two lines intersect, the opposite or vertical angles are equal.

$$\angle 7 = \angle 6 = 65° \quad \text{Ans}$$

Solve for ∠8. *Apply Principle 4.*
Two angles are either equal or supplementary if their corresponding sides are parallel.

$$\angle 8 = \angle 2 = 82° \quad \text{Ans}$$

Solve for ∠9. Since a straight angle (straight line) contains 180°, ∠8 and ∠9 are supplementary.

$$\angle 9 = 180° - \angle 8 = 180° - 82° = 98° \quad \text{Ans}$$

APPLICATION

Naming Angles

1. Name each of the following angles in three additional ways.

    a. ∠1 _____

    b. ∠2 _____

    c. ∠C _____

    d. ∠D _____

    e. ∠E _____

    f. ∠F _____

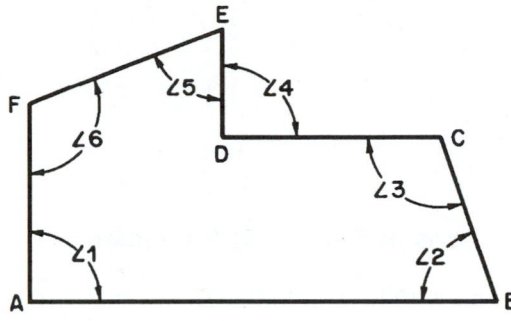

2. Name each of the following angles in two additional ways.

    a. ∠1 _____

    b. ∠CBF _____

    c. ∠3 _____

    d. ∠ECB _____

    e. ∠5 _____

    f. ∠BCD _____

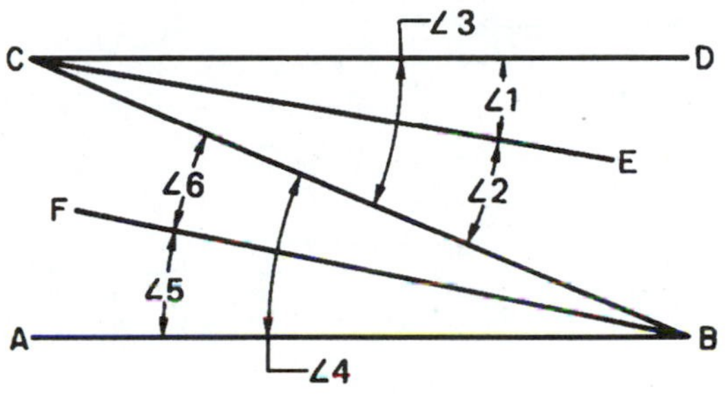

Types of Angles

3. Identify each of the following angles as acute, obtuse, right, straight, or reflex.

    a. ∠BAF _____

    b. ∠ABF _____

    c. ∠CBF _____

    d. ∠DCA _____

    e. ∠BFA _____

    f. ∠BFD _____

    g. ∠ABC _____

    h. ∠BCD _____

    i. ∠AED _____

    j. ∠1 _____

    k. ∠DFA _____

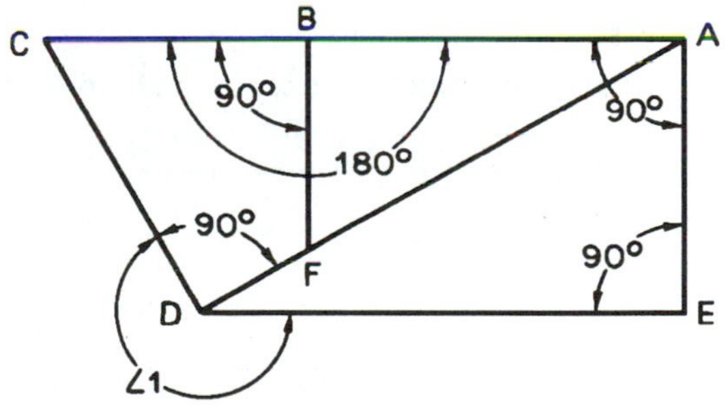

4. Name all pairs of adjacent angles shown in the figure.

    _____

    _____

    _____

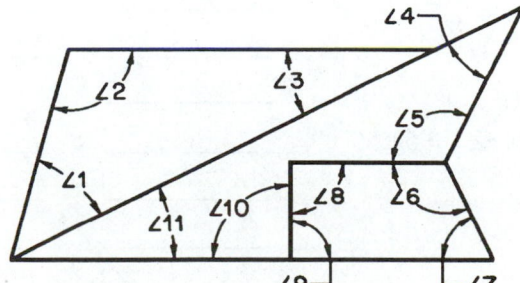

5. Alternate interior angles and corresponding angles are shown in the figure.

   a. Name all pairs of alternate interior angles. _____

   b. Name all pairs of corresponding angles. _____

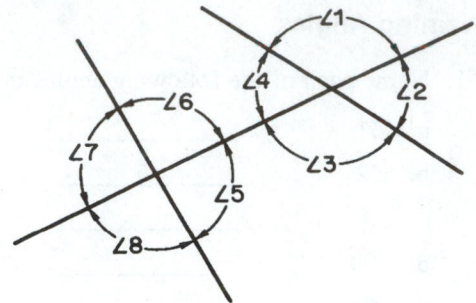

Applications of Geometric Principles

Solve the following problems.

6. Determine the values of ∠1 through ∠5.

   _____

   _____

   _____

   _____

   _____

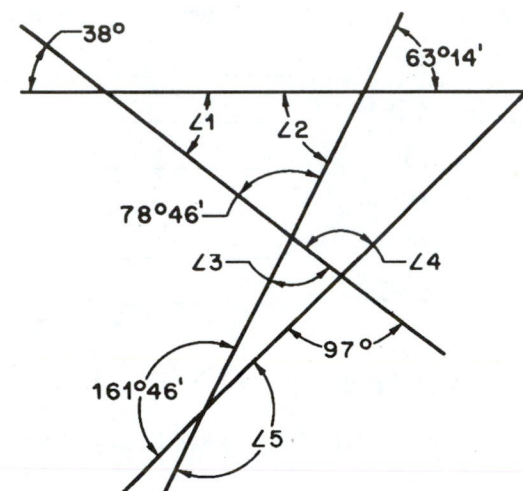

7. Determine the values of ∠2, ∠3, and ∠4 for these given values of ∠1.

   a. ∠1 = 32°   _____   _____

   _____

   b. ∠1 = 35°19′   _____   _____

   _____

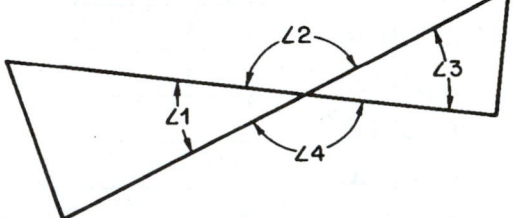

8. Given: AB ∥ CD. Determine the values of ∠2 through ∠8 for these given values of ∠1.

   a. ∠1 = 68°   _____   _____

   _____   _____

   _____   _____

   _____

   b. ∠1 = 52°55′   _____   _____

   _____   _____

   _____   _____

   _____

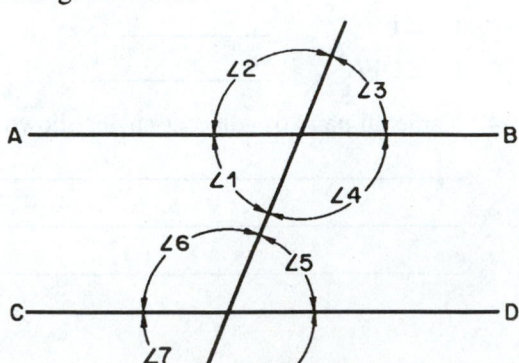

9. Given: Hole centerlines EF ‖ GH and MP ‖ KL. Determine the values of ∠1 through ∠15 for these values of ∠16.

   a. ∠16 = 71° _____

   _____

   b. ∠16 = 86°52′ _____

   _____

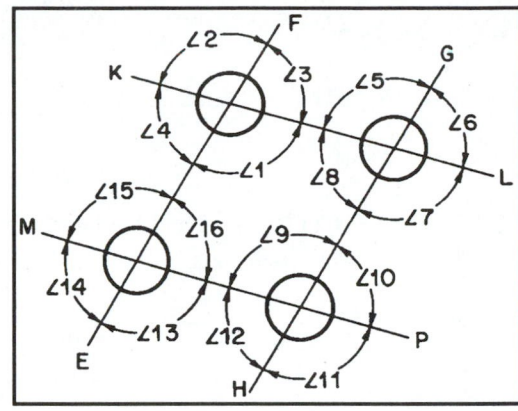

10. Given: Hole centerlines AB ‖ CD and EF ‖ GH. Determine the values of ∠1 through ∠22 for these given values of ∠23, ∠24, and ∠25.

   a. ∠23 = 97°, ∠24 = 34°, and ∠25 = 102°

   b. ∠23 = 112°23′, ∠24 = 27°53′, and ∠25 = 95°18′

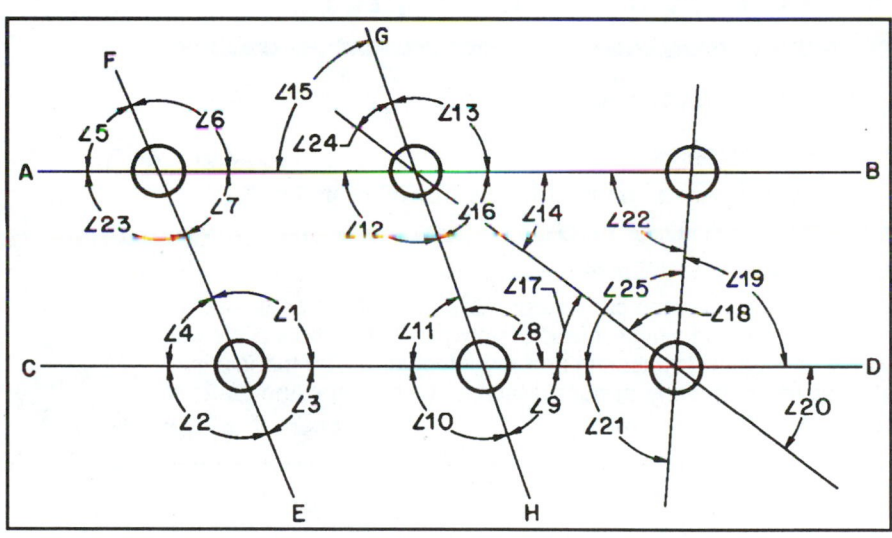

11. Given: AB ‖ CD, AC ‖ ED. Determine the value of ∠2 and ∠3 for these values of ∠1.

   a. ∠1 = 67°   _____   _____

   b. ∠1 = 74°12′   _____   _____

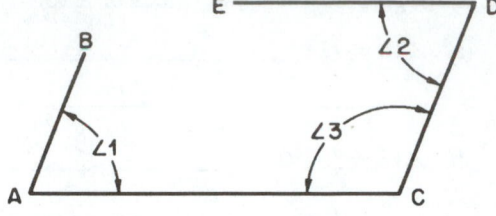

12. Given: FH ‖ GS ‖ KM and FG ‖ HK. Determine the values of ∠1, ∠2, and ∠3 for these values of ∠4.

   a. ∠4 = 116°   _____   _____   _____

   b. ∠4 = 107°43′   _____   _____   _____

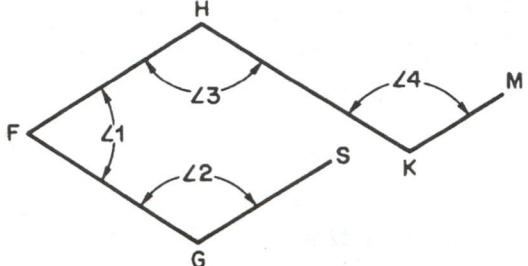

UNIT 47 Introduction to Triangles

Objectives   After studying this unit you should be able to

- **Identify different types of triangles.**

- **Determine unknown angles based on the principle that all triangles contain 180°.**

- **Identify corresponding parts of triangles.**

A *polygon* is a closed plane figure formed by three or more straight line segments. A *triangle* is a three-sided polygon; it is the simplest kind of polygon. The symbol △ means triangle. Triangles are widely applied in engineering and manufacturing. The triangle is a rigid figure which is the basic figure in many designs. Machine technicians and drafters require a knowledge of triangles in laying out work.

Types of Triangles

A *scalene triangle* has three unequal sides. It also has three unequal angles. Triangle ABC is scalene. Sides AB, AC, and BC are unequal and angles A, B, and C are unequal.

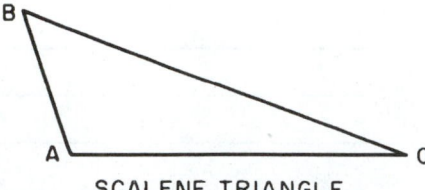

SCALENE TRIANGLE

An *isosceles triangle* has two equal sides. The equal sides are called *legs*. It also has two equal base angles. *Base angles* are the angles that are opposite the legs. In isosceles triangle RST, side RT = side ST and ∠R = ∠S.

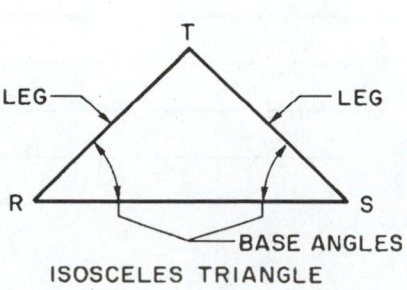

ISOSCELES TRIANGLE

An *equilateral triangle* has three equal sides. It also has three equal angles. In equilateral triangle DEF, sides DE = DF = EF and ∠D = ∠E = ∠F.

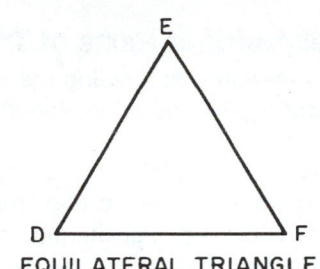

EQUILATERAL TRIANGLE

A *right triangle* has one right or 90° angle. The symbol ⌐ shown at the vertex of an angle means a right angle. The side opposite the right angle is called the *hypotenuse*. The other two sides are called *legs*. In right triangle HJK, ∠H = 90° and JK is the hypotenuse.

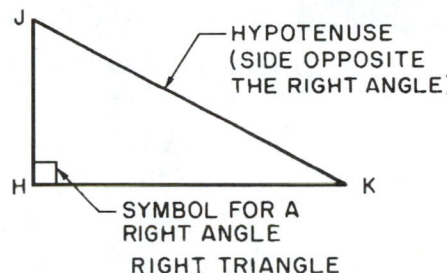

HYPOTENUSE
(SIDE OPPOSITE
THE RIGHT ANGLE)

SYMBOL FOR A
RIGHT ANGLE

RIGHT TRIANGLE

Angles of a Triangle

➤ **Principle 6**
The sum of the angles of any triangle is equal to 180°.

Example 1   Angles A, B, and C are hole centerline angles.
Angle A = 48°35′52″ and Angle C = 87°55′27″.
Determine ∠B.

$$∠B = 180° - (48°35′52″ + 87°55′27″)$$

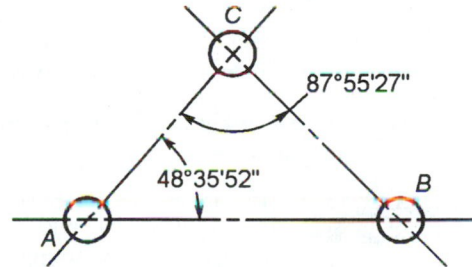

87°55'27"

48°35'52"

$180\ \boxed{-}\ \boxed{(}\ 48\ \boxed{° ′ ″}\ 35\ \boxed{° ′ ″}\ 52\ \boxed{° ′ ″}\ \boxed{+}\ 87\ \boxed{° ′ ″}\ 55\ \boxed{° ′ ″}\ 27\ \boxed{)}\ \boxed{=}\ \boxed{SHIFT}\ \boxed{←}$
→ 43°28′41″   Ans

or $180\ \boxed{-}\ \boxed{(}\ 48\ \boxed{.}\ 3552\ \boxed{2nd}\ \boxed{▸DD}\ \boxed{+}\ 87\ \boxed{.}\ 5527\ \boxed{2nd}\ \boxed{▸DD}\ \boxed{)}\ \boxed{=}\ \boxed{3rd}$
$\boxed{▸DMS}$ → 43°28′41″   Ans

Example 2   In isosceles triangle EFG, EF = EG and ∠E = 33°18′. Determine ∠F and ∠G.

∠E + ∠F + ∠G = 180°
180° − ∠E = ∠F + ∠G
180° − 33°18′ = ∠F + ∠G
146°42′ = ∠F + ∠G

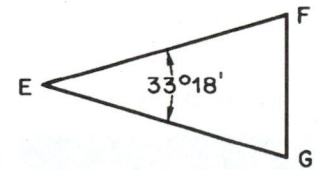

33°18'

Since ∠F = ∠G, ∠F and ∠G each = $\dfrac{146°42′}{2}$ = 73°21′   Ans

Example 3   Triangle HJK is equilateral. Determine ∠H, ∠J, and ∠K.

∠H + ∠J + ∠K = 180°

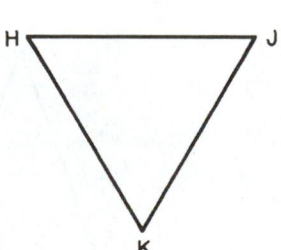

Since ∠H = ∠J = ∠K, each angle = $\dfrac{180°}{3}$ = 60°   Ans

Corresponding Parts of Triangles

It is essential to develop the ability to identify corresponding angles and sides of two or more triangles. Corresponding sides and angles between triangles are not determined by the positions of the triangles. The smallest angle of a triangle lies opposite the shortest side and the largest angle of a triangle lies opposite the longest side. *Corresponding angles* between two triangles are determined by comparing the lengths of the sides which lie opposite the angles. *Corresponding sides* between two triangles are determined by comparing the sizes of the angles which lie opposite the sides.

Example 1   In triangle ABC, determine the longest, next longest, and shortest sides.

The longest side is CB since it lies opposite the largest angle, 107°.   Ans

The next longest side is AB since it lies opposite the next largest angle, 43°.   Ans

The shortest side is AC since it lies opposite the smallest angle, 30°.   Ans

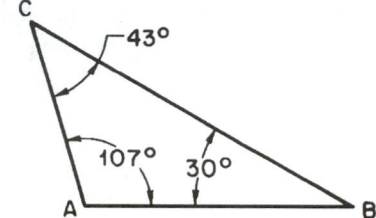

Example 2   In triangle DEF, determine the largest, next largest, and smallest angle. All dimensions are in inches.

The largest angle is ∠E since it lies opposite the longest side, 10 inches.   Ans

The next largest angle is ∠D since it lies opposite the next longest side, 7 inches.   Ans

The smallest angle is ∠F since it lies opposite the shortest side, 4 inches.   Ans

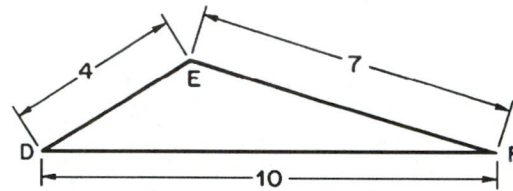

Example 3   In triangles ABC and FED, determine the pairs of corresponding angles between the two triangles. All dimensions are in millimeters.

Angle C corresponds to ∠D since each angle lies opposite the longest side of each triangle.

Angle B corresponds to ∠E since each angle lies opposite the next longest side of each triangle.

Angle A corresponds to ∠F since each angle lies opposite the shortest side of each triangle.

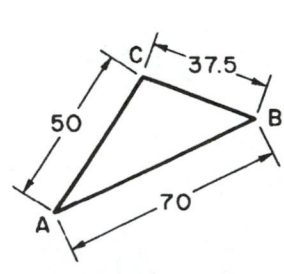
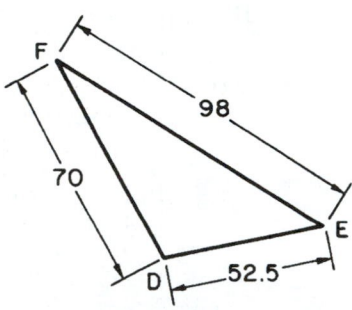

APPLICATION

Identify each of the triangles 1–8 as scalene, isosceles, equilateral, or right.

1. _____

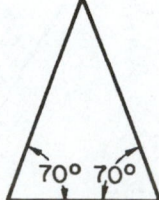

2. All dimensions are in inches.   _____

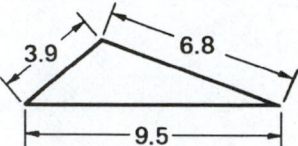

3. _____

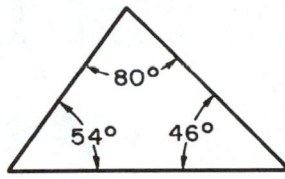

6. All dimensions are in millimeters. _____

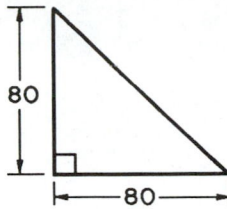

4. All dimensions are in millimeters. _____

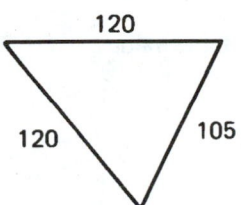

7. _____

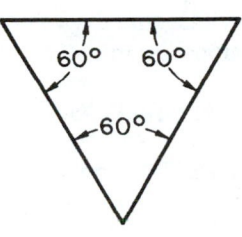

5. _____

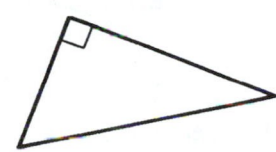

8. All dimensions are in inches. _____

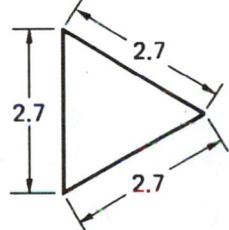

Angles of a Triangle

Solve the following problems.

9. Find the value of $\angle A + \angle B + \angle C$. _____

10. Find the value of the unknown angles for these given angle values.

   a. If $\angle 1 = 56°$ and $\angle 2 = 86°$, find $\angle 3$. _____

   b. If $\angle 2 = 81°$ and $\angle 3 = 46°$, find $\angle 1$. _____

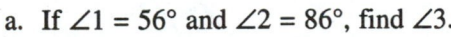

11. Find the value of the unknown angles for these given angle values.

   a. If $\angle 4 = 32°43'$ and $\angle 5 = 119°17'$, find $\angle 6$. _____

   b. If $\angle 5 = 123°17'13''$ and $\angle 6 = 27°$, find $\angle 4$. _____

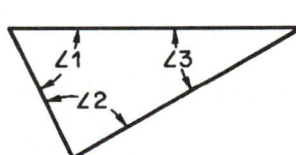

12. Find the value of the unknown angles for these given angle values.

   a. If $\angle A = 19°43'$, find $\angle B$. _____

   b. If $\angle B = 67°58'$, find $\angle A$. _____

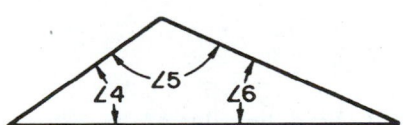
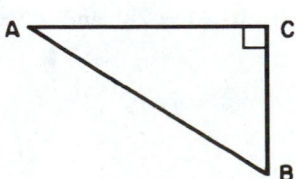

13. In triangle ABC, BC = 17.3 inches.

   a. Find AB.   _____

   b. Find AC.   _____

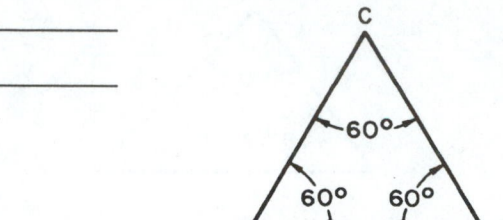

14. In triangle EFG, find the value of the unknown angles for these given angle values. All dimensions are in inches.

   a. If ∠E = 81°, find ∠G.   _____

   b. If ∠G = 83°27′, find ∠F.   _____

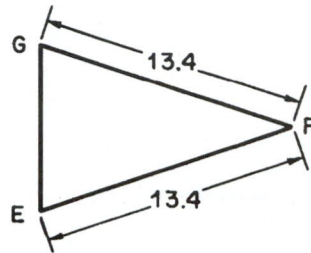

15. Find the value of the unknown angles for these given angle values. All dimensions are in millimeters.

   a. If ∠3 = 17°, find ∠1.   _____

   b. If ∠3 = 25°19′, find ∠2.   _____

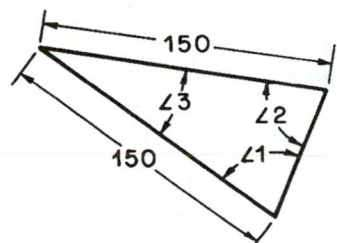

16. All dimensions are in inches.

   a. Find ∠3.   _____

   b. Find ∠4.   _____

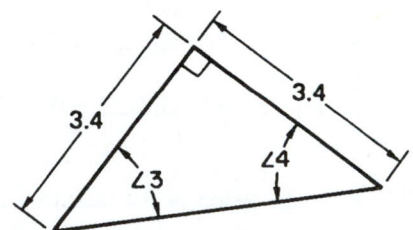

17. Find the value of the unknown angles for these given angle values.

   a. If ∠1 = 26° and ∠3 = 48°, find ∠2.   _____

   b. If ∠1 = 28° and ∠2 = 15°, find ∠3.   _____

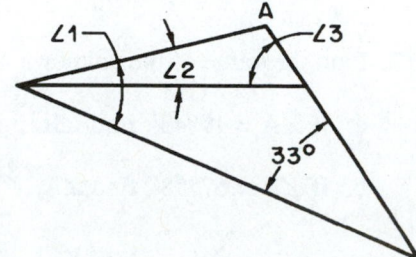

18. Hole centerlines AB ∥ CD.

    a. If ∠1 = 86°32′, find ∠2. _____

    b. If ∠2 = 67°47′, find ∠1. _____

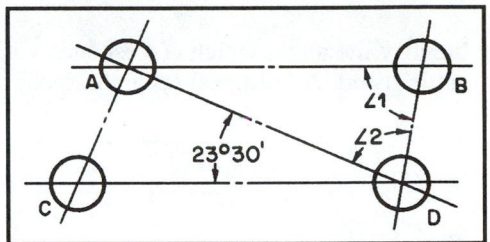

19. Find the value of the unknown angles listed.

    a. ∠3 _____

    b. ∠4 _____

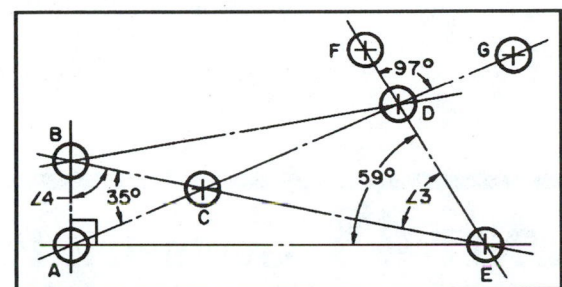

20. AB ∥ DE, BC is an extension of AB.

    a. If ∠E = 66°43′, find ∠A. _____

    b. If ∠A = 19°07′, find ∠E. _____

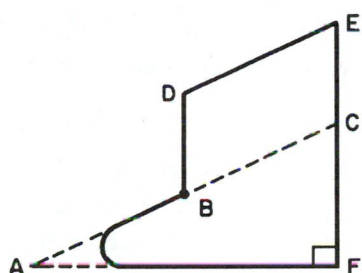

Corresponding Parts of Triangles

Determine the answers to the following problems which are based on corresponding parts.

21. All dimensions are in inches.

    a. Find the largest angle. _____

    b. Find the next largest angle. _____

    c. Find the smallest angle. _____

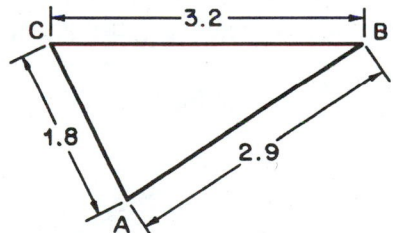

22. Refer to triangle EFG.

    a. Find the shortest side. _____

    b. Find the next shortest side. _____

    c. Find the longest side. _____

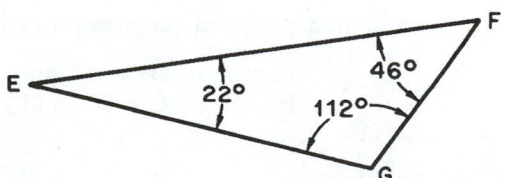

23. Identify the angle which corresponds with each angle listed. All dimensions are in millimeters.

    a. ∠A _____

    b. ∠B _____

    c. ∠1 _____

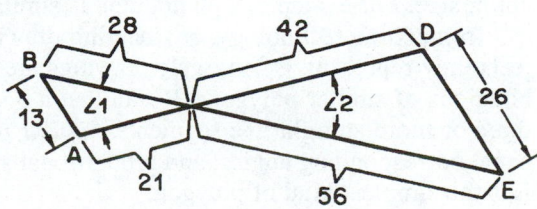

24. Identify the angle which corresponds with each angle listed. All dimensions are in inches.

   a. ∠F   _____

   b. ∠G   _____

   c. ∠H   _____

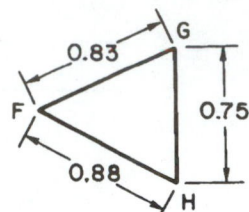 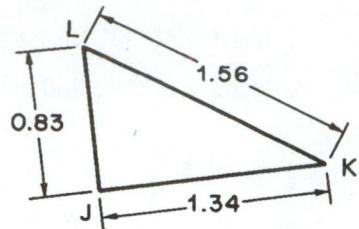

UNIT 48   Geometric Principles for Triangles and Other Common Polygons

Objectives   After studying this unit you should be able to

- Identify similar triangles and compute unknown angles and sides.
- Compute angles and sides of isosceles, equilateral, and right triangles.
- Determine interior angles of any polygon.

Congruent Triangles

Two triangles are *congruent* if they are identical in size and shape. If one triangle is placed on top of the other, they fit together exactly. The symbol ≅ means congruent. Corresponding parts of congruent triangles are equal.

Example   12△ABC ≅ 12△DEF.

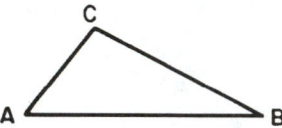 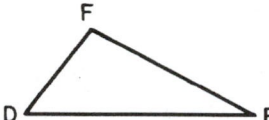

Corresponding parts of congruent triangles are equal.

   ∠A = ∠D, ∠B = ∠E and ∠C = ∠F.
   AB = DE, AC = DF, and BC = EF.

Similar Figures

Stated in a general way, similar figures are figures that are alike in shape but different in size. For example, a photograph is similar to the object that is photographed.

In machine technology, engineering drawings made to scale are similar to the objects they represent. Often, scale drawings are in the form of *similar polygons* or combinations of similar polygons. Recall that a *polygon* is a closed plane figure formed by three or more straight line segments. *Similar polygons* have the same number of sides, equal corresponding angles, and proportional sides. A triangle is a three-sided polygon; it is the simplest kind of polygon.

Similar Triangles

Two triangles are *similar* if their corresponding angles are equal. The symbol ~ means similar. Corresponding sides of similar triangles are proportional.

Example 1   Triangles ABC and DEF have equal corresponding angles.

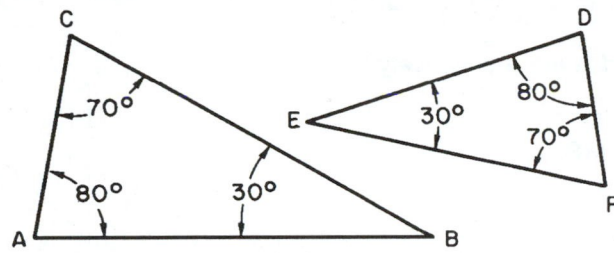

Two triangles are *similar* if their corresponding angles are equal.

12△ABC ~ △DEF

Example 2   The lengths of the sides of triangles HJK and LMN are given in inches.

$$\frac{HJ}{LM} = \frac{JK}{MN} = \frac{HK}{LN}$$

$$\frac{2}{4} = \frac{4}{8} = \frac{5}{10}$$

$$\frac{1}{2} = \frac{1}{2} = \frac{1}{2}$$

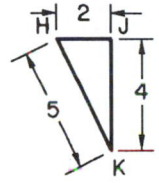 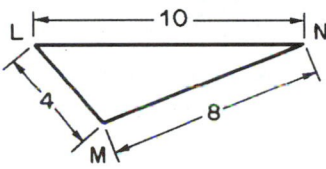

The corresponding sides are proportional.

△HJK ~ △LMN

Example 3   △PRS ~ △TWY.
All linear dimensions are in millimeters.

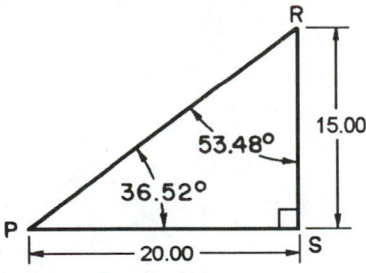 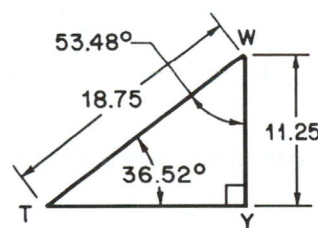

a.   Determine the length of side PR.

b.   Determine the length of side TY.

Set up proportions and solve for the unknown sides, PR and TY.

a. $\dfrac{WY}{RS} = \dfrac{WT}{PR}$

$$\frac{11.25 \text{ mm}}{15.00 \text{ mm}} = \frac{18.75 \text{ mm}}{PR}$$

$$11.25 \text{ mm } PR = 15.00 \text{ mm } (18.75 \text{ mm})$$

$$PR = \frac{15.00 \text{ mm } (18.75 \text{ mm})}{11.25 \text{ mm}}$$

$$PR = 25.00 \text{ mm} \qquad \text{Ans}$$

b. $\dfrac{WY}{RS} = \dfrac{TY}{PS}$

$$\frac{11.25 \text{ mm}}{15.00 \text{ mm}} = \frac{TY}{20.00 \text{ mm}}$$

$$15.00 \text{ mm } TY = 20.00 \text{ mm } (11.25 \text{ mm})$$

$$TY = \frac{20.00 \text{ mm } (11.25 \text{ mm})}{15.00 \text{ mm}}$$

$$TY = 15.00 \text{ mm} \qquad \text{Ans}$$

➤ **Principle 7**
Two triangles are similar if their sides are respectively parallel.

Given: AB ∥ DE, AC ∥ DF, and BC ∥ EF.

Conclusion: △ABC ~ △DEF.

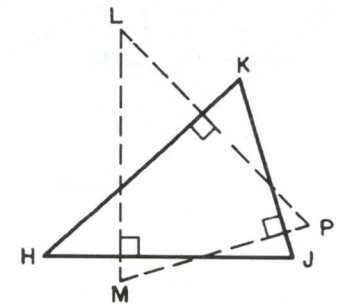

➤ **Two triangles are similar if their sides are respectively perpendicular.**

Given: HJ ⊥ LM, HK ⊥ LP, and JK ⊥ MP.

Conclusion: △HJK ~ △LMP.

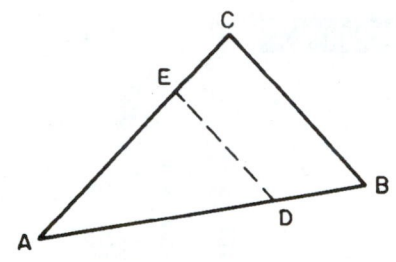

➤ **Within a triangle, if a line is drawn parallel to one side, the triangle formed is similar to the original triangle.**

Given: DE ∥ BC.

Conclusion: △ADE ~ △ABC.

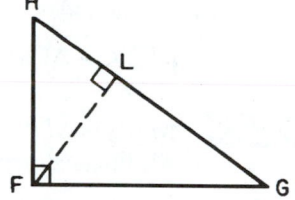

➤ **In a right triangle, if a line is drawn from the vertex of the right angle perpendicular to the opposite side, the two triangles formed and the original triangle are similar.**

Given: In rt △HFG, FL ⊥ HG.

Conclusion: △FLH ~ △GLF ~ △GFH.

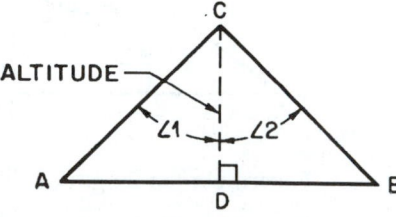

Isosceles, Equilateral, and Right Triangles

➤ **Principle 8**
In an isosceles triangle, an altitude to the base bisects the base and the vertex angle.

An *altitude* is a line drawn from a vertex perpendicular to the opposite side.

To *bisect* means to divide into two equal parts.

Given: Isosceles △ABC with AC = CB and line CD the altitude to base AB.

Conclusion: AD = BD and ∠1 = ∠2.

➤ **In an equilateral triangle, an altitude to any side bisects the side and the vertex angle.**

Given: Equilateral △EFG with EH the altitude to FG.

Conclusion: FH = GH and ∠3 = ∠4.

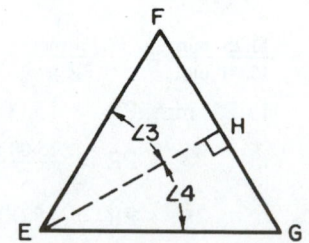

➤ **Principle 9**
In a right triangle, the square of the hypotenuse is equal to the sum of the squares of the other two sides or legs. If two sides of a right triangle are known, the third side can be calculated.

This principle, called the *Pythagorean Theorem,* is often used for solving machine problems.

Example 1   In right $\triangle ABC$, dimensions a and b are given; the centerlines meet at right angles at C. To inspect the distance between holes A and B, distance c must be computed. Dimensions are given in inches.

Side c is the hypotenuse. Substitute the given values for sides a and b and solve for distance c.

$$C^2 = a^2 + b^2$$
$$C = \sqrt{a^2 + b^2}$$
$$C = \sqrt{(6.027\,\text{in})^2 + (8.139\,\text{in})^2}$$
$$C = \sqrt{36.3247\,\text{in}^2 + 66.2433\,\text{in}^2}$$
$$C = \sqrt{102.5680\,\text{in}^2}$$
$$C = 10.128\,\text{in} \qquad \text{Ans (rounded)}$$

$$C = \sqrt{(6.027\,\text{in})^2 + (8.139\,\text{in})^2}$$

(6.027 X^2 + 8.139 X^2) $\sqrt{x}$ → 10.12758856,
10.128 in   Ans

or $\sqrt{}$ (6.027 X^2 + 8.139 X^2) = 10.12758856,
10.128 in   Ans

Example 2   In right $\triangle EFG$, $f = 5.800$ inches and hypotenuse $g = 7.200$ inches. Determine side e.

Side g is the hypotenuse. Substitute the given values, rearrange the equation, and solve for e.

$$g^2 = e^2 + f^2$$
$$(7.200\,\text{in})^2 = e^2 + (5.800\,\text{in})^2$$
$$51.840\,\text{sq in} = e^2 + 33.640\,\text{sq in}$$
$$18.200\,\text{sq in} = e^2$$
$$\sqrt{18.200\,\text{sq in}} = e$$
$$e = 4.266\,\text{in} \qquad \text{Ans (rounded)}$$

$$7.200^2 = e^2 + 5.800^2$$

Rearrange the equation in terms of e:

$$\sqrt{7.200^2 - 5.800^2} = e$$

(7.2 X^2 − 5.8 X^2) $\sqrt{x}$ → 4.2661458,
$e = 4.266$ in   Ans (rounded)

or $\sqrt{}$ (7.2 X^2 − 5.8 X^2) = 4.2661458,
$e = 4.266$ in   Ans (rounded)

Polygons

The types of polygons most common to machine trade applications in addition to triangles are squares, rectangles, parallelograms, and regular hexagons. A *regular polygon* is one which has equal sides and equal angles.

A *square* is a regular four-sided polygon. Each angle equals 90°. In the square ABCD shown, AB = BC = CD = AD and ∠A = ∠B = ∠C = ∠D = 90°.

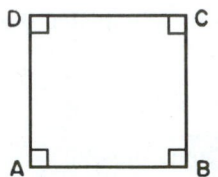

A *rectangle* is a four-sided polygon with opposite sides parallel and equal. Each angle equals 90°. In the rectangle EFGH shown, EF ∥ GH, EH ∥ FG; EF ∥ GH, EH ∥ FG; ∠E = ∠F = ∠G = ∠H = 90°.

A *parallelogram* is a four-sided polygon with opposite sides parallel and equal. Opposite angles are equal. In the parallelogram ABCD shown, AB ∥ CD, AD ∥ BC; AB = CD, AD = BC; ∠A = ∠C, ∠B = ∠D.

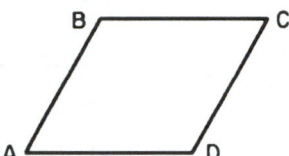

A *regular hexagon* is a six-sided figure with all sides equal and all angles equal. In the regular hexagon ABCDEF shown, AB = BC = CD = DE = EF = AF, and ∠A = ∠B = ∠C = ∠D = ∠E = ∠F.

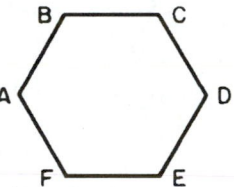

➤ **Principle 10**
The sum of the interior angles of a polygon of N sides is equal to (N – 2) times 180°.

Example 1 In figure EFGH, ∠E = 72°, ∠F = 95°, ∠G = 108°. Determine ∠H.

Since EFGH has 4 sides, N = 4.
The sum of the 4 angles = (4 − 2)180° = 2(180°) = 360°.

Add the 3 given angles and subtract from 360° to find ∠H.

$$∠H = 360° − (∠E + ∠F + ∠G)$$
$$∠H = 360° − (72° + 95° + 108°)$$
$$∠H = 360° − 275°$$
$$∠H = 85° \quad \text{Ans}$$

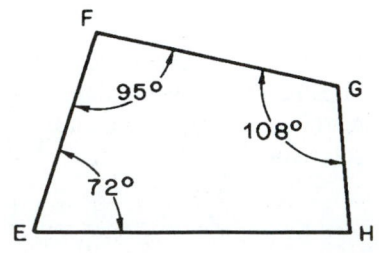

Example 2 Refer to figure ABCDEF and determine ∠1.

Since ABCDEF has 6 sides, N = 6.
The sum of the 6 angles = (6 − 2)180° = 4(180°) = 720°.

Find ∠2.

$$∠2 = 360° − 114.02° = 245.98°$$

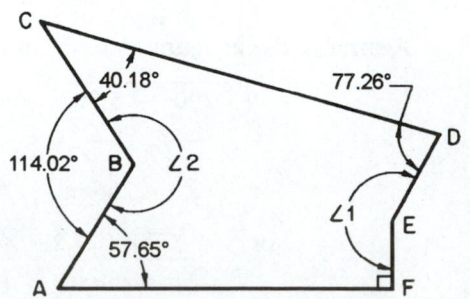

Add the 5 known interior angles and subtract from 720° to find ∠1.

$$\angle 1 = 720° - (57.65° + 245.98° + 40.18° + 77.26° + 90°)$$
$$\angle 1 = 720° - 511.07°$$
$$\angle 1 = 208.93° \quad \text{Ans}$$

720 [−] [(] 57.65 [+] 245.98 [+] 40.18 [+] 77.26 [+] 90 [)] [=] 208.93
$$\angle 1 = 208.93° \quad \text{Ans}$$

APPLICATION

Similar Triangles

1. Determine which of the following pairs of triangles (A–F) are similar. All linear dimensions are in inches.

   The similar pairs of triangles are _____ .

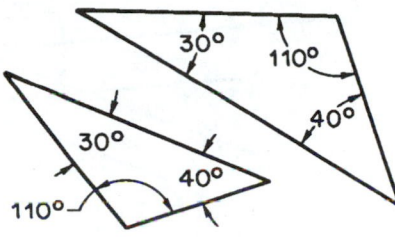

PAIR A

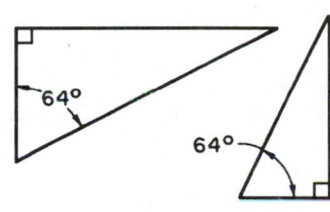

PAIR B

PAIR C

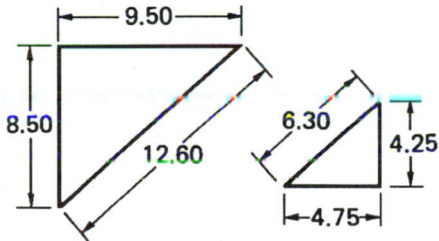

PAIR D

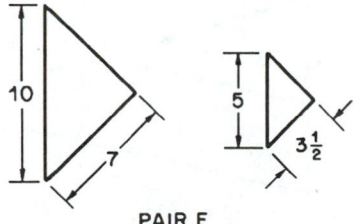

PAIR E

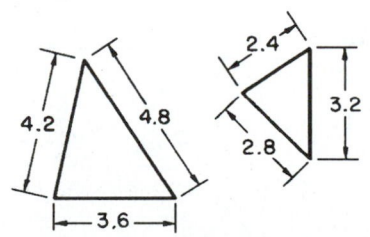

PAIR F

Solve the following problems.

2. In △ABC and △DEF, ∠A = ∠D, ∠B = ∠E, ∠C = ∠F.
   All dimensions are in inches.

   a. Find AC.   _____

   b. Find DE.   _____

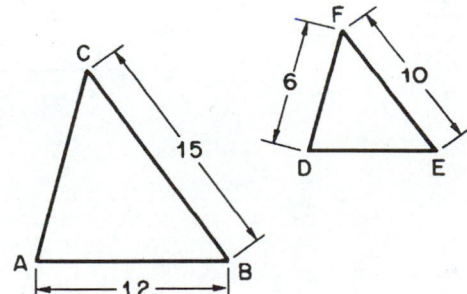

3. In the figure, ∠H = ∠P, ∠J = ∠M, ∠K = ∠L. All dimensions are in millimeters. Round the answers to 2 decimal places.

   a. Find HK. _____

   b. Find LM. _____

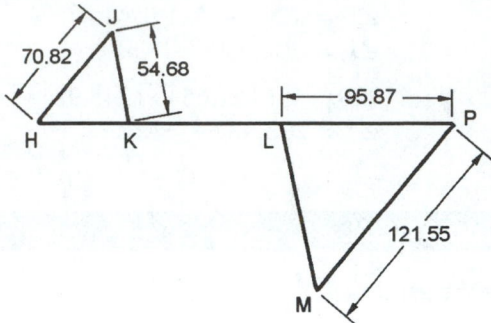

4. In △ABC and △DEF, AB ‖ DE, AC ‖ DF, BC ‖ EF.

   a. Find ∠A. _____

   b. Find ∠F. _____

   c. Find ∠B. _____

   d. Find ∠E. _____

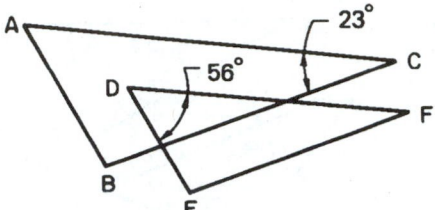

5. Use the figure to find the value of the following angles.

   a. Find ∠1. _____

   b. Find ∠2. _____

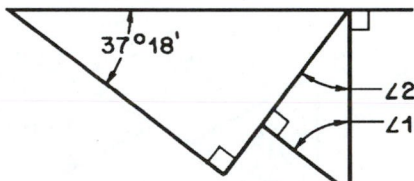

6. In △HJK, PM ‖ JK.

   a. Find ∠HPM. _____

   b. Find ∠PMK. _____

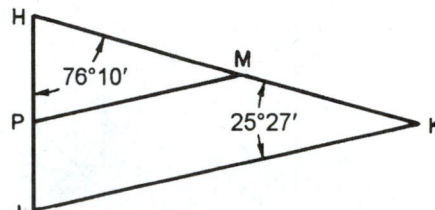

7. Refer to the figure to find these angles.

   a. ∠1 _____

   b. ∠2 _____

   c. ∠3 _____

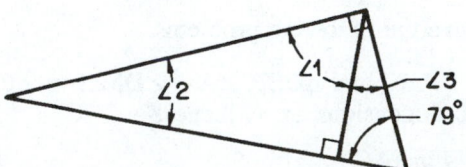

8. Refer to the figure to find these dimensions. All dimensions are in millimeters. Round the answers to 1 decimal place.

   a. Find dimension A.   _____

   b. Find dimension B.   _____

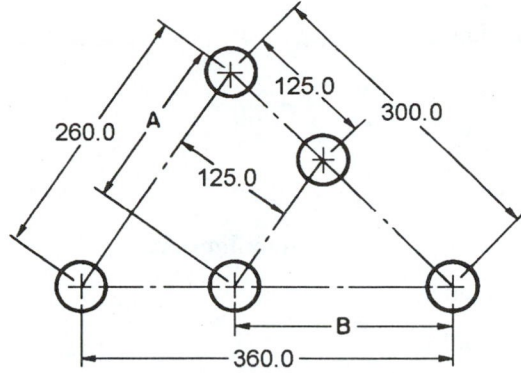

9. In this figure, AB ‖ DE and CB ‖ EF. All dimensions are in inches. Round the answers to 3 decimal places.

   a. Find x.   _____

   b. Find y.   _____

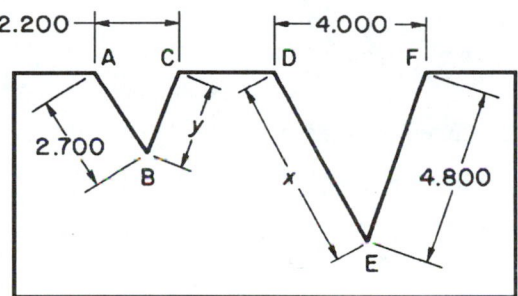

Isosceles, Equilateral, and Right Triangles

Solve the following problems.

10. All dimensions are in inches.

   a. Find x.   _____

   b. Find $\angle 1$.   _____

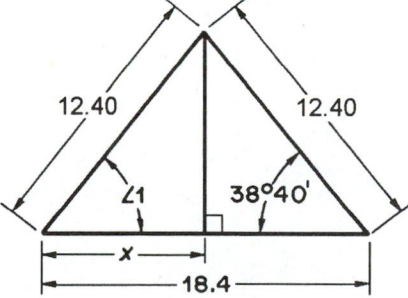

11. All dimensions are in millimeters.

   a. Find x.   _____

   b. Find y.   _____

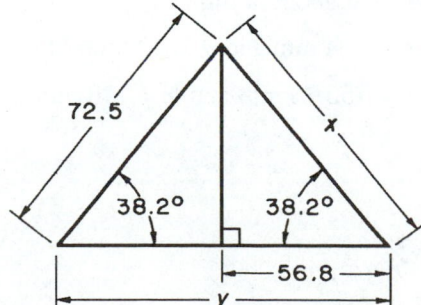

12. All dimensions are in inches.

    a. Find ∠1. _____

    b. Find ∠2. _____

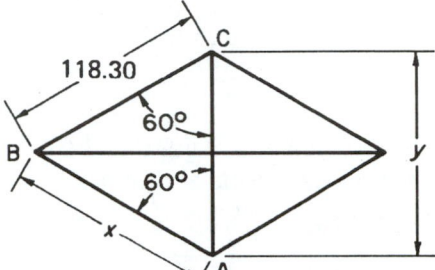

13. All dimensions are in millimeters.

    a. Find x. _____

    b. Find y. _____

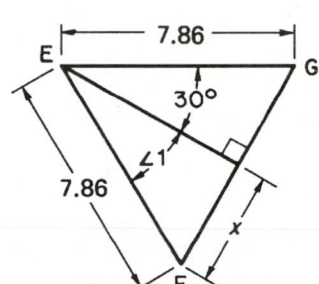

14. All dimensions are in inches.

    a. Find ∠1. _____

    b. Find x. _____

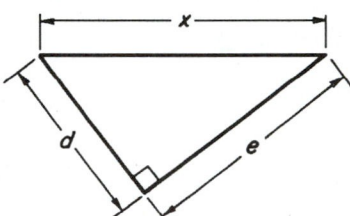

15. Refer to this figure. Using the given values, find the values of x.

    a. If $d = 9''$ and $e = 12''$, find x. _____

    b. If $d = 3''$ and $e = 4''$, find x. _____

16. Using the figure and these given values, find the values of y. Round the answers to the nearest whole millimeter.

    a. If $g = 108$ mm and $m = 123$ mm, find y. _____

    b. If $g = 153.70$ mm and $m = 170$ mm, find y. _____

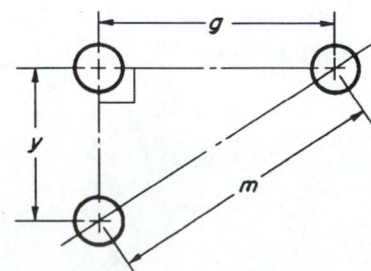

17. Using the figure and these given values, find the values of y.

   a. If Radius A = 360.00 mm and
      x = 480.00 mm, find y.   _____

   b. If Radius A = 216.00 mm and
      x = 288.00 mm, find y.   _____

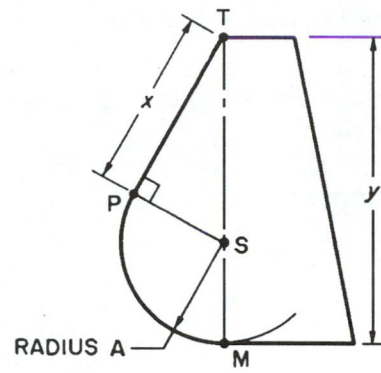

18. Three holes are drilled in the plate shown. All dimensions are in inches. Determine
    dimensions A and B to 3 decimal places.

   a. _____

   b. _____

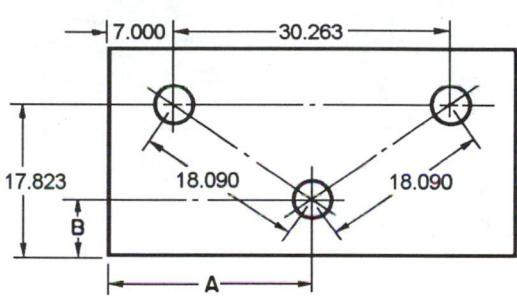

19. All dimensions are in inches. Round the answers to 3 decimal places.

   a. If y = 2.800″, find x.   _____
   b. If y = 3.000″, find x.   _____

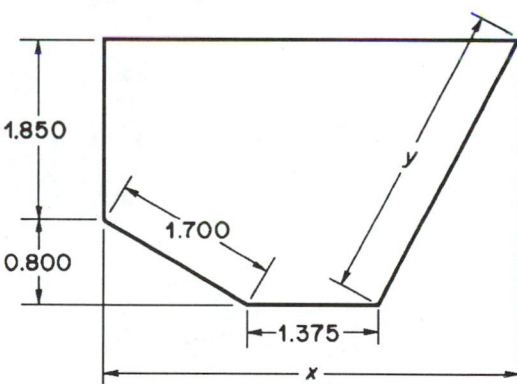

20. All dimensions are in inches. Round the answers to 3 decimal places.

   a. If y = 2.145″, find x.   _____
   b. If y = 2.265″, find x.   _____

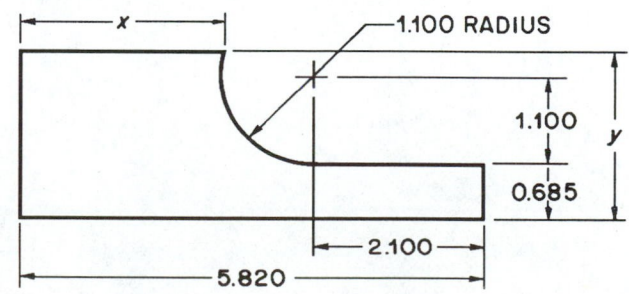

Polygons

Solve the following problems.

21. A template is shown. All dimensions are in millimeters. Determine length x and length y to two decimal places.

    x _____

    y _____

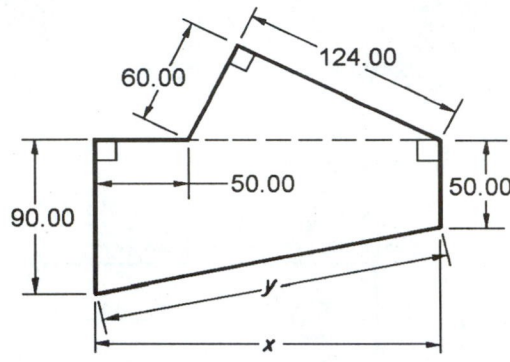

22. Refer to polygon ABCD.

    a. If $\angle 2 = 87.0°$, find $\angle 1$. _____

    b. If $\angle 1 = 114.0°$, find $\angle 2$. _____

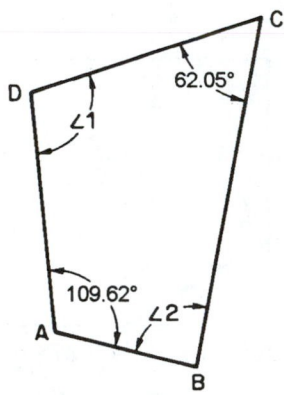

23. Use the angle values given.

    a. If $\angle 1 = 114°$, find $\angle 2$. _____

    b. If $\angle 2 = 83°$, find $\angle 1$. _____

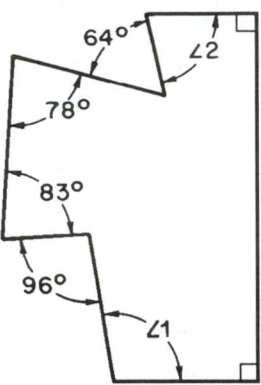

24. Use the angle values given to find $\angle 2$.

    a. If $\angle 1 = 37°$, find $\angle 2$. _____

    b. If $\angle 1 = 29°$, find $\angle 2$. _____

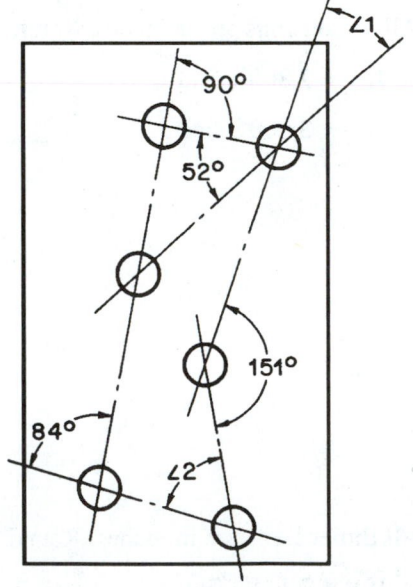

UNIT **49** Introduction to Circles

Objectives After studying this unit you should be able to

- Identify parts of a circle.
- Solve problems by using geometric principles which involve chords, arcs, central angles, perpendiculars, and tangents.

Definitions

A *circle* is a closed curve of which every point on the curve is equally distant from a fixed point called the center.

Refer to (1) for the following definitions:

The *circumference* is the length of the curved line which forms the circle.

A *chord* is a straight line segment that joins two points on the circle. AB is a chord.

A *diameter* is a chord that passes through the center of a circle. CD is a diameter.

A *radius* (plural radii) is a straight line segment that connects the center of the circle with a point on the circle. The radius is one-half the diameter. OE is a radius.

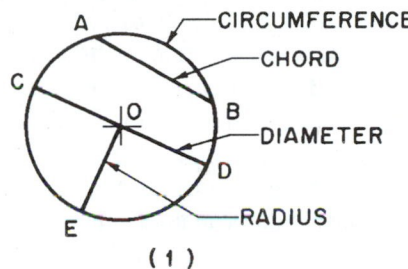

(1)

Refer to (2) for the following definitions:

An *arc* is that part of a circle between any two points on the circle. The symbol ⌢ written above the letters means arc. $\overset{\frown}{AB}$ is an arc.

A *tangent* to a circle is a straight line that touches the circle at one point only. The point on the circle touched by the tangent is called the *point of tangency* or *tangent point*. CD is a tangent and point P is a tangent point.

A *secant* is a straight line that passes through a circle and intersects the circle at two points. EF is a secant.

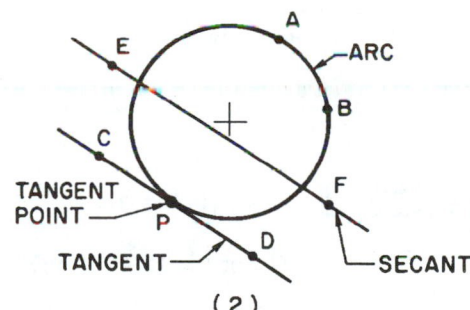

(2)

Refer to (3) for the following definitions:

A *segment* is that part of a circle which is bounded by a chord and its arc. The shaded figure ABC is a segment.

A *sector* is that part of a circle which is bounded by two radii and the arc intercepted by the radii. The shaded figure EOF is a sector.

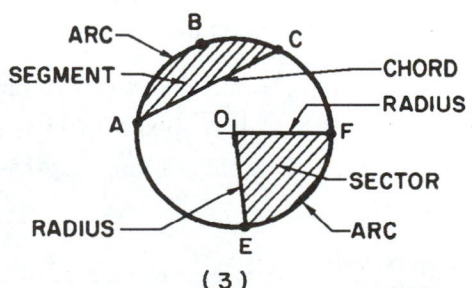

(3)

Refer to (4) for the following definitions:

A *central angle* is an angle whose vertex is at the center of a circle and whose sides are radii. Angle MON is a central angle.

An *inscribed angle* is an angle whose vertex is on the circle and whose sides are chords. Angle SRT is an inscribed angle.

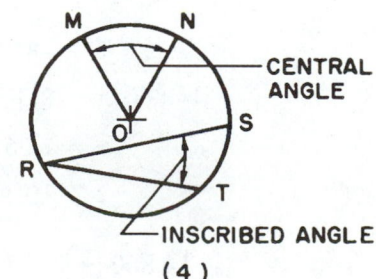

(4)

Circumference Formula

A polygon is *inscribed* in a circle when each vertex of the polygon is a point of the circle. In the figure shown, regular polygons are inscribed in circles. As the number of sides increases, the perimeter increases and approaches the circumference.

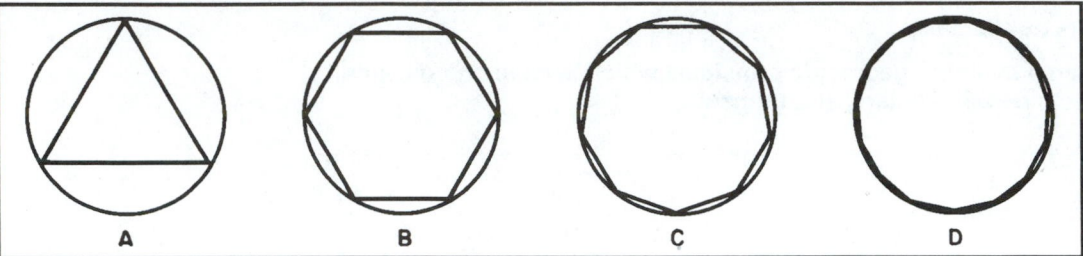

An important relationship exists between the circumference and the diameter of a circle. As the number of sides of an inscribed polygon increases, the perimeter approaches a certain number times the diameter. This number is called *pi*. The symbol for pi is π. No matter how many sides an inscribed polygon has, the value of π cannot be expressed exactly with digits. Pi is called an irrational number.

The circumference of a circle is equal to pi (π) times the diameter. Generally, for the degree of precision required in machining applications, a value of 3.1416 is used for π if a calculator is not available.

$$C = \pi d$$
or
$$C = 2\pi r$$

where C = circumference
π = pi
d = diameter
r = radius

Example 1   Compute the circumference of a circle with a 50.70 mm diameter.

$$C = \pi d = 3.1416(50.70 \text{ mm}) = 159.28 \text{ mm} \quad \text{Ans}$$

As presented on page 158, depressing the pi key ($\boxed{\pi}$) enters the value of pi to 10 digits (3.141592654) on most calculators. Recall that $\boxed{\pi}$ is the second or third function on many calculators.

$$C = \pi d, C = \pi (50.70 \text{ mm})$$
$$\boxed{\pi}\ \boxed{\text{x}}\ 50.70\ \boxed{=}\ 159.27875$$
$$C = 159.28 \text{ mm} \quad \text{Ans}$$

Example 2   Determine the radius of a circle which has a circumference of 14.860 inches.

$$C = 2\pi r$$
$$14.860 \text{ in} = 2(3.1416)(r)$$
$$r = 2.365 \text{ in} \quad \text{Ans}$$
$$r = \frac{14.860 \text{ in}}{2\pi}$$

$$14.860\ \boxed{\div}\ \boxed{(}\ 2\ \boxed{\text{x}}\ \boxed{\pi}\ \boxed{)}\ \boxed{=}\ 2.365042454, 2.365 \text{ in} \quad \text{Ans}$$

Geometric Principles

➤ **Principle 11**
In the same circle or in equal circles, equal chords cut off equal arcs.

Given: Circle A = Circle B and chords CD = EF = GH = MS.

Conclusion: $\overset{\frown}{CD} = \overset{\frown}{EF} = \overset{\frown}{GH} = \overset{\frown}{MS}$.

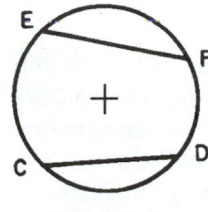
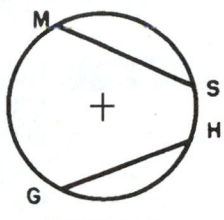

CIRCLE A CIRCLE B

➤ **Principle 12**
In the same circle or in equal circles, equal central angles cut off equal arcs.

Given: Circle D = Circle E and $\angle 1 = \angle 2 = \angle 3 = \angle 4$.

Conclusion: $\overset{\frown}{AB} = \overset{\frown}{FG} = \overset{\frown}{HK} = \overset{\frown}{MP}$.

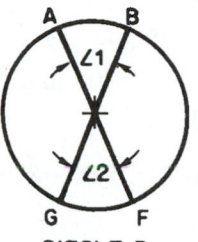
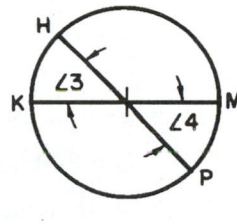

CIRCLE D CIRCLE E

➤
Principle 13
In the same circle or in equal circles, two central angles have the same ratio as the arcs which are cut off by the angles.

Example   Circle A = Circle B. If $\angle COD = 90°$, $\angle EOF = 50°$, $\overset{\frown}{CD} = 1.400''$, and $\overset{\frown}{GH} = 2.100''$, determine (a) the length of $\overset{\frown}{EF}$ and (b) the size of $\angle GOH$. All dimensions are in inches.

a.   Set up a proportion between $\overset{\frown}{CD}$ and $\overset{\frown}{EF}$ with their respective central angles. Solve for $\overset{\frown}{EF}$.

$$\frac{\angle COD}{\angle EOF} = \frac{\overset{\frown}{CD}}{\overset{\frown}{EF}}$$

$$\frac{90°}{50°} = \frac{1.400''}{\overset{\frown}{EF}}$$

$$9\overset{\frown}{EF} = 5(1.400'')$$

$$\overset{\frown}{EF} = \frac{5(1.400'')}{9}$$

$$\overset{\frown}{EF} = 0.778'' \text{Ans}$$

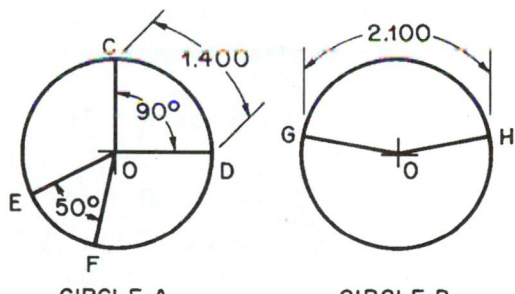

CIRCLE A CIRCLE B

b.   Set up a proportion between $\overset{\frown}{CD}$ and $\overset{\frown}{GH}$ with their central angles. Solve for $\angle GOH$.

$$\frac{\angle COD}{\angle GOH} = \frac{\overset{\frown}{CD}}{\overset{\frown}{GH}}$$

$$\frac{90°}{\angle GOH} = \frac{1.400''}{2.100''}$$

$$1.400(\angle GOH) = 90°(2.100)$$

$$\angle GOH = \frac{90°(2.100)}{1.400}$$

$$\angle GOH = 135° \text{Ans}$$

➤ **Principle 14**
A line drawn from the center of a circle perpendicular to a chord bisects the chord and the arc cut off by the chord. The perpendicular bisector of a chord passes through the center of a circle.

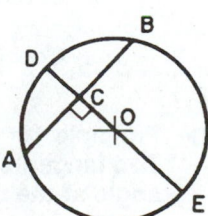

Given: Diameter DE ⊥ chord AB.

Conclusion: AC = BC and $\overset{\frown}{AD} = \overset{\frown}{BD}$ and $\overset{\frown}{AE} = \overset{\frown}{BE}$.

The use of Principle 14 with the Pythagorean Theorem (Principle 9) has wide practical application in the machine trades.

Example Holes A, B, and C are to be drilled in the plate shown. The centers of holes A and C lie on a 280.00-mm diameter circle. The center of hole B lies on the intersection of chord AC and segment OB which is perpendicular to AC. Compute working dimensions F, G, and H. All dimensions are in millimeters.

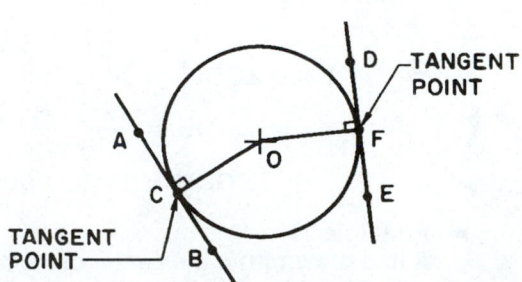

Compute dimension F: Applying Principle 14, AC is bisected by OB.

$$AB = BC = 250.00 \text{ mm} \div 2 = 125.00 \text{ mm}$$
$$F = 200.00 \text{ mm} - 125.00 \text{ mm} = 75.00 \text{ mm} \quad \text{Ans}$$

Compute dimension G.

$$G = 200.00 \text{ mm} + 125.00 \text{ mm} = 325.00 \text{ mm} \quad \text{Ans}$$

Compute dimension H.

In right $\triangle ABO$, AB = 125.00 mm, AO = 280.00 mm $\div$ 2 = 140.00 mm

Compute OB by applying the Pythagorean Theorem (Principle 9).

$$AO^2 = OB^2 + AB^2$$
$$(140.00 \text{ mm})^2 = OB^2 + (125.00 \text{ mm})^2$$
$$OB = 63.05 \text{ mm}$$

$$H = 180.00 \text{ mm} + 63.05 \text{ mm} = 243.05 \text{ mm} \quad \text{Ans}$$

$$H = 180.00 \text{ mm} + \sqrt{(140.00 \text{ mm})^2 - (125.00 \text{ mm})^2}$$

180 $\boxed{+}$ $\boxed{\sqrt{}}$ $\boxed{(}$ 140 $\boxed{x^2}$ $\boxed{-}$ 125 $\boxed{x^2}$ $\boxed{)}$ $\boxed{=}$
243.0476011, 243.05 mm   Ans

or   180 $\boxed{+}$ $\boxed{(}$ 140 $\boxed{x^2}$ $\boxed{-}$ 125 $\boxed{x^2}$ $\boxed{)}$ $\boxed{\sqrt{x}}$ $\boxed{=}$
243.0476011, 243.05 mm   Ans

➤ **Principle 15**
A line perpendicular to a radius at its extremity is tangent to the circle. A tangent is perpendicular to a radius at its tangent point.

Example 1 Given: Line AB ⊥ to a radius CO at point C.

Conclusion: Line AB is a tangent.

Example 2 Given: Tangent DE passes through point F of radius FO.

Conclusion: Tangent DE ⊥ radius FO.

➤ **Principle 16**
Two tangents drawn to a circle from a point outside the circle are equal. The angle at the outside point is bisected by a line drawn from the point to the center of the circle.

Example 1 Given: Tangents AP and BP are drawn to the circle from point P.
Conclusion: AP = BP.

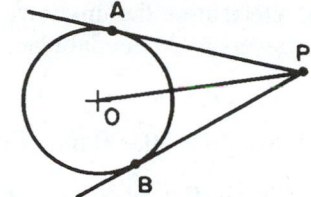

Example 2 Given: Line OP which extends from outside point P to center O.
Conclusion: ∠APO = ∠BPO.

➤ **Principle 17**
If two chords intersect inside a circle, the product of the two segments of one chord is equal to the product of the two segments of the other chord.

Example 1 Given: Chords AC and DE intersect at point B.
Conclusion: AB(BC) = BD(BE).

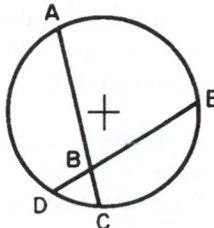

Example 2 If AB = 7.5 inches, BC = 2.8 inches, and BD = 2.1 inches, determine the length of BE.

$$AB(BC) = BD(BE)$$
$$7.5(2.8) = 2.1(BE)$$
$$21.0 = 2.1BE$$
$$BE = 10.0 \text{ inches} \qquad \text{Ans}$$

APPLICATION

Definitions

Name each of the parts of circles for the following problems.

1. a. AB _____
   b. CD _____
   c. EO _____
   d. Point O _____

3. a. M _____
   b. P _____
   c. SO _____
   d. TO _____
   e. RW _____
   f. $\overgroup{RW}$ _____

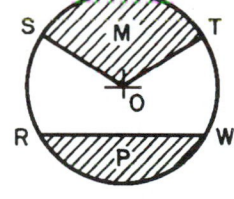

2. a. $\overgroup{GF}$ _____
   b. HK _____
   c. LM _____
   d. GF _____
   e. Point P _____

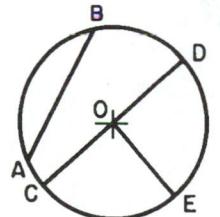

4. a. ∠1 _____
   b. ∠2 _____
   c. AO _____
   d. CD _____
   e. CE _____
   f. $\overgroup{AB}$ _____

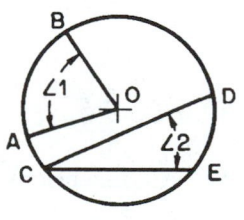

Circumference Formula

Use $C = \pi d$ or $C = 2\pi r$ where C = circumference
π = 3.1416 if not using calculator
d = diameter
r = radius

5. Determine the unknown value for each of the following problems. Round the answers to 3 decimal places.

   a. If $d = 6.500''$, find C. _____   e. If $C = 35.000''$, find d. _____

   b. If $d = 30.000$ mm, find C. _____   f. If $C = 218.000$ mm, find d. _____

   c. If $r = 18.600$ mm, find C. _____   g. If $C = 327.000$ mm, find r. _____

   d. If $r = 2.930''$, find C. _____   h. If $C = 7.680''$, find r. _____

6. Determine the length of wire, in feet, in a coil of 60.0 turns. The average diameter of the coil is 30.0 inches. Round the answer to the nearest whole foot. _____

7. A pipe with a wall thickness of 6.00 millimeters has an outside diameter of 79.20 millimeters. Compute the inside circumference of the pipe. Round the answer to 2 decimal places. _____

8. The flywheel of a machine has a 0.80-meter diameter and revolves 240.0 times per minute. How many meters does a point on the outside of the flywheel rim travel in 5.0 minutes? Round the answer to the nearest whole meter. _____

Geometric Principles

Solve the following problems. These problems are based on principles 11–14, although a problem may require the application of two or more of any of the principles. Round the answers to 3 decimal places where necessary unless otherwise stated.

9. Determine the length of belt required to connect the two pulleys shown. All dimensions are in inches. Round the answer to 2 decimal places. _____

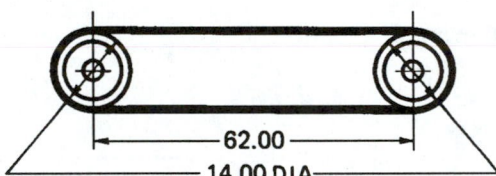

10. $\triangle ABC$ is equilateral. All dimensions are in inches.

    a. Find $\overset{\frown}{AB}$. _____

    b. Find $\overset{\frown}{BC}$. _____

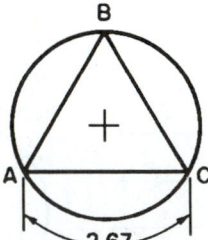

11. All dimensions are in inches.

    a. Find $\overset{\frown}{AB}$. _____

    b. Find $\overset{\frown}{BC}$. _____

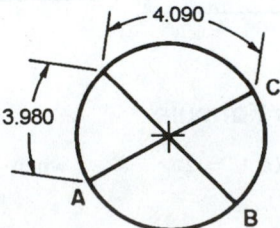

12. a. If $\overset{\frown}{EF}$ = 160 mm, find $\overset{\frown}{HP}$.   _____

    b. If $\overset{\frown}{HP}$ = 284 mm, find $\overset{\frown}{EF}$.   _____

    Round the answer to the nearest whole millimeter.

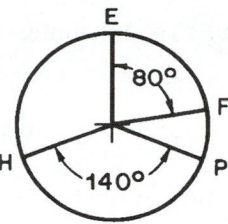

13. a. If $\overset{\frown}{SW}$ = 4.800″ and $\overset{\frown}{TM}$ = 5.760″, find ∠1.   _____

    b. If $\overset{\frown}{TM}$ = 4.128″ and $\overset{\frown}{SW}$ = 2.064″, find ∠1.   _____

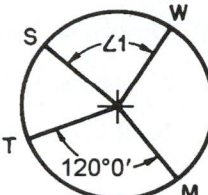

14. a. If AB = 5.378″ and $\overset{\frown}{AC}$ = 3.782″, find (1) DB and (2) $\overset{\frown}{ACB}$.   _____

    b. If DB = 3.017″ and $\overset{\frown}{ACB}$ = 7.308″, find (1) $\overset{\frown}{AB}$ and (2) $\overset{\frown}{CB}$.   _____

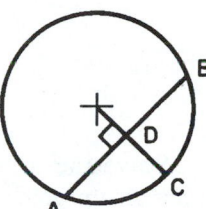

15. Find $\overset{\frown}{HK}$ when $\overset{\frown}{EF}$ = 21.23 mm.   _____

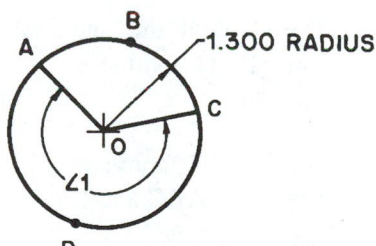

16. All dimensions are in inches.

    a. If ∠1 = 240°0′, find $\overset{\frown}{ABC}$.   _____

    b. If $\overset{\frown}{ABC}$ = 2.300″, find ∠1.   _____

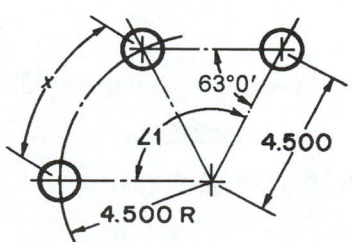

17. All dimensions are in inches.

    a. If x = 5.100″, find ∠1.   _____

    b. If x = 4.750″, find ∠1.   _____

18. a. If radius x = 7.500″ and y = 4.500″, find PM.   _____

    b. If radius x = 8.000″ and y = 4.800″, find PM.   _____

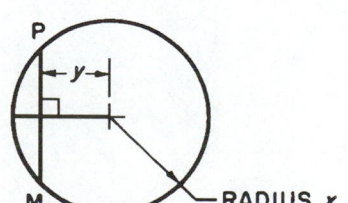

19. The circumference of this circle is 14.400″.

    a. If x = 3.200″, find $\angle 1$.   _____

    b. If $\angle 1$ = 36°0′, find x.   _____

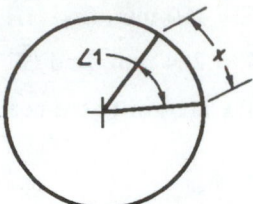

20. Determine the centerline distance between hole A and hole B for these values.

    a. Radius x = 8.000″ and DO = 2.100″.   _____

    b. Radius x = 1.200″ and DO = 0.700″.   _____

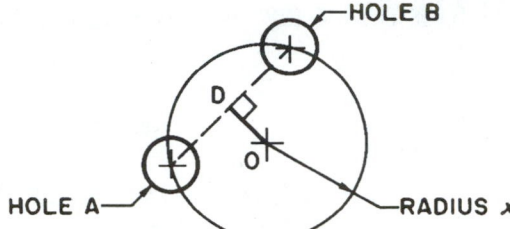

Solve the following problems. These problems are based on principles 15–17, although a problem may require the application of two or more of any of the principles. Round the answers to 3 decimal places where necessary unless otherwise stated.

21. Point P is a tangent point and $\angle 1$ = 107°18′.

    a. If $\angle 2$ = 41°21′, find (1) $\angle E$ and (2) $\angle F$.

    (1) _____   (2) _____

    b. If $\angle 2$ = 48°20′, find (1) $\angle E$ and (2) $\angle F$.

    (1) _____   (2) _____

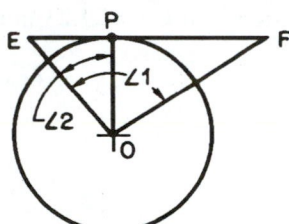

22. AB and CB are tangents.

    a. If y = 137.20 mm and $\angle ABC$ = 67.0°, find (1) $\angle 1$ and (2) x.

    (1) _____   (2) _____

    b. If x = 207.70 mm and $\angle 1$ = 33.8°, find (1) $\angle ABC$ and (2) y.

    (1) _____   (2) _____

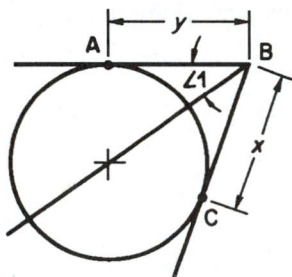

23. Point A is a tangent point. All dimensions are in inches.

    a. If y = 1.400″, find x.   _____

    b. If y = 1.800″, find x.   _____

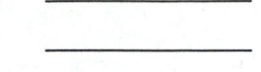

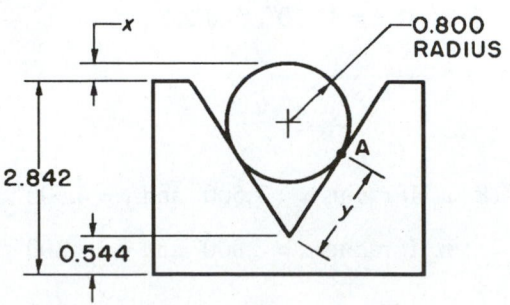

24. Points E, G, and F are tangent points.

    a. If ∠1 = 109°, find ∠2. _____

    b. If ∠1 = 118°45′, find ∠2. _____

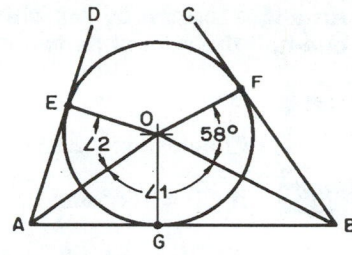

25. All dimensions are in millimeters. Round the answers to 2 decimal places.

    a. If EK = 150.00 mm, find GK. _____

    b. If GK = 120.00 mm, find EK. _____

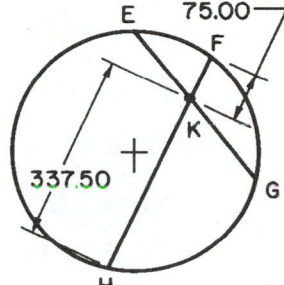

26. All dimensions are in inches.

    a. If PT = 1.800″, find x. _____

    b. If PT = 2.000″, find x. _____

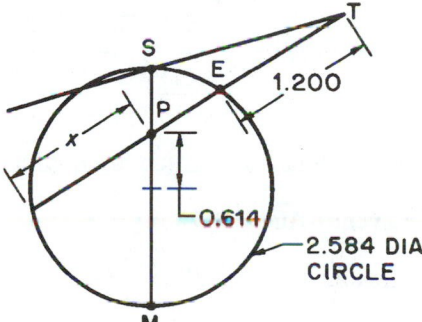

UNIT 50 Arcs and Angles of Circles

Objectives   After studying this unit you should be able to

- Solve problems by using geometric principles which involve angles formed inside, on, and outside a circle.
- Solve problems by using geometric principles which involve internally and externally tangent circles.

Angles Formed Inside a Circle

➤ **Principle 18**
  A central angle is equal to its intercepted arc.

(An *intercepted arc* is an arc which is cut off by a central angle.)

Given: $\overset{\frown}{AB} = 78°$.

Conclusion: ∠AOB = 78°.

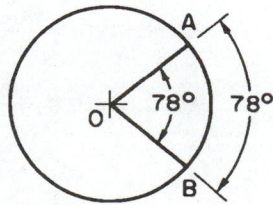

➤ **An angle formed by two chords which intersect inside a circle is equal to one-half the sum of its two intercepted arcs.**

Example 1   Given: Chords CD and EF intersect at point P.

Conclusion: $\angle EPD = \frac{1}{2}(\overset{\frown}{CF} + \overset{\frown}{DE})$.

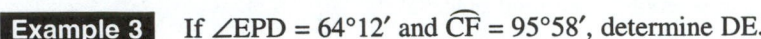

Example 2   If $\overset{\frown}{CF} = 106°$ and $\overset{\frown}{ED} = 42°$, determine $\angle EPD$.

$\angle EPD = \frac{1}{2}(106° + 42°) = 74°$ Ans

Example 3   If $\angle EPD = 64°12'$ and $\overset{\frown}{CF} = 95°58'$, determine DE.

$\angle EPD = \frac{1}{2}(\overset{\frown}{CF} + \overset{\frown}{DE})$

$64°12' = \frac{1}{2}(95°58' + \overset{\frown}{DE})$

$64°12' = 47°59' + \frac{1}{2}\overset{\frown}{DE}$

$16°13' = \frac{1}{2}\overset{\frown}{DE}$

$\overset{\frown}{DE} = 32°26'$ Ans

$DE = 2 \times 64°12' - 95°58'$

 $2\;\boxed{\times}\;64\;\boxed{°\,'\,''}\;12\;\boxed{°\,'\,''}\;\boxed{-}\;95\;\boxed{°\,'\,''}\;58\;\boxed{°\,'\,''}\;\boxed{=}$
$\boxed{\text{SHIFT}}\;\boxed{\longleftarrow}\; \rightarrow 32°26'$ Ans

or $2\;\boxed{\times}\;64\;\boxed{\cdot}\;12\;\boxed{\text{2nd}}\;\boxed{\overset{▸DD}{=}}\;\boxed{-}\;95\;\boxed{\cdot}\;58\;\boxed{\text{2nd}}\;\boxed{\overset{▸DD}{=}}\;\boxed{=}$
$\boxed{\text{3rd}}\;\boxed{▸\text{DMS}}\; \rightarrow 32°26'$ Ans

➤ **An inscribed angle is equal to one-half of its intercepted arc.**

Given: $\overset{\frown}{AC} = 105°$.

Conclusion: $\angle ABC = \frac{1}{2}\overset{\frown}{AC} = \frac{1}{2}(105°) = 52°30'$.

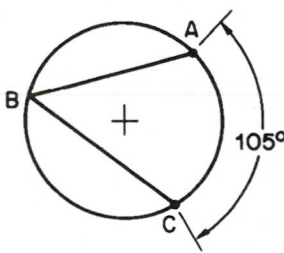

Arc Length Formula

Consider a complete circle as an arc of 360°. The ratio of the number of degrees of an arc to 360° is the fractional part of the circumference that is used to find the length of an arc. **The length of an arc equals the ratio of the number of degrees of the arc to 360° times the circumference.**

$$\text{Arc Length} = \frac{\text{Arc Degrees}}{360°}(2\pi r)$$

or

$$\text{Arc Length} = \frac{\text{Central Angle}}{360°}(2\pi r)$$

Example 1 $\overset{\frown}{ABC} = 130.00°$ and the radius is 120.00 mm. Determine the arc length $\overset{\frown}{ABC}$ to 2 decimal places.

$$\text{Arc Length} = \frac{\text{Arc Degrees}}{360°}(2\pi r)$$

$$\text{Arc Length} = \frac{130.00°}{360°}[2(3.1416)(120.00 \text{ mm})]$$

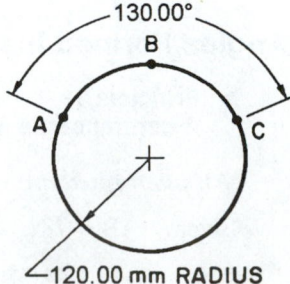

Arc Length = 272.27 mm   Ans (rounded)

130 ➕ 360 ✖ 2 ✖ π ✖ 120 🟰 272.27136,
272.27 mm   Ans (rounded)

Example 2 The arc length of $\overset{\frown}{DEF}$ is 8.426″ and the radius is 5.021″. Determine ∠1. All dimensions are in inches. Give the answer in degrees and minutes.

$$\text{Arc Length} = \frac{\text{Central Angle}}{360°} (2\pi r)$$

$$8.426'' = \frac{\angle 1}{360°} [2(3.1416)(5.021'')]$$

$$\angle 1 = 96°09' \quad \text{Ans}$$

$$\frac{8.426'' (360°)}{2\pi (5.021'')} = \text{Central Angle}$$

8.426 ✖ 360 ➕ (2 ✖ π ✖ 5.021 🟰
SHIFT ← → 96°09′   Ans

or 8.426 ✖ 360 ➕ (2 ✖ π ✖ 5.021 🟰
3rd ►DMS → 96°09′   Ans

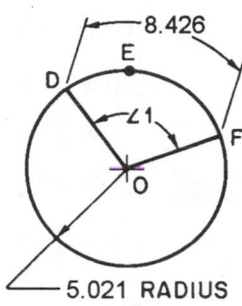

Angles Formed on a Circle

➤ **Principle 19**
An angle formed by a tangent and a chord at the tangent point is equal to one-half of its intercepted arc.

Example 1 Tangent CD meets chord AB at tangent point A and $\overset{\frown}{AEB}$ = 110°. Determine ∠CAB.

$$\angle CAB = \tfrac{1}{2} \overset{\frown}{AEB} = \tfrac{1}{2} (110°) = 55° \quad \text{Ans}$$

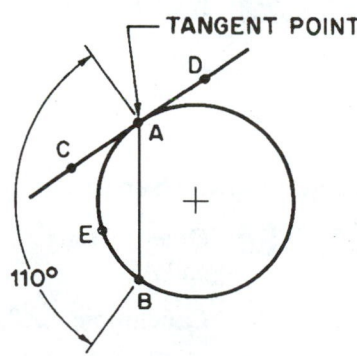

Example 2 The centers of 3 holes lie on line ABC. Line ABC is tangent to circle O at hole-center B. The hole-center D, of a fourth hole, lies on the circle. Determine ∠ABD.

A central angle is equal to its intercepted arc (Principle 18).

$$\overset{\frown}{DEB} = \angle DOB = 132°$$

Apply Principle 19.

$$\angle ABD = \tfrac{1}{2} \overset{\frown}{DEB} = \tfrac{1}{2} (132°) = 66° \quad \text{Ans}$$

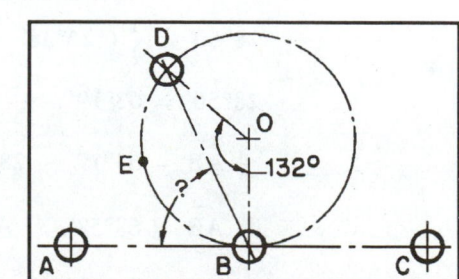

Angles Formed Outside a Circle

➤ **Principle 20**
An angle formed at a point outside a circle by two secants, two tangents, or a secant and a tangent is equal to one-half the difference of the intercepted arcs.

Two Secants

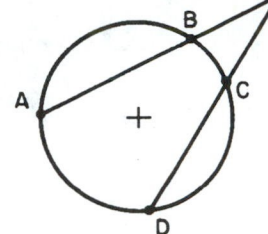

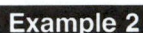 Given: Secants AP and DP meet at point P and intercept $\overset{\frown}{BC}$ and $\overset{\frown}{AD}$.

Conclusion: $\angle P = \frac{1}{2}(\overset{\frown}{AD} - \overset{\frown}{BC})$.

Example 2 If $\overset{\frown}{AD} = 85°40'0''$ and $\overset{\frown}{BC} = 39°17'0''$, find $\angle P$.

$\angle P = \frac{1}{2}(\overset{\frown}{AD} - \overset{\frown}{BC}) = \frac{1}{2}(85°40'0'' - 39°17'0'') = \frac{1}{2}(46°23'0'')$
$\quad\quad = 23°11'30''$ Ans

Example 3 If $\angle P = 28°$ and $\overset{\frown}{BC} = 40°$, determine $\overset{\frown}{AD}$.

$\angle P = \frac{1}{2}(\overset{\frown}{AD} - \overset{\frown}{BC})$

$28° = \frac{1}{2}(\overset{\frown}{AD} - 40°)$

$\overset{\frown}{AD} = 96°$ Ans

Two Tangents

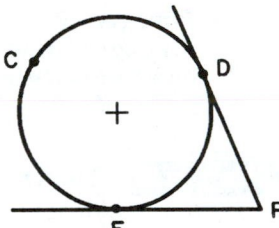

Example 1 Given: Tangents DP and EP meet at point P and intercept $\overset{\frown}{DE}$ and $\overset{\frown}{DCE}$.

Conclusion: $\angle P = \frac{1}{2}(\overset{\frown}{DCE} - \overset{\frown}{DE})$.

Example 2 If $\overset{\frown}{DCE} = 253°37'$ and $\overset{\frown}{DE} = 106°23'$, determine $\angle P$.

$\angle P = \frac{1}{2}(\overset{\frown}{DCE} - \overset{\frown}{DE}) = \frac{1}{2}(253°37' - 106°23') = \frac{1}{2}(147°14')$
$\quad\quad = 73°37'$ Ans

A Tangent and a Secant

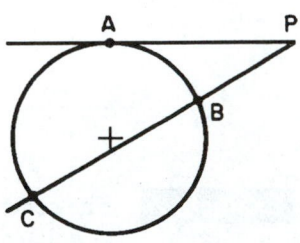

Example 1 Given: Tangent AP and secant CP meet at point P and intercept $\overset{\frown}{AC}$ and $\overset{\frown}{AB}$.

Conclusion: $\angle P = \frac{1}{2}(\overset{\frown}{AC} - \overset{\frown}{AB})$.

Example 2 If $\overset{\frown}{AC} = 126°38'$ and $\angle P = 28°50'$, determine $\overset{\frown}{AB}$.

$\angle P = \frac{1}{2}(\overset{\frown}{AC} - \overset{\frown}{AB})$

$28°50' = \frac{1}{2}(126°38' - \overset{\frown}{AB})$

$28°50' = 63°19' - \frac{1}{2}\overset{\frown}{AB}$

$\frac{1}{2}\overset{\frown}{AB} = 63°19' - 28°50'$

$\overset{\frown}{AB} = 68°58'$ Ans

$$\overset{\frown}{AB} = 2(63°19' - 28°50')$$

2 $\boxed{x}$ $\boxed{(}$ 63 $\boxed{° ' ''}$ 19 $\boxed{° ' ''}$ $\boxed{-}$ 28 $\boxed{° ' ''}$ 50 $\boxed{° ' ''}$ $\boxed{)}$ $\boxed{=}$
$\boxed{\text{SHIFT}}$ $\boxed{\longleftarrow}$ $\rightarrow 68°58'$ Ans

or 2 $\boxed{x}$ $\boxed{(}$ 63 $\boxed{.}$ 19 $\boxed{\text{2nd}}$ $\boxed{\blacktriangleright\text{DD}}$ $\boxed{-}$ 28 $\boxed{.}$ 50 $\boxed{\text{2nd}}$ $\boxed{\blacktriangleright\text{DD}}$ $\boxed{=}$
$\boxed{\text{3rd}}$ $\boxed{\blacktriangleright\text{DMS}}$ $\rightarrow 68°58'$ Ans

Internally and Externally Tangent Circles

Two circles that are tangent to the same line at the same point are tangent to each other. Circles are either internally or externally tangent.

Internally tangent—Two circles are internally tangent if both circles are on the same side of the common tangent line.

Externally tangent—Two circles are externally tangent if the circles are on opposite sides of the common tangent line.

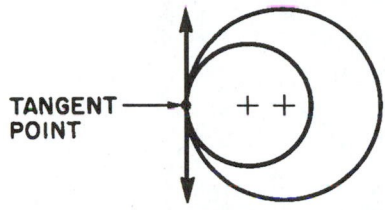

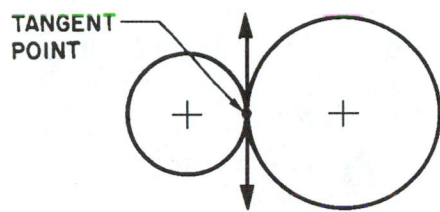

INTERNALLY TANGENT CIRCLES   EXTERNALLY TANGENT CIRCLES

➤ **Principle 21**
If two circles are either internally or externally tangent, a line connecting the centers of the circles passes through the point of tangency and is perpendicular to the tangent line.

Internally Tangent Circles

Example Given: Circle D and Circle E are internally tangent at point C. D is the center of Circle D and E is the center of Circle E. Line AB is tangent to both circles at point C.

Conclusion: An extension of line DE passes through tangent point C and line CDE ⊥ tangent line AB.

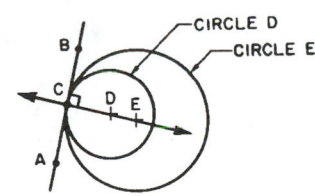

Principle 21 is often used as the basis for computing dimensions of parts on which two or more radii blend to give a smooth curved surface. This type of application is illustrated by the following example.

Example A part is to be machined as shown. The proper locations of the two radii will result in a smooth curve from point A to point B.

➤ **Note:** The curve from A to B is not an arc of one circle; it is made up of two different size circles. In order to make the part, the location to the center of the 12.000-inch radius (dimension *x*) must be determined. Compute *x*. All dimensions are in inches.

Refer to the figure on the next page.

The 12.000″ radius arc and the 25.000″ radius arc are internally tangent. Apply Principle 21. A line connecting arc centers F and H passes through tangent point C.

Tangent point C is the endpoint of the 25.000″ radius, CFH = 25.000″.

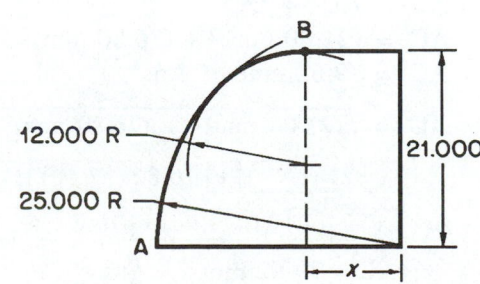

Tangent point C is the endpoint of the 12.000″
radius, CF = 12.000″.

FH = 25.000″ − 12.000″ = 13.000″

Since BFE is vertical and AEH is horizontal, ∠FEH
= 90°. In right △FEH, FH = 13.000″, FE = 21.000″
− BF = 9.000″.

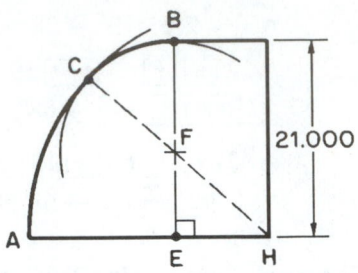

Apply the Pythagorean Theorem (Principle 9) to compute EH.

$$FH^2 = EH^2 + FE^2$$
$$(13.000 \text{ in})^2 = EH^2 + (9.000 \text{ in})^2$$
$$EH = 9.381 \text{ in}$$
$$x = EH = 9.381 \text{ in} \quad \text{Ans}$$

Externally Tangent Circles

Example 1   Given: Circle D and Circle E are externally tangent at point C. D is
the center of Circle D and E is the center of Circle E. Line AB is tangent to both circles at point C.

Conclusion: Line DE passes through tangent point
C and line DE ⊥ tangent line AB at point C.

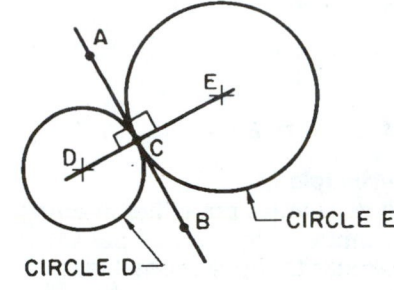

Example 2   Three holes are to be bored in a steel plate as shown. The 42.00-mm
and 61.40-mm diameter holes are tangent at point D. CD is the common tangent line. Determine the distances between hole centers (AB,
AC, and BC). All dimensions are in millimeters. Round the answers to
2 decimal places.

Compute AB.   Apply Principle 21.
Since AB connects the centers of two tangent circles,
AB passes through tangent point D.

$$AB = AD + DB = 21.00 \text{ mm} + 30.70 \text{ mm}$$
$$= 51.70 \text{ mm} \quad \text{Ans}$$

Compute AC and BC.
Since AB connects the centers of two tangent circles,
AB ⊥ tangent line DC. Triangle ADC and triangle
BDC are right triangles. Apply the Pythagorean
Theorem (Principle 9).

In right △ADC, AD = 21.00 mm and DC = 76.80 mm.

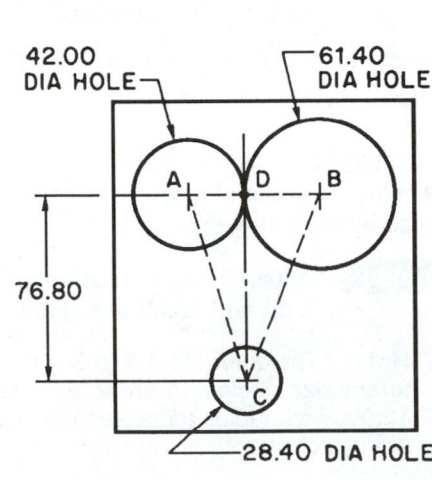

$$AC^2 = AD^2 + DC^2$$
$$AC^2 = (21.00 \text{ mm})^2 + (76.80 \text{ mm})^2$$
$$AC = 79.62 \text{ mm} \quad \text{Ans}$$

$$AC = \sqrt{(21.00 \text{ mm})^2 + (76.80 \text{ mm})^2}$$

(21 $\boxed{X^2}$ + 76.8 $\boxed{X^2}$) $\boxed{\sqrt{x}}$ → 79.61934438

or $\boxed{\sqrt{}}$ (21 $\boxed{X^2}$ + 76.8 $\boxed{X^2}$) $\boxed{=}$ 79.61934438

AC = 79.62 mm   Ans

In right $\triangle BDC$, DB = 30.70 mm and DC = 76.80 mm.

$$BC^2 = DB^2 + DC^2$$

$$BC^2 = (30.70 \text{ mm})^2 + (76.80 \text{ mm})^2$$

$$BC = 82.71 \text{ mm} \qquad \text{Ans}$$

$$BC = \sqrt{(30.70 \text{ mm})^2 + (76.80 \text{ mm})^2}$$

$$(\boxed{30.7} \boxed{x^2} \boxed{+} \boxed{76.8} \boxed{x^2} \boxed{)} \boxed{\sqrt{x}} \rightarrow 82.70870571$$

or $\boxed{\sqrt{}} \boxed{(} 30.7 \boxed{x^2} \boxed{+} 76.8 \boxed{x^2} \boxed{)} \boxed{=} 82.70870571$

$$BC = 82.71 \text{ mm} \qquad \text{Ans}$$

APPLICATION

Arc Length Formula

Determine the unknown value for each of the following problems. Round the answers to 3 decimal places.

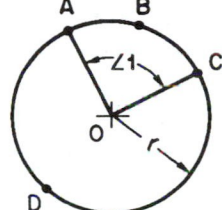

1. $\overset{\frown}{ABC} = 90°0'$ and $r = 3.500$ in. Find arc length $\overset{\frown}{ABC}$.   _____

2. $\overset{\frown}{ABC} = 85.00°$ and $r = 60.000$ mm. Find arc length $\overset{\frown}{ADC}$.   _____

3. Arc length $\overset{\frown}{ABC} = 510.000$ mm and $r = 120.000$ mm. Find $\angle 1$.   _____

4. Arc length $\overset{\frown}{ADC} = 22.700$ in and $r = 5.200$ in. Find $\angle 1$.   _____

5. Arc length $\overset{\frown}{ABC} = 18.750$ in and $\angle 1 = 72°0'$. Find r.   _____

6. Arc length $\overset{\frown}{ABC} = 620.700$ mm and $\angle 1 = 69.30°$. Find r.   _____

Geometric Principles

Solve the following problems. These problems are based on principles 18–21, although a problem may require the application of two or more of any of the principles. Where necessary, round linear answers in inches to 3 decimal places and millimeters to 2 decimal places. Round angular answers in decimal degrees to 2 decimal places and degrees and minutes to the nearest minute.

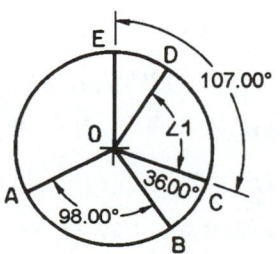

7. a. If $\angle 1 = 76.00°$, find

    (1) $\overset{\frown}{DC}$ _____

    (2) $\angle EOD$ _____

    (3) $\overset{\frown}{ABC}$ _____

  b. If $\angle 1 = 63.76°$, find

    (1) $\overset{\frown}{DC}$ _____

    (2) $\angle EOD$ _____

    (3) $\overset{\frown}{BCD}$ _____

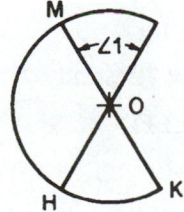

8. a. If $\angle 1 = 63°$, find

    (1) $\overset{\frown}{HK}$ _____

    (2) $\overset{\frown}{HM}$ _____

  b. If $\angle 1 = 59°47'$, find

    (1) $\overset{\frown}{HK}$ _____

    (2) $\overset{\frown}{HM}$ _____

9. a. If $\overset{\frown}{PS}$ = 46°, find

    (1) ∠1   _____

    (2) ∠2   _____

  b. If $\overset{\frown}{PS}$ = 39°, find

    (1) ∠1   _____

    (2) ∠2   _____

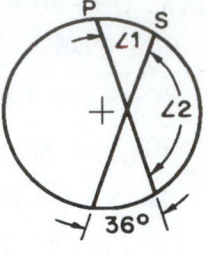

10. a. If $\overset{\frown}{DC}$ = 35°, find $\overset{\frown}{AB}$.   _____

  b. If $\overset{\frown}{AB}$ = 127°, find $\overset{\frown}{DC}$.   _____

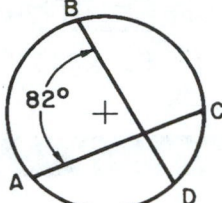

11. a. If ∠3 = 47° and $\overset{\frown}{GH}$ = 32°, find

    (1) $\overset{\frown}{EF}$ _____

    (2) ∠4   _____

  b. If ∠4 = 17°53′ and $\overset{\frown}{EF}$ = 103°, find

    (1) ∠3   _____

    (2) $\overset{\frown}{GH}$ _____

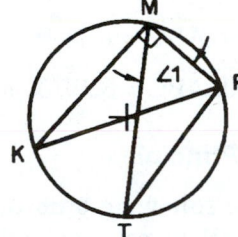

12. a. If ∠1 = 25° and $\overset{\frown}{MPT}$ = 95°, find

    (1) $\overset{\frown}{KTP}$ _____

    (2) $\overset{\frown}{PT}$ _____

    (3) $\overset{\frown}{MP}$ _____

  b. If ∠1 = 17°30′ and $\overset{\frown}{MPT}$ = 103°, find

    (1) $\overset{\frown}{KTP}$ _____

    (2) $\overset{\frown}{PT}$ _____

    (3) $\overset{\frown}{MP}$ _____

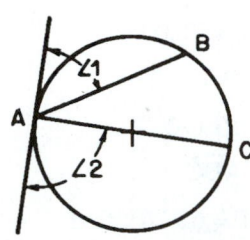

13. a. If $\overset{\frown}{AB}$ = 116°, find

    (1) ∠1   _____

    (2) ∠2   _____

  b. If $\overset{\frown}{AB}$ = 112°56′, find

    (1) ∠1   _____

    (2) ∠2   _____

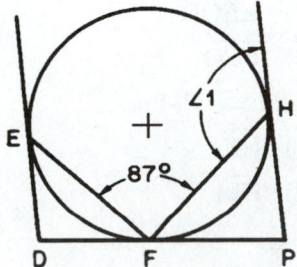

14. a. If $\overset{\frown}{EF}$ = 84°, find

    (1) ∠EFD   _____

    (2) $\overset{\frown}{HF}$ _____

    (3) ∠1   _____

  b. If EF = 79°, find

    (1) ∠EFD   _____

    (2) $\overset{\frown}{HF}$ _____

    (3) ∠1   _____

15. a. If $\overset{\frown}{ST}$ = 20°18′ and $\overset{\frown}{SM}$ = 38°07′, find

    (1) ∠1   _____

    (2) ∠2   _____

  b. If $\overset{\frown}{ST}$ = 25°17′ and $\overset{\frown}{SM}$ = 35°24′, find

    (1) ∠1   _____

    (2) ∠2   _____

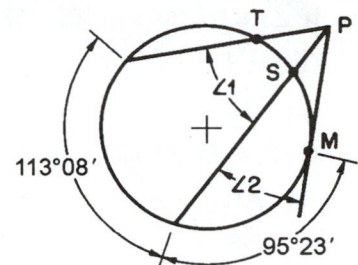

16. a. If $\overset{\frown}{AB}$ = 72°20′ and $\overset{\frown}{CD}$ = 50°18′, find

    (1) ∠1   _____

    (2) ∠2   _____

    (3) ∠3   _____

  b. If $\overset{\frown}{CD}$ = 43°15′ and $\overset{\frown}{AD}$ = 106°05′, find

    (1) ∠1   _____

    (2) ∠2   _____

    (3) ∠3   _____

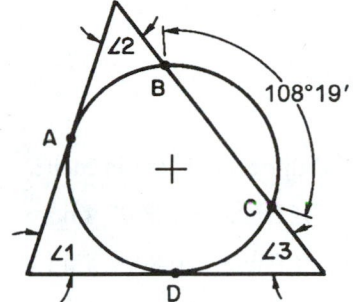

17. a. If ∠1 = 24.00° and ∠2 = 60.00°, find

    (1) $\overset{\frown}{DH}$ _____

    (2) $\overset{\frown}{EDH}$ _____

  b. If ∠1 = 29.00° and ∠2 = 64.00°, find

    (1) $\overset{\frown}{DH}$ _____

    (2) $\overset{\frown}{EDH}$ _____

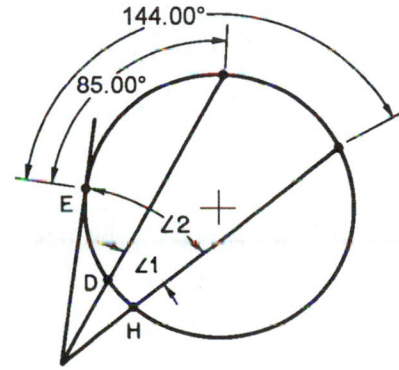

18. a. If Dia A = 3.756″ and Dia B = 1.622″, find x.   _____

  b. If x = 0.975″ and Dia B = 1.026″, find Dia A.   _____

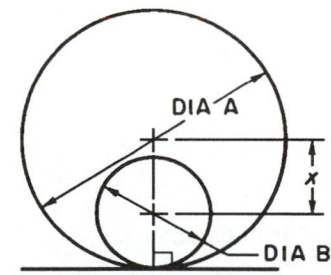

19. a. If x = 24.93 mm and y = 28.95 mm, find Dia A.   _____

  b. If x = 78.36 mm and y = 114.48 mm, find Dia A.   _____

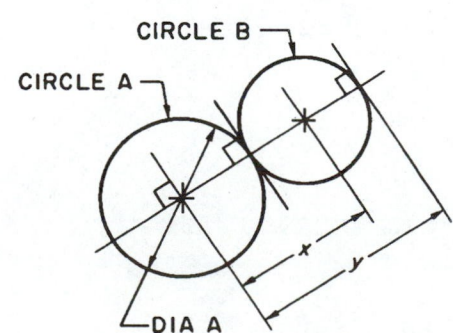

20. a. If $\angle 1 = 67°00'$ and $\angle 2 = 93°00'$, find

    (1) $\overset{\frown}{AB}$ _____

    (2) $\overset{\frown}{DE}$ _____

  b. If $\angle 1 = 75°00'$ and $\angle 2 = 85°00'$, find

    (1) $\overset{\frown}{AB}$ _____

    (2) $\overset{\frown}{DE}$ _____

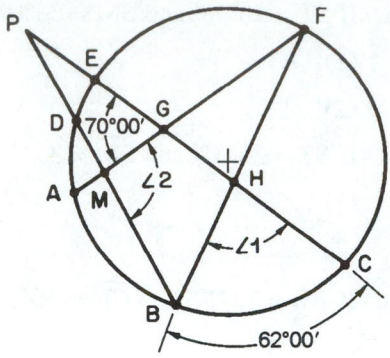

21. All dimensions are in inches.

  a. If Dia A = 1.000″, find x.   _____

  b. If Dia A = 0.800″, find x.   _____

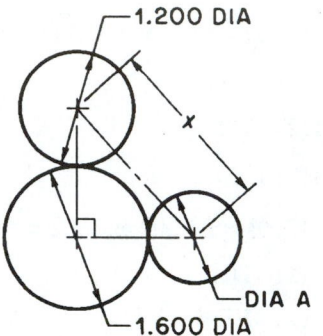

22. All dimensions are in inches.

  a. If $y = 0.350″$, find x.   _____

  b. If $y = 0.410″$, find x.   _____

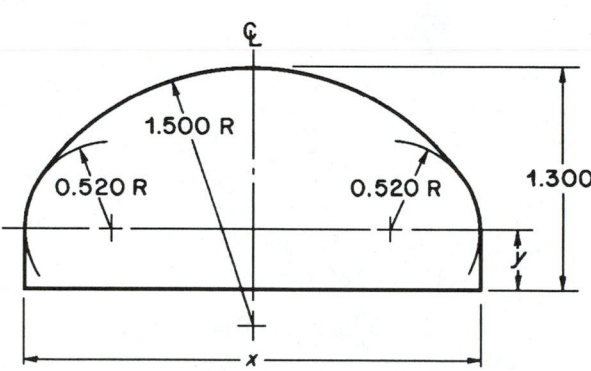

23. AC is a diameter.

  a. If $\angle 2 = 22°00'$, find $\angle 1$.   _____

  b. If $\angle 2 = 30°54'$, find $\angle 1$.   _____

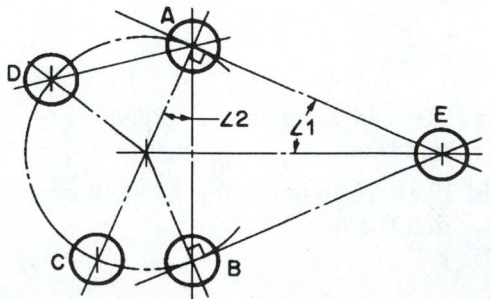

24. Three posts are mounted on the fixture shown. Each post is tangent to the arc made by the 0.650-inch radius. Determine (a) dimension A and (b) dimension B.

> **Note:** The fixture is symmetrical (identical) on each side of the horizontal centerline (℄).

All dimensions are in inches. a. _____

b. _____

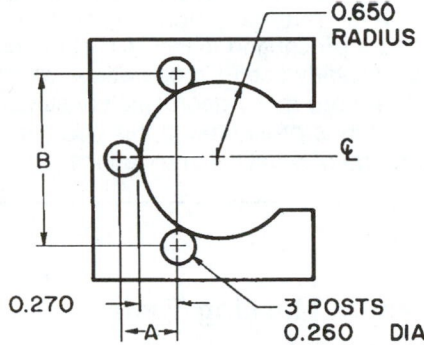

25. Points A, B, C, D, and E are tangent points.

a. If $\overset{\frown}{AB} = 46.00°$ and $\overset{\frown}{DE} = 66.00°$, find $\angle 1$. _____
b. If $\overset{\frown}{AB} = 53.00°$ and $\overset{\frown}{DE} = 70.00°$, find $\angle 1$. _____

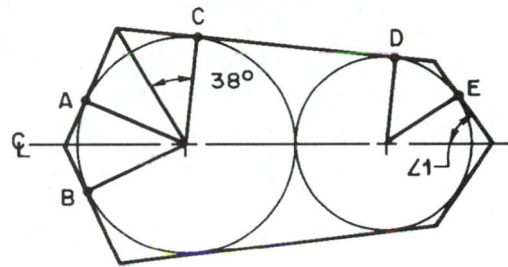

26. Three holes are to be located on the layout shown. The 72.40 mm and 30.80 mm diameter holes are tangent at point T and TA is the common tangent line between the two holes. Determine (a) dimension C and (b) dimension D.

a. _____

b. _____

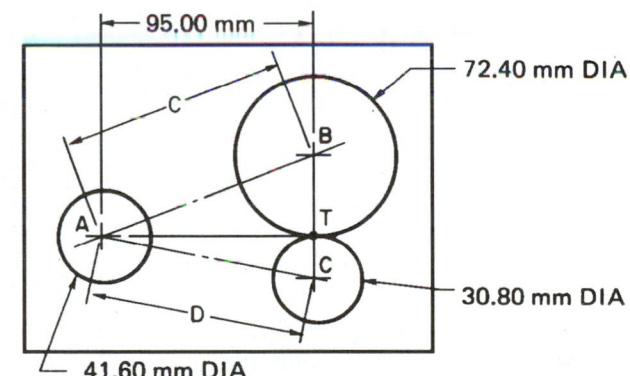

UNIT **51** Fundamental Geometric Constructions

Objectives   After studying this unit you should be able to

- **Make constructions which are basic to the machine trades.**
- **Lay out typical machine shop problems using the methods of construction.**

A knowledge of basic geometric constructions done with a compass or dividers and a steel rule is required of a machinist in laying out work. The constructions are used in determining stock allowances and reference locations on castings, forgings, and sheet stock.

For certain jobs where wide dimensional tolerances are permissible, the most practical and efficient way of producing a part may be by scribing and centerpunching locations. Layout dimensions are sometimes used as a reference for machining complex parts which require a high degree of precision. Locations lightly scribed on a part are used as a precaution to insure that the part or table movement is in the proper direction. It is particularly useful in operations which require part rotation or repositioning.

There are many geometric constructions, some of which are relatively complex. The constructions presented in this book are those which are most basic and common to a wide range of practical applications.

Common Marking Tools

Pocket Scriber
(Courtesy of L. S. Starrett Company)

Center Punch
(Courtesy of L. S. Starrett Company)

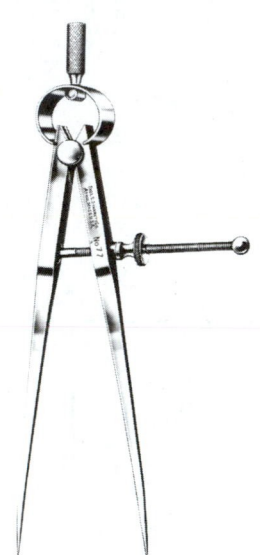

Dividers
(Courtesy of L. S. Starrett Company)

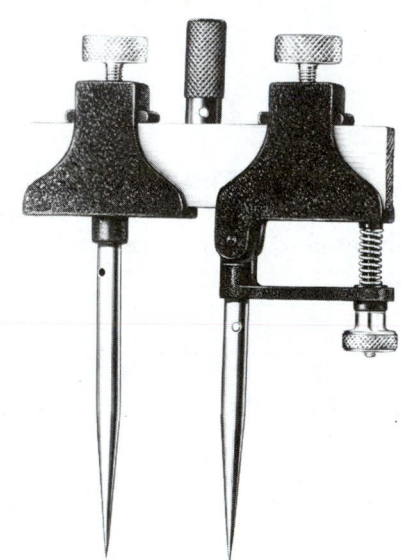

Trammels
(Courtesy of L. S. Starrett Company)

Construction 1   To Construct a Perpendicular Bisector of a Line Segment

Required: Construct a perpendicular bisector to line segment AB.

Procedure   Refer to the next page.

- With endpoint A as a center and using a radius equal to more than half AB, draw arcs above and below AB.

- With endpoint B as a center and with the same radius used at A, draw arcs above and below AB which intersect the first pair of arcs.

- Draw a connecting line between the intersection of the arcs above and below AB. Line CD is perpendicular to AB and point O is the midpoint of AB.

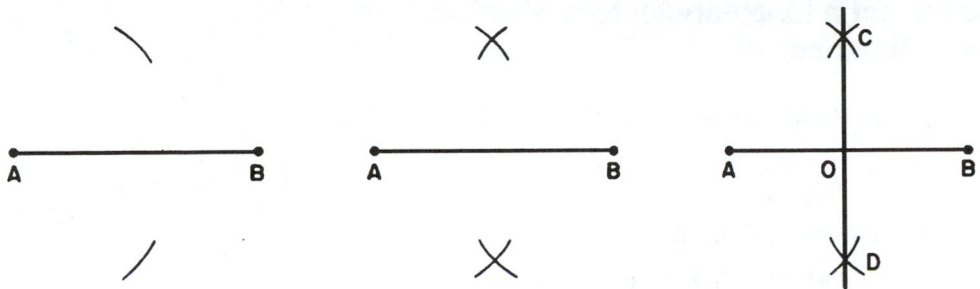

Practical Application

Locate the center of a circle.

Solution: The perpendicular bisector of a chord passes through the center of the circle. The center of a circle is located by drawing two chords and constructing a perpendicular bisector to each chord. The intersection of the two perpendicular bisectors locates the center of the circle. The construction lines are shown in this figure.

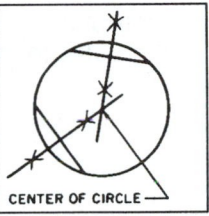

CENTER OF CIRCLE

Construction 2   To Construct a Perpendicular to a Line Segment at a Given Point on the Line Segment

Required: Construct a perpendicular at point O on line segment AB.

Procedure

- With given point O as a center, and with a radius of any convenient length, draw arcs intersecting AB at points C and D.

- With C as a center, and with a radius greater than OC, draw an arc. With D as a center, and with the same radius used at C, draw an arc which intersects the first arc at E.

- Draw a line connecting point E and point O. Line EO is perpendicular to line AB at point O.

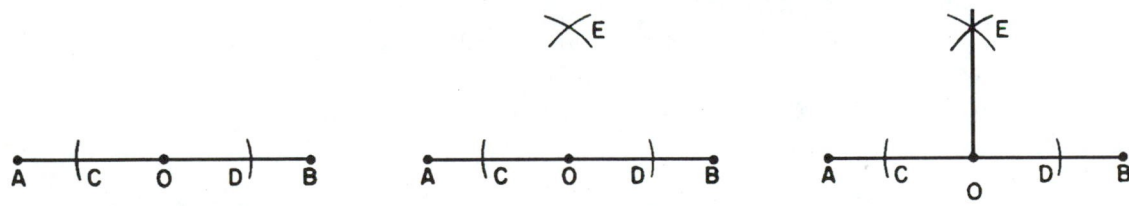

Practical Application

A triangular piece is to be scribed and cut. The piece is laid out as follows:

The 22-inch base is measured and marked off.

The $10\frac{7}{64}$-inch distance is measured and marked off at point A on the baseline.

From point A, a perpendicular to the baseline is constructed.

The construction lines are shown.

The $7\frac{1}{2}$-inch distance is measured and marked off at point B on the constructed perpendicular.

Lines are scribed connecting vertex B with the endpoints of the baseline.

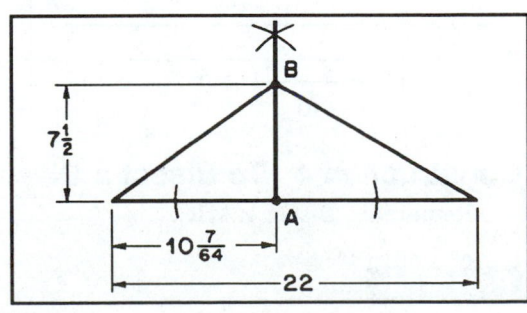

Construction 3 To Construct a Line Parallel to a Given Line at a Given Distance

Required: Construct a line parallel to line AB at a given distance of 1 inch.

Procedure

- Set the compass to the required distance (1 inch).

- With any points C and D as centers on AB, draw arcs with the given distance (1 inch) as the radius.

- Draw a line, EF, that touches each arc at one point (the tangent point). Line EF is parallel to line AB and EF is 1 inch from AB.

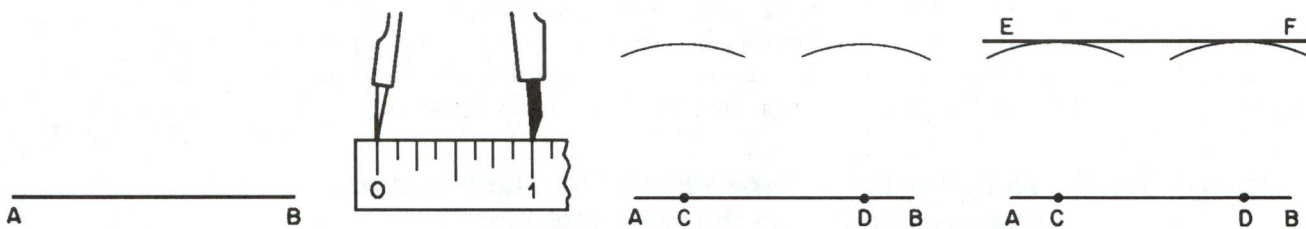

Practical Application

The cutout shown in the drawing is laid out on a sheet as follows:

All dimensions are in millimeters.

The centerline (℄) is scribed and the 310-mm distance is marked off. Points A and B are the endpoints of the 310-mm segment.

From points A and B perpendiculars are constructed. The perpendiculars are extended more than 70 mm (140 mm ÷ 2) above and below AB.

With points C and D as centers on AB, 70-mm radius arcs are drawn above and below AB. A line is scribed above and a line is scribed below AB touching the pairs of arcs. The lines are extended to intersect the perpendiculars constructed. The points of intersection are E, F, G, and H.

From point A and from point B on AB, 40-mm distances are marked off. Point J and point K are the endpoints.

Lines are scribed connecting J to E and G and connecting K to F and H. Scribed figure JEFKHG is the required cutout.

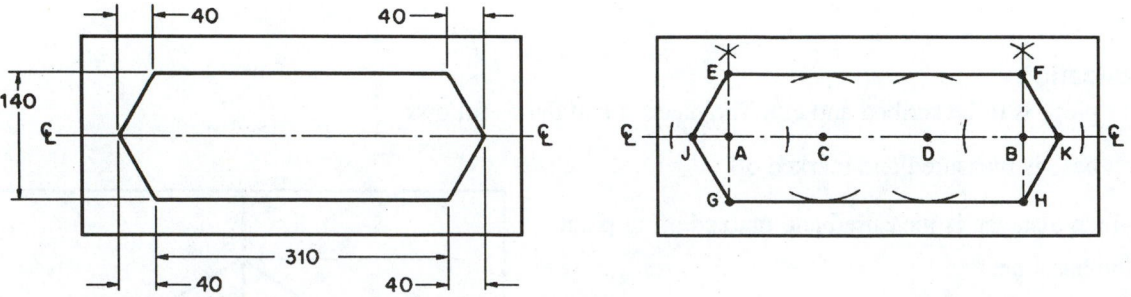

Construction 4 To Bisect a Given Angle

Required: Bisect ∠ABC.

Procedure

- With point B as the center draw an arc intersecting sides BA and BC at points D and E.

- With D as the center, and with a radius equal to more than half the distance DE, draw an arc. With E as the center, and with the same radius, draw an arc. The intersection of the two arcs is point F.

- Draw a line from point B to point F. Line BF is the bisector of ∠ABC.

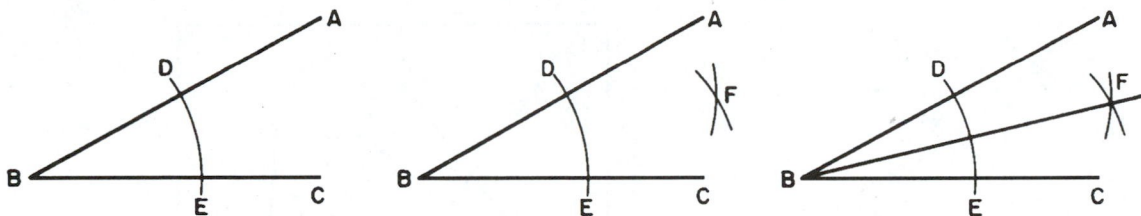

Practical Application

The centers of the three $\frac{1}{2}$-inch diameter holes in the mounting plate shown are located and center punched. Two $\frac{1}{4}$-inch diameter holes are located by constructing the bisector of ∠ABC as shown and marking and center punching the $1\frac{3}{8}$-inch and $4\frac{1}{4}$-inch hole center locations on the bisector.

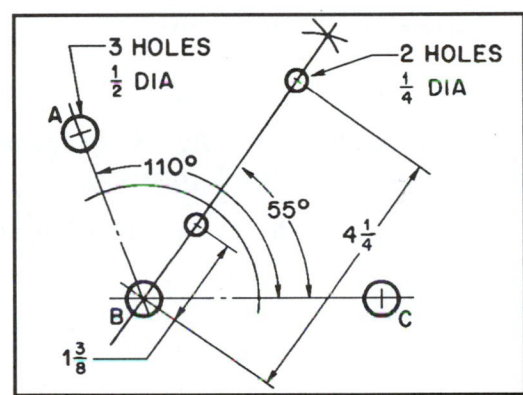

Construction 5   To Construct Tangents to a Circle from an Outside Point

Required: Construct tangents to given circle O from given outside point P.

Procedure

- Draw a line segment connecting center O and point P. Bisect OP. Point A is the midpoint of OP.

- With point A as the center and AP as a radius, draw arcs intersecting circle O at points B and C. Points B and C are tangent points.

- Connect points B and P, and C and P. Line segments BP and CP are tangents.

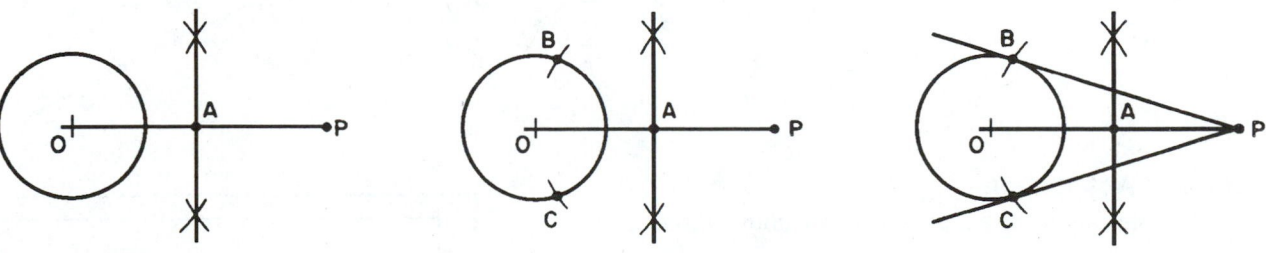

Practical Application

A piece is to be made as shown in the drawing on the next page. All dimensions are in millimeters. The piece is laid out as follows:

A baseline is scribed and AB (170 mm) is marked off.

Distance OA (152 mm) is set on dividers and with OA as the radius, an arc is scribed. Distance OB (104 mm) is set on dividers and with OB as the radius, an arc is scribed to intersect with the OA radius arc. The intersection of the arcs locates center O of the 42-mm radius circle.

Dividers are set to the 42-mm radius dimension, and the circle is scribed from center O.

Tangents to the circle from points A and B are constructed resulting in tangent points C and D and tangent line segments AC and BD. The piece is now laid out and ready to be cut to the scribed lines.

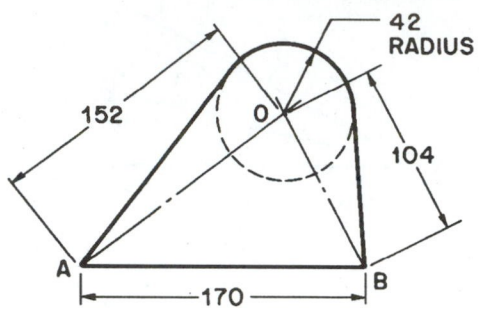
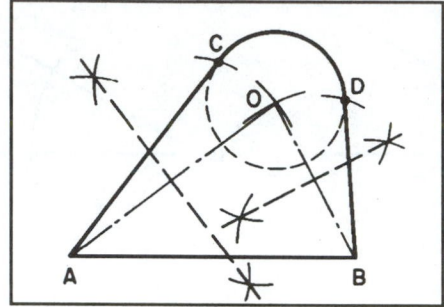

Construction 6 To Divide a Line Segment into a Given Number of Equal Parts

Required: Divide line segment AB into three equal parts.

Procedure

- From point A, draw line AC forming any convenient angle with AB.

- On AC, with a compass, lay off any three equal segments, AD, DE, and EF.

- Connect point F with point B. With centers at points F, E, and D, draw arcs of equal radii. The arc with a center at point F intersects AC at point G and BF at point H. Set distance GH on the compass and mark off this distance on the other two arcs. The points of intersection are K and M.

- Connect points E and K, and D and M, extending the lines past AB. Line AB is divided into three equal segments; AP = PS = SB.

- ➤ **Note:** Line segment AB can be divided into any required number of equal segments by laying off the required number of equal segments on AC and following the procedure given.

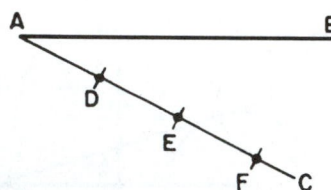
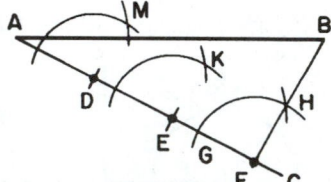
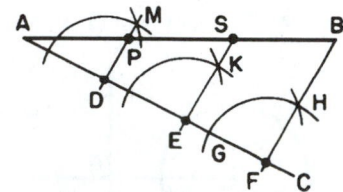

Practical Application

Six holes are to be equally spaced within a distance of $2\frac{11}{16}$ inches. Since six holes are required, there will be five equal spaces between holes. Dividing $2\frac{11}{16}$ inches by 5 results in fractional distances which are difficult to accurately measure or transfer, such as $\frac{8.6}{16}$ inch, $\frac{17.2}{32}$ inch, or $\frac{34.4}{64}$ inch. By careful construction, the hole centers are accurately located as shown in the figure.

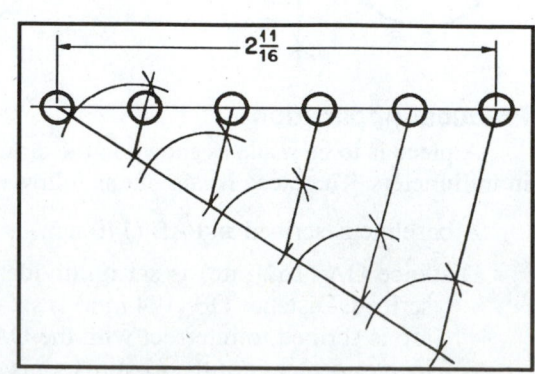

APPLICATION

Construction 1 and 2 Applications

Show construction lines and arcs for each of these problems.

1. Trace each line segment in problems a–d and construct perpendicular bisectors to each segment.

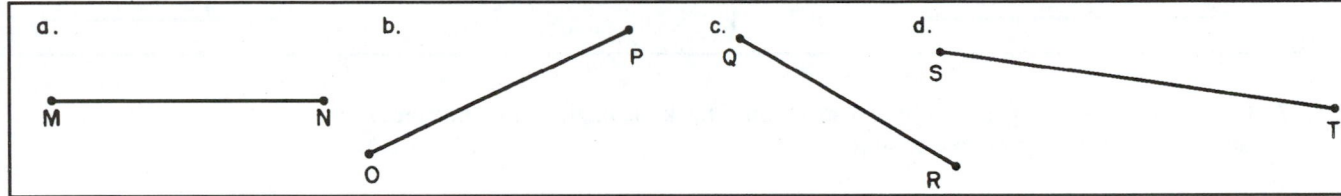

2. Trace each line in problems a–c and construct perpendiculars to each line at the given points on the lines.

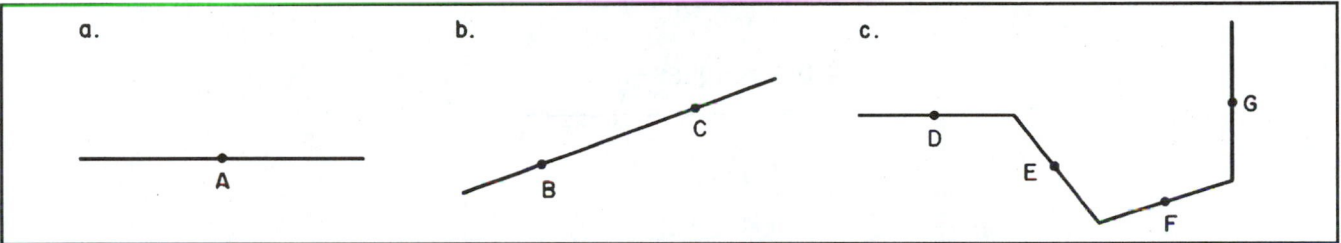

3. With a compass, draw a circle 2 inches in diameter. By construction, locate the center of the circle.

4. Lay out a figure as follows:

   a. Draw a horizontal line and mark off a distance of $2\frac{1}{2}$ inches. Label the left endpoint of the $2\frac{1}{2}$-inch line segment point A and label the right endpoint point D.

   b. From point A and above point A, construct a perpendicular to AD. Mark off a distance of $1\frac{7}{8}$ inches on the perpendicular from point A. Label the top endpoint point B.

   c. From point B and to the right of point B, construct a perpendicular to AB. Mark off a distance of $2\frac{1}{2}$ inches on the perpendicular from point B. Label the right endpoint point C.

   d. From point C and below point C, construct a perpendicular to BC. Mark off a distance of $1\frac{7}{8}$ inches on the perpendicular from point C. If your constructions are accurate the $1\frac{7}{8}$-inch distance marked off coincides with point D. What kind of a figure is formed by this construction? _____

Construction 3 and 4 Applications

Show construction lines and arcs for each of these problems.

5. Trace each of the lines in problems a–d and construct a line parallel to each line at a distance of $1\frac{1}{2}$ inches.

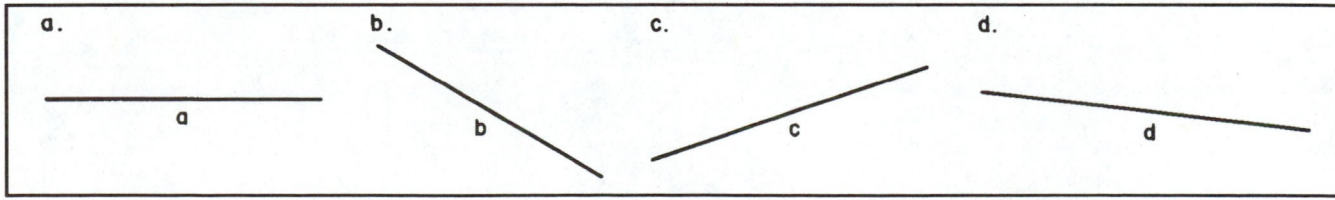

6. Trace each of the angles a–c and construct a bisector to each.

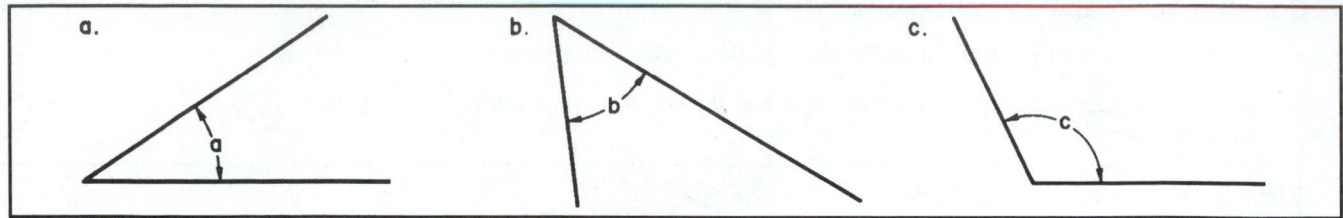

7. Lay out the following angles by construction. Check the angle with a protractor but do not lay out angles with a protractor.

   a. 45° b. 22°30′ c. 67°30′ d. $157\frac{1}{2}°$ e. $168\frac{3}{4}°$

8. Lay out the plate shown. Make the layout full size using construction methods. Use a protractor only for checking. All dimensions are in inches.

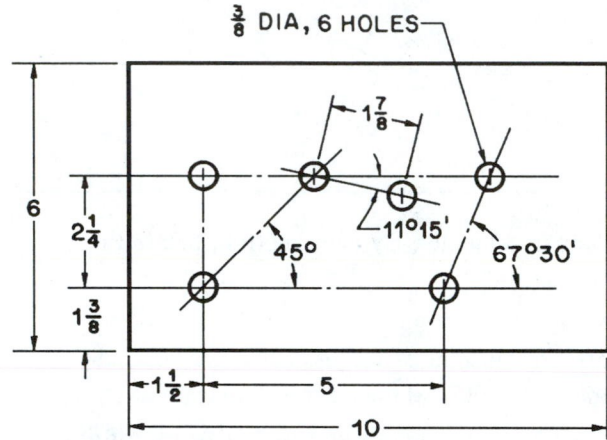

9. Lay out the gage shown. Make the layout full size using construction methods. Use a protractor only for checking. All dimensions are in inches.

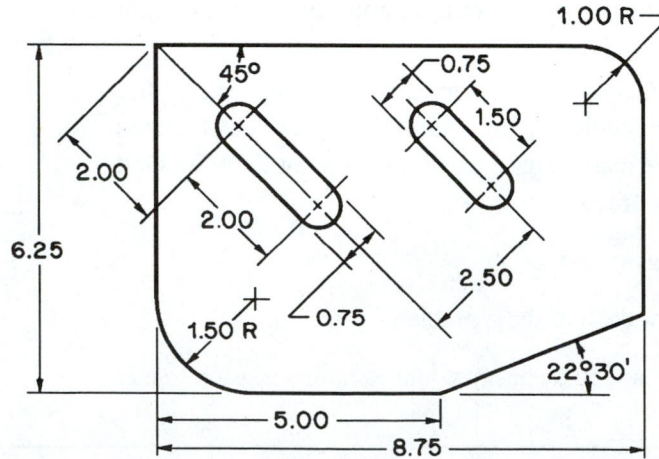

Construction 5 and 6 Applications

Show construction lines and arcs for each of these problems.

10. Trace each circle and point in problems a–c and construct tangents to the circles from the given points.

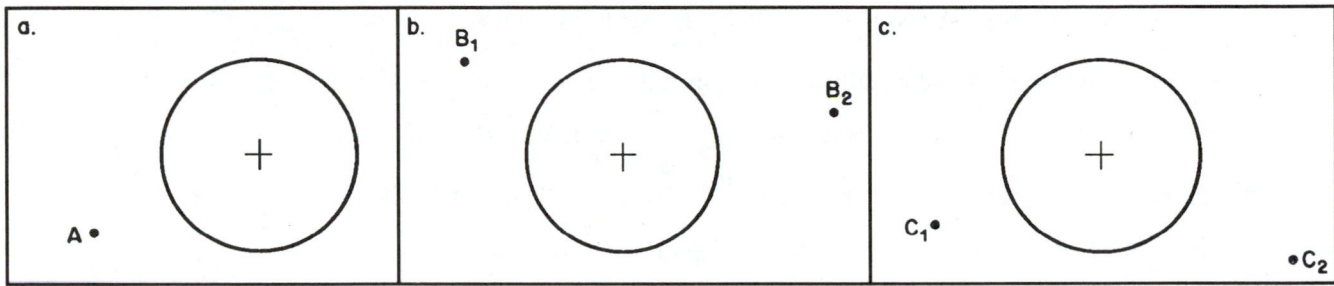

11. Trace each line segment of problems a, b, and c. Divide the given lines into the designated number of segments by means of construction.

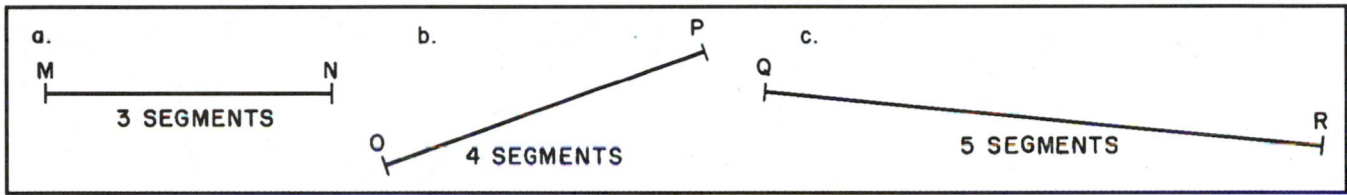

12. Lay out the template shown. Make the layout full size using construction methods. All dimensions are in inches.

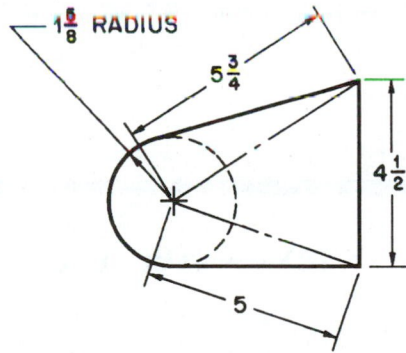

13. Lay out the cutout shown. Make the layout full size using construction methods. All dimensions are in inches.

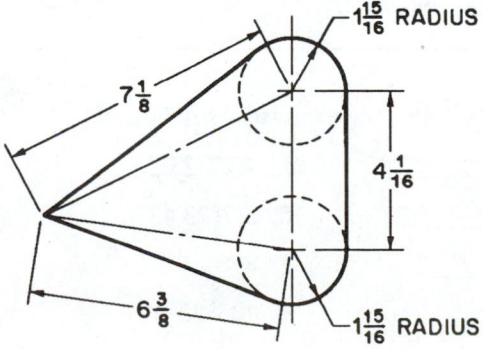

14. Trace the plate shown. Lay out three sets of holes by construction. Follow the given directions.

Directions:

- Bisect ∠A and construct 4 equally spaced $\frac{3}{16}$-inch diameter holes. Make the first hole $\frac{7}{8}$ inch from point A and the last hole $2\frac{7}{16}$ inches from point A.

- Bisect ∠B and construct 8 equally spaced $\frac{1}{4}$-inch diameter holes. Make the first hole $\frac{3}{4}$ inch from point B and the last hole $3\frac{11}{16}$ inches from the first hole.

- Bisect ∠C and construct 4 equally spaced $\frac{3}{16}$-inch diameter holes. Make the first hole $\frac{9}{16}$ inch from point C and the last hole $2\frac{11}{16}$ inches from point C.

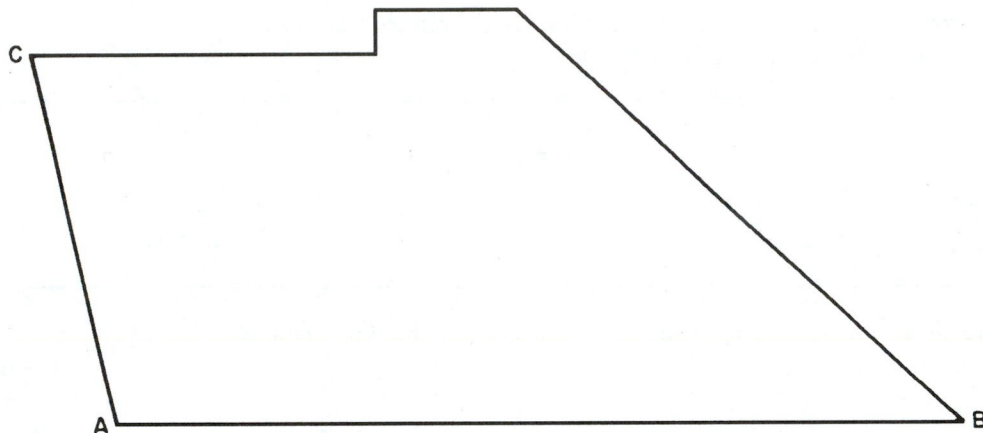

UNIT 52 Achievement Review—Section Four

Objective

You should be able to solve the exercises and problems in this Achievement Review by applying the principles and methods covered in units 44–51.

---

1. Add, subtract, multiply, or divide each of the following exercises as indicated.

a. 37°18′ + 86°23′ _____

b. 38°46′ + 23°43′ _____

c. 136°36′28″ − 94°17′15″ _____

d. 58°14′ − 44°58′ _____

e. 4(27°23′) _____

f. 3(7°23′43″) _____

g. 87° ÷ 2 _____

h. 103°20′ ÷ 4 _____

2. Determine ∠A. _____

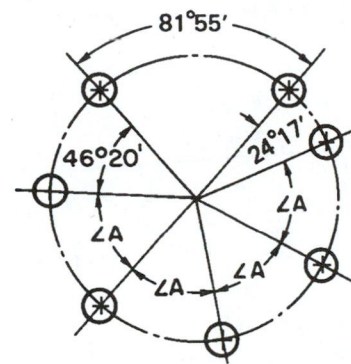

3. Given: The sum of all angles = 720°00′00″
   ∠3 = ∠4 = ∠5 = ∠6.
   ∠1 = ∠2 = 68°42′18″.

   Determine ∠3. _____

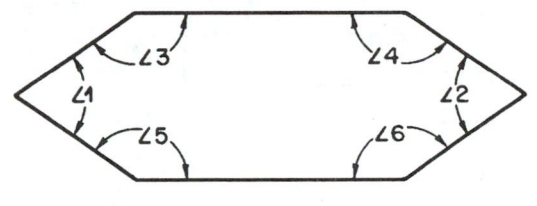

4. Express 68.85° as degrees and minutes. _____

5. Express 64.1420° as degrees, minutes, and seconds. _____

6. Express 37°23′ as decimal degrees to 2 decimal places. _____

7. Express 103°38′43″ as decimal degrees to 4 decimal places. _____

8. Using a simple protractor, measure each of the angles, 1–7 to the nearer degree. It may be necessary to extend sides of angles.

   ∠1 = _____

   ∠2 = _____

   ∠3 = _____

   ∠4 = _____

   ∠5 = _____

   ∠6 = _____

   ∠7 = _____

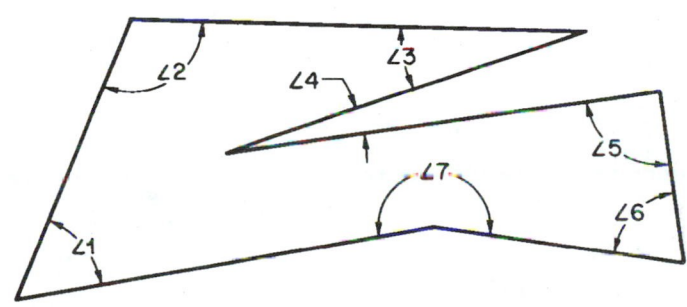

9. Write the values of the settings shown in the following vernier protractor scales.

   a. _____   b. _____   c. _____

10. Write the complement of each of the following angles.

    a. 67° _____   b. 17°41′ _____   c. 54°47′53″ _____

11. Write the supplement of each of the following angles.

    a. 41° _____   b. 99°32′ _____   c. 103°03′27″ _____

12. Given: AB ‖ CD and EF ‖ GH. Determine the value of each angle, $\angle 1 - \angle 10$ to the nearer minute.

$\angle 1 =$ _____

$\angle 2 =$ _____

$\angle 3 =$ _____

$\angle 4 =$ _____

$\angle 5 =$ _____

$\angle 6 =$ _____

$\angle 7 =$ _____

$\angle 8 =$ _____

$\angle 9 =$ _____

$\angle 10 =$ _____

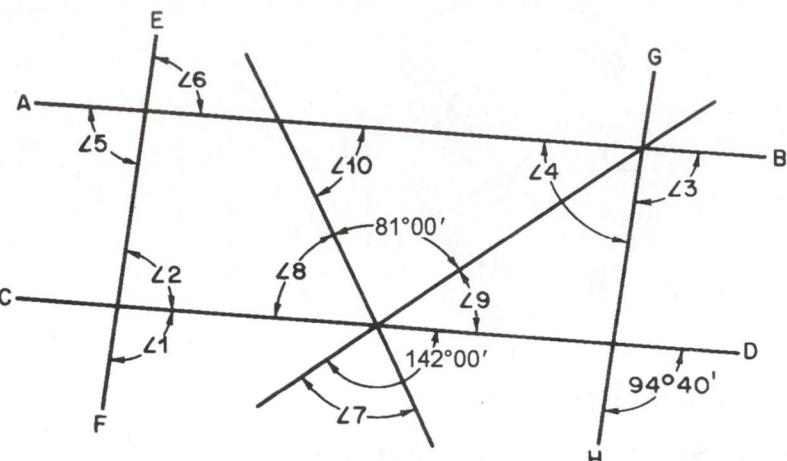

13. a. Determine:

    (1) $\angle 1$ _____

    (2) Side a _____

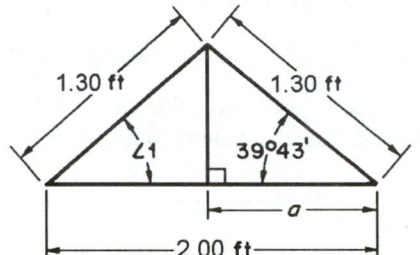

  b. Determine:

    (1) $\angle 1$ _____

    (2) Side b _____

    (3) Side c _____

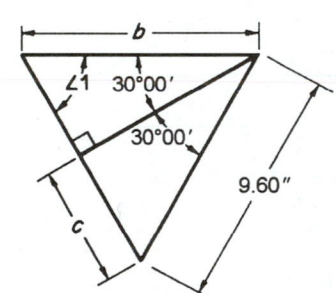

  c. Determine:

    (1) $\angle 1$ _____

    (2) $\angle 2$ _____

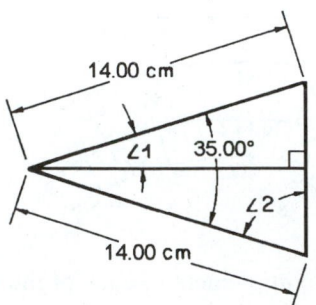

14. a. Given: $a = 8.400''$ and $b = 9.200''$. Find c.   _____

  b. Given: $b = 90.00$ mm and $c = 150.00$ mm. Find a.   _____

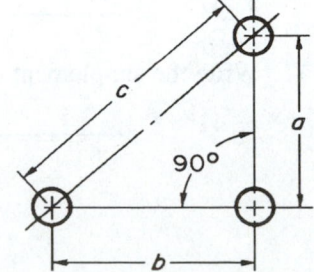

15. Compute ∠1. _____

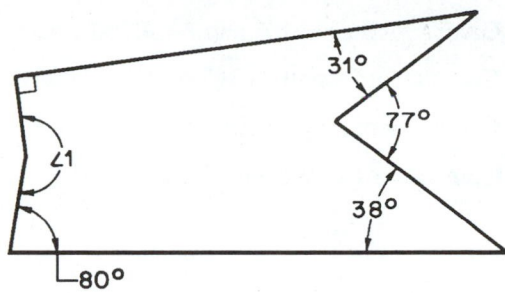

16. Determine the circumference of a circle which has a 5.360-inch radius. Round the answer to 3 decimal places. _____

17. Determine the diameter of a circle which has a 360.00-millimeter circumference. Round the answer to 2 decimal places. _____

18. a. Given: CD = 184 mm and $\overparen{CE}$ = 118 mm.

   Determine CF and $\overparen{CED}$.

   CF = _____

   $\overparen{CED}$ = _____

   b. Given: FD = 26 mm and $\overparen{CED}$ = 78 mm.

   Determine CD and $\overparen{ED}$.

   CD = _____

   $\overparen{ED}$ = _____

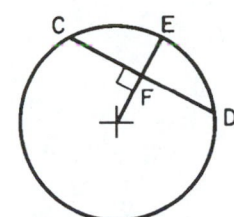

19. a. Given: EB = 5.150″.

   Determine AE. _____

   b. Given: AE = 4.200″.

   Determine AB. _____

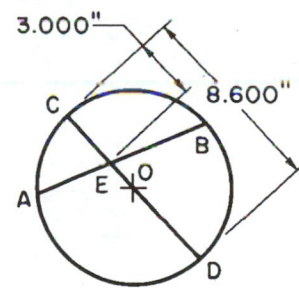

20. Given: Points A and E are tangent points. EB is a diameter. $\overparen{AFE}$ = 156°, $\overparen{CDE}$ = 140°, and $\overparen{ED}$ = 60°. Determine angles 1–10.

   ∠1 = _____

   ∠2 = _____

   ∠3 = _____

   ∠4 = _____

   ∠5 = _____

   ∠6 = _____

   ∠7 = _____

   ∠8 = _____

   ∠9 = _____

   ∠10 = _____

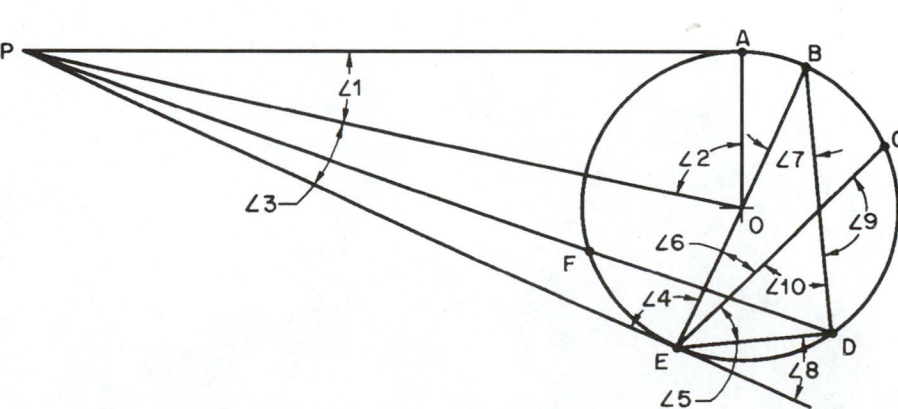

21. a. Given: $\overarc{ABC} = 110°$ and $r = 4.700''$.

    Compute arc length $\overarc{ABC}$ to 3 decimal places. _____

    b. Given: Arc length $\overarc{ABC} = 478.60$ mm and $r = 105.00$ mm.

    Compute $\angle 1$ to 2 decimal places. _____

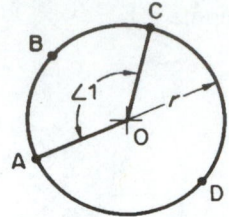

22. a. Given: Dia $H = 14.520''$ and $d = 8.300''$.

    Compute Dia M. _____

    b. Given: Dia $M = 36.900''$, $e = 15.840''$, and $d = 12.620''$.

    Compute f. _____

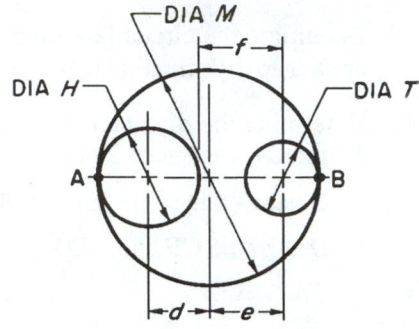

23. Lay out the template shown. Make the layout full size using construction methods. Do not use a protractor. All dimensions are in inches.

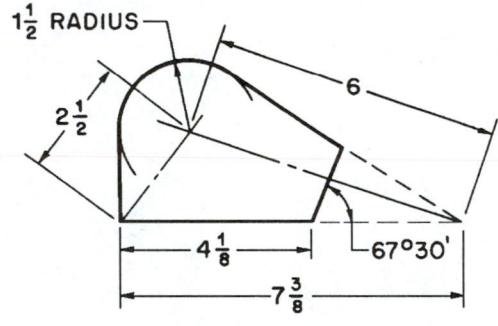

Section Five
Trigonometry

UNIT 53 Introduction to Trigonometric Functions

Objectives   After studying this unit you should be able to

- Identify the sides of a right triangle with reference to any angle.
- State the ratios of the six trigonometric functions in relation to given triangles.
- Find functions of angles given in decimal degrees and degrees, minutes, and seconds.
- Find angles in decimal degrees and degrees, minutes, and seconds of given functions.

Trigonometry is the branch of mathematics which is used to compute unknown angles and sides of triangles. Many problems that cannot be solved by the use of geometry alone are easily solved by trigonometry.

Practical machine shop problems are often solved by using a combination of elements of algebra, geometry, and trigonometry. Therefore, it is essential to develop the ability to analyze a problem in order to relate and determine the mathematical principles which are involved in its solution. Then the problem must be worked in clear orderly steps, based on mathematical facts.

When solving a problem, it is important to understand the trigonometric operations involved rather than to mechanically "plug in" values. Attempting to solve trigonometry problems without understanding the principles involved will prove to be unsuccessful, particularly in practical shop applications such as those found later in the text.

Ratio of Right Triangle Sides

In a right triangle, the ratio of two sides of the triangle determine the sizes of the angles, and the angles determine the ratio of two sides. Refer to the triangles shown. The size of angle A is determined by the ratio of side a to side b. When side $a = 1$ inch and side $b = 2$ inches, the ratio of a to b is 1:2 or 1/2. If side a is increased to 2 inches and side b remains 2 inches, the ratio of a to b is 1:1 or 1/1. Observe the increase in angle A as the ratio changed from 1/2 to 1/1.

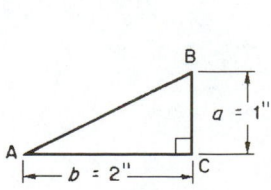
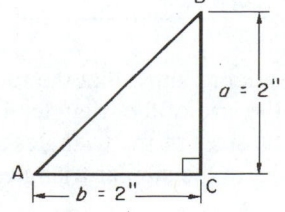
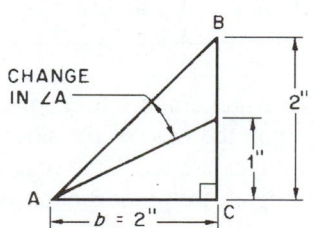

Identifying Right Triangle Sides by Name

The sides of a right triangle are named opposite side, adjacent side, and hypotenuse. The *hypotenuse* (hyp) is always the side opposite the right angle. It is always the longest side of a right triangle. The positions of the opposite and adjacent sides depend on the reference angle. The *opposite side* (opp side) is opposite the reference angle and the *adjacent side* (adj side) is next to the reference angle.

In the triangle showing $\angle A$ as the reference angle, side b is the adjacent side and side a is the opposite side. In the triangle showing $\angle B$ as the reference angle, side b is the opposite side and side a is the adjacent side. It is important to be able to identify the opposite and adjacent sides of right triangles in reference to any angle regardless of the positions of the triangles.

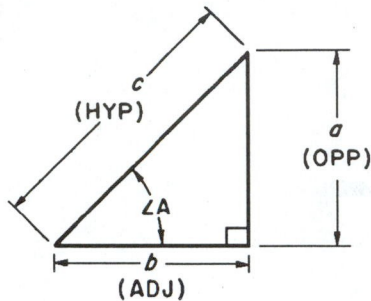 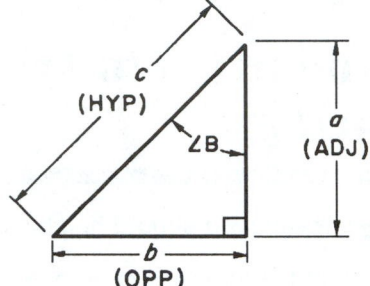

Trigonometric Functions: Ratio Method

There are two methods of defining trigonometric functions: the unity or unit circle method and the ratio method. Only the ratio method is presented in this book.

Since a triangle has three sides and a ratio is the comparison of any two sides, there are six different ratios. The names of the ratios are the sine, cosine, tangent, cotangent, secant, and cosecant.

The six trigonometric functions are defined in this table in relation to the triangle shown. The reference angle is A, the adjacent side is b, the opposite side is a, and the hypotenuse is c.

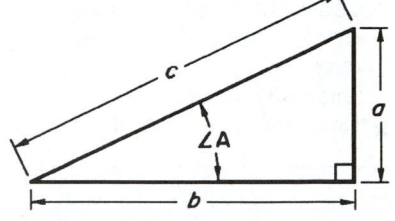

Function	Symbol	Definition of Function
sine of Angle A	sin A	$\sin A = \dfrac{\text{opp side}}{\text{hyp}} = \dfrac{a}{c}$
cosine of Angle A	cos A	$\cos A = \dfrac{\text{adj side}}{\text{hyp}} = \dfrac{b}{c}$
tangent of Angle A	tan A	$\tan A = \dfrac{\text{opp side}}{\text{adj side}} = \dfrac{a}{b}$
cotangent of Angle A	cot A	$\cot A = \dfrac{\text{adj side}}{\text{opp side}} = \dfrac{b}{a}$
secant of Angle A	sec A	$\sec A = \dfrac{\text{hyp}}{\text{adj side}} = \dfrac{c}{b}$
cosecant of Angle A	csc A	$\csc A = \dfrac{\text{hyp}}{\text{opp side}} = \dfrac{c}{a}$

To properly use trigonometric functions, it is essential to know that the function of an angle depends upon the ratio of the sides and **not** the size of the triangle. The functions of similar triangles are the same regardless of the sizes of the triangles since the sides of similar triangles are proportional. For example, in the similar triangles shown, the functions of angle A are the same for the three triangles. The equality of the tangent function is shown. Each of the other five functions have equal values for the three similar triangles.

$$\text{In } \triangle ABC,\ \tan \angle A = \frac{0.500}{1.000} = 0.500$$

$$\text{In } \triangle ADE,\ \tan \angle A = \frac{0.800}{1.600} = 0.500$$

$$\text{In } \triangle AFG,\ \tan \angle A = \frac{1.200}{2.400} = 0.500$$

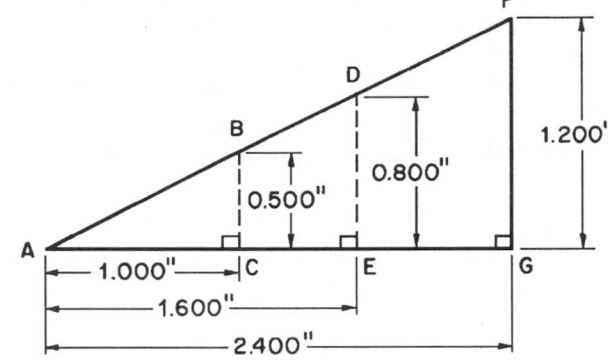

English and Metric Units of Angular Measure

As discussed in Unit 44, angular measure in the English system is generally expressed in degrees and minutes or in degrees, minutes, and seconds for very precise measurements. In the metric system, the decimal degree is the preferred unit of measure. Unless otherwise specified, degrees and minutes or degrees, minutes, and seconds are to be used when solving English system units of measure problems. Decimal degrees are to be used when solving metric system units of measure problems.

Determining Functions of Given Angles and Determining Angles of Given Functions

Calculator Applications

Determining functions of given angles or angles of given functions is readily accomplished using a calculator. As previously stated, calculator procedures vary among different makes of calculators. Also, different models of the same make calculator vary in some procedures. Generally, where procedures differ, there are basically two different procedures. Where relevant, both procedures are shown. However, because of the many makes and models of calculators, some procedures on your calculator may differ from the procedures shown. If so, it is essential that you refer to your user's guide or owner's manual.

The trigonometric keys, | sin |, | cos |, and | tan |, calculate the sine, cosine, and tangent of the angle in the display. An angle can be measured in degrees, radians, or gradients. **When calculating functions of angles measured in degrees, be certain that the calculator is in the degree mode.** A calculator is in the degree mode when the abbreviation DEG or D appears in the display when the calculator is turned on.

Depending on the make and model of the calculator, for most calculators there are essentially two ways of putting the calculator in the degree mode.

1. Pressing the | DRG | key changes the mode to radian, RAD, gradient, GRAD, or degree, DEG. Press | DRG | until DEG is displayed.

   ➤ **Note:** On some calculators DRG is the second or third function.

2. Press | MODE |, press | 4 | or press | MODE | twice, press | 1 |. The calculator is in the degree mode and DEG or D is displayed.

Procedure for Determining the Sine, Cosine, and Tangent Functions

The procedure for determining functions of angles varies with different calculators. Although there are exceptions, basically, there are two different procedures.

1. The value of the angle is entered first, and then the appropriate function key, $\boxed{\text{sin}}$, $\boxed{\text{cos}}$, or $\boxed{\text{tan}}$ is pressed.

2. The appropriate function key, $\boxed{\text{sin}}$, $\boxed{\text{cos}}$, or $\boxed{\text{tan}}$ is pressed first, and then the value of the angle is entered.

The following examples show both procedures.

➤ **Note:** As previously stated, on certain calculators $\boxed{=}$ is used instead of $\boxed{\text{EXE}}$.

Examples Round each answer to 5 decimal places.

1. Determine the sine of 43°.

   43 $\boxed{\text{sin}}$ → 0.68199836, 0.68200   Ans

   or $\boxed{\text{sin}}$ 43 $\boxed{\text{EXE}}$ 0.6819983601, 0.68200   Ans

2. Determine the cosine of 6.034°.

   6.034 $\boxed{\text{cos}}$ → 0.994459692, 0.99446   Ans

   or $\boxed{\text{cos}}$ 6.034 $\boxed{\text{EXE}}$ 0.9944596918, 0.99446   Ans

3. Determine the tangent of 51.9162°.

   51.9162 $\boxed{\text{tan}}$ → 1.276090171, 1.27609   Ans

   or $\boxed{\text{tan}}$ 51.9162 $\boxed{\text{EXE}}$ 1.276090171, 1.27609   Ans

4. Determine the sine of 61°49′.

   61.49 $\boxed{\text{2nd}}$ $\boxed{\text{►DD}}$ $\boxed{\text{sin}}$ → 0.881440874, 0.88144   Ans

   61 $\boxed{\circ\;\prime\;\prime\prime}$ 49 $\boxed{\circ\;\prime\;\prime\prime}$ $\boxed{\text{sin}}$ → 0.81440874

   or $\boxed{\text{sin}}$ 61 $\boxed{\circ\;\prime\;\prime\prime}$ 49 $\boxed{\circ\;\prime\;\prime\prime}$ $\boxed{\text{EXE}}$ 0.8814408742, 0.88144   Ans

5. Determine the tangent of 32°7′23″.

   32.0723 $\boxed{\text{2nd}}$ $\boxed{\text{►DD}}$ $\boxed{\text{tan}}$ → 0.627859699, 0.62786   Ans

   32 $\boxed{\circ\;\prime\;\prime\prime}$ 7 $\boxed{\circ\;\prime\;\prime\prime}$ 23 $\boxed{\circ\;\prime\;\prime\prime}$ $\boxed{\text{tan}}$ → 0.627859698

   or $\boxed{\text{tan}}$ 32 $\boxed{\circ\;\prime\;\prime\prime}$ 7 $\boxed{\circ\;\prime\;\prime\prime}$ 23 $\boxed{\circ\;\prime\;\prime\prime}$ $\boxed{\text{EXE}}$ 0.6278596985, 0.62786   Ans

Procedure for Determining the Cosecant, Secant, and Cotangent Functions

The cosecant, secant, and cotangent functions are reciprocal functions. The cosecant is the reciprocal of the sine.

$$\csc \angle A = \frac{1}{\sin \angle A}$$

The secant is the reciprocal of the cosine.

$$\sec \angle A = \frac{1}{\cos \angle A}$$

The cotangent is the reciprocal of the tangent.

$$\cot \angle A = \frac{1}{\tan \angle A}$$

Cosecants, secants, and cotangents are computed with the reciprocal key, $\boxed{1/x}$ or $\boxed{x^{-1}}$. On certain calculators, the reciprocal key is a second function.

As with sine, cosine, and tangent functions, basically there are two different procedures in determining reciprocal functions.

1. The value of the angle is entered first. Next press the appropriate function key, $\boxed{\text{sin}}$, $\boxed{\text{cos}}$, or $\boxed{\text{tan}}$; press $\boxed{\frac{1}{x}}$.

2. The appropriate function key, $\boxed{\text{sin}}$, $\boxed{\text{cos}}$, $\boxed{\text{tan}}$ is pressed first. The value of the angle is entered next,

   press $\boxed{\text{EXE}}$, press $\boxed{x^{-1}}$, press $\boxed{\text{EXE}}$.
   └─ or press $\boxed{=}$, press $\boxed{\frac{1}{x}}$.

The following examples show both procedures.

Examples   Round each answer to 5 decimal places.

1. Determine the cosecant of 57.16°.

   57.16 $\boxed{\text{sin}}$ $\boxed{\frac{1}{x}}$ → 1.190209506, 1.19021   Ans

   or $\boxed{\text{sin}}$ 57.16 $\boxed{\text{EXE}}$ $\boxed{x^{-1}}$ $\boxed{\text{EXE}}$ 1.190209506, 1.19021   Ans
   └─ or $\boxed{=}$ $\boxed{\frac{1}{x}}$ → 1.190209506

2. Determine the secant of 13.795°.

   13.795 $\boxed{\text{cos}}$ $\boxed{\frac{1}{x}}$ → 1.029701649, 1.02970   Ans

   or $\boxed{\text{cos}}$ 13.795 $\boxed{\text{EXE}}$ $\boxed{x^{-1}}$ $\boxed{\text{EXE}}$ 1.029701649, 1.02970 Ans
   └─ or $\boxed{=}$ $\boxed{\frac{1}{x}}$ → 1.029701649

3. Determine the cotangent of 78.63°.

   78.63 $\boxed{\text{tan}}$ $\boxed{\frac{1}{x}}$ → 0.20109054, 0.20109   Ans

   or $\boxed{\text{tan}}$ 78.63 $\boxed{\text{EXE}}$ $\boxed{x^{-1}}$ $\boxed{\text{EXE}}$ 0.2010905402, 0.20109   Ans
   └─ or $\boxed{=}$ $\boxed{\frac{1}{x}}$ → 0.2010905402

4. Determine the cosecant of 24°51′.

   24.51 $\boxed{\text{2nd}}$ $\boxed{\blacktriangleright\text{DD}}$ $\boxed{\text{sin}}$ $\boxed{\frac{1}{x}}$ → 2.379569353, 2.37957   Ans

   24 $\boxed{\circ\,\prime\,\prime\prime}$ 51 $\boxed{\circ\,\prime\,\prime\prime}$ $\boxed{\text{sin}}$ $\boxed{\frac{1}{x}}$ → 2.379569353

   or $\boxed{\text{sin}}$ 24 $\boxed{\circ\,\prime\,\prime\prime}$ 51 $\boxed{\circ\,\prime\,\prime\prime}$ $\boxed{\text{EXE}}$ $\boxed{x^{-1}}$ $\boxed{\text{EXE}}$ 2.379569353, 2.37957   Ans
   └─ or $\boxed{=}$ $\boxed{\frac{1}{x}}$ → 2.379569353

5. Determine the secant of 43°36′25″.

   43.3625 $\boxed{\text{2nd}}$ $\boxed{\blacktriangleright\text{DD}}$ $\boxed{\text{cos}}$ $\boxed{\frac{1}{x}}$ → 1.381047089, 1.38105   Ans

   43 $\boxed{\circ\,\prime\,\prime\prime}$ 36 $\boxed{\circ\,\prime\,\prime\prime}$ 25 $\boxed{\circ\,\prime\,\prime\prime}$ $\boxed{\text{cos}}$ $\boxed{\frac{1}{x}}$ → 1.381047089

   or $\boxed{\text{cos}}$ 43 $\boxed{\circ\,\prime\,\prime\prime}$ 36 $\boxed{\circ\,\prime\,\prime\prime}$ 25 $\boxed{\circ\,\prime\,\prime\prime}$ $\boxed{\text{EXE}}$ $\boxed{x^{-1}}$ $\boxed{\text{EXE}}$ 1.381047089, 1.38105   Ans
   └─ or $\boxed{=}$ $\boxed{\frac{1}{x}}$ → 1.381047089

Angles of Given Functions

Determining the angle of a given function is the inverse of determining the function of a given angle. When a certain function value is known, the angle can be found easily.

The term *arc* is often used as a prefix to any of the names of the trigonometric functions, such as arcsine, arctangent, etc. Such expressions are called inverse functions and they mean angles. For example, sin 30°15′ = 0.503774, then 30°15′ = arcsin 0.503774 or 30°15′ is the angle whose sine is 0.503774.

Arcsin is often written as $\sin^{-1}$, arccos is written as $\cos^{-1}$, and arctan is written as $\tan^{-1}$.

Procedure for Determining Angles of Given Functions

The procedure for determining angles of given functions varies somewhat with the make and model of calculator. With most calculators, the inverse functions are shown as second functions [sin⁻¹], [cos⁻¹], and [tan⁻¹] of function keys [sin], [cos], and [tan].

With some calculators, the function value is entered before the function key is pressed. With other calculators, the function key is pressed before the function value is entered.

The following examples show the procedure for determining angles of given functions. All examples show the procedures where [sin⁻¹], [cos⁻¹], and [tan⁻¹] are the second functions. Remember, for certain calculators it is necessary to substitute [=] in place of [EXE].

Examples

1. Find the angle whose tangent is 1.902. Round the answer to 2 decimal places.

    1.902 [2nd] (or [SHIFT]) [tan⁻¹] → 62.2662961, 62.27°   Ans

    or [SHIFT][tan⁻¹] 1.902 [EXE] 62.2662961, 62.27°   Ans

2. Find the angle whose sine is 0.21256. Round the answer to 2 decimal places.

    .21256 [2nd] (or [SHIFT]) [sin⁻¹] → 12.27241712, 12.27°   Ans

    or [SHIFT][sin⁻¹] .21256 [EXE] 12.27241712, 12.27°   Ans

3. Find the angle whose cosine is 0.732976. Give the answer in degrees, minutes, and seconds.

    .732976 [2nd][cos⁻¹][3rd][►DMS] → 42°51′48″7, 42°51′49″   Ans

    or .732976 [SHIFT][cos⁻¹][SHIFT][←] → 42°51′48.72″   Ans

    or [SHIFT][cos⁻¹] .732976 [EXE][SHIFT][←] → 42°51′48.72″, 42°51′49″   Ans

Angles for the reciprocal functions—cosecant, secant, and cotangent—are calculated using the reciprocal key, [1/x] or [x⁻¹].

Examples

1. Find the angle whose secant is 1.2263. Round the answer to 2 decimal places.

    1.2263 [1/x][2nd] (or [SHIFT]) [cos⁻¹] → 35.36701576, 35.37°   Ans

    or [SHIFT][cos⁻¹] 1.2263 [x⁻¹] (or [1/x]) [EXE] 35.36701576, 35.37°   Ans

2. Find the angle whose cotangent is 0.4166. Give the answer in degrees and minutes.

    .4166 [1/x][2nd][tan⁻¹][3rd][►DMS] 67°23′00″2, 67°23′   Ans

    or .4166 [1/x][SHIFT][tan⁻¹][SHIFT][←] → 67°23′0.2″

    or [SHIFT][tan⁻¹] .4166 [x⁻¹] (or [1/x]) [EXE][SHIFT][° ′ ″] → 67°23′0.2″, 67°23′   Ans

APPLICATION

Identifying Right Triangle Sides by Name

With reference to ∠1, name the sides of each of the following triangles as opposite, adjacent, or hypotenuse.

1. Name sides *r*, *x*, and *y*.   *r* = _____   2. Name sides *r*, *x*, and *y*.   *r* = _____

    x = _____   *x* = _____

    y = _____   *y* = _____

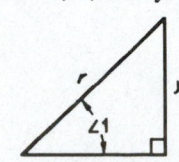

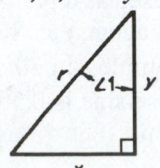

3. Name sides *a*, *b*, and *c*.

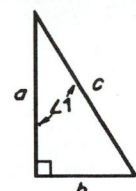

a = _____

b = _____

c = _____

10. Name sides *h*, *k*, and *l*.

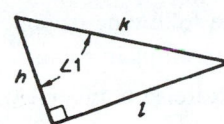

h = _____

k = _____

l = _____

4. Name sides *a*, *b*, and *c*.

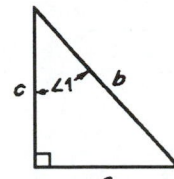

a = _____

b = _____

c = _____

11. Name sides *m*, *p*, and *s*.

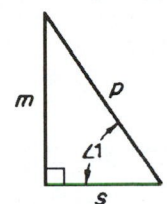

m = _____

p = _____

s = _____

5. Name sides *a*, *b*, and *c*.

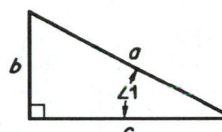

a = _____

b = _____

c = _____

12. Name sides *m*, *p*, and *s*.

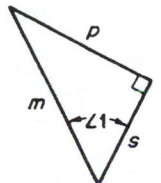

m = _____

p = _____

s = _____

6. Name sides *d*, *m*, and *p*.

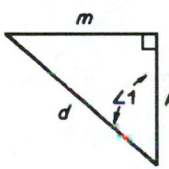

d = _____

m = _____

p = _____

13. Name sides *m*, *r*, and *t*.

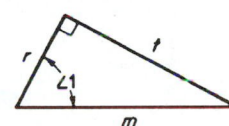

m = _____

r = _____

t = _____

7. Name sides *d*, *m*, and *p*.

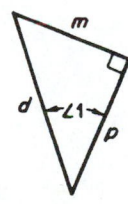

d = _____

m = _____

p = _____

14. Name sides *m*, *r*, and *t*.

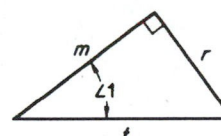

m = _____

r = _____

t = _____

8. Name sides *e*, *f*, and *g*.

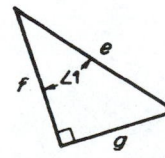

e = _____

f = _____

g = _____

15. Name sides *f*, *g*, and *h*.

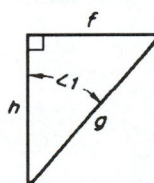

f = _____

g = _____

h = _____

9. Name sides *h*, *k*, and *l*.

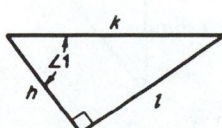

h = _____

k = _____

l = _____

16. Name sides *f*, *g*, and *h*.

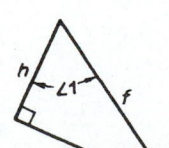

f = _____

g = _____

h = _____

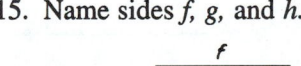

Trigonometric Functions

The sides of each of the following triangles are labeled with different letters. State the ratio of each of the 6 functions in relation to $\angle 1$ for each of the triangles. For example, for the triangle in exercise number 17, $\sin \angle 1 = \frac{y}{r}$, $\cos \angle 1 = \frac{x}{r}$, $\tan \angle 1 = \frac{y}{x}$, $\cot \angle 1 = \frac{x}{y}$, $\sec \angle 1 = \frac{r}{x}$, and $\csc \angle 1 = \frac{r}{y}$.

17. _____   _____   19. _____   _____   21. _____   _____

    _____   _____   _____   _____   _____   _____

    _____   _____   _____   _____   _____   _____

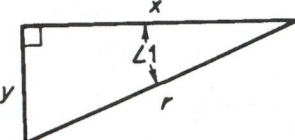 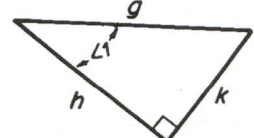 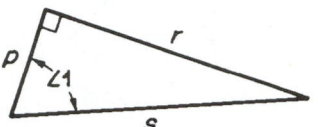

18. _____   _____   20. _____   _____   22. _____   _____

    _____   _____   _____   _____   _____   _____

    _____   _____   _____   _____   _____   _____

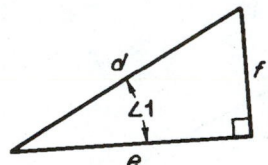 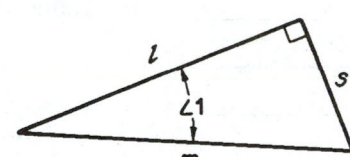 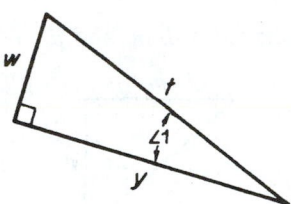

23. Three groups of triangles are given below. Each group consists of four triangles. Within each group, name the triangles—a, b, c, or d—in which angles A are equal.

Group 1 _____

a.   b.   c.   d.

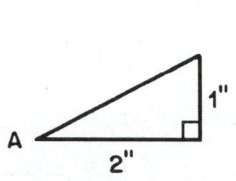 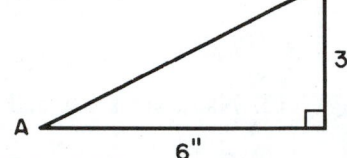 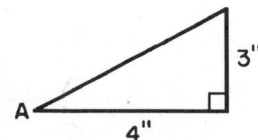 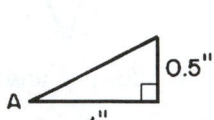

Group 2 _____

a.   b.   c.   d.

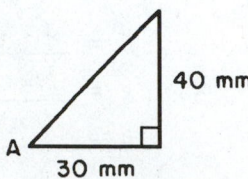 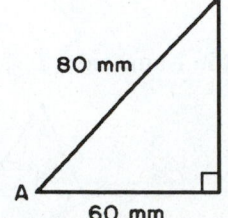 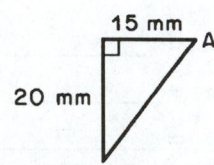 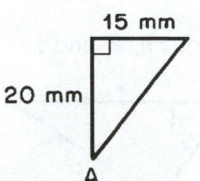

Group 3 _____

a.   b.   c.   d.

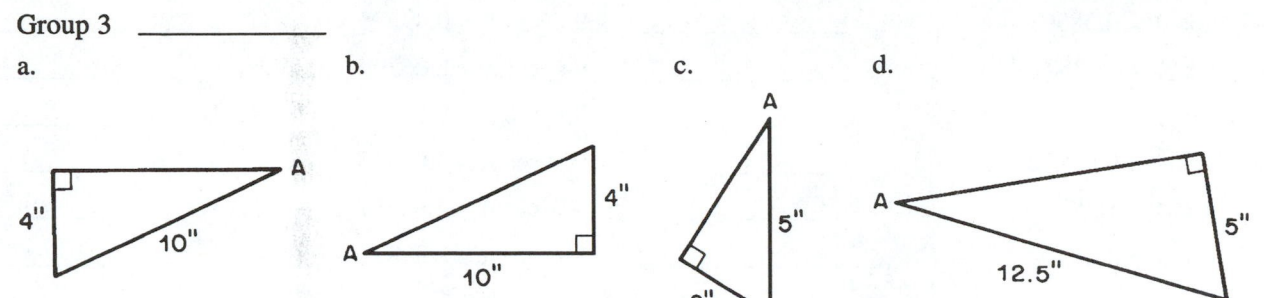

Determine the sine, cosine, or tangent functions of the following angles. Round the answers to 5 decimal places.

24. sin 36°	_____	34. tan 73.86°	_____	44. cos 19°42′	_____
25. cos 53°	_____	35. sin 50.05°	_____	45. sin 71°59′	_____
26. tan 47°	_____	36. cos 16.77°	_____	46. tan 42°36′	_____
27. cos 18°	_____	37. sin 0.86°	_____	47. sin 20°28′	_____
28. sin 79°	_____	38. tan 59.89°	_____	48. cos 6°16′	_____
29. cos 4°	_____	39. cos 60.605°	_____	49. tan 37°26′12″	_____
30. tan 65.18°	_____	40. cos 77.144°	_____	50. tan 9°4′50″	_____
31. sin 27.06°	_____	41. tan 10°18′	_____	51. cos 86°30′38″	_____
32. tan 12.92°	_____	42. sin 26°29′	_____	52. sin 53°46′19″	_____
33. cos 4.63°	_____	43. sin 6°53′	_____	53. tan 70°51′44″	_____

Determine the cosecant, secant, or cotangent functions of the following angles. Round the answers to 5 decimal places.

54. csc 27°	_____	61. sec 77.08°	_____	68. cot 17°19′	_____
55. sec 56°	_____	62. sec 86.92°	_____	69. sec 80°51′	_____
56. cot 19°	_____	63. csc 44.077°	_____	70. sec 6°43′	_____
57. sec 48°	_____	64. csc 6.904°	_____	71. csc 76°0′15″	_____
58. csc 6.16°	_____	65. cot 31.081°	_____	72. cot 2°58′59″	_____
59. cot 18.85°	_____	66. sec 20°16′	_____	73. sec 55°16′32″	_____
60. cot 36.97°	_____	67. csc 46°27′	_____	74. csc 19°34′18″	_____

Determine the value of angle A in decimal degrees for each of the given functions. Round the answers to the nearest hundredth of a degree.

75. sin A = 0.83692	_____	81. sin A = 0.02539	_____
76. cos A = 0.23695	_____	82. tan A = 1.56334	_____
77. tan A = 0.59334	_____	83. tan A = 0.09632	_____
78. cos A = 0.97370	_____	84. cos A = 0.20893	_____
79. tan A = 3.96324	_____	85. cos A = 0.87736	_____
80. sin A = 0.77376	_____	86. sin A = 0.10532	_____

87. $\cos A = 0.38591$	_____	93. $\cot A = 0.89538$	_____
88. $\tan A = 0.67871$	_____	94. $\cot A = 6.06790$	_____
89. $\sin A = 0.63634$	_____	95. $\csc A = 5.93632$	_____
90. $\cos A = 0.05332$	_____	96. $\sec A = 1.02353$	_____
91. $\sec A = 1.58732$	_____	97. $\csc A = 4.93317$	_____
92. $\csc A = 2.08363$	_____	98. $\cot A = 2.89895$	_____

Determine the value of angle A in degrees and minutes for each of the given functions. Round the answers to the nearest minute.

99. $\cos A = 0.23076$	_____	111. $\cos A = 0.69304$	_____
100. $\tan A = 0.56731$	_____	112. $\tan A = 3.03030$	_____
101. $\sin A = 0.92125$	_____	113. $\sin A = 0.70705$	_____
102. $\tan A = 4.09652$	_____	114. $\cos A = 0.90501$	_____
103. $\cos A = 0.03976$	_____	115. $\csc A = 1.38630$	_____
104. $\sin A = 0.09741$	_____	116. $\sec A = 5.05377$	_____
105. $\sin A = 0.73204$	_____	117. $\cot A = 0.27982$	_____
106. $\tan A = 0.95300$	_____	118. $\csc A = 2.02103$	_____
107. $\cos A = 0.00495$	_____	119. $\sec A = 9.90778$	_____
108. $\cos A = 0.89994$	_____	120. $\cot A = 8.03012$	_____
109. $\sin A = 0.30536$	_____	121. $\csc A = 3.03539$	_____
110. $\tan A = 7.60385$	_____	122. $\sec A = 2.71177$	_____

UNIT 54 Analysis of Trigonometric Functions

Objectives   After studying this unit you should be able to

- **Determine the variations of functions as angles change.**
- **Compute cofunctions of complementary angles.**

Variation of Functions

As the size of an angle increases the sine, tangent, and secant functions increase while the cofunctions (cosine, cotangent, cosecant) decrease. As the reference angles approach 0° or 90°, the function variation can be shown. These examples illustrate variations of an increasing function and a decreasing function for a reference angle which is increasing in size.

➤ **Note:**   Use this figure for Examples 1–2.

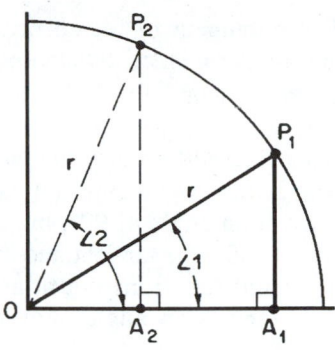

OP_1 and OP_2 are radii of the arc of a circle.

$$OP_1 = OP_2 = r$$

Example 1   Variation of an increasing function; the sine function.

The sine of an angle $= \dfrac{\text{opposite side}}{\text{hypotenuse}}$

$$\sin \angle 1 = \frac{A_1P_1}{r}$$

$$\sin \angle 2 = \frac{A_2P_2}{r}$$

A_2P_2 is greater than A_1P_1; therefore, $\sin \angle 2$ is greater than $\sin \angle 1$. Observe that if $\angle 1$ decreases to $0°$, side $A_1P_1 = 0$.

$$\sin 0° = \frac{0}{r} = 0$$

If $\angle 2$ increases to $90°$, size $A_2P_2 = r$.

$$\sin 90° = \frac{r}{r} = 1$$

Conclusion: As an angle increases from $0°$ to $90°$, the sine of the angle increases from 0 to 1.

Example 2   Variation of a decreasing function; the cosine function.

The cosine of an angle $= \dfrac{\text{adjacent side}}{\text{hypotenuse}}$

$$\cos \angle 1 = \frac{OA_1}{r}$$

$$\cos \angle 2 = \frac{OA_2}{r}$$

OA_2 is less than OA_1; therefore, $\cos \angle 2$ is less than $\cos \angle 1$. Observe that if $\angle 1$ decreases to $0°$, side $OA_1 = r$.

$$\cos 0° = \frac{r}{r} = 1$$

If $\angle 2$ increases to $90°$, size $OA_2 = 0$.

$$\cos 90° = \frac{0}{r} = 0$$

Conclusion: As an angle increases from $0°$ to $90°$, the cosine of the angle decreases from 1 to 0.

It is helpful to sketch figures for all functions in order to further develop an understanding of the relationship of angles and their functions. Particular attention should be given to functions of angles close to $0°$ and $90°$.

A summary of the variations taken from the table of trigonometric functions is shown for an angle increasing from $0°$ to $90°$.

As an angle increases from 0° to 90°	
sin increases from 0 to 1	cos decreases from 1 to 0
tan increases from 0 to ∞	cot decreases from ∞ to 0
sec increases from 1 to ∞	csc decreases from ∞ to 1

The cotangent of 0°, cosecant of 0°, tangent of 90°, and secant of 90° involve division by zero; since division by zero is not possible, these values are undefined. Although they are undefined, the values are often written as ∞. The symbol ∞ means infinity. Infinity is the quality of existing beyond or being greater than any countable value. It cannot be used for computations at this level of mathematics.

Rather than to attempt to treat ∞ as a value, think of the tangent and secant functions not at an angle of 90°, but at angles very close to 90°. Observe that as an angle approaches 90°, the tangent and secant functions get very large. Think of the cotangent and cosecant functions not at an angle of 0°, but as very small angles close to 0°. Observe that as an angle approaches 0° the cotangent and cosecant functions get very large.

Functions of Complementary Angles

Two angles are complementary when their sum is 90°. For example, 20° is the complement of 70° and 70° is the complement of 20°. In the triangle shown, $\angle A$ is the complement of $\angle B$ and $\angle B$ is the complement of $\angle A$. The six functions of the angle and the cofunctions of the complementary angle are shown.

sin 20° = cos 70° = 0.34202	cos 20° = sin 70° = 0.93969
tan 20° = cot 70° = 0.36397	cot 20° = tan 70° = 2.7475
sec 20° = csc 70° = 1.0642	csc 20° = sec 70° = 2.9238

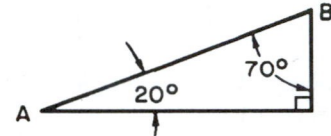

A function of an angle is equal to the cofunction of the complement of the angle.

The complement of an angle equals 90° minus the angle. The relationships of the six functions of angles and the cofunctions of the complementary angles are shown.

sin A = cos (90° − A)	cos A = sin (90° − A)
tan A = cot (90° − A)	cot A = tan (90° − A)
sec A = csc (90° − A)	csc A = sec (90° − A)

Examples   For each function of an angle, write the cofunction of the complement of the angle.

1. sin 30° = cos (90° − 30°) = cos 60°   Ans

2. cot 10° = tan (90° − 10°) = tan 80°   Ans

3. tan 72.53° = cot (90° − 72.53°) = cot 17.47°   Ans

4. sec 40°20′ = csc (90° − 40°20′) = csc (89°60′ − 40°20′) = csc 49°40′   Ans

5. cos 90° = sin (90° − 90°) = sin 0°   Ans

APPLICATION

Variation of Functions

Refer to this figure in answering exercises 1–7. It may be helpful to sketch figures.

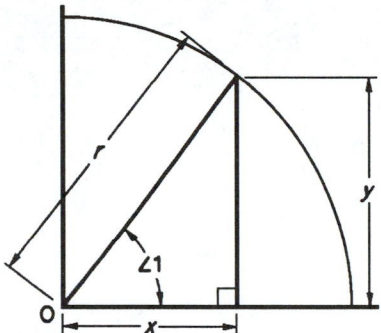

1. When $\angle 1$ is almost 90°:

   a. how does side y compare to side r?   _____

   b. how does side x compare to side r?   _____

   c. how does side x compare to side y?   _____

2. When $\angle 1$ is 90°:

   a. what is the value of side x?   _____

   b. how does side y compare to side r?   _____

3. When $\angle 1$ is slightly greater than 0°:

   a. how does side y compare to side r?   _____

   b. how does side x compare to side r?   _____

   c. how does side x compare to side y?   _____

4. When $\angle 1$ is 0°:

   a. what is the value of side y?   _____

   b. how does side x compare to side r?   _____

5. When side x = side y:

   a. what is the value of $\angle 1$?   _____

   b. what is the value of the tangent function?   _____

   c. what is the value of the cotangent function?   _____

6. When side x = side r:

   a. what is the value of the sine function? _____

   b. what is the value of the secant function? _____

   c. what is the value of the cosine function? _____

   d. what is the value of the tangent function? _____

7. When side y = side r:

   a. what is the value of the sine function? _____

   b. what is the value of the cotangent function? _____

   c. what is the value of the cosine function? _____

   d. what is the value of the cosecant function? _____

For each exercise, functions of two angles are given. Which of the functions of the two angles is greater? Do **not** use a calculator.

8. sin 38°; sin 43°	_____	14. tan 21°40′; tan 12°50′	_____
9. tan 17°; tan 18°	_____	15. cos 81°19′; cos 81°20′	_____
10. cos 78°; cos 85°	_____	16. sin 0.42°; sin 0.37°	_____
11. cot 40°; cot 36°	_____	17. csc 40.50°; 40.45°	_____
12. sec 5°; sec 8°	_____	18. cot 27°23′; cot 87°0′	_____
13. csc 22°; csc 25°	_____	19. sec 55°; sec 54°50′	_____

Functions of Complementary Angles

For each function of an angle, write the cofunction of the complement of the angle.

20. tan 23°	_____	30. cot 7°10′	_____
21. sin 49°	_____	31. sec 36°06′	_____
22. cos 26°	_____	32. csc 0°38′	_____
23. sec 82°	_____	33. sin 5.89°	_____
24. cot 35°	_____	34. cos 3.76°	_____
25. csc 51°	_____	35. cot 0°	_____
26. cos 90°	_____	36. tan 90°	_____
27. sin 0°	_____	37. sec 43°19′	_____
28. tan 57.5°	_____	38. cos 0.01°	_____
29. cos 12.2°	_____	39. sin 89°59′	_____

For each exercise, functions and cofunctions of two angles are given. Which of the functions or cofunctions of the two angles is greater? Do **not** use a calculator.

40. cos 48°; sin 18°	_____	43. tan 30°; cot 45°	_____
41. cos 55°; sin 40°	_____	44. sec 42°; csc 58°	_____
42. tan 30°; cot 65°	_____	45. sec 43°; csc 58°	_____

46. sin 14°; cos 78°   _____   49. cot 87°50′; tan 2°40′   _____

47. sin 12°; cos 75°   _____   50. sec 0.2°; csc 89.9°   _____

48. cot 89°10′; tan 1°20′   _____   51. sec 0.2°; csc 89.0°   _____

UNIT 55   Basic Calculations of Angles and Sides of Right Triangles

Objectives   After studying this unit you should be able to

- **Compute an unknown angle of a right triangle when two sides are known.**
- **Compute an unknown side of a right triangle when an angle and a side are known.**

Determining an Unknown Angle When Two Sides of a Right Triangle Are Known

In order to solve for an unknown angle of a right triangle where neither acute angle is known, at least two sides must be known. An understanding of the procedures required for solving for unknown angles is essential to the machinist.

- In relation to the desired angle, identify two given sides as adjacent, opposite, or hypotenuse.
- Determine the functions that are ratios of the sides identified in relation to the desired angle.

  ➤ **Note:**   Two of the six trigonometric functions are ratios of the two known sides. Either of the two functions can be used. Both produce the same value for the unknown.

- Choose one of the two functions; substitute the given sides in the ratio.
- Determine the angle that corresponds to the quotient of the ratio.

When sides are given in inches (English units), compute the angle to the nearer minute. When sides are given in millimeters (metric units), compute the angle to the nearer hundredth degree.

  ➤ **Note:**   With the calculator examples given in this unit and in the trigonometry units which follow, generally two basic procedures are shown for each example. If your calculator functions differ from these procedures you may find it helpful to refer back to Unit 53. Additional procedures are given in many examples.

Example 1   Determine ∠A of the right triangle shown to the nearer minute.

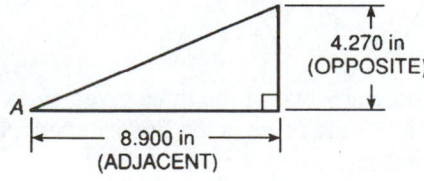

Solution In relation to $\angle A$, the 8.900-inch side is the adjacent side, and the 4.270-inch side is the opposite side.

Determine the two functions whose ratios consist of the adjacent and opposite sides. Then, tan $\angle A$ = opposite side/adjacent side, and cot $\angle A$ = adjacent side/opposite side. Either the tangent or cotangent function can be used.

Choosing the tangent function: tan $\angle A = \frac{4.270 \text{ in}}{8.900 \text{ in}}$.

Determine the angle whose tangent function is the quotient of $\frac{4.270}{8.900}$.

$\angle A$ = 4.27 [+] 8.9 [=] [2nd] [tan⁻¹] [3rd] [►DMS] 25°37′49″9, 25°38′   Ans

or $\angle A$ = [SHIFT] [tan⁻¹] [(] 4.27 [+] 8.9 [)] [EXE] [SHIFT] [◄—] → 25°37′49.95″, 25°38′   Ans

Example 2 Determine $\angle B$ of the right triangle shown to the nearest hundredth degree.

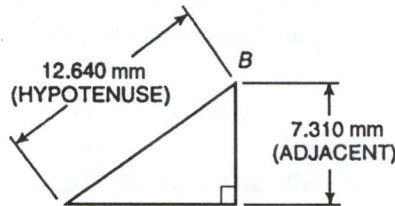

Solution In relation to $\angle B$, the 12.640-millimeter side is the hypotenuse, and the 7.310-millimeter side is the adjacent side.

Determine the two functions whose ratios consist of the adjacent side and the hypotenuse. Then, cos $\angle B$ = adjacent side/hypotenuse; and sec $\angle B$ = hypotenuse/adjacent side. Either the cosine or secant function can be used. Choosing the cosine function: cos $\angle B = \frac{7.310 \text{ mm}}{12.640 \text{ mm}}$.

Determine the angle whose cosine function is the quotient of $\frac{7.310}{12.640}$.

$\angle B$ = 7.31 [+] 12.64 [=] [2nd] [cos⁻¹] → 54.66733748, 54.67°   Ans

or $\angle B$ = [SHIFT] [cos⁻¹] [(] 7.31 [+] 12.64 [)] [EXE] 54.66733748, 54.67°   Ans

Example 3 Determine $\angle 1$ and $\angle 2$ of the triangle shown to the nearest minute.

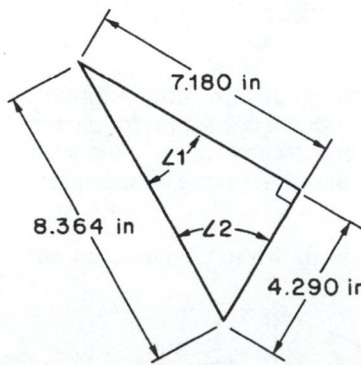

Solution Compute either $\angle 1$ or $\angle 2$. Choose any two of the three given sides for a ratio. In relation to $\angle 1$, the 4.290-inch side is the opposite side, and the 8.364-inch side is the hypotenuse.

Determine the two functions whose ratios consist of the opposite side and the hypotenuse. Then, sin $\angle 1$ = opposite side/hypotenuse, and csc $\angle 1$ = hypotenuse/opposite. Either the sine or cosecant can be used.

Choosing the sine function: sin $\angle 1 = \frac{4.290 \text{ in}}{8.364 \text{ in}}$.

Determine the angle whose sine function is the quotient of $\frac{4.290}{8.364}$.

$\angle 1$ = 4.29 $\boxed{+}$ 8.364 $\boxed{=}$ $\boxed{\text{2nd}}$ $\boxed{\sin^{-1}}$ $\boxed{\text{3rd}}$ $\boxed{\blacktriangleright\text{DMS}}$ 30°51′28″8, 30°51′   Ans

or $\angle 1$ = $\boxed{\text{SHIFT}}$ $\boxed{\sin^{-1}}$ $\boxed{(}$ 4.29 $\boxed{+}$ 8.364 $\boxed{)}$ $\boxed{\text{EXE}}$ $\boxed{\text{SHIFT}}$ $\boxed{\leftarrow}$ → 30°51′28.89″, 30°51′   Ans

Since $\angle 1 + \angle 2 = 90°$, $\angle 2 = 90° - 30°51′$, $\angle 2 = 59°9′$ Ans

Determining an Unknown Side When an Acute Angle and One Side of a Right Triangle Are Known

In order to solve for an unknown side of a right triangle, at least an acute angle and one side must be known.

Procedure   To determine an unknown side when an acute angle and one side of a right triangle are known

- Identify the given side and the unknown side as adjacent, opposite, or hypotenuse in relation to the given angle.

- Determine the trigonometric functions that are ratios of the sides identified in relation to the given angle.

  > **Note:** Two of the six functions will be found as ratios of the two identified sides. Either of the two functions can be used. Both produce the same value for the unknown. If the unknown side is made the numerator of the ratio, the problem is solved by multiplication. If the unknown side is made the denominator of the ratio, the problem is solved by division.

- Choose one of the two functions and substitute the given side and given angle.

- Solve as a proportion for the unknown side.

Example 1   Determine side x of the right triangle shown. Round the answer to 3 decimal places.

Solution   In relation to the 61°50′ angle, the 5.410-inch side is the adjacent side and side x is the opposite side.

Determine the two functions whose ratios consist of the adjacent and opposite sides. Tan 61°50′ = opposite side/adjacent side, and cot 61°50′ = adjacent side/opposite side. Either the tangent or cotangent function can be used.

Choosing the tangent function: tan 61°50′ = $\frac{x}{5.410 \text{ in}}$.

Solve as a proportion.

$$\frac{\tan 61°50′}{1} = \frac{x}{5.410 \text{ in}}$$

$$x = \tan 61°50′ \, (5.410 \text{ in})$$

x = 61.50 $\boxed{\text{2nd}}$ $\boxed{\blacktriangleright\text{DD}}$ $\boxed{\tan}$ $\boxed{\times}$ 5.41 $\boxed{=}$ 10.10371739, 10.100 in   Ans

or x = $\boxed{\tan}$ 61 $\boxed{°\,′\,″}$ 50 $\boxed{°\,′\,″}$ $\boxed{\times}$ 5.41 $\boxed{\text{EXE}}$ 10.10371739, 10.100 in   Ans

Example 2   Determine side *r* of the right triangle shown. Round the answer to 3 decimal places.

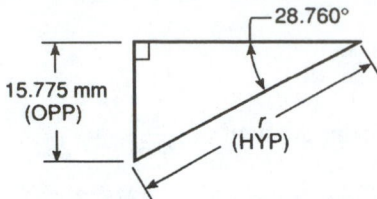

Solution   In relation to the 28.760° angle, the 15.775-millimeter side is the opposite side and side *r* is the hypotenuse.

Determine the two functions whose ratios consist of the opposite side and the hypotenuse. Sin 28.760° = opposite side/hypotenuse, and csc 28.760° = hypotenuse/opposite side. Either the sine or cosecant function can be used.

Choosing the sine function: sin 28.760° = $\frac{15.775 \text{ mm}}{r}$.

Solve as a proportion.

$$\frac{\sin 28.760°}{1} = \frac{15.775 \text{ mm}}{r}$$

$$r = \frac{15.775 \text{ mm}}{\sin 28.760°}$$

$r = 15.775$ | + | 28.76 | sin | = | 32.78659364, 32.787 mm   Ans

or $r = 15.775$ | + | | sin | 28.76 | EXE | 32.78659364, 32.787 mm   Ans

Example 3   Determine side *x*, side *y*, and ∠1 of the right triangle shown. Round the answer to 3 decimal places.

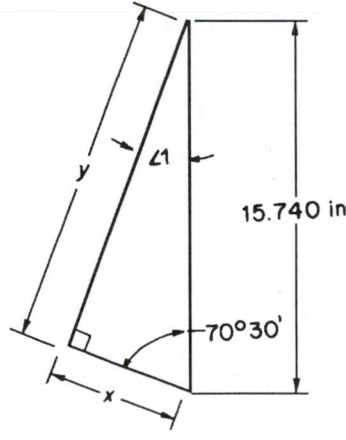

Solution   Compute either side *x* or side *y*. Choosing side *x*, in relation to the 70°30′ angle, side *x* is the adjacent side. The 15.740-inch side is the hypotenuse.

Determine the two functions whose ratios consist of the adjacent side and the hypotenuse. Either the cosine or secant function can be used.

Choosing the cosine function: cos 70°30′ = $\frac{x}{15.740}$ in.

Solve as a proportion.

$$\frac{\cos 70°30′}{1} = \frac{x}{15.740 \text{ in}}$$

$$x = \cos 70°30′ \,(15.740 \text{ in})$$

$x = 70.30$ [2nd] [►DD] [cos] [x] 15.74 [=] 5.254119964, 5.254 in   Ans

or $x =$ [cos] 70 [° ′ ″] 30 [° ′ ″] [x] 15.74 [EXE] 5.254119964, 5.254 in   Ans

Solve for side y by using either a trigonometric function or the Pythagorean Theorem. If the Pythagorean Theorem is used to determine y, then $y^2 = (15.740)^2 - (5.254)^2$ and $y = \sqrt{(15.740)^2 - (5.254)^2}$. In cases like this, it is generally more convenient to solve for the side by using a trigonometric function. In relation to the 70°30′ angle, side y is the opposite side. The 15.740-inch side is the hypotenuse.

Determine the two functions whose ratios consist of the opposite side and the hypotenuse. Either the sine or cosecant function can be used.

Choosing the sine function: $\sin 70°30' = y/15.740$ in.

➤ **Note:** Since side x has been calculated, it can be used with the 70°30′ angle to determine side y. However, it is better to use the given 15.740-inch hypotenuse rather than the calculated side x. Whenever possible, use given values rather than calculated values when solving problems. The calculated values could have been incorrectly computed or improperly rounded off resulting in an incorrect answer.

Solve as a proportion.

$$\frac{\sin 70°30'}{1} = \frac{y}{15.740 \text{ in}}$$

$y = 70.30$ [2nd] [►DD] [sin] [x] 15.74 [=] 14.83717707, 14.837 in   Ans

or $y =$ [sin] 70 [° ′ ″] 30 [° ′ ″] [x] 15.74 [EXE] 14.83717707, 14.837 in   Ans

Determine $\angle 1$: $\angle 1 = 90° - 70°30' = 19°30'$ Ans

APPLICATION

Determining an Unknown Angle When Two Sides of a Right Triangle Are Known

Solve the following problems. Compute angles to the nearer minute in triangles with English unit sides. Compute angles to the nearer hundredth degree in triangles with metric unit sides.

1. Determine ∠A. _____

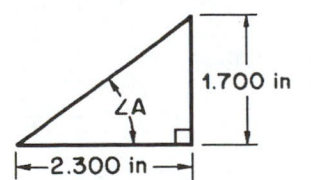

3. Determine ∠1. _____

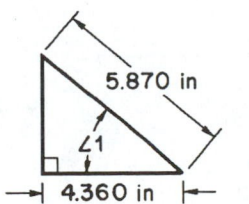

5. Determine ∠1. _____

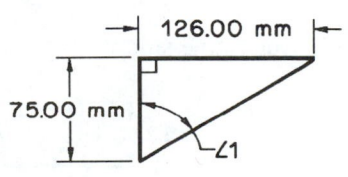

2. Determine ∠B. _____

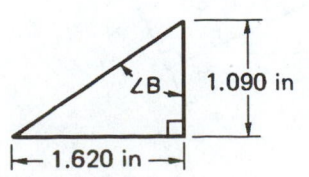

4. Determine ∠x. _____

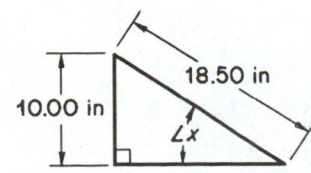

6. Determine ∠A. _____

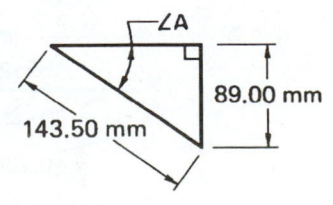

7. Determine $\angle y$. _____

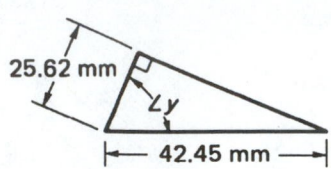

8. Determine $\angle B$. _____

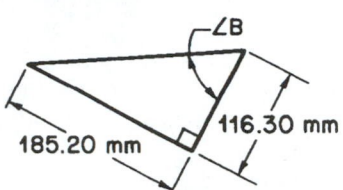

9. a. Determine $\angle 1$. _____
   b. Determine $\angle 2$. _____

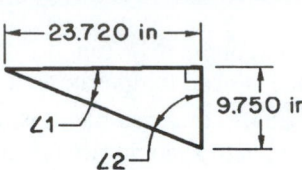

10. a. Determine $\angle A$. _____
    b. Determine $\angle B$. _____

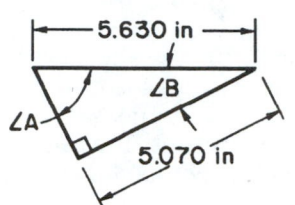

11. a. Determine $\angle x$. _____
    b. Determine $\angle y$. _____

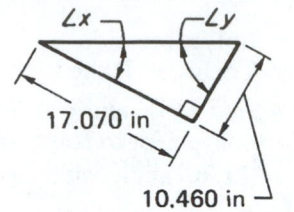

12. a. Determine $\angle C$. _____
    b. Determine $\angle D$. _____

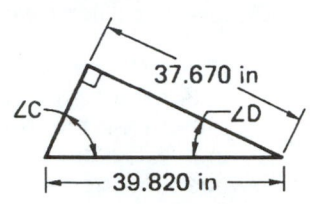

Determining an Unknown Side When an Acute Angle and One Side of a Right Triangle Are Known

Solve the following problems. Compute the sides to 3 decimal places in triangles dimensioned in English units. Compute the sides to 2 decimal places in triangles dimensioned in metric units.

13. Determine side b. _____

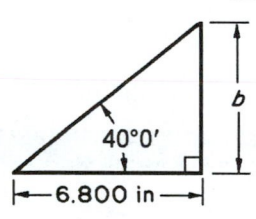

14. Determine side c. _____

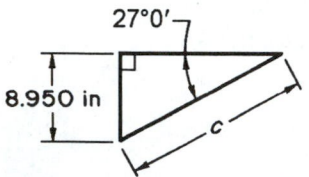

15. Determine side x. _____

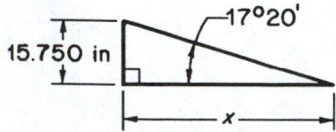

16. Determine side d. _____

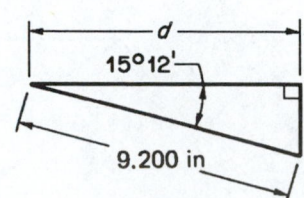

17. Determine side y. _____

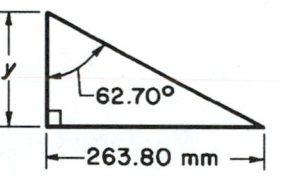

18. Determine side f. _____

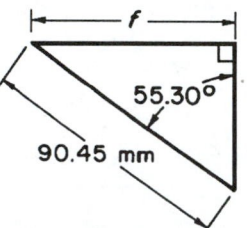

19. Determine side p. _____

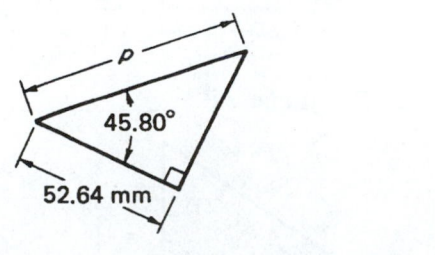

20. Determine side *y*. _____

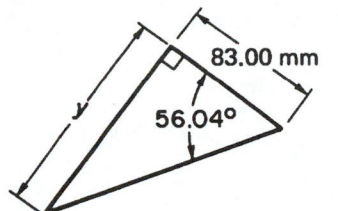

21. a. Determine side *d*. _____
    b. Determine side *e*. _____

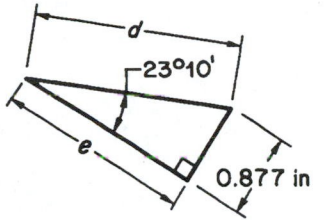

22. a. Determine side *s*. _____
    b. Determine side *t*. _____

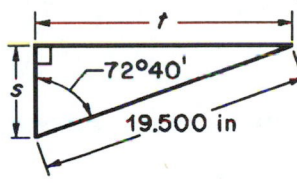

23. a. Determine side *x*. _____
    b. Determine side *y*. _____

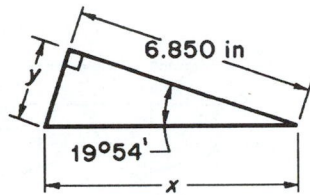

24. a. Determine side *p*. _____
    b. Determine side *n*. _____

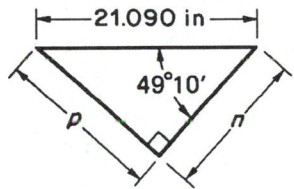

Determining Unknown Sides and Angles

Solve the following problems. For triangles dimensioned in English units, compute the sides to 3 decimal places and the angles to the nearer minute. For triangles dimensioned in metric units, compute the sides to 2 decimal places and the angles to the nearer hundredth degree.

25. a. Determine ∠B. _____
    b. Determine side *x*. _____
    c. Determine side *y*. _____

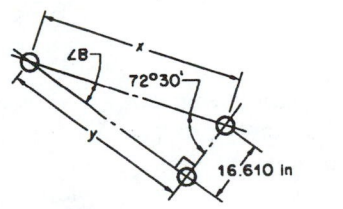

26. a. Determine ∠1. _____
    b. Determine ∠2. _____
    c. Determine side *a*. _____

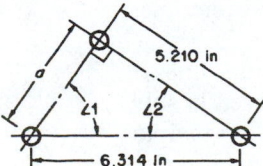

27. a. Determine side *a*. _____
    b. Determine side *b*. _____
    c. Determine ∠2. _____

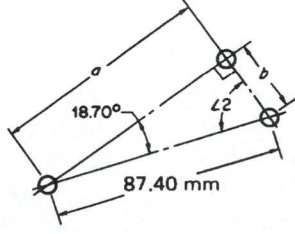

28. a. Determine ∠A. _____
    b. Determine ∠B. _____
    c. Determine side *r*. _____

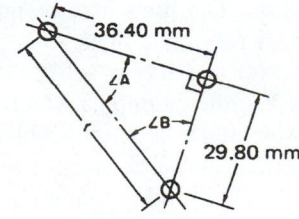

29. a. Determine ∠B. _____

    b. Determine side *b*. _____

    c. Determine side *c*. _____

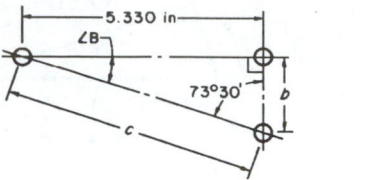

30. a. Determine ∠D. _____

    b. Determine ∠E. _____

    c. Determine side *m*. _____

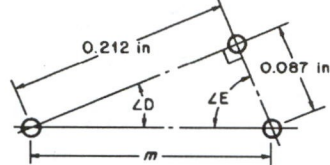

31. a. Determine ∠1. _____

    b. Determine side *g*. _____

    c. Determine side *h*. _____

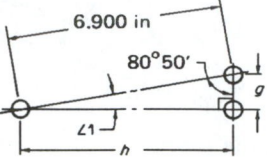

UNIT 56 Simple Practical Machine Applications

Objective   **After studying this unit you should be able to**

- **Solve simple machine technology problems which require the projection of auxiliary lines and the use of geometric principles and trigonometric functions.**

Method of Solution

In the previous unit, you solved for unknown angles and sides of right triangles. Emphasis was placed on developing an understanding and the ability to apply proper procedures in solving for angles and sides. No attempt was made to show the many practical applications of right-angle trigonometry.

The examples discussed in this unit are simple practical shop applications of right angle trigonometry, although they may not be given directly in the form of right triangles. To solve most of the examples, it is necessary to project auxiliary lines to produce a right triangle. The unknown, or a dimension required to compute the unknown, is part of the triangle. The auxiliary lines may be projected between given points, or from given points. The lines may be projected parallel or perpendicular to centerlines, tangents, or other reference lines.

It is important to study carefully the procedures and the use of auxiliary lines as they are applied to the examples which follow. The same basic method is used in solving many similar machine shop problems. A knowledge of both geometric principles and trigonometric functions and the ability to relate and apply them to specific situations are required in solving many machine shop problems.

Sine Bar and Sine Plate

Sine bars and sine plates are used to measure angles which have been cut in parts and to position parts which are to be cut at specified angles. One end of the sine bar or plate is raised with gage blocks in order to set a desired angle. The most common sizes of bars and plates are 5 inches and 10 inches between rolls. In setting angles, the sine bar or the top plate of the sine plate is the hypotenuse of a right triangle, and the gage blocks are the opposite side in reference to the desired angle.

Example 1 Determine the gage block height x which is required to set an angle of 24°20′ with a 5-inch sine bar as shown.

$$\sin 24°20' = \frac{\text{gage block height } x}{\text{sine bar length}}$$

$$\frac{\sin 24°20'}{1} = \frac{x}{5 \text{ in}}$$

$x = (\sin 24°21')5$

$x = 24.21$ [2nd] [►DD] [sin] [×] 5 [=] 2.061547756,

  2.0615   Ans

or $x =$ [sin] 24 [° ′ ″] 21 [° ′ ″] [EXE] [×] 5 [EXE]

  2.061547756, 2.0615   Ans

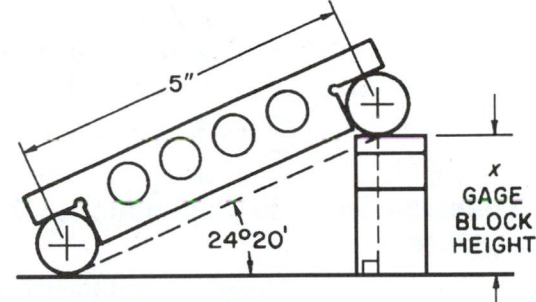

5″

24°20′

x
GAGE
BLOCK
HEIGHT

Example 2 Determine the angle set on a 10-inch sine plate using a gage block height of 3.0625 inches.

$$\sin \angle x = \frac{3.0625 \text{ in}}{10 \text{ in}}$$

Determine the angle whose sine function is the quotient of $\frac{3.0625}{10}$.

$\angle x = 3.0625$ [+] 10 [=] [2nd] [sin⁻¹] [3rd] [►DMS]

  → 17°50′   Ans

or $\angle x =$ [SHIFT] [sin⁻¹] [(] 3.0625 [+] 10 [)] [EXE] [SHIFT]

  [←] → 17°50′   Ans

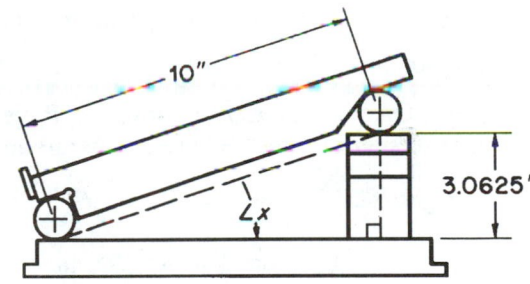

10″

$\angle x$

3.0625″

Tapers and Bevels

Example 1 Determine the included taper angle of the shaft shown. All dimensions are in inches.

The problem must be solved by using a figure in the form of a right triangle. Therefore, project line AB from point A parallel to the centerline. Right △ABC is formed in which ∠BAC is one-half the included taper angle. Side AB = 10.500″.

Side BC $= \frac{1.800'' - 0.700''}{2} = 0.550''$

Using sides AB and BC, solve for ∠BAC.

tan ∠BAC $= \frac{BC}{AB} = \frac{0.550''}{10.500''}$

Determine the angle whose tangent function is the quotient of $\frac{0.550}{10.500}$.

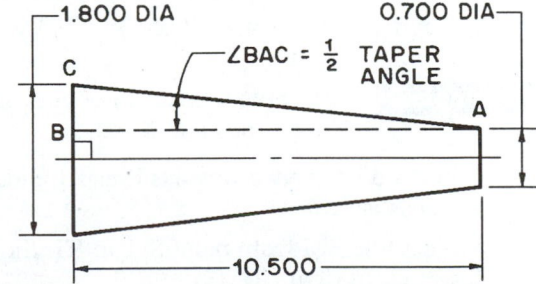

1.800 DIA

0.700 DIA

∠BAC = ½ TAPER ANGLE

C

B

A

10.500

  $\angle x =$ .55 [+] 10.5 [=] [2nd] [tan⁻¹] [3rd] [►DMS] → 2°59′55″, 3°0′

or $\angle x =$ [SHIFT] [tan⁻¹] [(] .55 [+] 10.5 [)] [EXE] [SHIFT] [←] → 2°59′55″, 3°0′

The included taper angle = 2(3°0′) = 6°0′   Ans

Example 2 Determine diameter x of the part shown. All dimensions are in millimeters.

Project line DE from point D parallel to the centerline, in order to form right $\triangle$DEF.

Side DE = 21.80 mm − 7.50 mm = 14.30 mm

$$\angle EDF = 32.50°$$

Using side DE and $\angle$EDF, solve for side EF.

$$\tan \angle EDF = \frac{EF}{DE}$$

$$\tan 32.50° = \frac{EF}{14.30 \text{ mm}}$$

EF = tan 32.50°(14.30 mm)

EF = 32.5 ⬛tan⬛ ⬛x⬛ 14.3 ⬛=⬛ 9.11010473, 9.11 mm

or EF = ⬛tan⬛ 32.5 ⬛x⬛ 14.3 ⬛EXE⬛ 9.11010473, 9.11 mm

Dia x = 26.25 mm − 2(9.11 mm) = 8.03 mm   Ans

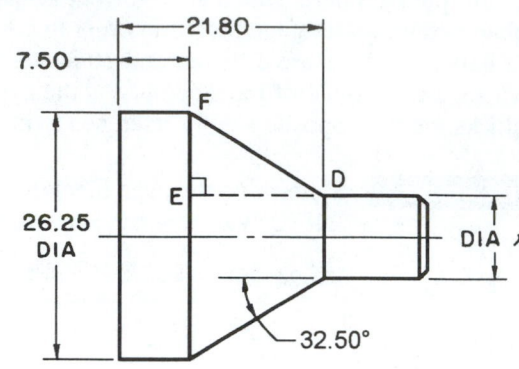

Isosceles Triangle Applications: Distance Between Holes and V-Slots

The solutions to many practical trigonometry problems are based on recognizing figures as isosceles triangles. In an isosceles triangle, an altitude to the base bisects the base and the vertex angle.

Example 1 In this figure, five holes are equally spaced on a 5.200-inch diameter circle. Determine the straight line distance between two consecutive holes.

Project radii from center O to hole centers A and B.

Project a line from A to B. $\angle AOB = \dfrac{360°}{5} = 72°$

Since OA = OB, $\triangle$AOB is isosceles. Project line OC $\perp$ to AB from center O. Line OC bisects $\angle$AOB and side AB.

In right $\triangle$AOC, $\angle AOC = \dfrac{72°}{2} = 36°$

$$AO = \frac{5.200 \text{ in}}{2} = 2.600 \text{ in}$$

Solve for side AC. $\sin \angle AOC = \dfrac{AC}{AO}$

$$\sin 36° = \frac{AC}{2.600 \text{ in}}$$

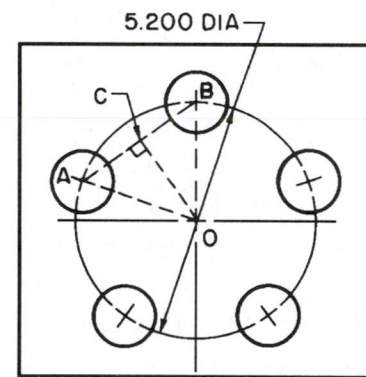

AC = sin 36°(2.600 in)

AC = 36 ⬛sin⬛ ⬛x⬛ 2.6 ⬛=⬛ 1.528241656, 1.528 in

or AC = ⬛sin⬛ 36 × 2.6 ⬛EXE⬛ 1.528241656, 1.528 in

AB = 2(1.528 in) = 3.056 in   Ans

Example 2 Determine the depth of cut x required to machine the V-slot shown. All dimensions are in inches.

Connect a line between points R and T. Sides RS = TS; therefore, $\triangle$RST is isosceles.

Project line SM from point S $\perp$ to RT. Side RT and $\angle$RST are bisected. In right $\triangle$RMS,

$$\angle RSM = \frac{62°46'}{2} = 31°23'$$

$$RM = \frac{3.856 \text{ in}}{2} = 1.928 \text{ in}$$

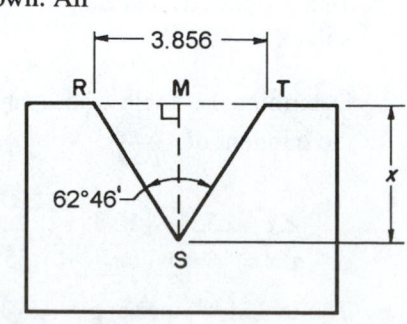

Solve for depth of cut MS.

$$\tan \angle RSM = \frac{RM}{MS}$$

$$\tan 31°23' = \frac{1.928 \text{ in}}{MS}$$

MS = 1.928 in ÷ tan 31°23′

MS = 1.928 $\boxed{+}$ 31.23 $\boxed{\text{2nd}}$ $\boxed{\blacktriangleright\text{DD}}$ $\boxed{\tan}$ $\boxed{=}$ 3.160638084, 3.161 in

or MS = 1.928 $\boxed{+}$ $\boxed{\tan}$ 31 $\boxed{°\,'\,''}$ 23 $\boxed{°\,'\,''}$ $\boxed{\text{EXE}}$ 3.160638084, 3.161 in

x = MS = 3.161   Ans

Tangents to Circles Applications: V-Blocks, Thread Wire Checking Dimensions, Dovetails, and Angle Cuts

A tangent is perpendicular to a radius of a circle at its tangent point. Solutions to many applied trigonometry problems are based on this principle.

Example 1 A 75.00-millimeter diameter pin is used to inspect the groove machined in the block shown. Determine dimension x. The sides of the groove are equal. All dimensions are in millimeters.

Project a line from center O to point B. Project radius AO from center O to tangent point A. Since a radius is ⊥ to a tangent line at the point of tangency, △AOB is a

right triangle. In right △AOB, OA = $\frac{75.00 \text{ mm}}{2}$ = 37.50 mm

Since the angle formed by two tangents to a circle from an outside point is bisected by a line from the point to

the center of the circle, $\angle ABO = \frac{37.00°}{2} = 18.50°$

Solve for side OB.

$$\sin \angle ABO = \frac{OA}{OB}$$

$$\sin 18.50° = \frac{37.50 \text{ mm}}{OB}$$

OB = 37.50 mm ÷ sin 18.50°

OB = 37.5 $\boxed{+}$ 18.5 $\boxed{\sin}$ $\boxed{=}$ 118.1829489, 118.18 mm

or OB = 37.5 $\boxed{+}$ $\boxed{\sin}$ 18.5 $\boxed{\text{EXE}}$ 118.1829489, 118.18 mm

Find the height from the base of the block to the top of the pin.

87.50 mm + OB + radius of pin =

87.50 mm + 118.18 mm + 37.50 mm = 243.18 mm

x = 243.18 mm − 222.50 mm = 20.68 mm   Ans

Example 2 An internal dovetail is shown. Two pins or balls are used to check the dovetail for both location and angular accuracy. Calculate check dimension x. All dimensions are in inches.

Project line HO from point H to the pin center O; HO bisects the 72°20′ angle. Project a radius from point O to the point of tangency K; $\angle HKO$ is a right angle since a radius is perpendicular to a tangent at the point of tangency.

In right △HOK, $\angle HKO = \frac{72°20'}{2} = 36°10'$

$$KO = \frac{1.000 \text{ in}}{2} = 0.500 \text{ in}$$

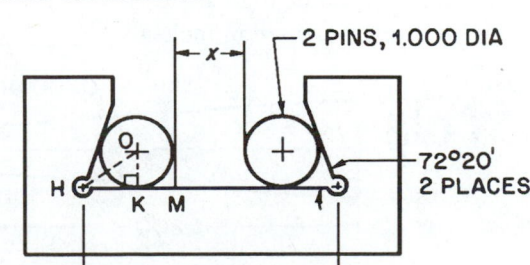

Solve for side HK. $\cot \angle KHO = \dfrac{HK}{KO}$

$$\cot 36°10' = \dfrac{HK}{0.500}$$

HK = cot 36°10′ (0.500)

HK = 36.10 ⌷2nd⌷ ⌷▶DD⌷ ⌷tan⌷ ⌷1/x⌷ ⌷x⌷ .5 ⌷=⌷ 0.683997962, 0.684 in

or   HK = tan 36 ⌷° ′ ″⌷ 10 ⌷° ′ ″⌷ ⌷EXE⌷ ⌷1/x⌷ ⌷x⌷ .5 ⌷EXE⌷ 0.683997962, 0.684 in

        └─ or ⌷x⁻¹⌷

    HK = 0.684 in

    KM = pin radius = 0.500 in

    HM = HK + KM = 0.684 in + 0.500 in = 1.184 in

x = 6.312 in − 2(HM) = 6.312 in − 2(1.184 in)

x = 6.312 ⌷−⌷ 2 ⌷x⌷ 1.184 ⌷=⌷ 3.944 in   Ans

APPLICATION

Sine Bars and Sine Plates

1. Determine the height of gage blocks required to set the following angles on a 10″ sine plate.

   a. 35° _____   d. 9°44′ _____   g. 0°20′ _____

   b. 13°10′ _____   e. 28°32′ _____   h. 2°26′ _____

   c. 36°50′ _____   f. 44°20′ _____   i. 19°51′ _____

2. Determine the height of gage blocks required to set the following angles with a 5″ sine bar.

   a. 40°40′ _____   d. 0°30′ _____   g. 39°12′ _____

   b. 7° _____   e. 21°57′ _____   h. 44°50′ _____

   c. 12°10′ _____   f. 13°18′ _____   i. 8°17′ _____

Tapers and Bevels

Solve the following problems. For English unit dimensioned problems, calculate angles to the nearer minute and lengths to the nearer thousandths inch. For metric unit dimensioned problems, calculate angles to the nearer hundredth degree and lengths to the nearer hundredth millimeter.

3. Find the included taper $\angle x$.
   All dimensions are in inches. _____

4. Find length x.
   All dimensions are in inches. _____

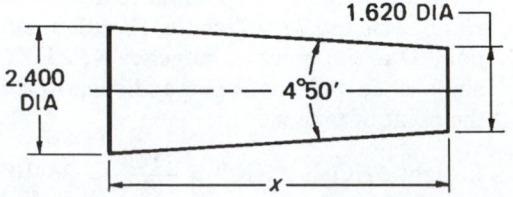

5. Find diameter *y*.
All dimensions are in millimeters. _____

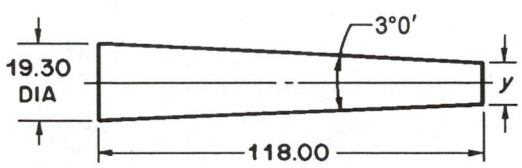

6. Find diameter *x*.
All dimensions are in inches. _____

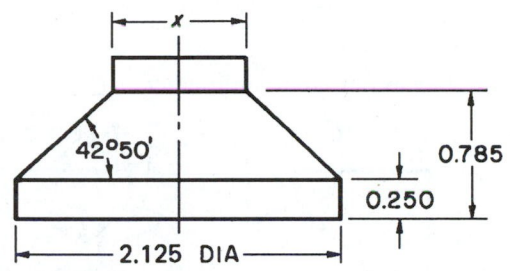

Distance Between Holes and V-Slots

9. Find center distance *y*.
All dimensions are in millimeters. _____

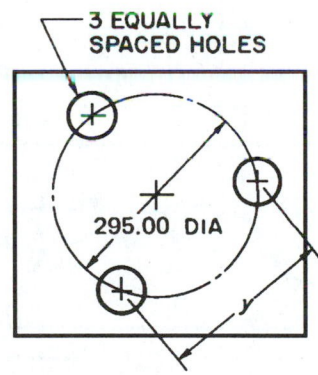

10. Find inside caliper dimension *x*.
All dimensions are in inches. _____

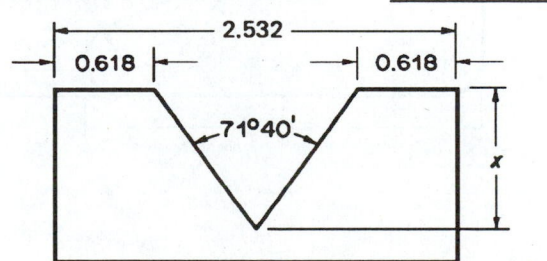

7. Find ∠*x*.
All dimensions are in inches. _____

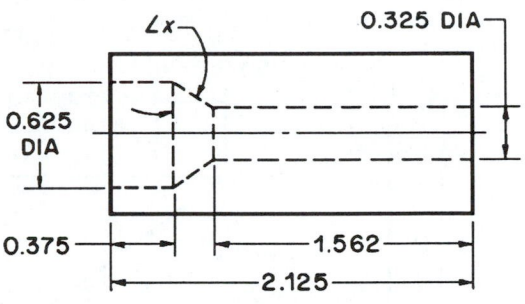

8. Find dimension *y*.
All dimensions are in inches. _____

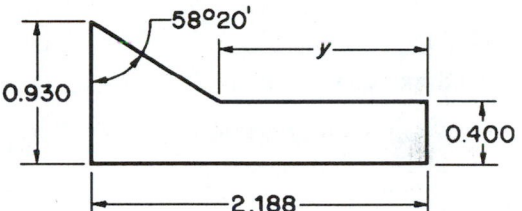

11. Find radius *r*.
All dimensions are in millimeters. _____

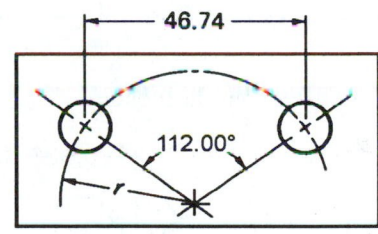

12. Find arc dimension *x*.
All dimensions are in inches. _____

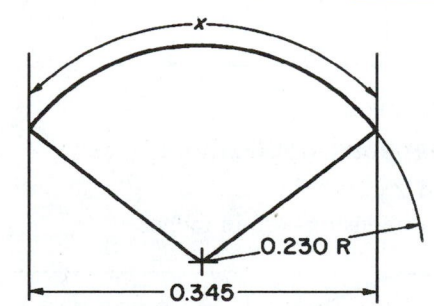

13. Find the depth of cut *x*.
All dimensions are in inches. _____

V-Blocks, Thread Wire Checking Dimensions, Dovetails, and Angle Cuts

14. Find $\angle x$.

All dimensions are in inches. _____

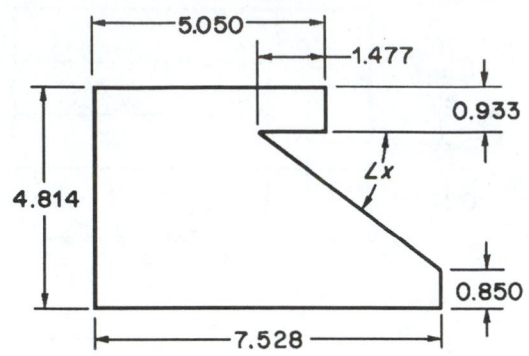

15. Find gage dimension y.

All dimensions are in millimeters. _____

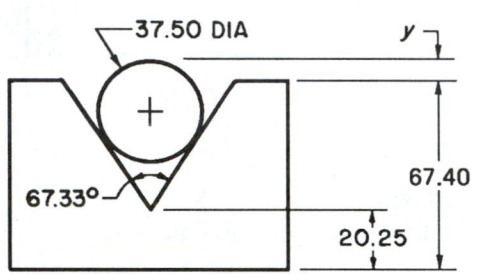

16. Find $\angle y$.

All dimensions are in inches. _____

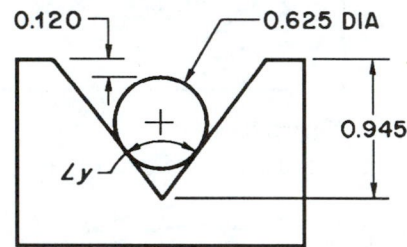

Miscellaneous Applications

20. Find $\angle y$.

All dimensions are in inches. _____

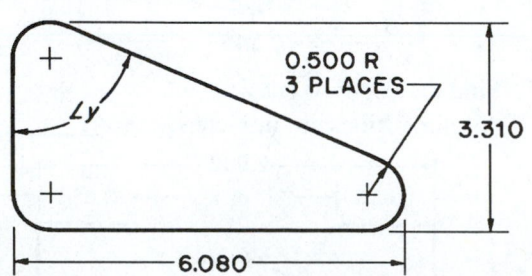

17. Find gage dimension x.

All dimensions are in inches. _____

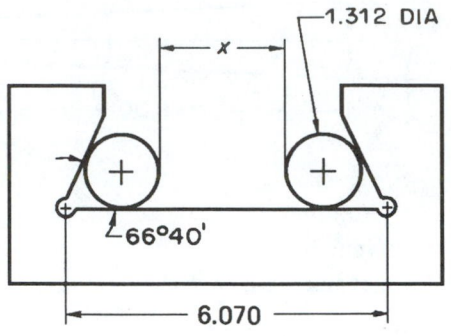

18. Find $\angle x$.

All dimensions are in millimeters. _____

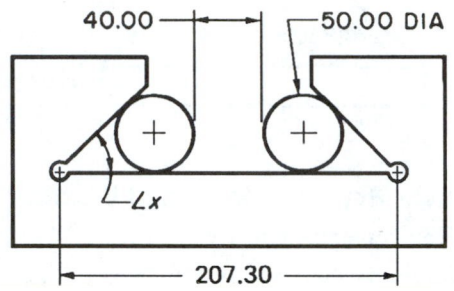

19. Find dimension y.

All dimensions are in inches. _____

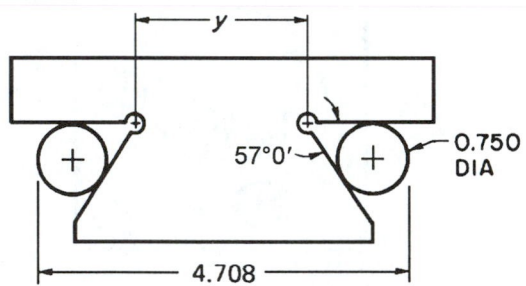

21. Find dimension x.

All dimensions are in millimeters. _____

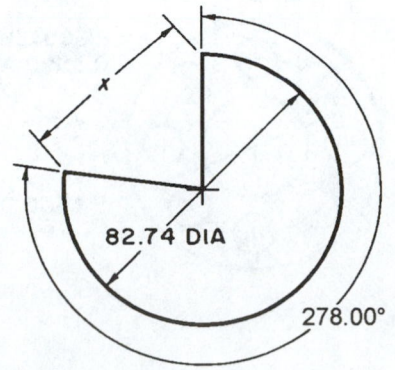

22. Find ∠x.
    All dimensions are in inches. _____

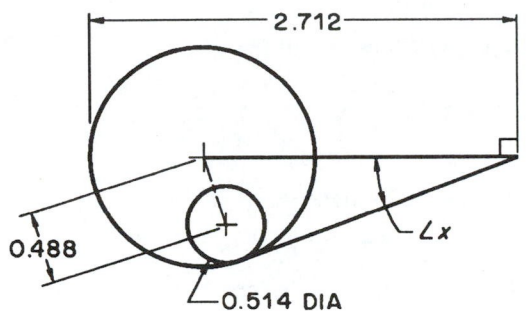

23. Find distance y.
    All dimensions are in millimeters. _____

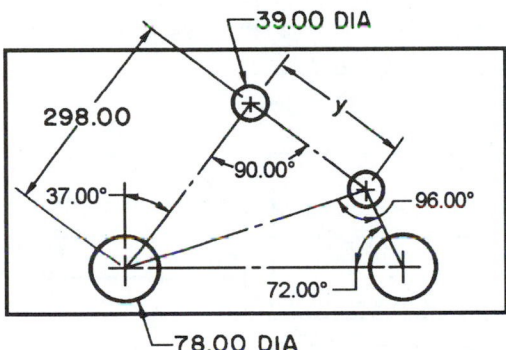

24. Find dimension y.
    All dimensions are in inches. _____

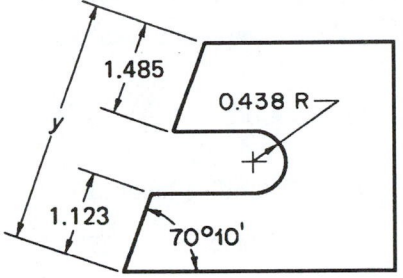

UNIT 57 Complex Practical Machine Applications

Objective   After studying this unit you should be able to

- **Solve complex applied machine technology problems which require forming two or more right triangles by the projection of auxiliary lines.**

The problems in this unit are more challenging than those in the last unit and are typical of those found in actual practice when working directly from engineering drawings. The solutions of these problems require the projection of auxiliary lines to form two or more right triangles.

Study the procedures which are given in detail for solving the examples. There is a common tendency to begin writing computations before analyzing the problem. This tendency must be avoided. As problems become more complex, a greater proportion of time and effort is required in the analyses. The written computations must be developed in clear and orderly steps.

Method of Solution

Analyze the problem before writing computations.

- Relate given dimensions to the unknown and determine whether other dimensions in addition to the given dimensions are required in the solution.

- Determine the auxiliary lines which are required to form right triangles which contain dimensions that are needed for the solution.

- Determine whether sufficient dimensions are known to obtain required values within the right triangles. If enough information is not available for solving a triangle, continue the analysis until enough information is obtained.

- Check each step in the analysis to verify that there are no gaps or false assumptions.

Write the computations.

Example 1   Determine length x of the part shown. All dimensions are in inches.

Analyze the problem:

Project auxiliary lines to form right $\triangle$ABF and right $\triangle$CDE. If distances AB and CD can be determined, length x can be computed.

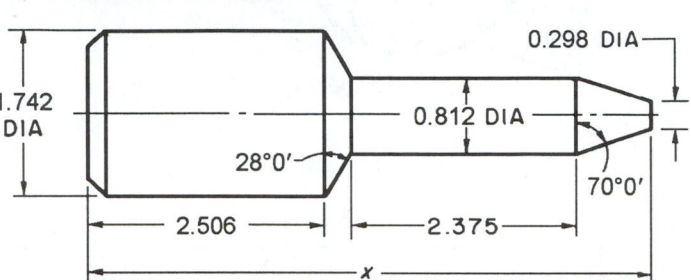

$$x = 2.506 \text{ in} + AB + 2.375 \text{ in} + CD$$

Determine whether enough information is given to solve for AB. In right $\triangle$ABF:
$\angle$FAB = 90° − 28° = 62° (complementary angles)

$$BF = \frac{1.742 \text{ in} - 0.812 \text{ in}}{2} = 0.465 \text{ in}$$

There is enough information to determine AB.

Determine whether enough information is given to solve for CD. In right $\triangle$CDE:
$\angle$ECD = 90° − 70° = 20° (complementary angles)

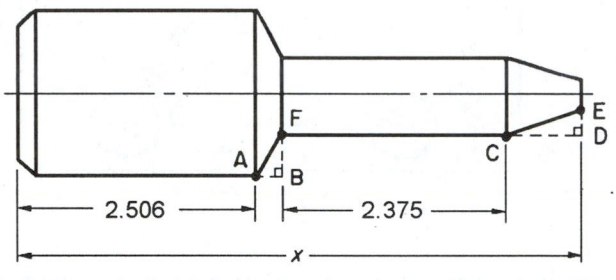

$$DE = \frac{0.812 \text{ in} - 0.298 \text{ in}}{2} = 0.257 \text{ in}$$

There is enough information to determine CD.

Computations:

Solve for AB.

$$\tan \angle FAB = \frac{BF}{AB}$$

$$\tan 62°0' \frac{0.465 \text{ in}}{AB}$$

AB = 0.465 in ÷ tan 62°0′
.465 [÷] 62 [tan] [=] 0.247244886
or   .465 [÷] [tan] 62 [EXE] 0.247244886
AB = 0.2422 in

Solve for CD.

$$\tan CD = \frac{DE}{CD}$$

$$\tan 20°0' = \frac{0.257 \text{ in}}{CD}$$

$$CD = 0.257 \text{ in} \div \tan 20°0'$$

.257 [÷] 20 [tan] [=] 0.706101697
or   .257 [÷] [tan] 20 [EXE] 0.706101696
CD = 0.7061 in

Solve for x.

$$x = 2.506 \text{ in} + AB + 2.375 \text{ in} + CD$$
$$x = 2.506 \text{ in} + 0.2422 \text{ in} + 2.375 \text{ in} + 0.7061 \text{ in}$$
$$x = 5.829 \text{ in} \quad \text{Ans}$$

Example 2 Determine $\angle x$ of the plate shown. All dimensions are in millimeters.

➤ **Note:** Generally, when solving problems which involve an arc which is tangent to one or more lines, it is necessary to project the radius of the arc to the tangent point and to project a line from the vertex of the unknown angle to the center of the arc.

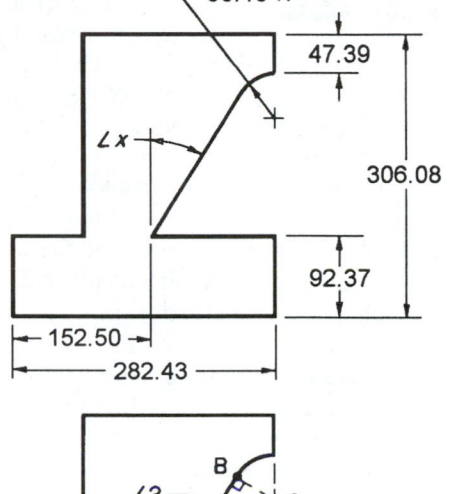

Analyze the problem:

Project auxiliary lines between the points A and O, from point O to the tangent point B, and from point O to point C. Right $\triangle ACO$ and right $\triangle ABO$ are formed. If $\angle 1$ and $\angle 2$ can be computed, $\angle x$ can be determined. $\angle x = 90° - (\angle 1 + \angle 2)$

Determine whether enough information is given to solve for $\angle 1$.
In right $\triangle ACO$:

$$AC = 282.43 \text{ mm} - 152.50 \text{ mm} = 129.93 \text{ mm}$$

$$CO = 306.08 \text{ mm} - (92.37 \text{ mm} + 50.18 \text{ mm} + 47.39 \text{ mm})$$
$$= 116.14 \text{ mm}$$

There is enough information to determine $\angle 1$.

Determine whether enough information is given to solve for $\angle 2$.

In right $\triangle ABO$: $\quad$ BO = 50.18 mm
$\qquad\qquad\qquad$ AO can be determined after solving for $\angle 1$.
$\qquad\qquad\qquad$ There is enough information to determine $\angle 2$.

Computations:

Solve for $\angle 1$. $\qquad\qquad\qquad \tan \angle 1 = \dfrac{CO}{AC} = \dfrac{116.14 \text{ mm}}{129.93 \text{ mm}}$

$\angle 1 = 116.14 \boxed{+} 129.93 \boxed{=} \boxed{\text{2nd}} \boxed{\tan^{-1}} \to 41.79244435$
or $\angle 1 = \boxed{\text{SHIFT}} \boxed{\tan^{-1}} \boxed{(} 116.14 \boxed{+} 129.93 \boxed{)} \boxed{=} 41.79244435$
$\angle 1 = 41.79244°$

Solve for AO.

$$\sin \angle 1 = \frac{CO}{AO}$$

$$\sin 41.79244° = \frac{116.14 \text{ mm}}{AO}$$

$$AO = \frac{116.14 \text{ mm}}{\sin 41.79244°}$$

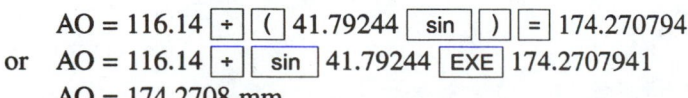

$AO = 116.14 \boxed{+} \boxed{(} 41.79244 \boxed{\sin} \boxed{)} \boxed{=} 174.2707941$
or $AO = 116.14 \boxed{+} \boxed{\sin} 41.79244 \boxed{\text{EXE}} 174.2707941$
$AO = 174.2708 \text{ mm}$

Solve for $\angle 2$. $\qquad\qquad\qquad \sin \angle 2 = \dfrac{BO}{AO} = \dfrac{50.18 \text{ mm}}{174.2708 \text{ mm}}$

$\angle 2 = 50.18 \boxed{+} 174.2708 \boxed{=} \boxed{\text{2nd}} \boxed{\sin^{-1}} \to 16.73482688$
or $\angle 2 = \boxed{\text{SHIFT}} \boxed{\sin^{-1}} \boxed{(} 50.18 \boxed{+} 174.2708 \boxed{)} \boxed{\text{EXE}} 16.73482688$
$\angle 2 = 16.73483°$

Solve for $\angle x$. $\angle x = 90° - (\angle 1 + \angle 2)$

$\angle x = 90° - (41.79244° + 16.73483°) = 31.47273°$
$\angle x = 31.47°$ Ans

Example 3 The front view of a piece with a V-groove is shown. A 1.250-inch diameter pin is used to check the cut for depth and angular accuracy. Compute check dimension x. All dimensions are in inches.

Analyze the problem:

Dimension x is determined by the pin size, the points of tangency where the pin touches the groove, the angle of the V-groove, and the depth of the groove. Therefore, these dimensions and locations must be part of the calculations.

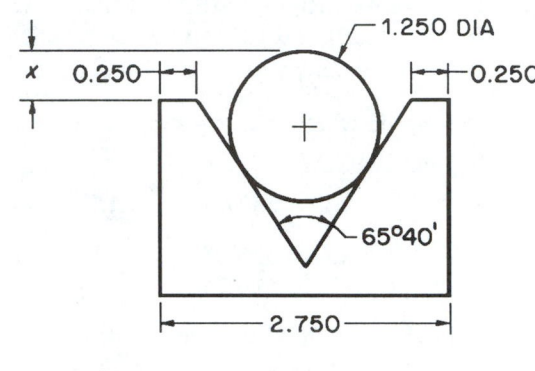

Project auxiliary lines from point A through the center of the pin O, from point O to the tangent point P, and from point B horizontally intersect vertical line AD at point C. Right $\triangle$APO and right $\triangle$ACB are formed. If AO and AC can be determined, check dimension x can be computed.

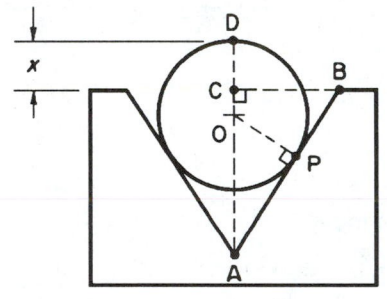

$$DO = \text{radius of pin} = 0.625 \text{ in}$$
$$x = (AO + DO) - AC$$

Determine whether enough information is given to solve for AO.

In right $\triangle$APO: $PO = \dfrac{1.250 \text{ in}}{2} = 0.625 \text{ in}$

$$\angle OAP = \dfrac{65°40'}{2} = 32°50'$$

There is enough information to determine AO.

Determine whether enough information is given to solve for AC.

In right $\triangle$ACB: $BC = \dfrac{2.750 \text{ in}}{2} - 0.250 \text{ in} = 1.125 \text{ in}$

$$\angle BAC = 32°50'$$

There is enough information to determine AC.

Computations:

Solve for AO. $\sin \angle OAP = \dfrac{PO}{AO}$

$$\sin 32°50' = \dfrac{0.625 \text{ in}}{AO}$$

$$AO = \dfrac{0.625 \text{ in}}{\sin 32°50'}$$

AO = .625 $\boxed{\div}$ 32.50 $\boxed{\text{2nd}}$ $\boxed{\blacktriangleright\text{DD}}$ $\boxed{\sin}$ $\boxed{=}$ 1.152717256
or   AO = .625 $\boxed{\div}$ $\boxed{\sin}$ 32 $\boxed{° ' ''}$ 50 $\boxed{° ' ''}$ $\boxed{\text{SHIFT}}$ $\boxed{\longleftarrow}$ $\boxed{\text{EXE}}$ 1.152717256
AO = 1.1527 in

Solve for AC.

$$\tan \angle BAC = \frac{BC}{AC}$$

$$\tan 32°50' = \frac{1.125 \text{ in}}{AC}$$

$$AC = \frac{1.125 \text{ in}}{\tan 32°50'}$$

AC = 1.125 ⊞ 32.50 ⌈2nd⌉ ⌈►DD⌉ ⌈tan⌉ ⌈=⌉ 1.743429928

or   AC = 1.125 ⊞ ⌈tan⌉ 32 ⌈° ′ ″⌉ 50 ⌈° ′ ″⌉ ⌈SHIFT⌉ ⌈◄—⌉ ⌈EXE⌉ 1.743429928

AC = 1.7434 in

Solve for check dimension *x*.

$x = (AO + DO) - AC$

$x = (1.1527 \text{ in} + 0.625 \text{ in}) - 1.7434 \text{ in} = 0.034 \text{ in}$ Ans

Example 4   Determine $\angle x$ in the series of holes shown in this plate. All dimensions are in inches.

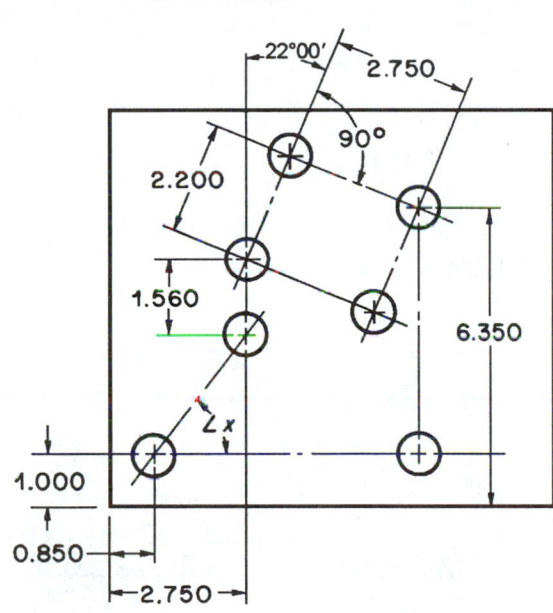
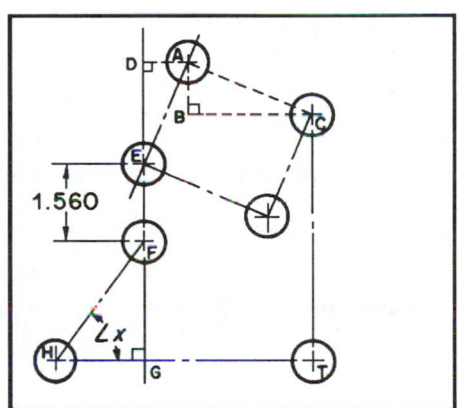

Project auxiliary lines AD, AB, BC. Right △ABC, right △ADE, and right △AGH are formed. If HG and FG can be determined, $\angle x$ can be computed.

HG = 2.750 in − 0.850 in = 1.900 in

FG = (TC + AB) − (DE + 1.560 in)

FG = [(6.350 in − 1.000 in) + AB] − (DE + 1.560 in)

Solve for AB.

In right △ABC: AC = 2.750 in

                      $\angle ACB = 22°00'$ (Two angles whose corresponding sides are perpendicular and equal.)

$$\sin 22°00' = \frac{AB}{2.750 \text{ in}}$$

AB = sin 22°00′ (2.750 in)

AB = 22 ⌈sin⌉ ⌈x⌉ 2.75 ⌈=⌉ 1.030168132

or   ⌈sin⌉ 22 ⌈x⌉ 2.75 ⌈EXE⌉ 1.030168132

AB = 1.0302 in

Solve for DE.

In right △ADE:
$$\angle DEA = 22°00'$$
$$AE = 2.200 \text{ in}$$
$$\cos 22°00' = \frac{DE}{2.200 \text{ in}}$$

DE = cos 22°00′ (2.200 in)

DE = 22 $\boxed{\cos}$ $\boxed{\times}$ 2.2 $\boxed{=}$ 2.03980448

or $\boxed{\cos}$ 22 $\boxed{\times}$ 2.2 $\boxed{\text{EXE}}$ 2.03980448

DE = 2.0398 in

Solve for FG.

FG = [(6.350 in − 1.000 in) + AB] − (DE + 1.560 in)
FG = (5.350 in + 1.0302 in) − (2.0398 in + 1.560 in) = 2.7804 in

Solve for ∠x. $\qquad \tan \angle x = \dfrac{FG}{HG} = \dfrac{2.7804 \text{ in}}{1.900 \text{ in}}$

∠x = 2.7804 $\boxed{+}$ 1.9 $\boxed{=}$ $\boxed{\text{2nd}}$ $\boxed{\tan^{-1}}$ $\boxed{\text{3rd}}$ $\boxed{\text{▶DMS}}$ → 55°39′11″

or ∠x = $\boxed{\text{SHIFT}}$ $\boxed{\tan^{-1}}$ $\boxed{(}$ 2.7804 $\boxed{+}$ 1.9 $\boxed{)}$ $\boxed{\text{EXE}}$ $\boxed{\text{SHIFT}}$ $\boxed{\longleftarrow}$ → 55°39′11″

∠x = 55°39′　　Ans

Example 5　Determine dimension *x* of the template shown. All dimensions are in inches.

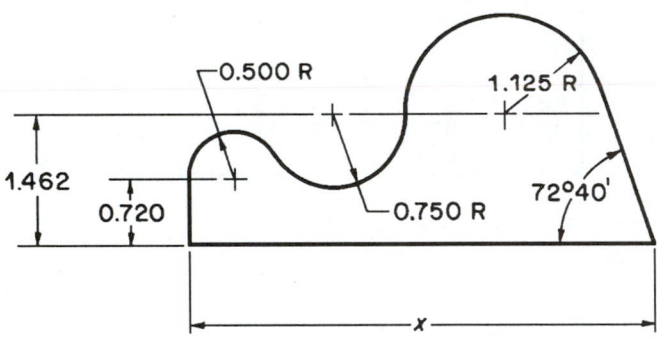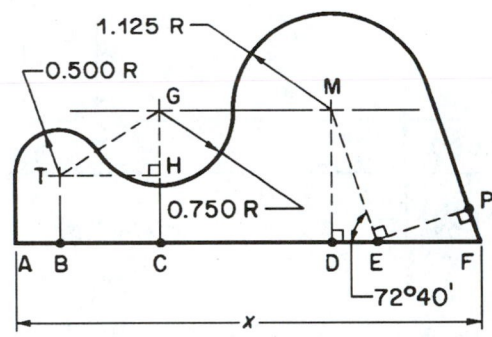

Project auxiliary lines to form right △GHT, right △DEM, and right △EFP.

$$x = AB + BC + CD + DE + EF$$

$$AB = 0.500 \text{ in}$$

$$CD = GM = 0.750 \text{ in} + 1.125 \text{ in} = 1.875 \text{ in}$$

(A line connecting the centers of two externally tangent circles passes through the point of tangency.)

If BC, DE, and EF can be determined, *x* can be computed.

Solve for BC.　(BC = TH)

In right △GHT:
$$GH = 1.462 \text{ in} - 0.720 \text{ in} = 0.742 \text{ in}$$
$$GT = 0.500 \text{ in} + 0.750 \text{ in} = 1.250 \text{ in}$$
(GT passes through the point of tangency.)

$$\sin \angle GTH = \frac{GH}{GT} = \frac{0.742 \text{ in}}{1.250 \text{ in}}$$

∠GTH = .742 | + | 1.25 | = | | 2nd | | sin⁻¹ | → 36.4128935

or | SHIFT | | sin⁻¹ | | (| .742 | + | 1.25 |) | | EXE | 36.4128935

∠GTH = 36.41289°

$$\tan 36.41289° = \frac{GH}{TH}$$

$$\tan 36.41289° = \frac{0.742 \text{ in}}{TH}, \; TH = \frac{0.742 \text{ in}}{\tan 36.41289°}$$

TH = .742 | + | 36.41289 | tan | | = | 1.005950425

or .742 | + | tan 36.41289 | EXE | 1.005950425

TH = 1.0060 in

BC = 1.0060 in

Solve for DE.

In right △DEM:

∠DEM = 72°40′

DM = 1.462 in

$$\tan \angle DEM = \frac{DM}{DE}$$

$$\tan 72°40′ = \frac{1.462 \text{ in}}{DE}$$

$$DE = \frac{1.462 \text{ in}}{\tan 72°40′}$$

DE = 1.462 | + | 72 | · | 40 | 2nd | | ►DD | | tan | | = | 0.456295531

or 1.462 | + | tan | 72 | ° ′ ″ | 40 | ° ′ ″ | | SHIFT | | ← | | EXE | 0.456295531

DE = 0.4563 in

Solve for EF.

In right △EFP:

∠F = 72°40′

EP = 1.125 in

(1.125 radius is ⊥ to tangent line at the point of tangency.)

$$\sin \angle F = \frac{EP}{EF}$$

$$\sin 72°40′ = \frac{1.125 \text{ in}}{EF}$$

$$EF = \frac{1.125 \text{ in}}{\sin 72°40′}$$

EF = 1.125 | + | 72 | · | 40 | 2nd | | ►DD | | sin | | = | 1.178519354

or 1.125 | + | sin | 72 | ° ′ ″ | 40 | ° ′ ″ | | SHIFT | | ← | | EXE | 1.178519354

EF = 1.1785 in

Solve for x.

x = AB + BC + CD + DE + EF

x = 0.500 in + 1.0060 in + 1.875 in + 0.4563 in + 1.1785 in = 5.016 in   Ans

APPLICATION

Complex Practical Machine Applications

Solve the following problems. For English unit dimensioned problems, calculate angles to the nearer minute and lengths to the nearer thousandth inch. For metric unit dimensioned problems, calculate angles to the nearer hundredth degree and lengths to the nearer hundredth millimeter.

1. Find length x.
   All dimensions are in inches. _____

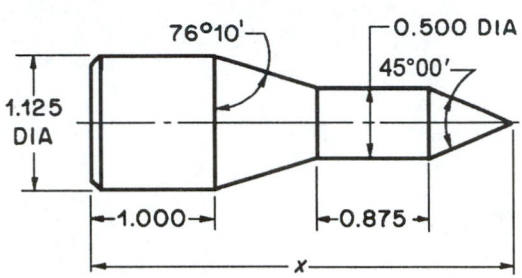

2. Find $\angle x$.
   All dimensions are in millimeters. _____

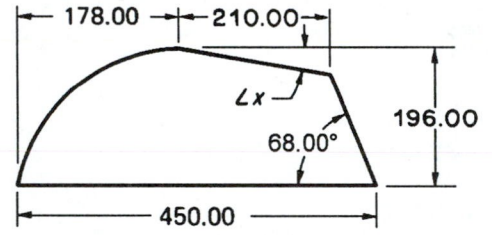

3. Find $\angle x$.
   All dimensions are in inches. _____

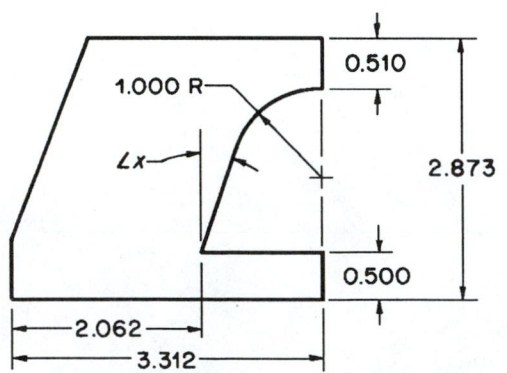

4. Find $\angle y$.
   All dimensions are in millimeters. _____

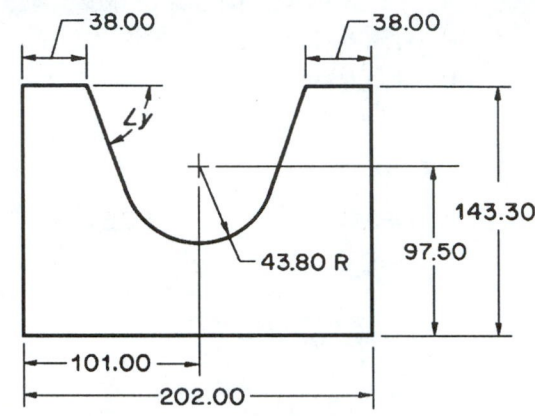

5. Find gage dimension y.
   All dimensions are in inches. _____

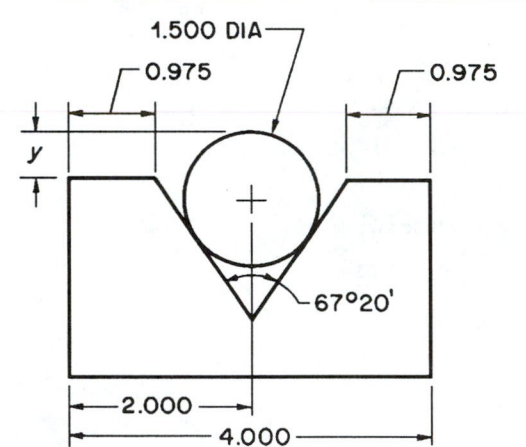

6. Find dimension x.
   All dimensions are in inches. _____

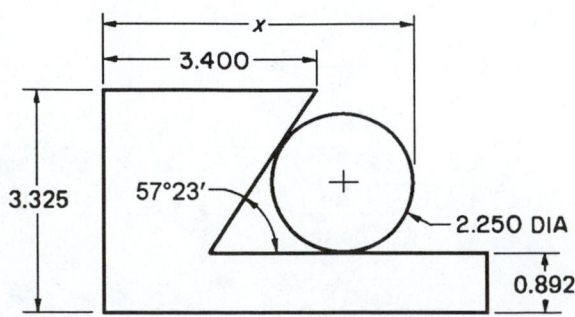

7. Find ∠x.
   All dimensions are in inches. _____

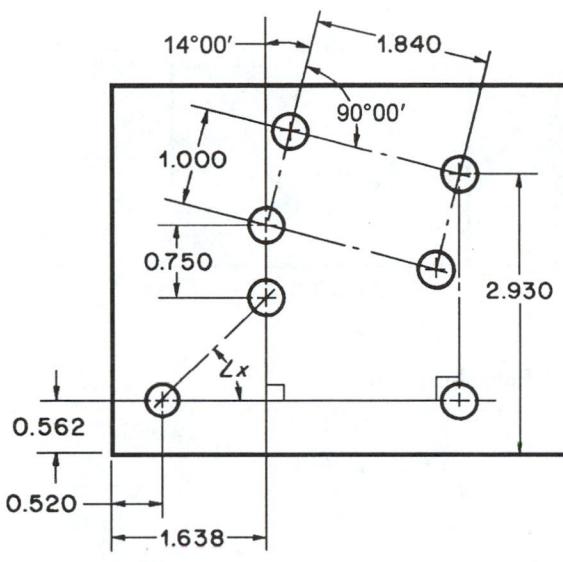

8. Find ∠y.
   All dimensions are in inches. _____

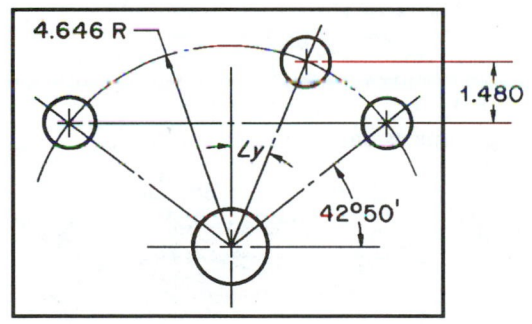

9. Find length x.
   All dimensions are in millimeters. _____

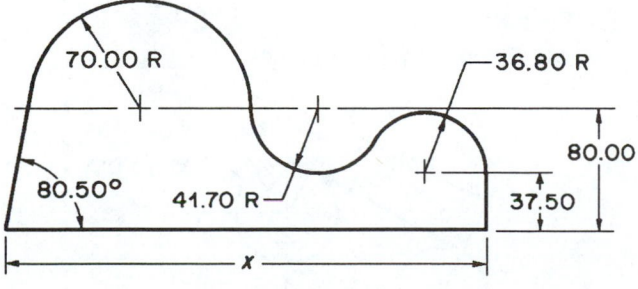

10. Find ∠y.
    All dimensions are in inches. _____

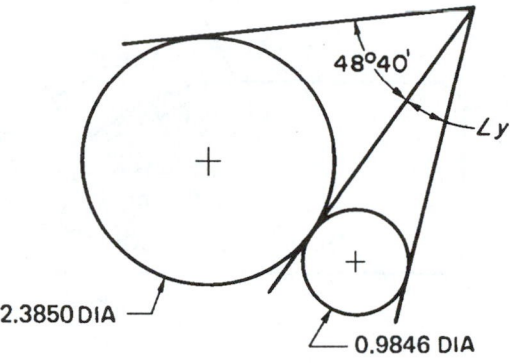

11. Find dimension x.
    All dimensions are in inches. _____

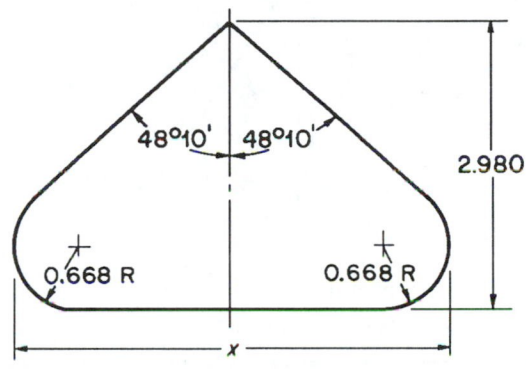

12. Find dimension y.
    All dimensions are in inches. _____

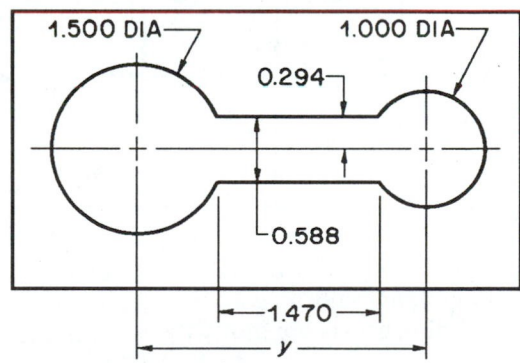

13. Find dimension y.
    All dimensions are in inches. _____

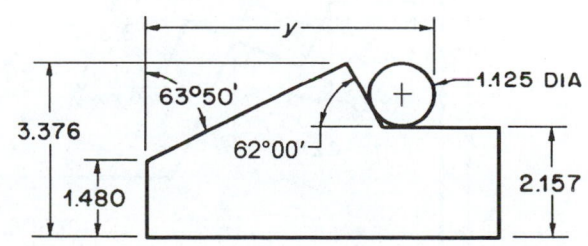

14. Find ∠x.
All dimensions are in millimeters. _____

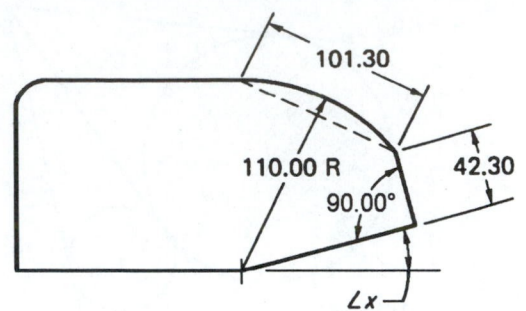

15. Find ∠x.
All dimensions are in inches. _____

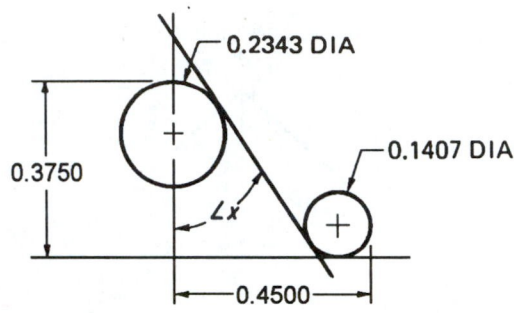

16. Find dimension y.
All dimensions are in inches. _____

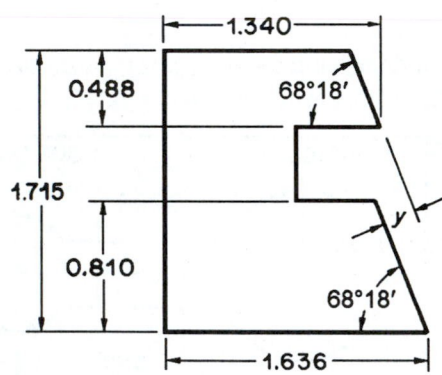

17. Find dimension y.
All dimensions are in inches. _____

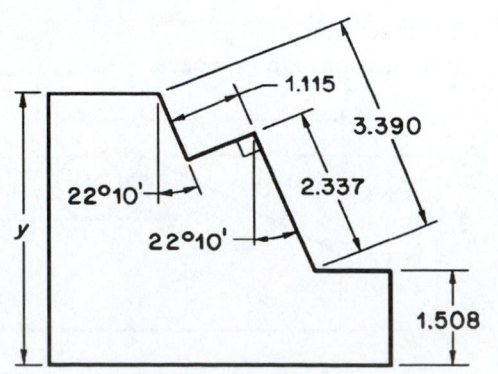

18. Find dimension x.
All dimensions are in inches. _____

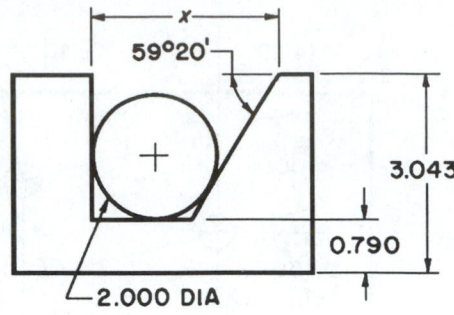

19. Find dimension x.
All dimensions are in inches. _____

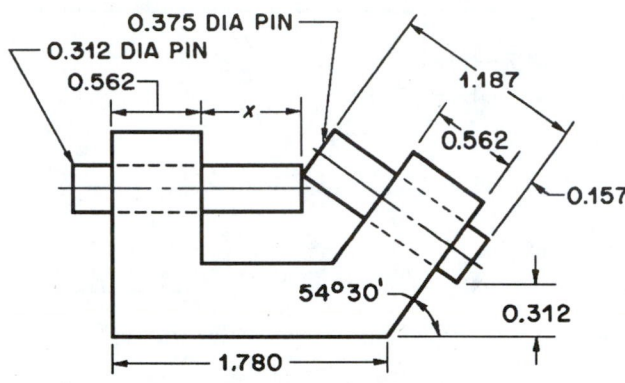

20. Find ∠x.
All dimensions are in millimeters. _____

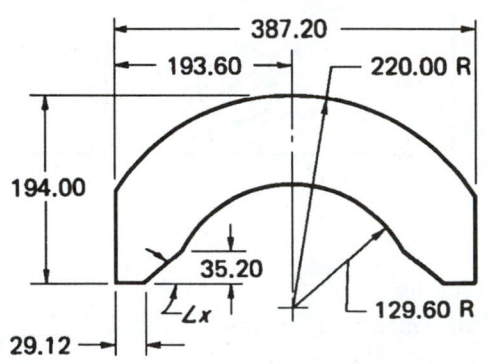

21. Find ∠x.
All dimensions are in inches. _____

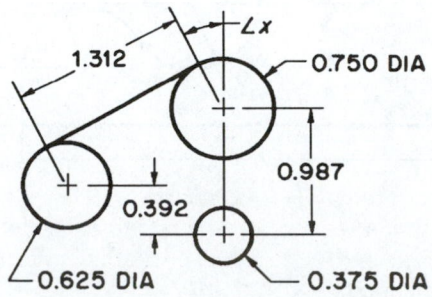

22. Find check dimension *y*.
    All dimensions are in inches.   _____

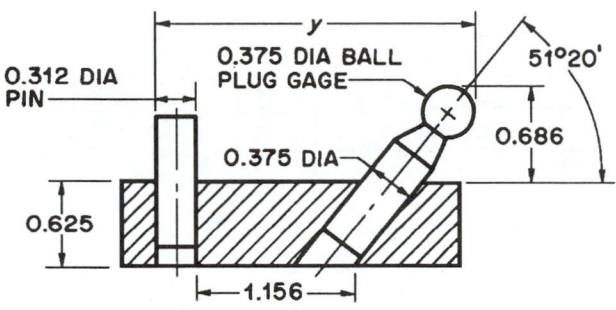

23. Find dimension *y*.
    All dimensions are in inches.   _____

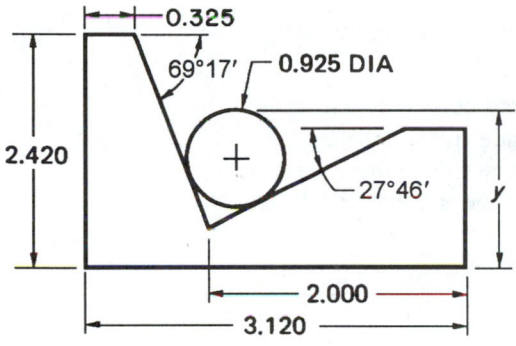

24. Find ∠*x*.
    All dimensions are in millimeters.   _____

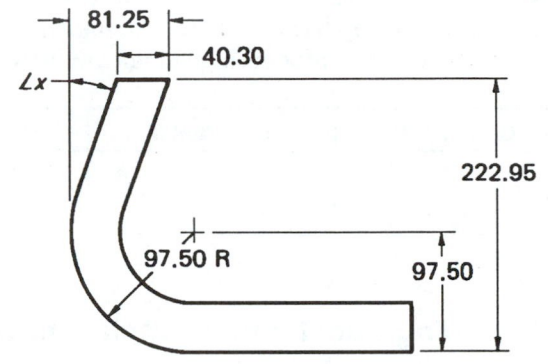

UNIT **58** The Cartesian Coordinate System

Objective   **After studying this unit you should be able to**

● **Compute functions of angles greater than 90°.**

Cartesian (Rectangular) Coordinate System

It is sometimes necessary to determine functions of angles greater than 90°. In a triangle that is not a right triangle, one of the angles can be greater than 90°. Computations using functions of angles greater than 90° are often required in order to solve oblique triangle problems.

Functions of any angles are easily described in reference to the Cartesian Coordinate System. A fixed point (O) called the *origin* is located at the intersection of a vertical and horizontal axes. The horizontal axis is the *x*-axis and the vertical axis is the *y*-axis. The *x* and *y* axes divide a plane into four parts which are called *quadrants*. Quadrant I is the upper right section. In a counterclockwise direction from Quadrant I are Quadrants II, III, and IV.

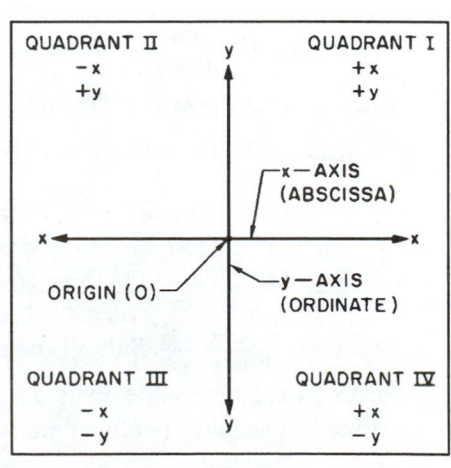

All points located to the right of the *y*-axis have positive (+) *x* values; all points to the left of the *y*-axis have negative (–) *x* values. All points above the *x*-axis have positive (+) *y* values; all points below the *x*-axis have negative (–) *y* values. The *x* value is called the *abscissa* and the *y* value is called the *ordinate*.

The *x* and *y* values for each quadrant are listed in the table.

Quadrant I	Quadrant II	Quadrant III	Quadrant IV
$+x$	$-x$	$-x$	$+x$
$+y$	$+y$	$-y$	$-y$

Determining Functions of Angles in Any Quadrant

As a ray is rotated through any of the four quadrants, functions of an angle are determined as follows:

- The ray is rotated in a counterclockwise direction with its vertex at the origin (O). Zero degrees is on the *x*-axis in quadrant I.

- From a point on the rotated ray, a line segment is projected perpendicular to the *x*-axis. A right triangle is formed of which the rotated side (ray) is the hypotenuse, the projected line segment is the opposite side, and the side on the *x*-axis is the adjacent side. The *reference angle* is the acute angle of the triangle which has the vertex at the origin (O).

- The sign of the functions of a reference angle is determined by noting the signs (+ or –) of the opposite and adjacent sides of the right triangle. The hypotenuse (*r*) is always positive in all four quadrants.

These examples illustrate the method of determining functions of angles greater than 90° in the various quadrants.

Example 1   Determine the sine and cosine functions of 115°.

With the endpoint of the ray (*r*) at the origin (O), the ray is rotated 115° in a counterclockwise direction.

From a point on *r*, side *y* is projected perpendicular to the *x*-axis. In the right triangle formed, in relation to the reference angle ($\angle x$), *r* is the hypotenuse, *y* is the opposite side, and *x* is the adjacent side.

$$\angle x = 180° - 115° = 65°$$

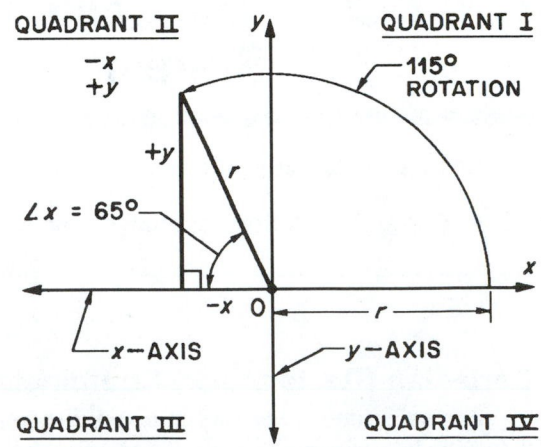

Sin $\angle x = \frac{\text{opposite side}}{\text{hypotenuse}}$. In Quadrant II, *y* is positive and *r* is always positive. Therefore, sin $\angle x = \frac{+y}{+r}$. In Quadrant II, the sine is a positive (+) function.

$$\sin 115° = \sin (180° - 115°) = \sin 65°$$
$$\sin 115° = 115 \boxed{\text{sin}} \rightarrow 0.906307787 \quad \text{Ans}$$
$$\text{or} \quad \sin 115° = \boxed{\text{sin}} \; 115 \boxed{\text{EXE}} \; 0.906307787 \quad \text{Ans}$$

Cos $\angle x = \frac{\text{adjacent side}}{\text{hypotenuse}}$. Side *x* is negative (–); therefore, cos $\angle x = \frac{-x}{+r}$. Since the quotient of a negative value divided by a positive value is negative, in Quadrant II, the cosine is a negative (–) function.

$\cos 115° = -\cos(180° - 115°) = -\cos 65°$

$\cos 115° = 115 \boxed{\cos} \rightarrow -0.422618262$ Ans

or $\cos 115° = \boxed{\cos} 115 \boxed{EXE} -0.422618261$ Ans

> **Note:** A negative function of an angle does **not** mean that the angle is negative; it is a negative function of a positive angle. For example, $-\cos 65°$ does not mean $\cos(-65°)$.

Example 2   Determine the tangent and secant functions of 218°.

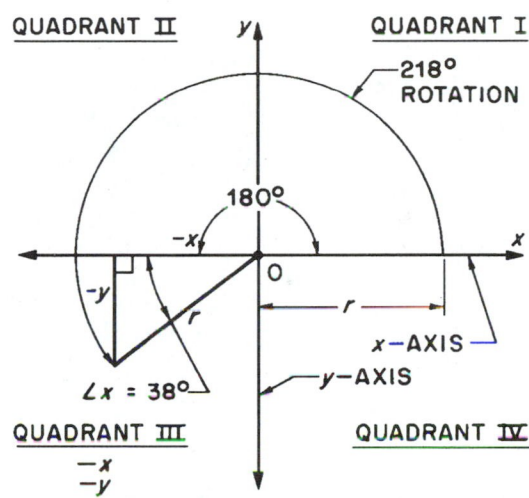

Rotate r 218° in a counterclockwise direction.

Project side $y \perp$ to the x-axis.

Reference $\angle x = 218° - 180° = 38°$

$$\tan \angle x = \frac{-y}{-x} = + \text{function}$$

$\tan 218° = \tan 38°$

$\tan 218° = 218 \boxed{\tan} \rightarrow 0.781285627$ Ans

or $\tan 218° = \boxed{\tan} 218 \boxed{EXE} 0.781285626$ Ans

$$\sec \angle x = \frac{+r}{-x} = - \text{function}$$

$\sec 218° = -\sec 38°$

$\sec 218° = 218 \boxed{\cos} \boxed{1/x} \rightarrow 1.269018215$ Ans

or $\sec 218° = \boxed{\cos} 218 \boxed{EXE} \boxed{1/x}$ or $\boxed{x^{-1}} \boxed{EXE} 1.269018215$ Ans

Example 3   Determine the cotangent and cosecant functions of 310°.

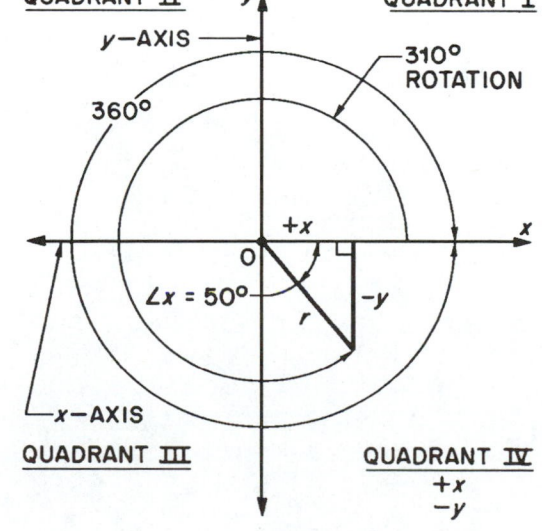

Rotate 310° counterclockwise.

Project side $y \perp$ to the x-axis.

Reference $\angle x = 360° - 310° = 50°$

$$\cot \angle x = \frac{+x}{-y} = - \text{function}$$

$\cot 310° = -\cot 50°$

$310 \boxed{\tan} \boxed{1/x} \rightarrow -0.8390996$ Ans

$$\csc \angle x = \frac{+r}{-y} = - \text{function}$$

$\csc 310° = -\csc 50°$

$310 \boxed{\sin} \boxed{1/x} \rightarrow -1.3054073$ Ans

APPLICATION

Determining Functions of Angles in Any Quadrant

Determine the sine, cosine, tangent, cotangent, secant, and cosecant functions for each of these angles. For each angle, sketch a right triangle. Label the sides of the triangles + or −. Determine the reference angles and functions of the angles. Round the answers to 4 decimal places.

1. 120°

2. 207°

3. 260°

4. 172°

5. 300°

6. 350°

7. 208°50′

8. 96°42′

9. 146°10′

10. 199.40°

11. 313.17°

12. 179.90°

UNIT 59 Oblique Triangles: Law of Sines and Law of Cosines

Objectives   After studying this unit you should be able to

- Solve simple oblique triangles using the Law of Sines and the Law of Cosines.

- Solve practical shop problems by applying the Law of Sines and the Law of Cosines.

Oblique Triangles

An *oblique triangle* is one that does not contain a right angle. An oblique triangle may be either acute or obtuse. In an acute triangle, each of the three angles is acute or less than 90°. In an obtuse triangle, one of the angles is obtuse or greater than 90°. The machinist must often solve practical machine shop problems which involve oblique triangles. These problems can be reduced to a series of right triangles, but the process can be cumbersome and time consuming. Two formulas, the Law of Sines and the Law of Cosines, can be used to simplify such computations. In order to use either formula, three parts of an oblique triangle must be known; at least one part must be a side.

Law of Sines

The Law of Sines states that in any triangle, the sides are proportional to the sines of the opposite angles. In reference to the triangle shown, the formula is stated:

$$\frac{a}{\sin A} = \frac{b}{\sin B} = \frac{c}{\sin C}$$

The Law of Sines is used to solve the following two kinds of problems:

- Problems where any two angles and any one side of an oblique triangle are known.

- Problems where any two sides and an angle opposite one of the given sides of an oblique triangle are known.

Solving Oblique Triangle Problems Given Two Angles and a Side

Example 1   Given two angles and a side, determine side x of the oblique triangle shown. All dimensions are in inches.

Since side x is opposite the 36° angle and the 3.500 inch side is opposite the 58° angle, the proportion is set up as:

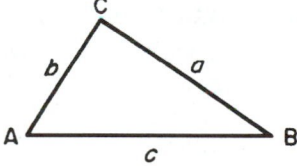

$$\frac{x}{\sin 36°} = \frac{3.500 \text{ in}}{\sin 58°}$$

$$x = \frac{\sin 36° \,(3.500 \text{ in})}{\sin 58°}$$

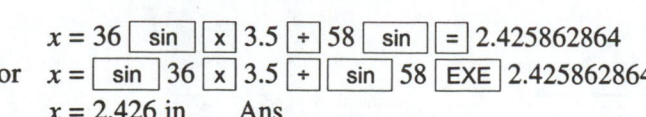

$x = 36 \boxed{\;\sin\;} \boxed{\times} 3.5 \boxed{+} 58 \boxed{\;\sin\;} \boxed{=} 2.425862864$

or $x = \boxed{\;\sin\;} 36 \boxed{\times} 3.5 \boxed{+} \boxed{\;\sin\;} 58 \boxed{\text{EXE}} 2.425862864$

$x = 2.426 \text{ in}$ Ans

Example 2 Given two angles and a side of the oblique triangle shown. All dimensions are in millimeters.

a. Determine ∠A.
b. Determine side a.
c. Determine side b.

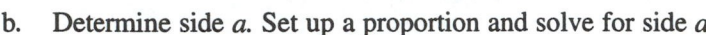

a. Determine ∠A.

$$\angle A = 180° - (37.3° + 24.5°)$$

$\angle A = 180 \boxed{-} \boxed{(} 37.3 \boxed{+} 24.5 \boxed{)} \boxed{=} 118.2$
$\angle A = 118.2°$ Ans

b. Determine side a. Set up a proportion and solve for side a.

$$\frac{a}{\sin 118.2°} = \frac{108.60 \text{ mm}}{\sin 37.3°}$$

$$a = \frac{\sin 118.2° \, (108.60 \text{ mm})}{\sin 37.3°}$$

$a = 118.2 \boxed{\sin} \boxed{\times} 108.6 \boxed{+} 37.3 \boxed{\sin} \boxed{=} 157.9395824$

or $a = \boxed{\sin} 118.2 \boxed{\times} 108.6 \boxed{+} \boxed{\sin} 37.3 \boxed{\text{EXE}} 157.9395824$

$a = 157.94$ mm   Ans

c. Determine side b. Set up a proportion and solve for side b.

$$\frac{b}{\sin 24.5°} = \frac{108.60 \text{ mm}}{\sin 37.3°}$$

$$b = \frac{\sin 24.5° \, (108.60 \text{ mm})}{\sin 37.3°}$$

$b = 24.5 \boxed{\sin} \boxed{\times} 108.6 \boxed{+} 37.3 \boxed{\sin} \boxed{=} 74.31773633$

or $b = \boxed{\sin} 24.5 \boxed{\times} 108.6 \boxed{+} \boxed{\sin} 37.3 \boxed{\text{EXE}} 74.31773633$

$b = 74.32$ mm

Solving Oblique Triangle Problems Given Two Sides and an Angle Opposite One of the Given Sides

A special condition exists when solving certain problems in which two sides and an angle opposite one of the sides is given. If triangle data are given in word form or if a triangle is inaccurately sketched, there may be two solutions to a problem.

It is possible to have two different triangles with the same two sides and the same angle opposite one of the given sides. A situation of this kind is called an ambiguous case. The following example illustrates the ambiguous case or a problem with two solutions.

Example   *(The Ambiguous Case or 2 solutions)*   A triangle has a 1.5-inch side, a 2.5-inch side, and an angle of 32° which is opposite the 1.5-inch side.

Using the given data, a figure is accurately drawn. Observe that two different triangles are constructed using identical given data. Both △BCA and △DCA have a 1.5-inch side, a 2.5-inch side, and a 32° angle opposite the 1.5-inch side. The two different triangles are shown.

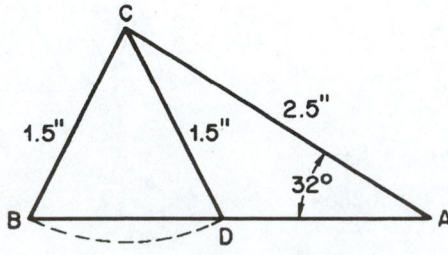

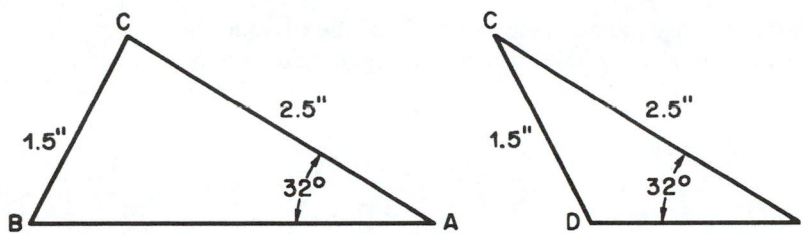

The only conditions under which a problem can have two solutions is when the given angle is acute and the given side opposite the given angle is smaller than the other given side. For example, in the problem illustrated the 32° angle is acute, and the 1.5-inch side opposite the 32° angle is smaller than the 2.5-inch side.

In most problems you do not get involved with two solutions. Even under the conditions in which there can be two solutions, if the problem is shown in picture form as an accurately drawn triangle, it can readily be observed that there is only one solution.

Example 1 Given two sides and an opposite angle of the oblique triangle shown. All dimensions are in inches.

a. Determine $\angle x$.

b. Determine side y.

The 6.000-inch side opposite the 63°50′ angle is larger than the 4.500-inch side; therefore, there is only one solution.

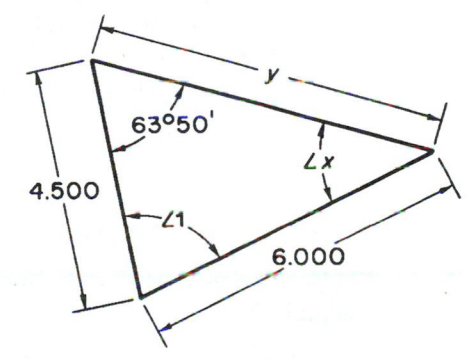

a. Determine $\angle x$.

$$\frac{4.500 \text{ in}}{\sin \angle x} = \frac{6.000 \text{ in}}{\sin 63°50'}$$

$$\sin \angle x = \frac{\sin 63°50' \ (4.500 \text{ in})}{6.000 \text{ in}}$$

$\sin \angle x =$ 63 $\boxed{\cdot}$ 50 $\boxed{\text{2nd}}$ $\boxed{\blacktriangleright \underset{=}{\text{DD}}}$ $\boxed{\sin}$ $\boxed{\times}$ 4.5 $\boxed{\div}$ 6 $\boxed{=}$ 0.673136307

or $\sin \angle x =$ $\boxed{\sin}$ 63 $\boxed{°\,'\,''}$ 50 $\boxed{°\,'\,''}$ $\boxed{\times}$ 4.5 $\boxed{\div}$ 6 $\boxed{\text{EXE}}$ 0.673136306

$\angle x =$ .673136307 $\boxed{\text{2nd}}$ $\boxed{\sin^{-1}}$ $\boxed{\text{3rd}}$ $\boxed{\blacktriangleright\text{DMS}}$ 42°18′34″

or $\angle x =$ $\boxed{\text{SHIFT}}$ $\boxed{\sin^{-1}}$ .673136307 $\boxed{\text{EXE}}$ $\boxed{\text{SHIFT}}$ $\boxed{\longleftarrow}$ 42°18′34″

$\angle x =$ 42°19′   Ans

b. Determine side y.

$\angle 1 = 180° - (63°50' + \angle x) = 180° - 106°9' = 73°51'$

$$\frac{6.000 \text{ in}}{\sin 63°50'} = \frac{y}{\sin 73°51'}$$

$$y = \frac{\sin 73°51' \ (6.000 \text{ in})}{\sin 63°50'}$$

$y =$ 73 $\boxed{\cdot}$ 51 $\boxed{\text{2nd}}$ $\boxed{\blacktriangleright \underset{=}{\text{DD}}}$ $\boxed{\sin}$ $\boxed{\times}$ 6 $\boxed{+}$ 63 $\boxed{\cdot}$ 50 $\boxed{\text{2nd}}$ $\boxed{\blacktriangleright \underset{=}{\text{DD}}}$ $\boxed{\sin}$

$\boxed{=}$ 6.421307977

or $y =$ $\boxed{\sin}$ 73 $\boxed{°\,'\,''}$ 51 $\boxed{°\,'\,''}$ $\boxed{\times}$ 6 $\boxed{\div}$ $\boxed{\sin}$ 63 $\boxed{°\,'\,''}$ 50 $\boxed{°\,'\,''}$ $\boxed{\text{EXE}}$ 6.421307977

$y =$ 6.421 in   Ans

Example 2 Given two sides and an opposite angle, determine $\angle x$ of the oblique triangle shown. All dimensions are in millimeters. The figure is drawn accurately to scale.

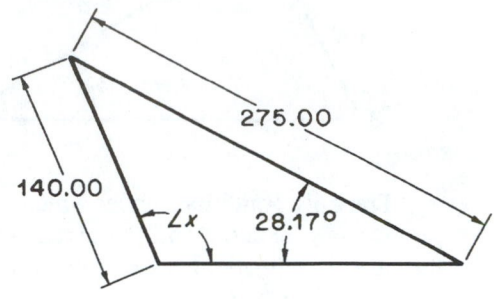

$$\frac{140.00 \text{ mm}}{\sin 28.17°} = \frac{275.00 \text{ mm}}{\sin \angle x}$$

$$\sin \angle x = \frac{\sin 28.17° \, (275.00 \text{ mm})}{140.00 \text{ mm}}$$

$\sin \angle x = $ 28.17 $\boxed{\sin}$ $\boxed{\times}$ 275 $\boxed{\div}$ 140 $\boxed{=}$ 0.927318171

or $\sin \angle x = $ $\boxed{\sin}$ 28.17 $\boxed{\times}$ 275 $\boxed{\div}$ 140 $\boxed{\text{EXE}}$ 0.927318171

.927318171 $\boxed{\text{2nd}}$ $\boxed{\sin^{-1}}$ $\rightarrow$ 68.02055272

or $\boxed{\text{SHIFT}}$ $\boxed{\sin^{-1}}$ .927318171 $\boxed{\text{EXE}}$ 68.02055272

The angle that corresponds to the sine function 0.927318171 is 68.02°. Because $\angle x$ is greater than 90°, $\angle x =$ the supplement of 68.02°.

$\angle x = 180° - 68.02° = 111.98°$ Ans

Law of Cosines (Given Two Sides and the Included Angle)

In any triangle, the square of any side is equal to the sum of the squares of the other two sides minus twice the product of these two sides multiplied by the cosine of their included angle.

In reference to the triangle shown the formula is stated:

$$a^2 = b^2 + c^2 - 2bc(\cos A)$$

$$b^2 = a^2 + c^2 - 2ac(\cos B)$$

$$c^2 = a^2 + b^2 - 2ab(\cos C)$$

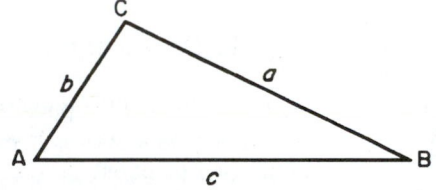

The Law of Cosines, as stated in the formulas, is used to solve the following kind of oblique triangle problems.

- Problems where two sides and the included angle of an oblique triangle are known.

➤ **Note:** An angle of an oblique triangle may be greater than 90°. Therefore, you must often determine the cosine of an angle greater than 90° and less than 180°. These angles lie in quadrant II of the Cartesian coordinate system. Recall that the cosine of an angle between 90° and 180° equals the negative (–) cosine of the supplement of the angle. For example, the cosine of 118°10′ = –cos (180° – 118°10′) = –cos 61°50′.

Solving Oblique Triangle Problems Given Two Sides and the Included Angle

Example 1 Given two sides and the included angle, determine side x of the oblique triangle shown. All dimensions are in millimeters. Observe that 36.83° is included between the 62.00 mm and 56.00 mm sides.

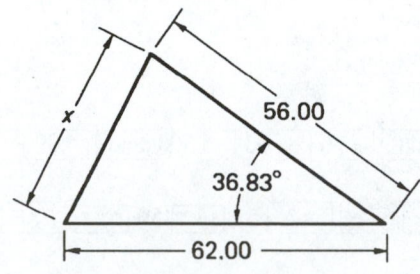

Substitute the values in their appropriate places in the formula and solve for x.

$$x^2 = (56.00 \text{ mm})^2 + (62.00 \text{ mm})^2 - 2(56.00 \text{ mm})(62.00 \text{ mm})(\cos 36.83°)$$

$$x = \sqrt{(56.00 \text{ mm})^2 + (62.00 \text{ mm})^2 - 2(56.00 \text{ mm})(62.00 \text{ mm})(\cos 36.83°)}$$

$x = $ ⏍ 56 ⏍X^2 ⏍ + 62 ⏍X^2 ⏍ − 2 ⏍×⏍ 56 ⏍×⏍ 62 ⏍×⏍ 36.83 ⏍cos⏍ ⏍)⏍ ⏍√x⏍
→ 37.70809057

or $x = $ ⏍√ ⏍ ⏍(⏍ 56 ⏍X^2 ⏍ + 62 ⏍X^2 ⏍ − 2 ⏍×⏍ 56 ⏍×⏍ 62 ⏍×⏍ cos 36.83 ⏍)⏍
⏍EXE⏍ 37.70809057

$x = 37.71 \text{ mm}$ Ans

Example 2   Given two sides and the included angle of the oblique triangle shown. All dimensions are in inches.

a. Determine side a.

b. Determine $\angle B$.

c. Determine $\angle C$.

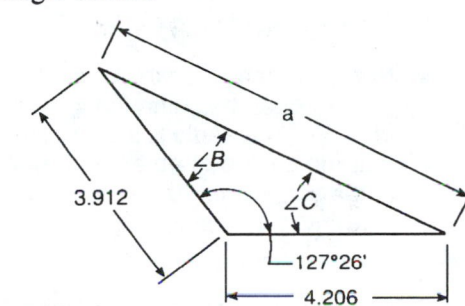

a.   Solve for a, using the Law of Cosines.

$$a^2 = (3.912 \text{ in})^2 + (4.206 \text{ in})^2 - 2(3.912 \text{ in})(4.206 \text{ in})(\cos 127°26')$$

$$a = \sqrt{(3.912 \text{ in})^2 + (4.206 \text{ in})^2 - 2(3.912 \text{ in})(4.206 \text{ in})(\cos 127°26')}$$

$a = $ ⏍(⏍ 3.912 ⏍X^2 ⏍ + 4.206 ⏍X^2 ⏍ − 2 ⏍×⏍ 3.912 ⏍×⏍ 4.206 ⏍×⏍ 127.26 ⏍2nd⏍
⏍▸DD⏍ ⏍cos⏍ ⏍=⏍ ⏍)⏍ ⏍√x⏍ → 7.27988697

or $a = $ ⏍√ ⏍ ⏍(⏍ 3.912 ⏍X^2 ⏍ + 4.206 ⏍X^2 ⏍ − 2 ⏍×⏍ 3.912 ⏍×⏍ 4.206 ⏍×⏍ ⏍cos⏍
127 ⏍° ′ ″⏍ 26 ⏍° ′ ″⏍ ⏍)⏍ ⏍EXE⏍ 7.27988697

$a = 7.280 \text{ in}$ Ans

Solve for $\angle B$, using the Law of Sines.

$$\frac{4.206 \text{ in}}{\sin \angle B} = \frac{7.280 \text{ in}}{\sin 127°26'}$$

$$\sin \angle B = \frac{4.206 \text{ in} (\sin 127°26')}{7.280 \text{ in}}$$

$\sin \angle B = $ 4.206 ⏍×⏍ 127.26 ⏍2nd⏍ ⏍▸DD⏍ ⏍sin⏍ ⏍÷⏍ 7.280 ⏍=⏍ 0.458766636

or $\sin \angle B = $ 4.206 ⏍×⏍ ⏍sin⏍ 127 ⏍° ′ ″⏍ 26 ⏍° ′ ″⏍ ⏍÷⏍ 7.280 ⏍EXE⏍ 0.4587666359

$\angle B = $ .458766636 ⏍2nd⏍ ⏍sin⁻¹⏍ ⏍3rd⏍ ⏍▸DMS⏍ 27°18′27″1

or $\angle B = $ ⏍SHIFT⏍ ⏍sin⁻¹⏍ .458766636 ⏍EXE⏍ ⏍SHIFT⏍ ⏍◄—⏍ → 27°18′27.18″

$\angle B = 27°18'$ Ans

Solve for $\angle C$.

$$\angle C = 180° - (127°26' + 27°18') = 25°16'   \text{Ans}$$

Law of Cosines (Given Three Sides)

In any triangle, the cosine of an angle is equal to the sum of the squares of the two adjacent sides minus the square of the opposite side, divided by twice the product of the two adjacent sides.

In reference to the triangle shown:

$$\cos A = \frac{b^2 + c^2 - a^2}{2bc}$$

$$\cos B = \frac{a^2 + c^2 - b^2}{2ac}$$

$$\cos C = \frac{a^2 + b^2 - c^2}{2ab}$$

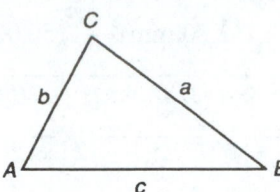

➤ **Note:** These formulas, which are stated in terms of the cosines of angles, are rearrangements of the formulas on page 378, which are stated in terms of the squares of the sides.

The Law of Cosines, as stated in the formulas, is used to solve the following kind of oblique triangle problems.

• Problems where three sides of an oblique triangle are known.

➤ **Note:** When an unknown angle is determined, its cosine function may be negative. A negative cosine function means that the angle being computed is greater than 90°. The angle lies in quadrant II of the Cartesian coordinate system. Recall that the cosine of an angle between 90° and 180° equals the negative cosine of the supplement of the angle. For example, the cosine of 147°40′ = −cos (180° − 147°40′) = −cos 32°20′.

Solving Oblique Triangle Problems Given Three Sides

Examples

1. Given three sides, determine ∠A of the oblique triangle shown. All dimensions are in inches.

$$\cos \angle A = \frac{(6.400 \text{ in})^2 + (7.800 \text{ in})^2 - (4.700 \text{ in})^2}{2(6.400 \text{ in})(7.800 \text{ in})}$$

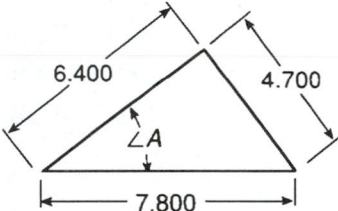

cos ∠A = 6.4 $\boxed{X^2}$ $\boxed{+}$ 7.8 $\boxed{X^2}$ $\boxed{-}$ 4.7 $\boxed{X^2}$ $\boxed{=}$ $\boxed{÷}$ $\boxed{(}$ 2
$\boxed{×}$ 6.4 $\boxed{×}$ 7.8 $\boxed{)}$ $\boxed{=}$ 0.798377404

∠A = .798377404 $\boxed{2\text{nd}}$ $\boxed{\cos^{-1}}$ $\boxed{3\text{rd}}$ $\boxed{\blacktriangleright\text{DMS}}$ 37°01′28″4

or   ∠A = $\boxed{\text{SHIFT}}$ $\boxed{\cos^{-1}}$ .798377404 $\boxed{\text{EXE}}$ $\boxed{\text{SHIFT}}$ $\boxed{\longleftarrow}$ 37°1′28.44″

∠A = 37°01′   Ans

2. Given three sides, determine ∠P of the oblique triangle shown. All dimensions are in millimeters.

$$\cos \angle P = \frac{(8.323 \text{ mm})^2 + (9.745 \text{ mm})^2 - (15.118 \text{ mm})^2}{2(8.323 \text{ mm})(9.745 \text{ mm})}$$

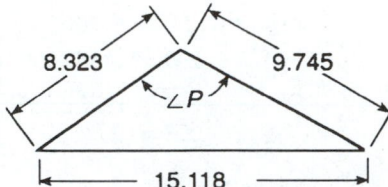

cos ∠P = 8.323 $\boxed{X^2}$ $\boxed{+}$ 9.745 $\boxed{X^2}$ $\boxed{-}$ 15.118 $\boxed{X^2}$ $\boxed{=}$ $\boxed{÷}$ $\boxed{(}$ 2 $\boxed{×}$ 8.323 $\boxed{×}$
9.745 $\boxed{)}$ $\boxed{=}$ −0.396488999

∠P = .39648899 $\boxed{+/-}$ $\boxed{2\text{nd}}$ $\boxed{\cos^{-1}}$ → 113.3588715

or   ∠P = $\boxed{\text{SHIFT}}$ $\boxed{\cos^{-1}}$ $\boxed{-}$ .396488999 $\boxed{\text{EXE}}$ 113.3588715

          └─ or .39648899 $\boxed{+/-}$ $\boxed{=}$ 113.3588715

∠P = 113.36°   Ans

APPLICATION

For English unit dimensioned problems, calculate angles to the nearer minute and lengths to the nearer thousandth inch. For metric unit dimensioned problems, calculate angles to the nearer hundredth degree and lengths to the nearer hundredth millimeter.

Law of Sines

Solve the following problems using the Law of Sines.

1. Find side x.
   All dimensions are in inches.   _____

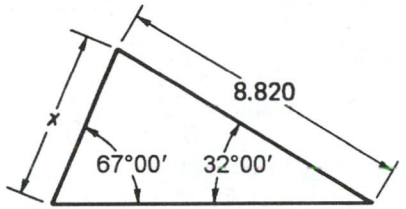

2. Find side x.
   All dimensions are in inches.   _____

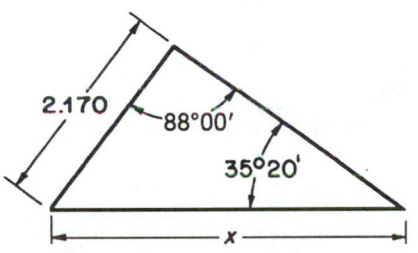

3. Find side x.
   All dimensions are in inches.   _____

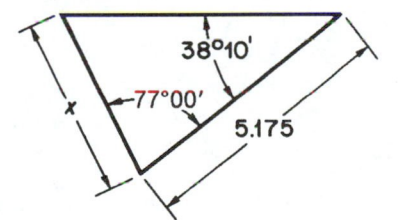

4. Find side x.
   All dimensions are in millimeters.   _____

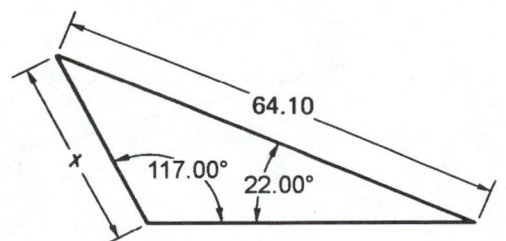

5. Find $\angle x$.
   All dimensions are in inches.   _____

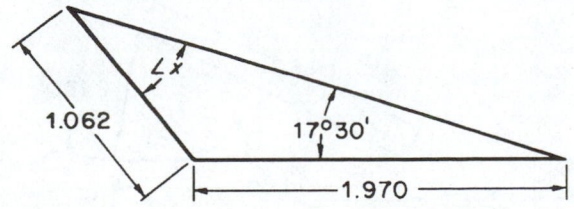

6. Find $\angle x$.
   All dimensions are in millimeters.   _____

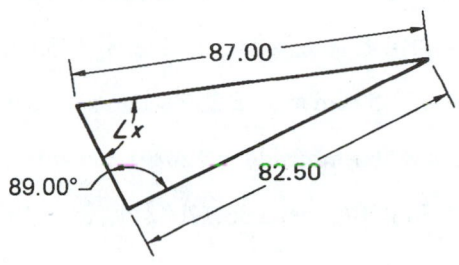

7. Find $\angle x$.
   All dimensions are in inches.   _____

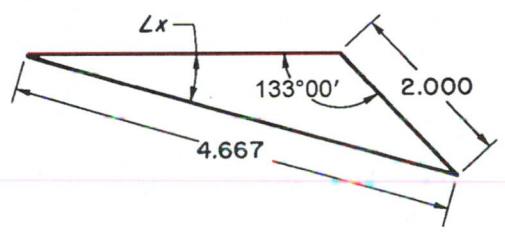

8. Find $\angle x$.
   All dimensions are in inches.   _____

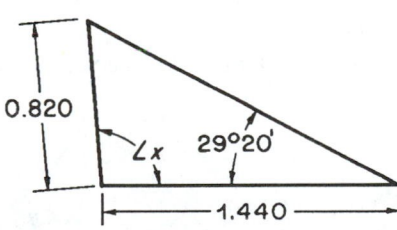

9. Find side x.
   All dimensions are in millimeters.   _____

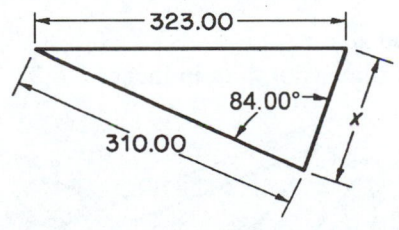

Identifying Problems with One or Two Solutions

Two sides and an angle opposite one of the sides of triangles are given in the following problems. Identify each problem as to whether it has one or two solutions. Do **not** solve the problems for angles and sides.

10. A 4″ side, a 5″ side, a 37° angle opposite the 4″ side. _____

11. A 95.00-mm side, a 98.00-mm side, a 75° angle opposite the 95.00-mm side. _____

12. A 21-mm side, a 29-mm side, a 41° angle opposite the 29-mm side. _____

13. A 0.943″ side, a 1.612″ side, and an 82°15′ angle opposite the 0.943″ side. _____

14. A 2.10-ft side, a 3.05-ft side, a 29°30′ angle opposite the 3.05-ft side. _____

15. A 16.35-mm side, a 23.86-mm side, a 115° angle opposite the 23.86-mm side. _____

16. An 87.60-mm side, a 124.80-mm side, a 12.90° angle opposite the 87.60-mm side. _____

17. A 34.090″ side, a 35.120″ side, a 46°18′ angle opposite the 34.090″ side. _____

Law of Cosines

Solve the following problems using the Law of Cosines.

18. Find side x.
    All dimensions are in inches. _____

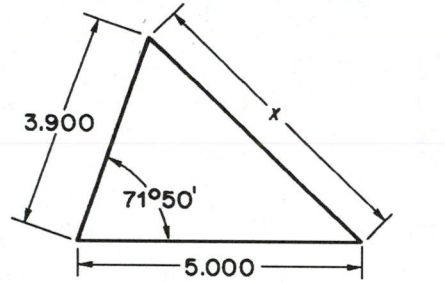

21. Find $\angle x$.
    All dimensions are in millimeters. _____

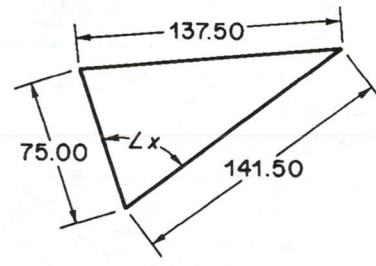

19. Find side x.
    All dimensions are in millimeters. _____

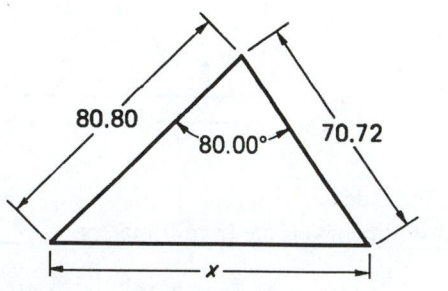

22. Find $\angle x$.
    All dimensions are in inches. _____

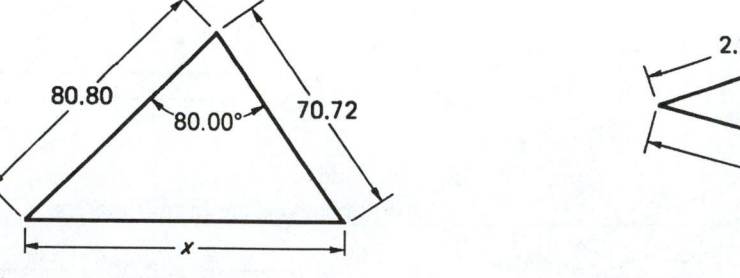

20. Find side x.
    All dimensions are in inches. _____

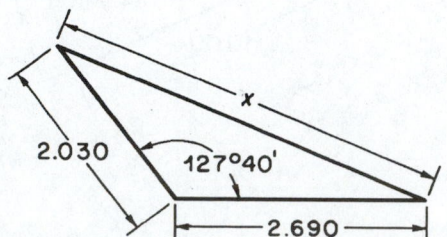

23. Find $\angle x$.
    All dimensions are in inches. _____

24. Find ∠x.
    All dimensions are in inches.   _____

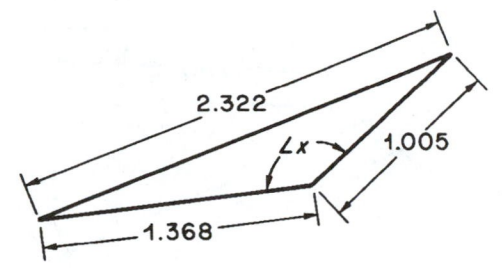

26. Find side x.
    All dimensions are in inches.   _____

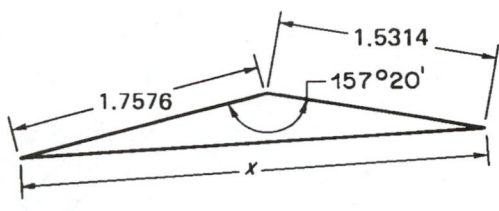

25. Find side x.
    All dimensions are in millimeters.   _____

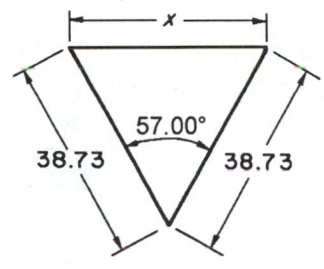

Combination of the Law of Cosines and the Law of Sines

Solve the following problems using a combination of the Law of Cosines and the Law of Sines.

27. All dimensions are in inches.
    a. Find side x.   _____
    b. Find ∠y.   _____

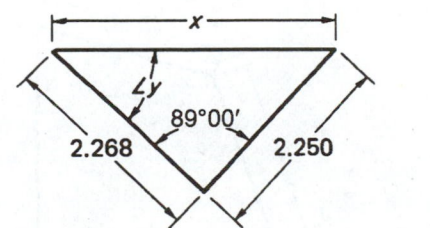

28. All dimensions are in inches.
    a. Find side x.   _____
    b. Find ∠y.   _____

29. All dimensions are in millimeters.
    a. Find side x.   _____
    b. Find ∠y.   _____

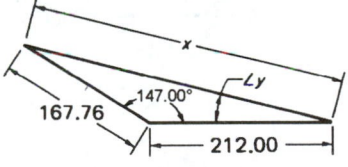

30. All dimensions are in inches.
    a. Find ∠x.   _____
    b. Find ∠y.   _____

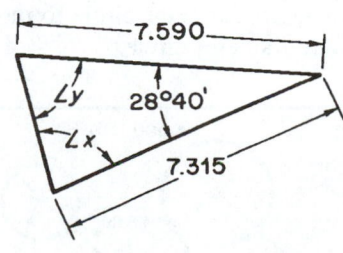

31. All dimensions are in inches.
    a. Find $\angle x$. _____
    b. Find $\angle y$. _____

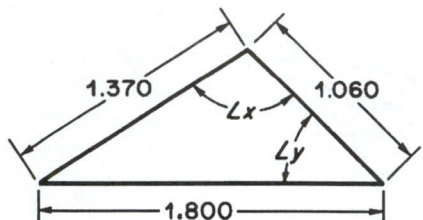

Practical Machine Shop Problems

Solve the following machine shop problems.

33. Find $\angle x$.
    All dimensions are in millimeters. _____

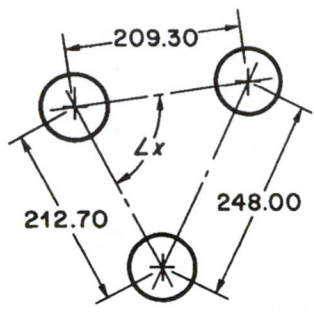

34. Find distance y.
    All dimensions are in inches. _____

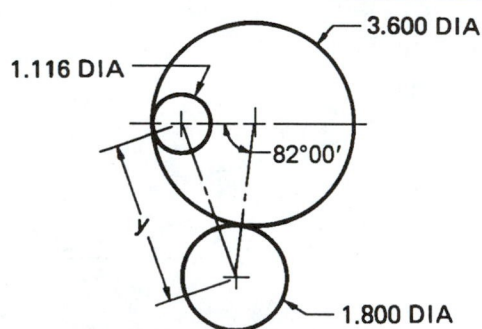

35. Three circles are to be bored in the plate shown. The 4.000-inch diameter and 5.500-inch diameter circles are each tangent to the 7.500-inch diameter circle. Determine the distance from the center of the 4.000-inch diameter circle to the center of the 5.500-inch diameter circle.

_____

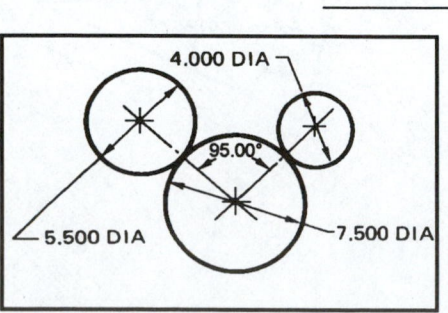

32. All dimensions are in millimeters.
    a. Find $\angle x$. _____
    b. Find $\angle y$. _____

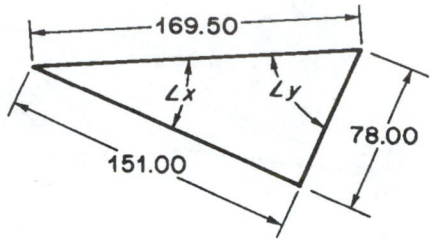

36. Find distance x.
    All dimensions are in inches. _____

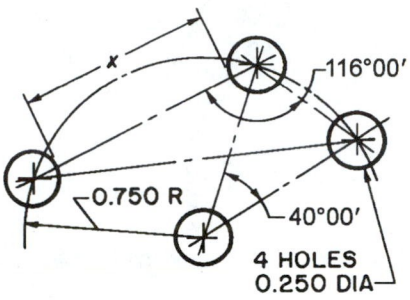

37. Find $\angle x$.
    All dimensions are in inches. _____

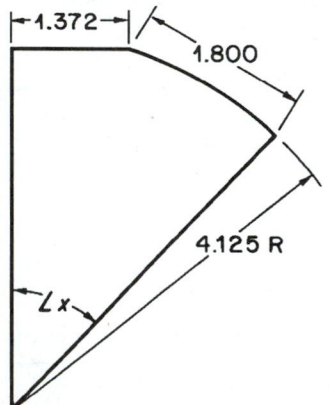

38. Find dimension y.
    All dimensions are in millimeters. _____

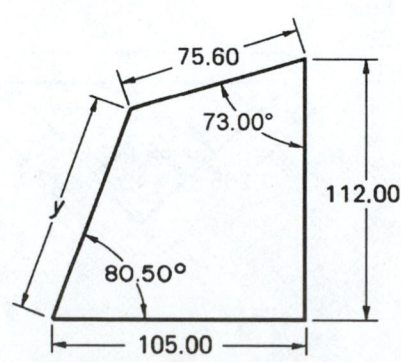

39. Find dimension y.
    All dimensions are in inches. _____

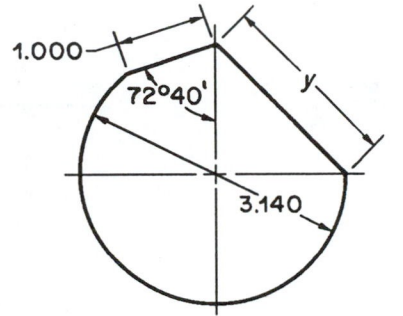

40. Find dimension y.
    All dimensions are in inches. _____

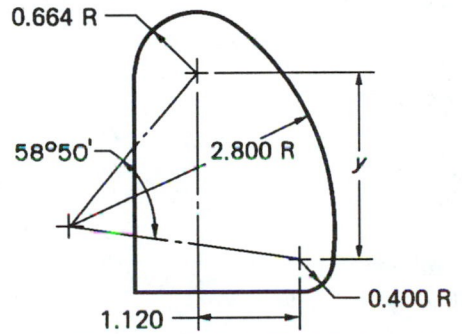

41. Find ∠x.
    All dimensions are in inches. _____

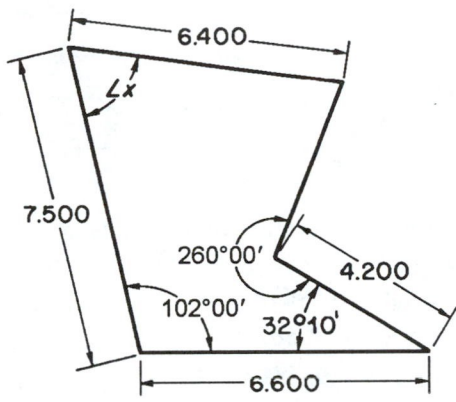

42. Find ∠y.
    All dimensions are in millimeters. _____

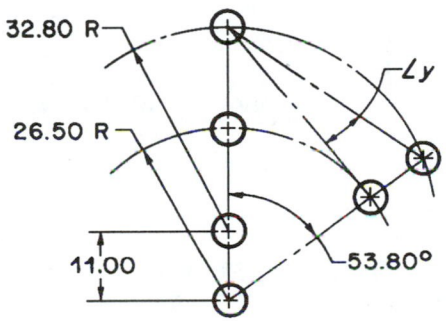

UNIT 60 Achievement Review— Section Five

Objective

You should be able to solve the exercises and problems in this Achievement Review by applying the principles and methods covered in units 53–59.

1. With reference to ∠1, name the sides of each of the following triangles as opposite, adjacent, or hypotenuse.

   a. _____   b. _____   c. _____   d. _____

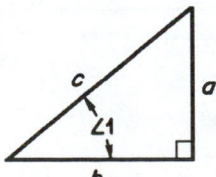

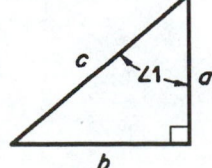

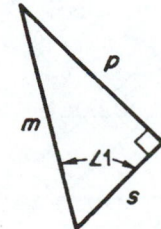

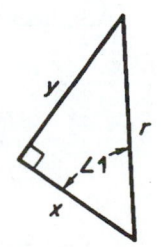

2. Determine the functions of the following angles. Round the answers to 4 decimal places.
   a. sin 22° _____
   b. cot 46°20′ _____
   c. sec 37°50′ _____
   d. tan 0°21′ _____
   e. cos 63°18′ _____
   f. tan 14°24′ _____
   g. sin 7.43° _____
   h. csc 57.82° _____

3. Determine the values of ∠A in degrees and minutes that correspond to the following functions.
   a. cos A = 0.69675 _____
   b. tan A = 0.50587 _____
   c. sin A = 0.98531 _____
   d. cot A = 1.1340 _____
   e. sec A = 1.5753 _____
   f. cos A = 0.15902 _____

4. Determine the values of ∠A in decimal-degrees to 2 decimal places that correspond to the following functions.
   a. sin A = 0.72847 _____
   b. tan A = 1.3925 _____
   c. cos A = 0.34038 _____

5. For each of the following functions of angles, write the cofunction of the complement of the angle.
   a. sin 36° _____
   b. tan 48°19′ _____
   c. cos 16°53′ _____
   d. cot 80.47° _____

6. Solve the following problems. Compute angles to the nearer minute in triangles with English unit sides. Compute angles to the nearer hundredth degree in triangles with metric unit sides. Compute sides to 3 decimal places.

   a. Determine ∠A.
      All dimensions are in inches. _____

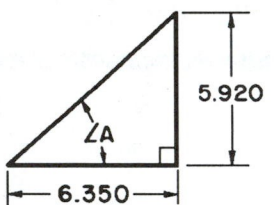

   b. Determine side a.
      All dimensions are in inches. _____

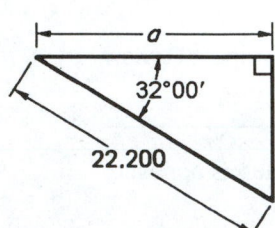

   c. Determine ∠D.
      All dimensions are in millimeters. _____

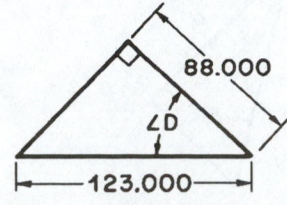

   d. Determine ∠1.
      All dimensions are in millimeters. _____

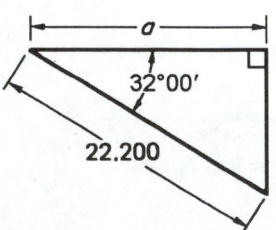

   e. All dimensions are in millimeters.
      (1) Determine side g. _____
      (2) Determine side h. _____
      (3) Determine ∠H. _____

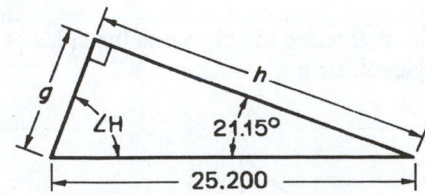

f. All dimensions are in inches.

   (1) Determine ∠A.   _____

   (2) Determine ∠B.   _____

   (3) Determine side *c*.   _____

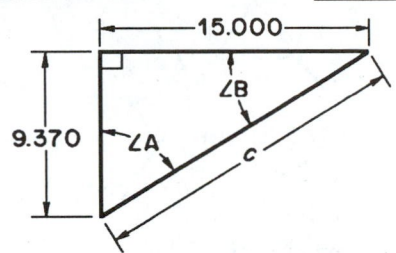

7. Solve the following applied right triangle problems. Compute linear values to 3 decimal places, English unit angles to the nearer minute, and metric angles to the nearer hundredth degree.

a. All dimensions are in inches.

   (1) Determine dimension *c*.   _____

   (2) Determine dimension *d*.   _____

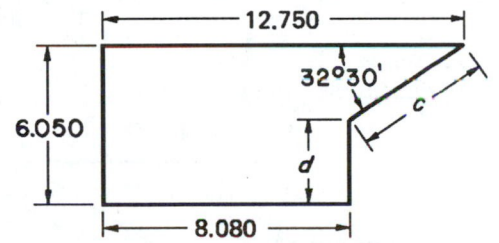

b. Determine ∠T.
All dimensions are in millimeters. _____

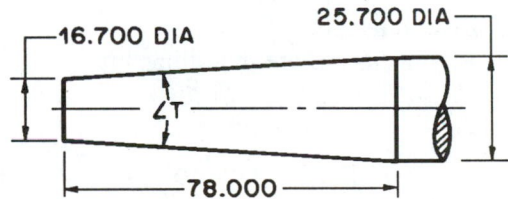

c. Determine dimension *x*.
All dimensions are in millimeters. _____

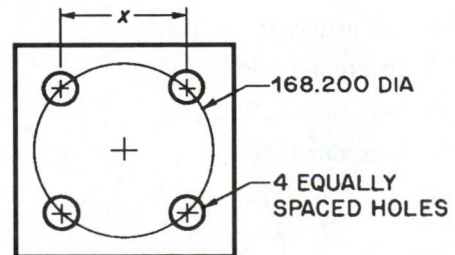

d. Determine dimension *d*.
All dimensions are in inches. _____

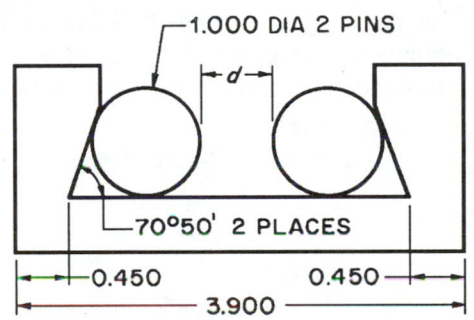

e. All dimensions are in millimeters.

   (1) Determine ∠A.   _____

   (2) Determine distance *x*.   _____

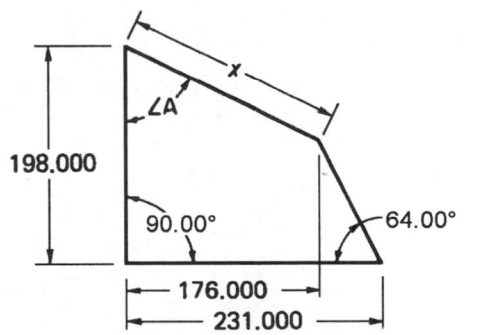

f. Determine check dimension *y*.
All dimensions are in inches. _____

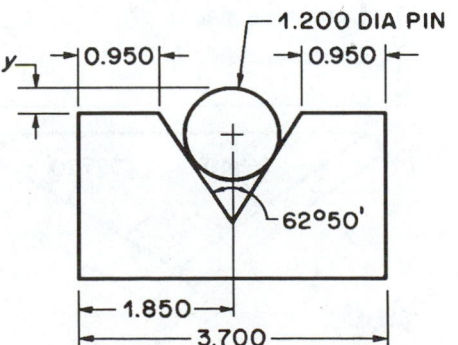

g. Determine ∠x.
All dimensions are in inches. _____

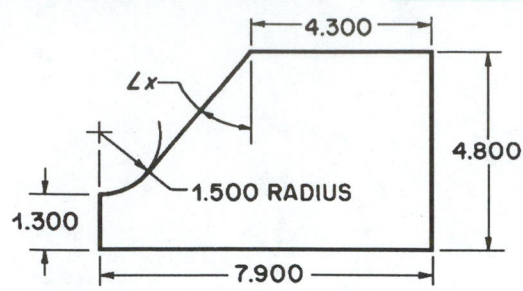

h. Determine ∠y.
All dimensions are in millimeters. _____

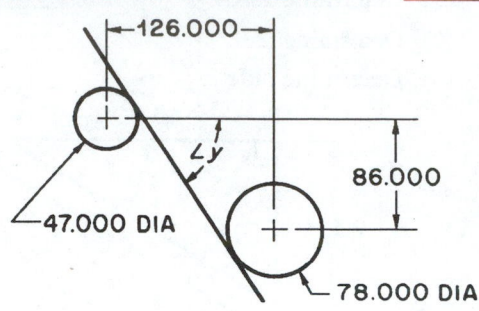

8. Determine the sine, cosine, tangent, cotangent, secant, and cosecant of each of the following angles.

   a. 115° _____

   b. 223° _____

   c. 310°30′ _____

9. Solve the following problems using the Law of Sines and/or the Law of Cosines. Compute side lengths to 3 decimal places, English unit angles to the nearer minute, and metric unit angles to the nearer hundredth degree.

   a. Determine side a.
   All dimensions are in millimeters. _____

   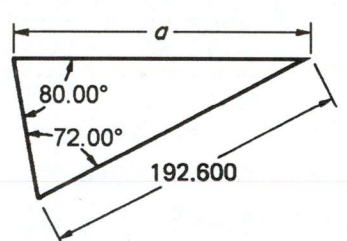

   d. Determine side d.
   All dimensions are in millimeters. _____

   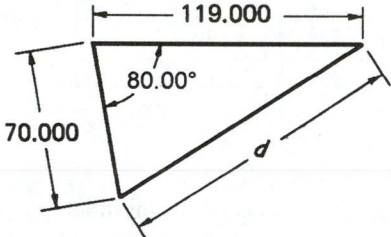

   b. Determine ∠D.
   All dimensions are in inches. _____

   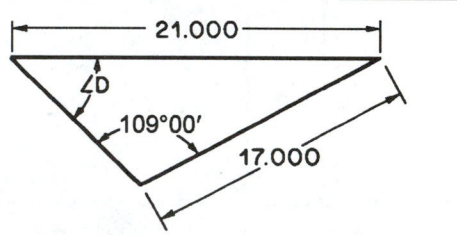

   e. Determine ∠E.
   All dimensions are in millimeters. _____

   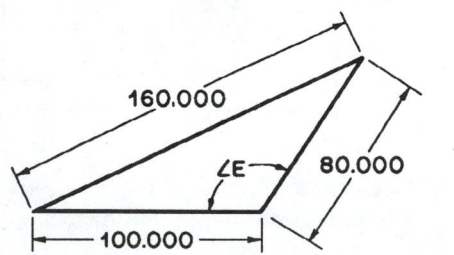

   c. All dimensions are in inches.

   (1) Determine ∠A. _____

   (2) Determine side a. _____

   (3) Determine side b. _____

   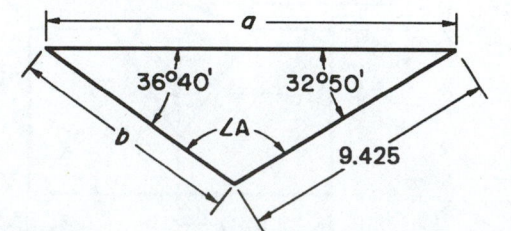

   f. All dimensions are in inches.

   (1) Determine side m. _____

   (2) Determine ∠N. _____

   (3) Determine ∠P. _____

   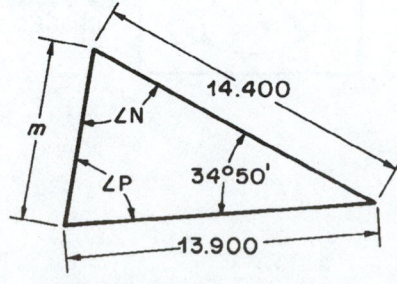

g. Determine dimension *d*.
   All dimensions are in inches. _____

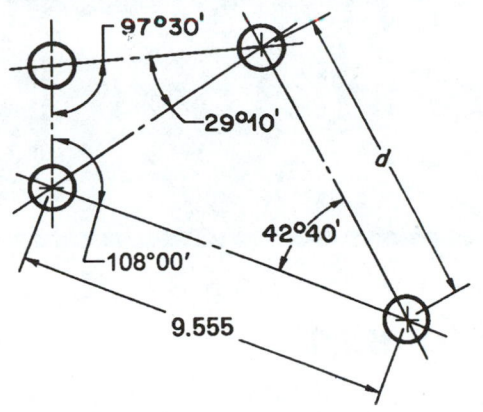

h. Determine ∠*x*.
   All dimensions are in millimeters. _____

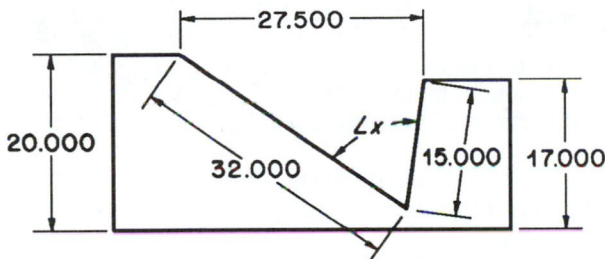

i. Determine ∠*x*.
   All dimensions are in inches. _____

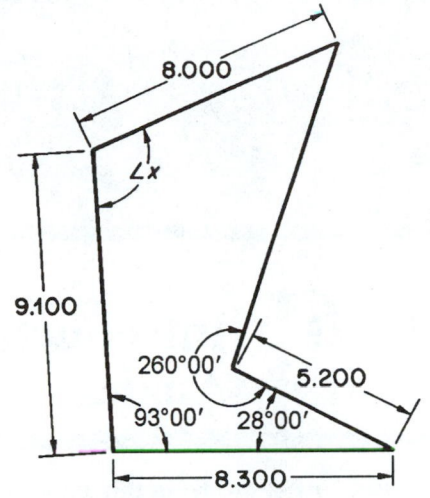

j. A piece of stock is to be machined as shown.
   Determine dimension *b*.

   All dimensions are in inches. _____

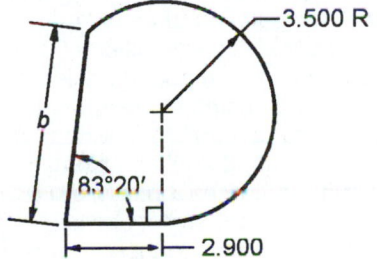

Section Six
Compound Angles

UNIT **61** Introduction to Compound Angles

Objectives   After studying this unit you should be able to

- Compute true lengths of diagonals of rectangular solids.
- Compute true angles of diagonals of rectangular solids.

In the machine trades, the application of principles of solid or three-dimensional trigonometry is commonly called *compound angles*. Generally compound angle problems require the computation of an unknown angle in a plane which is the resultant of two or more known angles lying in different planes.

Applications of compound angles are frequently required in machining fixture parts, die sections, and cutting tools. An understanding of compound angle procedures is necessary in setting up parts for drilling or boring compound-angular holes.

Often, compound angle problems are encountered when machining parts as shown on engineering drawings. Usually, the top, front, and right side views of orthographic projections are shown. Wherever applicable, compound angle examples and problems in this text are given in relation to these views.

Formulas for specific compound angle applications can be found in certain trade handbooks. These formulas are useful provided the particular compound angle applications are properly visualized and identified. There are variations in compound angle situations. Merely plugging in values in given formulas without fully visualizing the components of a problem can result in costly errors.

Certain basic compound angle situations are presented in this text. A comprehensive study of compound angles is not intended. An understanding of applications is emphasized. Visualization of a problem with its components is stressed.

Pictorial views of compound angle situations with their components located and identified in rectangular solids or pyramids are shown. The student should make sketches in pictorial form to develop understanding. Formulas should be used in the solution of a problem only after the problem has been clearly visualized.

Diagonal of a Rectangular Solid

A pictorial view of a rectangular solid with diagonal AB is shown. A rectangular solid has six rectangular faces.

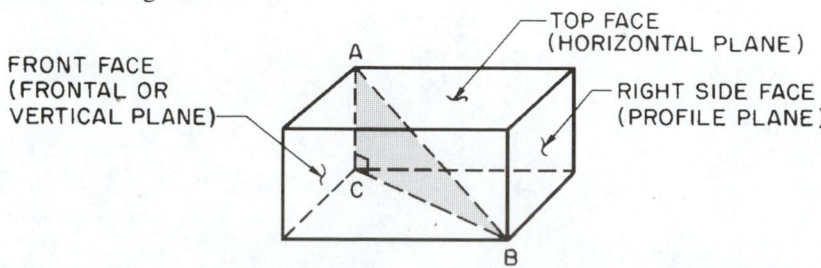

The top face (horizontal plane), front face (frontal or vertical plane), and right side face (profile plane) are identified. These faces correspond to the top, front, and right side views as they appear on an engineering drawing as shown.

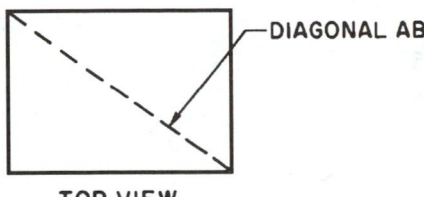

TOP VIEW

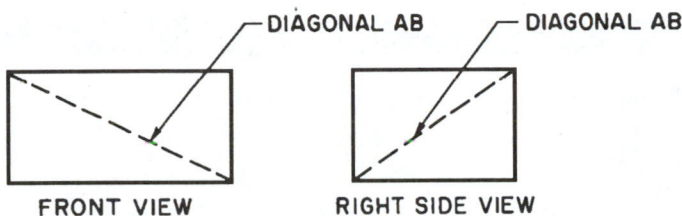

FRONT VIEW   RIGHT SIDE VIEW

Observe that although AB appears as a diagonal in each of the three views, it does not appear in its actual (true) length in any of the views. Neither does the true angle made by AB with either a vertical or horizontal plane appear in any of the three views. The true length of a line is shown when the line is contained in a plane which is viewed perpendicular to the line of sight.

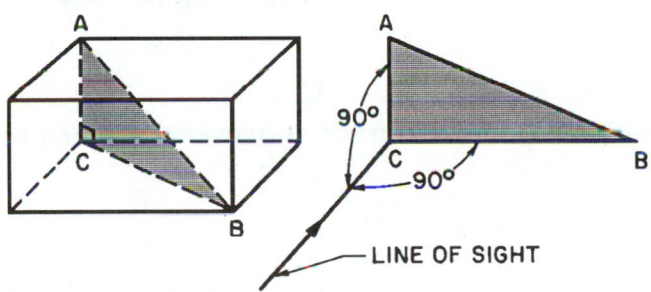

Computing True Lengths and True Angles

Example Compute true length AB and true $\angle$CAB shown. All dimensions are in inches.

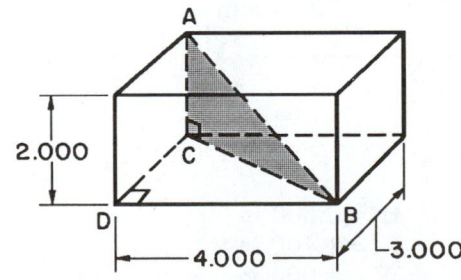

In right $\triangle$CDB:
Compute CB.
Apply the Pythagorean Theorem:

$$CB^2 = DB^2 + DC^2$$
$$CB^2 = (4.000 \text{ in})^2 + (3.000 \text{ in})^2$$
$$CB^2 = 16.000 \text{ in}^2 + 9.000 \text{ in}^2$$
$$CB^2 = 25.000 \text{ in}^2$$
$$CB = 5.000 \text{ in}$$

In right $\triangle$ACB:
Compute AB.

$$AB^2 = AC^2 + CB^2$$
$$AB^2 = (2.000 \text{ in})^2 + (5.000 \text{ in})^2$$
$$AB^2 = 4.000 \text{ in}^2 + 25.000 \text{ in}^2$$
$$AB^2 = 29.000 \text{ in}^2$$
$$AB = 5.385 \text{ in} \qquad \text{Ans}$$

Compute ∠CAB.

$$\tan \angle CAB = \frac{CB}{AC} = \frac{5.000 \text{ in}}{2.000 \text{ in}} = 2.5000$$

∠CAB = 2.5 [2nd] [tan⁻¹] [3rd] [▸DMS] → 68°11′55″

or ∠CAB = [SHIFT] [tan⁻¹] 2.5 [EXE] [SHIFT] [←] → 68°11′55″

∠CAB = 68°12′   Ans

APPLICATION

In each of the following problems a diagonal is shown within a rectangular solid.

a. Compute the true length of diagonal AB

b. Compute ∠CAB

1. Given: H = 1.500 in
           L = 2.700 in
           W = 2.000 in

   a. _____   b. _____

2. Given: H = 50.00 mm
           L = 100.00 mm
           W = 80.00 mm

   a. _____   b. _____

Use this figure for #1 and #2.

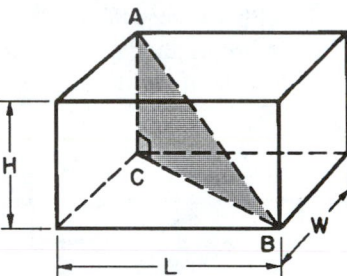

3. Given: H = 4.340 in
           L = 4.900 in
           W = 4.200 in

   a. _____   b. _____

4. Given: H = 75.00 mm
           L = 90.00 mm
           W = 70.00 mm

   a. _____   b. _____

Use this figure for #3 and #4.

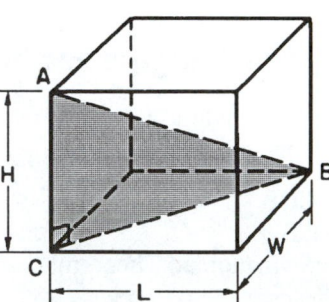

5. Given: H = 0.800 in
           L = 1.400 in
           W = 1.000 in

   a. _____   b. _____

6. Given: H = 18.00 mm
           L = 32.40 mm
           W = 25.20 mm

   a. _____   b. _____

Use this figure for #5 and #6.

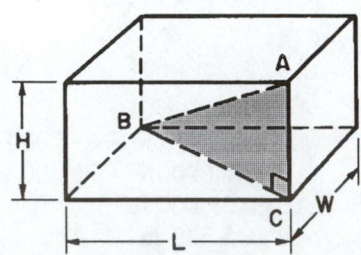

UNIT 62   Drilling and Boring Compound-Angular Holes: Computing Angles of Rotation and Tilt Using Given Lengths

Objective   After studying this unit you should be able to

- Compute the angles of rotation and angles of tilt of hole axes in given rectangular solids.

- Sketch, dimension, and label compound-angular components within rectangular solids and compute angles of rotation and angles of tilt.

Computing Angles of Rotation and Angles of Tilt for Drilling and Boring Compound-Angular Holes

A part is usually positioned on an angle plate when drilling or boring compound-angular holes. In order to position a part, the angle of rotation and the angle of tilt must be computed.

The *angle of rotation,* $\angle R$, is the angle that the piece is rotated so the hole axis is in a plane perpendicular to the pivot axis of the angle plate to which the piece is mounted.

The *angle of tilt,* $\angle T$, is the angle that the angle plate is raised to put the axis of the hole in a vertical position.

The following example shows the procedure, using given length dimensions, for finding the angle of rotation and the angle of tilt.

Example   Three views of a compound-angular hole are shown. All dimensions are in inches.

   a. Determine the angle of rotation.

   b. Determine the angle of tilt.

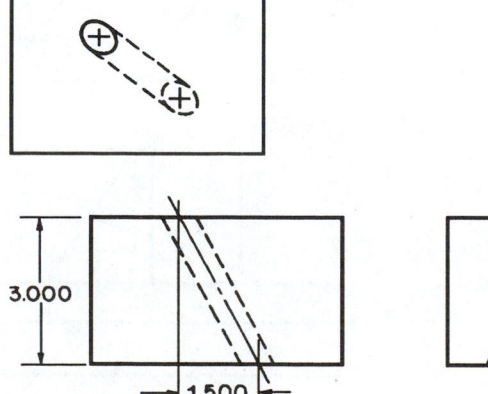
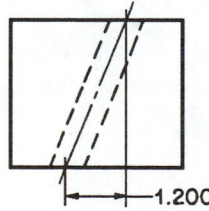

Sketch, dimension, and label a rectangular solid showing a right triangle within the solid which contains the hole axis as a side and the true angle.

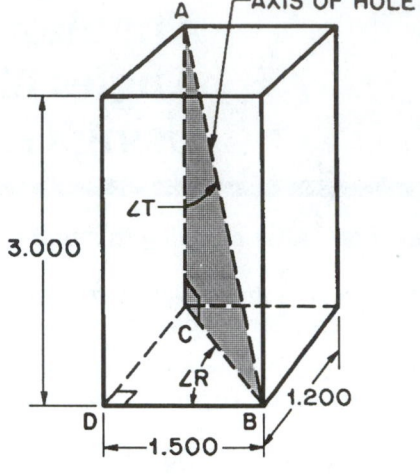

a. Compute the angle of rotation, $\angle R$. In right $\triangle BDC$:

$$\tan \angle R = \frac{DC}{DB} = \frac{1.200 \text{ in}}{1.500 \text{ in}} = 0.80000$$

$\angle R = .8$ |2nd| |tan⁻¹| |3rd| |▶DMS| $\rightarrow 38°39'35''$

or $\angle R =$ |SHIFT| |tan⁻¹| .8 |EXE| |SHIFT| |◄—| $\rightarrow 38°39'35''$

Angle of Rotation ($\angle R$) = 38°40'   Ans

b. Compute angle of tilt, $\angle T$.
   In right $\triangle BDC$:

$$\sin \angle R = \frac{DC}{CB}$$

$$\sin 38°40' = \frac{1.200 \text{ in}}{CB}$$

$$CB = \frac{1.200 \text{ in}}{\sin 38°40'}$$

CB = 1.2 |+| 38 |·| 40 |2nd| |▶DD| |sin| $\rightarrow 1.920649903$

or CB = 1.2 |+| |sin| 38 |° ' ''| 40 |° ' ''| |EXE| 1.920649903

CB = 1.92065 in

In right $\triangle ACB$:

$$\tan \angle T = \frac{CB}{AC}$$

$$\tan \angle T = \frac{1.92065 \text{ in}}{3.000 \text{ in}} = 0.64022$$

$\angle T = .64022$ |2nd| |tan⁻¹| |3rd| |▶DMS| $\rightarrow 32°37'41''$

or $\angle T =$ |SHIFT| |tan⁻¹| .64022 |EXE| |SHIFT| |◄—| $\rightarrow 32°37'41'$

Angle of Tilt ($\angle T$) = 32°38'   Ans

Procedure for Positioning the Part on an Angle Plate for Drilling

- Rotate the part to the angle of rotation, $\angle R$ as shown. Care must be taken as to whether the part is rotated to the computed $\angle R$ or the complement of $\angle R$. Rotate the part 38°40'.

  ➤ **Note:** The position of right $\triangle ACB$ is shown with hidden lines.

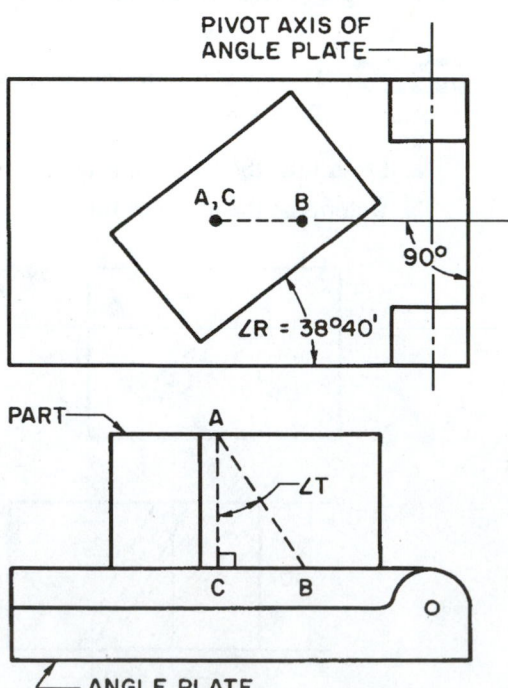

- Raise the angle plate to tilt angle, ∠T. Tilt to 32°38′ as shown. Care must be taken as to whether the part is tilted to the computed ∠T or the complement of ∠T. Observe that the position of hole axis AB is vertical.

With the part set to the angle of rotation and to the angle of tilt it is positioned to drill the hole on vertical axis AB.

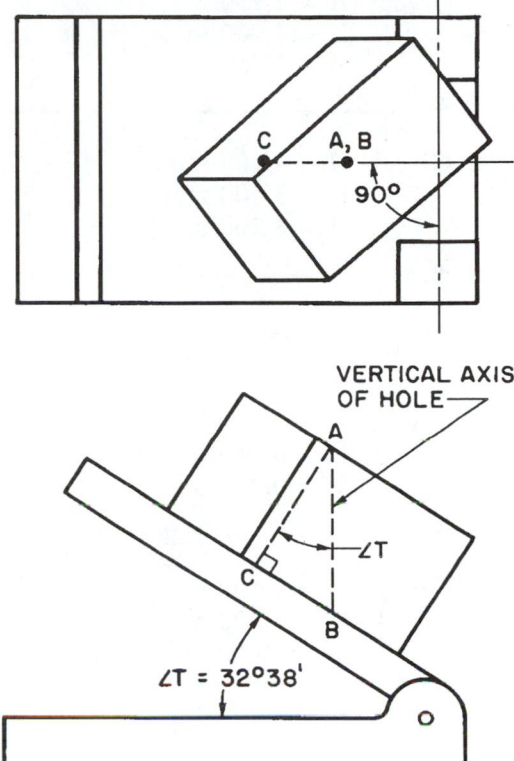

APPLICATION

In each of the following problems, the axis of a hole is shown in a rectangular solid. In order to position the hole axis for drilling, the angle of rotation and the angle of tilt must be determined. Compute angles to the nearer minute in triangles with English unit sides. Compute angles to the nearer hundredth degree in triangles with metric unit sides.

- a. Compute the angle of rotation, ∠R.
- b. Compute the angle of tilt, ∠T.

1. Given: H = 2.600 in
            L = 2.400 in
            W = 1.900 in

    a. _____   b. _____

2. Given: H = 55.00 mm
            L = 48.00 mm
            W = 30.00 mm

    a. _____   b. _____

Use this figure for #1 and #2.

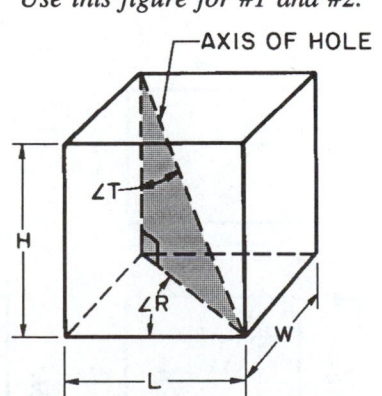

3. Given: H = 4.750 in
   L = 4.000 in
   W = 3.750 in

   a. _____   b. _____

4. Given: H = 42.00 mm
   L = 37.00 mm
   W = 32.00 mm

   a. _____   b. _____

Use this figure for #3 and #4.

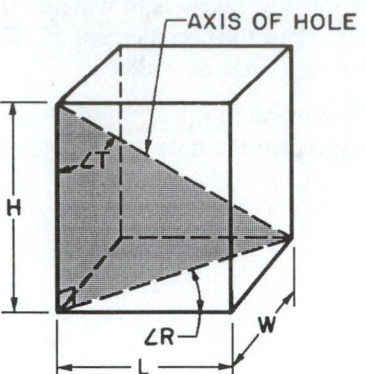

5. Given: H = 0.970 in
   L = 0.860 in
   W = 0.750 in

   a. _____   b. _____

6. Given: H = 22.00 mm
   L = 18.00 mm
   W = 15.00 mm

   a. _____   b. _____

Use this figure for #5 and #6.

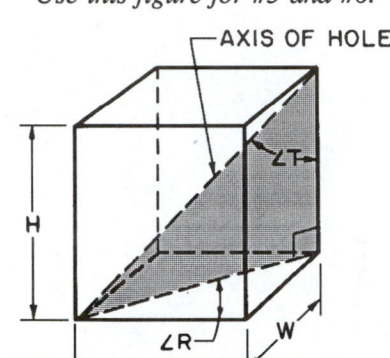

In each of the following problems, the top, front, and right side views of a compound-angular hole are shown. For each problem do the following:

a. Sketch, dimension, and label a rectangular solid. Within the solid, show the right triangle which contains the hole axis as a side and the angle of tilt. Show the position of the angle of rotation.

b. Compute the angle of rotation, ∠R.

c. Compute the angle of tilt, ∠T.

Compute angles to the nearer minute in triangles with English unit sides. Compute angles to the nearer hundredth degree in triangles with metric unit sides.

7. All dimensions are in inches.   a. *(sketch)*   8. All dimensions are in millimeters.   a. *(sketch)*

   b. _____   c. _____   b. _____   c. _____

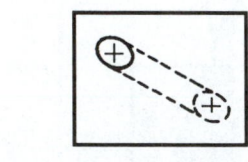

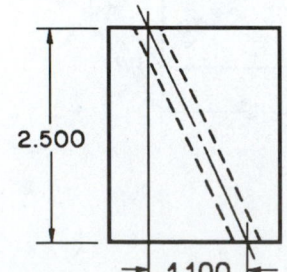

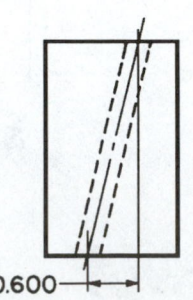

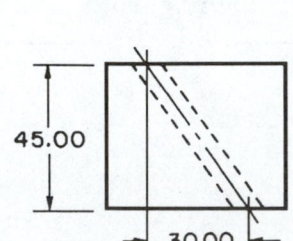

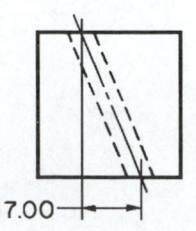

9. All dimensions are in inches. a. *(sketch)*   10. All dimensions are in millimeters. a. *(sketch)*

b. _____ c. _____   b. _____ c. _____

UNIT 63 Drilling and Boring Compound-Angular Holes: Computing Angles of Rotation and Tilt Using Given Angles

Objectives After studying this unit you should be able to

- Compute angles of rotation and angles of tilt of hole axes in rectangular solids. No length dimensions are known.

- Sketch, dimension, and label compound-angular components within rectangular solids and compute angles of rotation and angles of tilt. No length dimensions are known.

- Compute angles of rotation and angles of tilt by use of formulas.

- Compute front view and side view angles by use of formulas.

Computing Angles of Rotation and Angles of Tilt When No Length Dimensions Are Known

In certain compound angle problems no length dimensions are known; instead, angles in two different planes are known. In problems of this type where no length dimensions are known, it is necessary to assign a value of 1 (unity) to one of the sides in order to compute with trigonometric functions.

The side which is assigned a value of 1 must be a side which is common to two of the formed right triangles. One of the right triangles must have a known angle. The other right triangle must have either a known angle or an angle which is to be computed, $\angle R$ or $\angle T$.

Example   Three views of a compound-angular hole are shown. Hole angles are given in the front and right side views. No length dimensions are given.

a. Determine the angle of rotation.

b. Determine the angle of tilt.

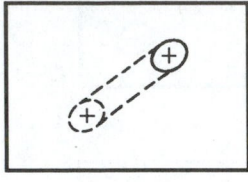

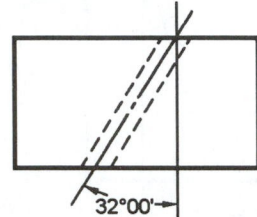 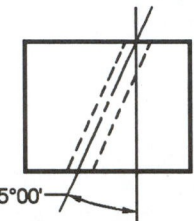

32°00′   25°00′

Sketch a rectangular solid. Project auxiliary lines which form right triangles containing the given angles, the axis of the hole, and the angles to be computed, ∠R and ∠T.

BC is a side of right △BCD which contains the given 25°00′ angle. BC is also a side of right △BCE which contains ∠R which is to be computed. Make BC = 1.

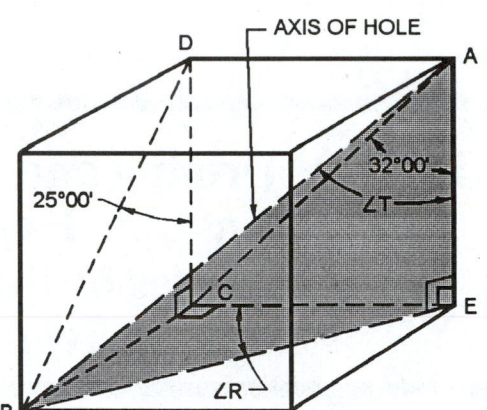

AXIS OF HOLE

a. Compute angle of rotation, ∠R.

In right △BCD: BC = 1, ∠D = 25°00′

$$\tan 25°00' = \frac{BC}{DC}$$

$$\tan 25°00' = \frac{1}{DC}, \quad DC = \frac{1}{\tan 25°00'}$$

DC = 1 [÷] 25 [tan] [=] 2.144506921

or DC = 1 [÷] [tan] 25 [EXE] 2.144506921

DC = 2.1445

In right △ACE: AE = DC = 2.1445, ∠A = 32°00′

$$\tan 32°00' = \frac{CE}{AE}$$

$$\tan 32°00' = \frac{CE}{2.1445}, \quad CE = \tan 32°00' \,(2.1445)$$

CE = 32 [tan] [×] 2.1445 [=] 1.340032325

or CE = [tan] 32 [×] 2.1445 [EXE] 1.340032325

CE = 1.3400

In right △BCE: BC = 1, CE = 1.3400

$$\tan ∠R = \frac{BC}{CE} = \frac{1}{1.3400} = 0.74627$$

∠R = .74627 [2nd] [tan⁻¹] [3rd] [▶DMS] 36°43′58″

or ∠R = [SHIFT] [tan⁻¹] .74627 [EXE] [SHIFT] [←] → 36°43′58″

∠R = 36°44′   Ans

b. Compute angle of tilt, $\angle T$.

In right $\triangle BCE$: $\angle R = 36°44'$, $BC = 1$

$$\sin 36°44' = \frac{BC}{BE}$$

$$\sin 36°44' = \frac{1}{BE}, BE = \frac{1}{\sin 36°44'}$$

$BE = 1 \boxed{\div} 36 \boxed{\cdot} 44 \boxed{2nd} \boxed{\blacktriangleright \underset{=}{DD}} \boxed{\sin} \boxed{=} 1.671984972$

or $BE = 1 \boxed{\div} \boxed{\sin} 36 \boxed{\circ\,\prime\,\prime\prime} 44 \boxed{\circ\,\prime\,\prime\prime} \boxed{EXE} 1.671984972$

$BE = 1.6720$

In right $\triangle AEB$: $AE = 2.1445$, $BE = 1.6720$

$$\tan \angle T = \frac{BE}{AE}$$

$$\tan \angle T = \frac{1.6720}{2.1445} = 0.77967$$

$\angle T = .77967 \boxed{2nd} \boxed{\tan^{-1}} \boxed{3rd} \boxed{\blacktriangleright DMS} \rightarrow 37°56'33''$

or $\angle T = \boxed{SHIFT} \boxed{\tan^{-1}} .77967 \boxed{EXE} \boxed{SHIFT} \boxed{\longleftarrow} \rightarrow 37°56'33''$

$\angle T = 37°57'$ Ans

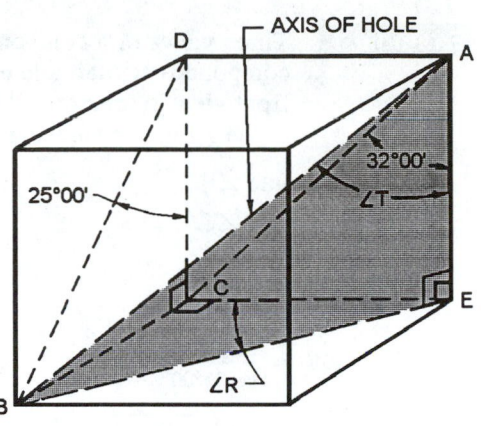

The part is rotated 36°44' ($\angle R$) on the angle plate and the angle plate is raised 37°57' ($\angle T$).

Formulas for Computing Angles of Rotation and Angles of Tilt Used in Drilling

Formulas for determining angles of rotation and angles of tilt have been computed. These formulas reduce the amount of computation required in solving compound angle problems. The formulas should **not** be used unless a problem is completely visualized and the method of solution shown in the previous example is fully understood.

To use formulas for the angles of rotation and angles of tilt, $\angle A$ and $\angle B$ shown in the figure must be identified.

$\angle A$ is the given angle in the front view (frontal plane) in relation to the vertical.

$\angle B$ is the given angle in the side view (profile plane) in relation to the vertical.

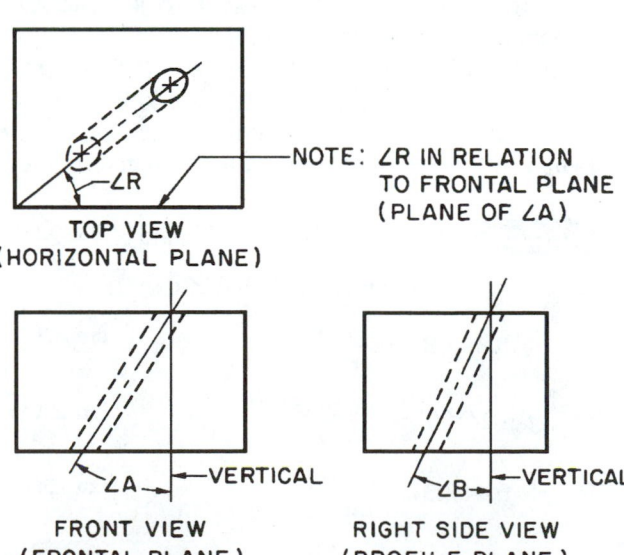

Formula for the Angle of Rotation in Relation to the Frontal Plane (Plane of $\angle A$) Used in Drilling

$$\tan \angle R = \frac{\tan \angle B}{\tan \angle A}$$

Formula for the Angle of Tilt Used in Drilling

$$\tan \angle T = \sqrt{\tan^2 \angle A + \tan^2 \angle B}$$

In using the formula given for the angle of rotation, $\angle R$ must be determined in relation to the frontal plane (plane of $\angle A$). If $\angle R$ is to be determined in relation to the profile plane (plane of $\angle B$), the complement of the computed formula $\angle R$ must be used.

Example 1   Three views of a compound-angular hole are shown. (This is the same compound-angular hole used in the previous example.) The angle in the front view in relation to the vertical is 32°00′ ($\angle A$ = 32°00′). The angle in the right side view in relation to the vertical is 25°00′ ($\angle B$ = 25°00′).

a. Compute $\angle R$.

b. Compute $\angle T$.

a. $\tan \angle R = \dfrac{\tan \angle B}{\tan \angle A}$

$\tan \angle R = \dfrac{\tan 25°00′}{\tan 32°00′}$

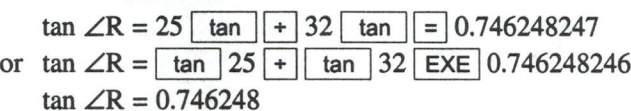

$\tan \angle R$ = 25 $\boxed{\tan}$ $\boxed{+}$ 32 $\boxed{\tan}$ $\boxed{=}$ 0.746248247

or $\tan \angle R$ = $\boxed{\tan}$ 25 $\boxed{+}$ $\boxed{\tan}$ 32 $\boxed{\text{EXE}}$ 0.746248246

$\tan \angle R$ = 0.746248

$\angle R$ = .746248 $\boxed{\text{2nd}}$ $\boxed{\tan^{-1}}$ $\boxed{\text{3rd}}$ $\boxed{\blacktriangleright\text{DMS}}$ → 36°43′55″

or $\angle R$ = $\boxed{\text{SHIFT}}$ $\boxed{\tan^{-1}}$ .746248 $\boxed{\text{EXE}}$ $\boxed{\text{SHIFT}}$ $\boxed{\longleftarrow}$

→ 36°43′55″

$\angle R$ = 36°44′   Ans

b. $\tan \angle T = \sqrt{\tan^2 \angle A + \tan^2 \angle B}$

$\tan \angle T = \sqrt{\tan^2 32°00′ + \tan^2 25°00′}$

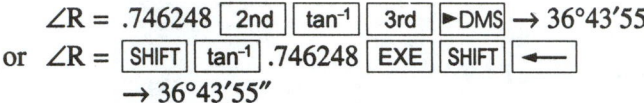

$\tan \angle T$ = $\boxed{(}$ 32 $\boxed{\tan}$ $\boxed{x^2}$ $\boxed{+}$ 25 $\boxed{\tan}$ $\boxed{x^2}$ $\boxed{)}$ $\boxed{\sqrt{x}}$ → 0.779682332

or $\tan \angle T$ = $\boxed{\sqrt{}}$ $\boxed{(}$ $\boxed{(}$ $\tan$ 32 $\boxed{)}$ x^2 $\boxed{+}$ $\boxed{(}$ $\tan$ 25 $\boxed{)}$ $\boxed{x^2}$ $\boxed{)}$

$\boxed{\text{EXE}}$ 0.779682332

$\angle T$ = .779682332 $\boxed{\text{2nd}}$ $\boxed{\tan^{-1}}$ $\boxed{\text{3rd}}$ $\boxed{\blacktriangleright\text{DMS}}$ → 37°56′34″

or $\angle T$ = $\boxed{\text{SHIFT}}$ $\boxed{\tan^{-1}}$ .779682332 $\boxed{\text{EXE}}$ $\boxed{\text{SHIFT}}$ $\boxed{\longleftarrow}$ → 37°56′34″

$\angle T$ = 37°57′   Ans

Observe that the values of $\angle R$ = 36°44′ and $\angle T$ = 37°57′ are the same as those computed in the previous example.

Occasionally a problem requires computing a front view angle when the side view angle and the angle of tilt or rotation are known. Also, it may be required to compute a side view angle in a problem when the front view angle and angle of tilt or rotation are known. The formulas for angle of rotation and tilt are used.

Example 2   Given: $\angle B$ = 20°00′, and $\angle R$ = 24°00′

Compute: $\angle A$ and $\angle T$

$\tan \angle R = \dfrac{\tan \angle B}{\tan \angle A}$ $\tan \angle T = \sqrt{\tan^2 \angle A + \tan^2 \angle B}$

$\tan 24°00′ = \dfrac{\tan 20°00′}{\tan \angle A}$ $\tan \angle T = \sqrt{\tan^2 39°16′ + \tan^2 20°}$

$0.44523 = \dfrac{0.36397}{\tan \angle A}$ $\tan \angle T = \sqrt{(0.81752)^2 + (0.36397)^2}$

$\tan \angle A = \dfrac{0.36397}{0.44523}$ $\tan \angle T = \sqrt{0.6683390 + 0.1324743}$

$\tan \angle A = 0.81749$ $\tan \angle T = \sqrt{0.8008133}$

$\angle A = 39°16′$ Ans $\tan \angle T = 0.89488$

$\angle T = 41°49′$ Ans

Example 3   Given: $\angle A = 40.00°$ and $\angle T = 42.50°$

Compute: $\angle B$ and $\angle R$

$$\tan \angle T = \sqrt{\tan^2 \angle A + \tan^2 \angle B}$$

$$\tan 42.50° = \sqrt{\tan^2 40.00° + \tan^2 \angle B}$$

$$0.91633 = \sqrt{(0.83910)^2 + \tan^2 \angle B}$$

$$0.91633^2 = 0.83910^2 + \tan^2 \angle B$$

$$\tan^2 \angle B = 0.13557$$
$$\tan \angle B = 0.36820$$

$$\angle B = 20.21° \qquad \text{Ans}$$

$$\tan \angle R = \frac{\tan \angle B}{\tan \angle A} = \frac{0.36820}{0.83910} = 0.43880$$

$$\angle R = 23.69° \qquad \text{Ans}$$

APPLICATION

For problems 1–16, compute angles to nearer minute or hundredth degree.

Computing Angles of Rotation and Tilt Without Using Drilling Formulas

In each of the following problems, 1–6, the axis of a hole is shown in a rectangular solid. In order to position the hole axis for drilling, the angle of rotation and the angle of tilt must be determined. Do **not** use drilling formulas in solving these problems.

a. Compute the angle of rotation, $\angle R$.
b. Compute the angle of tilt, $\angle T$.

1. Given: $\angle BDC = 35°00'$
            $\angle CAE = 42°00'$

a. _____   b. _____

2. Given: $\angle BDC = 27°00'$
            $\angle CAE = 33°50'$

a. _____   b. _____

Use this figure for #1 and #2.

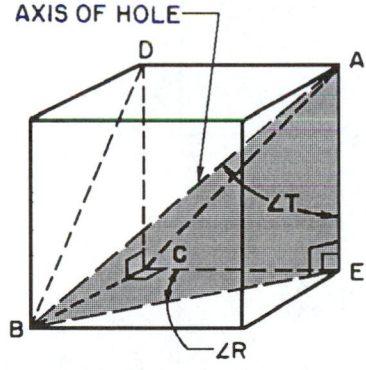

3. Given: $\angle EAC = 18.25°$
            $\angle CDB = 31.00°$

a. _____   b. _____

4. Given: $\angle EAC = 21°50'$
            $\angle CDB = 33°00'$

a. _____   b. _____

Use this figure for #3 and #4.

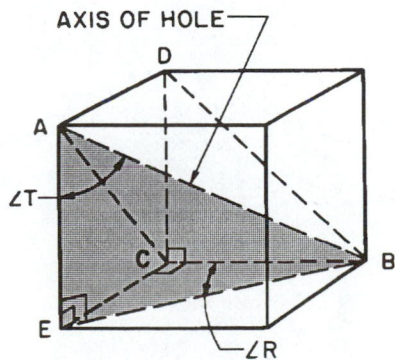

5. Given: ∠DAE = 30°00′
             ∠CAE = 42°10′

   a. _____   b. _____

6. Given: ∠DAE = 27.40°
             ∠CAE = 41.00°

   a. _____   b. _____

Use this figure for #5 and #6.

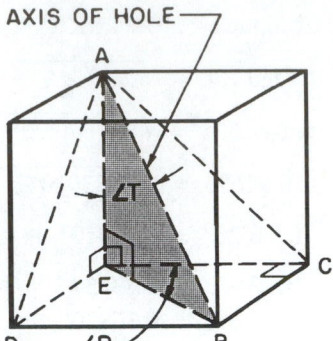

In each of the following problems, 7–10, the top, front, and right side views of a compound-angular hole are shown. Do **not** use drilling formulas in solving these problems. For each problem:

a. Sketch and label a rectangular solid. Within the solid, show the right triangle which contains the hole axis as a side and the angle of tilt. Show the position of the angle of rotation. Show the right triangles which contain the given angles.

b. Compute the angle of rotation, ∠R.

c. Compute the angle of tilt, ∠T.

7. a. *(sketch)*   8. a. *(sketch)*

    b. _____   c. _____   b. _____   c. _____

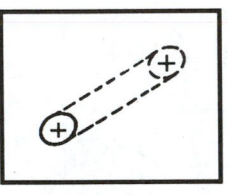

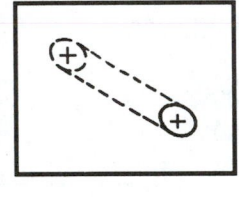

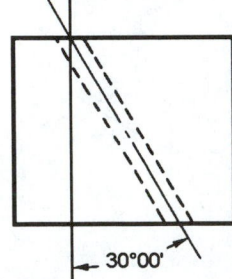

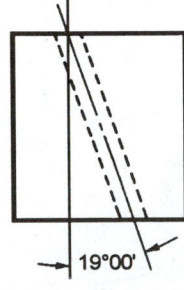

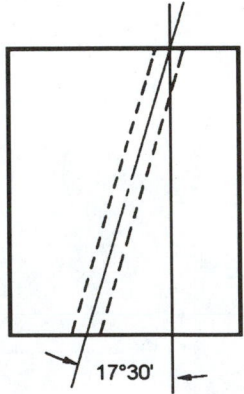

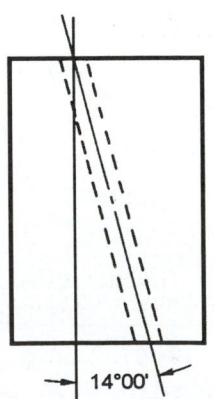

9. a. *(sketch)*

   b. _____   c. _____

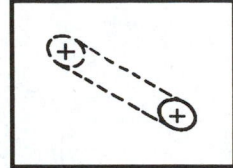

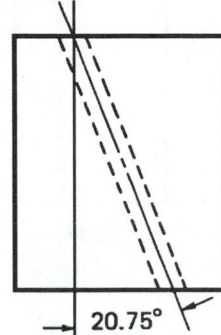 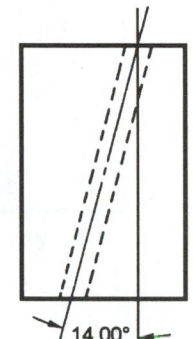

20.75°   14.00°

10. a. *(sketch)*

   b. _____   c. _____

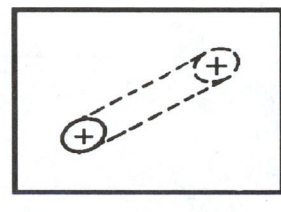

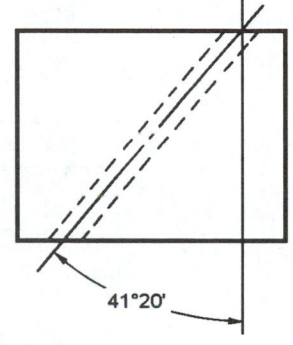 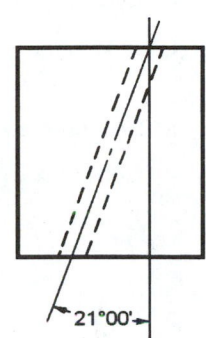

41°20'   21°00'

Computing Angles Using Drilling Formulas

In each of the following problems, the top, front, and right side views of a compound-angular hole are shown. Compute the required angles using these formulas.

$$\tan \angle R = \frac{\tan \angle B}{\tan \angle A}$$

$$\tan \angle T = \sqrt{\tan^2 \angle A + \tan^2 \angle B}$$

11. Given: $\angle A = 41°00'$
             $\angle B = 18°00'$

  a. Compute $\angle R$.   _____

  b. Compute $\angle T$.   _____

Use this figure for #11, #12, and #13.

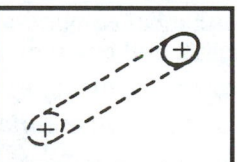

12. Given: $\angle B = 23°00'$
             $\angle R = 33°10'$

  a. Compute $\angle A$.   _____

  b. Compute $\angle T$.   _____

13. Given: $\angle A = 38.00°$
             $\angle T = 41.30°$

  a. Compute $\angle B$.   _____

  b. Compute $\angle R$.   _____

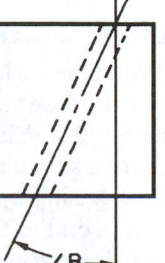

∠A   ∠B

14. Given: $\angle A = 25°00'$
$\angle B = 19°10'$

a. Compute $\angle R$. _____

b. Compute $\angle T$. _____

15. Given: $\angle B = 19.00°$
$\angle R = 32.10°$

a. Compute $\angle A$. _____

b. Compute $\angle T$. _____

16. Given: $\angle A = 23°20'$
$\angle T = 29°30'$

a. Compute $\angle B$. _____

b. Compute $\angle R$. _____

Use this figure for #14, #15, and #16.

UNIT 64 Machining Compound-Angular Surfaces: Computing Angles of Rotation and Tilt

Objectives **After studying this unit you should be able to**

- Compute angles of rotation and angles of tilt in angle plate positioning for machining compound-angular surfaces as given in rectangular solids.
- Sketch, dimension, and label compound-angular surface components within rectangular solids and compute angles of rotation and angles of tilt.
- Compute angles of rotation and angles of tilt by use of formulas.
- Compute front view and side view surface angles by use of formulas.

Machining Compound-Angular Surfaces

When the surface of a part appears as a diagonal in each of two conventional views, such as the front and right side views, setting up the part for machining involves compound angles. When just the surface (plane) must be considered in a compound angle problem, a single rotation and a single tilt are required.

The setting up of a part in a compound angle problem in which a surface (plane) and a line on the surface must both be considered is more complex. Setups of this type require single rotation and double tilt or single tilt and double rotation.

The presentation of compound-angular surfaces in this text is limited to problems which require only single rotation and single tilt. An understanding of the procedures shown will enable you to set up most compound angle surface cutting problems encountered. The procedures are also the basis for the solution of more complex compound angle problems which require double tilt or double rotation.

Example Three views of a rectangular solid block are shown in which a compound-angular surface is to be machined.

a. Determine the angle of rotation, ∠R.

b. Determine the angle of tilt, ∠T.

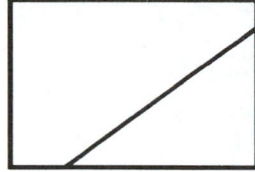

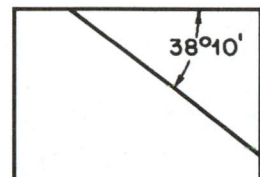

Sketch the rectangular solid and the pyramid ABCD formed by the surface ABC to be cut and the extended sides of the block. Project auxiliary lines which form right triangles containing the given angles and the angles to be computed, ∠R and ∠T.

In cutout (pyramid) ABCD, *DE must be projected perpendicular to AB*. Right △CDE contains the angle of tilt, ∠T. The angle of rotation, ∠R, is contained in right △AED. Observe that right △AED is contained in the horizontal plane and ∠R is given in reference to line AD which lies in both the horizontal and frontal planes.

Since no length dimensions are given, assign a value of 1 (unity) to a side which is common to two or more sides of the formed *right* triangles. One or more of the triangles must have a known angle. The other right triangle or triangles must have a known angle or an angle to be computed, ∠R or ∠T. Side DC is contained in the following three right triangles.

NOTE: DE MUST BE PROJECTED PERPENDICULAR TO AB

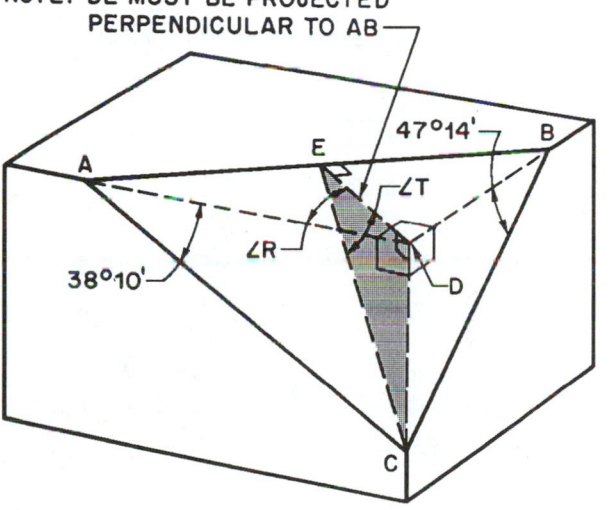

Right △ADC which contains the given angle 38°10′.

Right △BDC which contains the given angle 47°14′.

Right △CDE which contains ∠T.

Make DC = 1.

a. In right △BDC, DC = 1, ∠B = 47°14′. Compute DB.

$$\cot 47°14′ = \frac{DB}{DC}$$

$$\cot 47°14′ = \frac{DB}{1}$$

 DB = 47.14 $\boxed{\text{2nd}}$ $\boxed{\blacktriangleright\text{DD}}$ $\boxed{\tan}$ $\boxed{1/x}$ → 0.924930088

or   DB = $\boxed{\tan}$ 47 $\boxed{°\,′\,″}$ 14 $\boxed{°\,′\,″}$ $\boxed{\text{EXE}}$ $\boxed{1/x}$ or $\boxed{x^{-1}}$ → 0.924930088

DB = 0.92493

In right $\triangle ADC$, DC = 1, $\angle A = 38°10'$. Compute AD.

$$\cot 38°10' = \frac{AD}{DC}$$

$$\cot 38°10' = \frac{AD}{1}$$

AD = 38.10 $\boxed{\text{2nd}}$ $\boxed{\blacktriangleright \text{DD}}$ $\boxed{\text{tan}}$ $\boxed{1/x}$ → 1.272295718

or　AD = $\boxed{\text{tan}}$ 38 $\boxed{°\,'\,''}$ 10 $\boxed{°\,'\,''}$ $\boxed{\text{EXE}}$ $\boxed{1/x}$ or $\boxed{x^{-1}}$ → 1.272295718

AD = 1.2723

In right $\triangle ADB$, DB = 0.92493, AD = 1.2723. Compute $\angle A$.

$$\tan \angle A = \frac{DB}{AD} = \frac{0.92493}{1.2723} = 0.72697$$

$\angle A = .72697$ $\boxed{\text{2nd}}$ $\boxed{\text{tan}^{-1}}$ $\boxed{\text{3rd}}$ $\boxed{\blacktriangleright \text{DMS}}$ → 36°00'58"

or　$\angle A = \boxed{\text{SHIFT}}$ $\boxed{\text{tan}^{-1}}$ .72697 $\boxed{\text{EXE}}$ $\boxed{\text{SHIFT}}$ $\boxed{\longleftarrow}$ → 36°00'58"

$\angle A = 36°01'$

In right $\triangle AED$ compute the angle of rotation, $\angle R$.

$\angle R$ and $\angle A$ are complementary.

$\angle R = 90° - 36°01' = 53°59'$　　Ans

b.　In right $\triangle AED$, $\angle A = 36°01'$, AD = 1.2723. Compute DE.

$$\sin 36°01' = \frac{DE}{AD}$$

$$\sin 36°01' = \frac{DE}{1.2723}$$

$$DE = \sin 36°01' (1.2723)$$

DE = 36.01 $\boxed{\text{2nd}}$ $\boxed{\text{sin}}$ $\boxed{\times}$ 1.2723 $\boxed{=}$ 0.74813856

or　DE = $\boxed{\text{sin}}$ 36 $\boxed{°\,'\,''}$ 01 $\boxed{°\,'\,''}$ $\boxed{\text{EXE}}$ $\boxed{\times}$ 1.2723 $\boxed{\text{EXE}}$ 0.748138559

DE = 0.74814

In right $\triangle CDE$, CD = 1, DE = 0.74814. Compute the angle of tilt, $\angle T$.

$$\cot \angle T = \frac{DE}{DC} = \frac{0.74814}{1} = 0.74814$$

$\angle T = .74814$ $\boxed{1/x}$ $\boxed{\text{2nd}}$ $\boxed{\text{tan}^{-1}}$ $\boxed{\text{3rd}}$ $\boxed{\blacktriangleright \text{DMS}}$ → 53°11'54"

or　$\angle T = \boxed{\text{SHIFT}}$ $\boxed{\text{tan}^{-1}}$ .74814 $\boxed{1/x}$ or $\boxed{x^{-1}}$ $\boxed{\text{EXE}}$ $\boxed{\text{SHIFT}}$ $\boxed{\longleftarrow}$ → 53°11'54"

$\angle T = 53°12'$　　Ans

Procedure for Positioning the Part on an Angle Plate for Machining

- Rotate the part to the angle of rotation, $\angle R$, as shown. Care must be taken as to whether the part is rotated to the computed $\angle R$ or the complement of $\angle R$. Rotate the part 53°59′.

  ➤ **Note:** The position of right $\triangle CDE$ is shown with hidden lines.

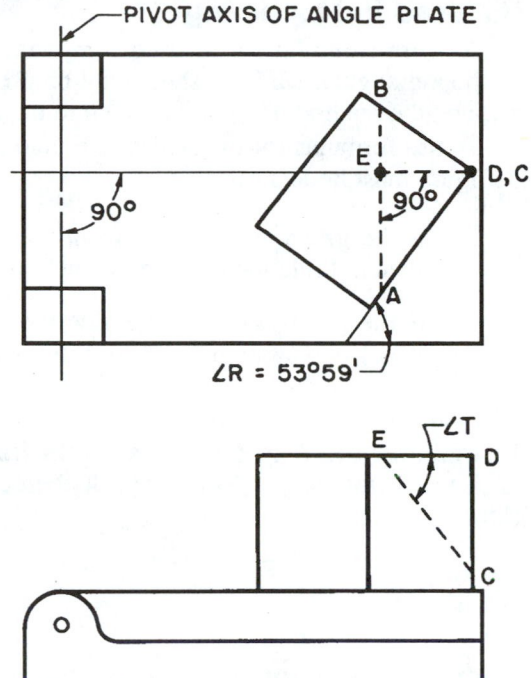

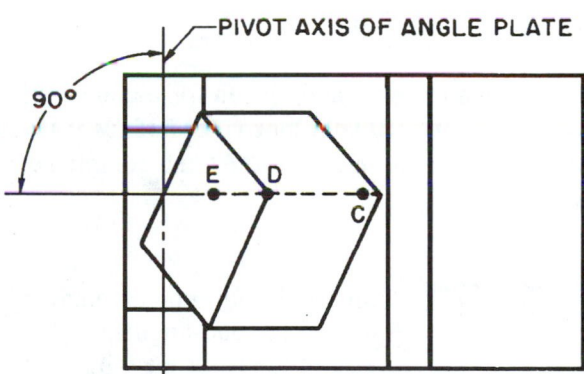

- Raise the angle plate to the tilt angle, $\angle T$. Tilt to 53°12′ as shown. Care must be taken as to whether the part is tilted to the computed $\angle T$ or the complement of $\angle T$. Observe that the position of the plane AEBC to be cut is horizontal.

With the part set to the angle of rotation and to the angle of tilt, it is positioned to machine the surface on the horizontal plane.

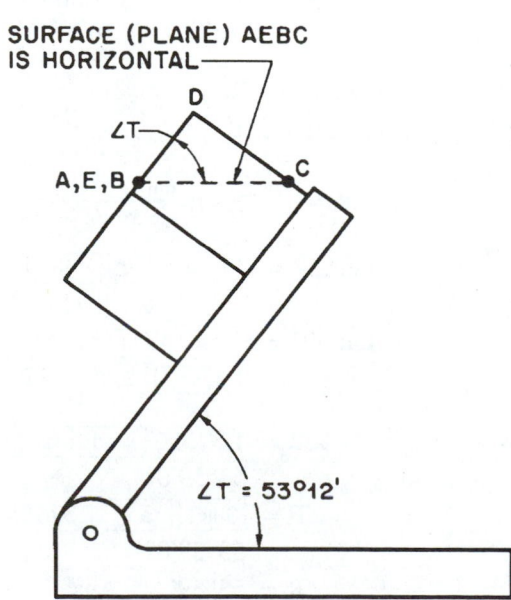

Formulas for Computing Angles of Rotation and Angles of Tilt Used in Machining

As with formulas for drilling compound-angular holes, formulas for machining compound-angular surfaces should not be used until the problem is completely visualized and the method of solution shown in the previous example is fully understood.

To use formulas for the angles of rotation and angles of tilt, ∠A and ∠B shown in the figure must be identified.

∠A is the given angle in the front view (frontal plane) in relation to the horizontal plane.

∠B is the given angle in the side view (profile plane) in relation to the horizontal plane.

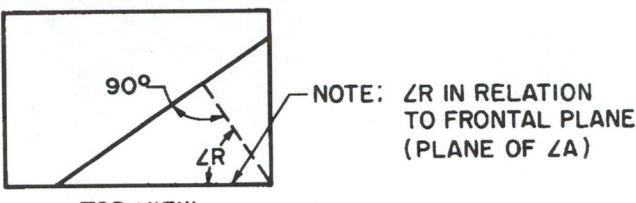

TOP VIEW
(HORIZONTAL PLANE)

Formula for the Angle of Rotation in Relation to the Frontal Plane (Plane of ∠A) Used in Machining

$$\tan \angle R = \frac{\tan \angle B}{\tan \angle A}$$

Formula for the Angle of Tilt Used in Machining

$$\tan \angle T = \frac{\tan \angle A}{\cos \angle R}$$

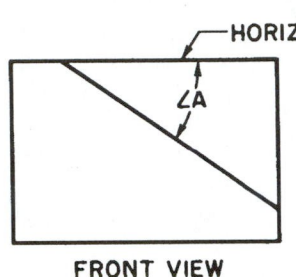

FRONT VIEW
(FRONTAL PLANE)

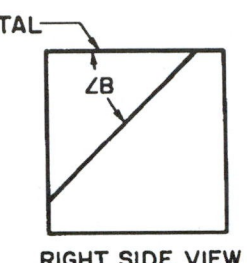

RIGHT SIDE VIEW
(PROFILE PLANE)

In using the formula given for the angle of rotation, ∠R must be determined in relation to the frontal plane (plane of ∠A). If ∠R is to be determined in relation to the profile plane (plane of ∠B), the complement of the computed formula ∠R must be used.

Example 1   Three views of a compound-angular surface are shown. (This is the same compound-angular surface used in the previous example.) The angle in the front view in relation to the horizontal is 38°10′ (∠A = 38°10′). The angle in the right side view in relation to the horizontal is 47°14′ (∠B = 47°14′).

   a. Compute ∠R.
   b. Compute ∠T.

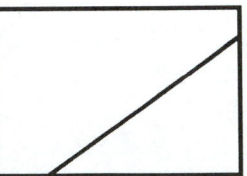

a. $\tan \angle R = \frac{\tan \angle B}{\tan \angle A} = \frac{\tan 47°14′}{\tan 38°10′}$

   tan ∠R = 47.14 [2nd] [▸DD] [tan] [+] 38.10 [2nd]
            [▸DD] [tan] [=] 1.375558795

   or tan ∠R = [tan] 47 [° ′ ″] 14 [° ′ ″] [+] [tan] 38 [° ′ ″]
              10 [° ′ ″] [EXE] 1.375558795

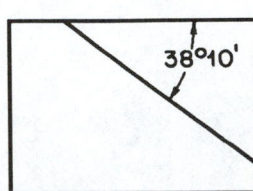

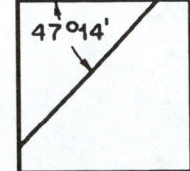

   ∠R = 1.37556 [2nd] [tan⁻¹] [3rd] [▸DMS] →
        53°59′01″

   or   ∠R = [SHIFT] [tan⁻¹] 1.37556 [EXE] [SHIFT] [◂─] →
             53°59′01″

   ∠R = 53°59′   Ans

b. $\tan \angle T = \dfrac{\tan \angle A}{\cos \angle R} = \dfrac{\tan 38°10'}{\cos 53°59'}$

$\tan \angle T = 38.10 \boxed{\text{2nd}} \boxed{\blacktriangleright\underline{\text{DD}}} \boxed{\tan} \boxed{+} 53.59 \boxed{\text{2nd}} \boxed{\blacktriangleright\underline{\text{DD}}} \boxed{\cos} \boxed{=} 1.336655293$

or $\tan \angle T = \boxed{\tan} 38 \boxed{° ' ''} 10 \boxed{° ' ''} \boxed{+} \boxed{\cos} 53 \boxed{° ' ''} 59 \boxed{° ' ''} \boxed{\text{EXE}} 1.336655293$

$\angle T = 1.33666 \boxed{\text{2nd}} \boxed{\tan^{-1}} \boxed{\text{3rd}} \boxed{\blacktriangleright\text{DMS}} \rightarrow 53°11'55''$

or $\angle T = \boxed{\text{SHIFT}} \boxed{\tan^{-1}} 1.33666 \boxed{\text{EXE}} \boxed{\text{SHIFT}} \boxed{\longleftarrow} \rightarrow 53°11'55''$

$\angle T = 53°12'$ Ans

Observe that the values of $\angle R = 53°59'$ and $\angle T = 53°12'$ are the same as those computed in the previous example.

The same formulas for angles of rotation and tilt are used to compute an unknown front view angle when a side view angle and an angle of rotation or tilt are known. An unknown side view angle may be computed if the front view angle and angle of rotation or tilt are known.

Example 2 Given: $\angle B = 18.15°$, and $\angle R = 27.45°$.
Compute: $\angle A$ and $\angle T$.

$\tan \angle R = \dfrac{\tan \angle B}{\tan \angle A}$ $\tan \angle T = \dfrac{\tan \angle A}{\cos \angle R} = \dfrac{0.63108}{0.88741} = 0.71115$

$\tan 27.45° = \dfrac{\tan 18.15°}{\tan \angle A}$ $\angle T = 35.42°$ Ans

$0.51946 = \dfrac{0.32782}{\tan \angle A}$

$\tan \angle A = \dfrac{0.32782}{0.51946}$

$\tan \angle A = 0.63108$

$\angle A = 32.25°$ Ans

APPLICATION

For problems 1–14, compute angles to the nearer minute or hundredth degree.

Computing Angles of Rotation and Tilt Without Using Machining Formulas

Three views of a rectangular solid block are shown in which a compound-angular surface is to be machined. A pictorial view of the block with auxiliary lines required for computations is also shown. Do **not** use machining formulas in solving these problems. For each of the following problems, 1–4:

a. Determine the angle of rotation, $\angle R$.

b. Determine the angle of tilt, $\angle T$.

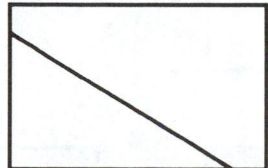

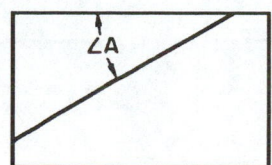

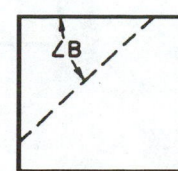

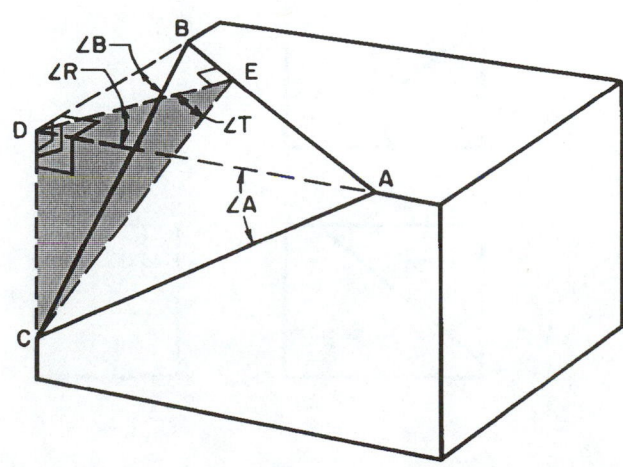

1. Given: ∠A = 32°00′
∠B = 44°00′

a. _____   b. _____

2. Given: ∠A = 27°00′
∠B = 39°00′

a. _____   b. _____

3. Given: ∠A = 18°10′
∠B = 27°50′

a. _____   b. _____

4. Given: ∠A = 23.20°
∠B = 37.10°

a. _____   b. _____

In each of the following problems, 5–8, the top, front, and right side views of a compound-angular surface are shown. Do **not** use machining formulas in solving these problems. For each problem:

   a. Sketch and label a rectangular solid and the pyramid formed by the surface to be cut and the extended sides of the block. Show the right triangles which contain ∠T and the right triangles which contain the given angles. Identify ∠T, ∠R, and the given angles.

   b. Compute the angle of rotation, ∠R.

   c. Compute the angle of tilt, ∠T.

5. a. *(sketch)*

   b. _____   c. _____

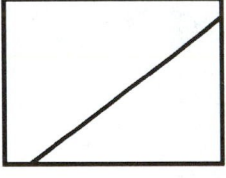
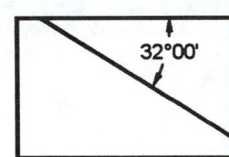
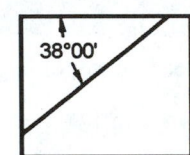

7. a. *(sketch)*

   b. _____   c. _____

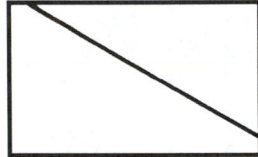
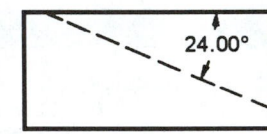
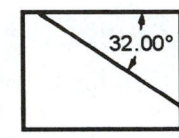

6. a. *(sketch)*

   b. _____   c. _____

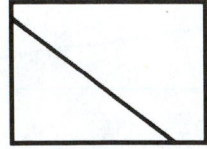
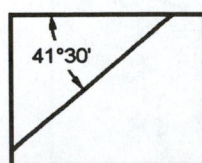
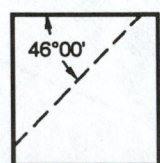

8. a. *(sketch)*

   b. _____   c. _____

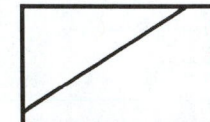
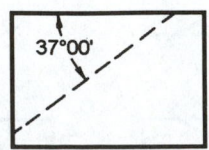
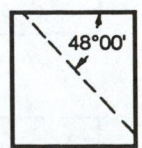

Computing Angles Using Machining Formulas

In each of the following problems, the top, front, and right side views of a compound-angular surface are shown. Compute the required angles using these formulas.

$$\tan \angle R = \frac{\tan \angle B}{\tan \angle A}$$

$$\tan \angle T = \frac{\tan \angle A}{\cos \angle R}$$

9. Given: $\angle A = 39°00'$
    $\angle B = 44°00'$

    a. Compute $\angle R$. _____

    b. Compute $\angle T$. _____

10. Given: $\angle B = 43°20'$
    $\angle R = 46°00'$

    a. Compute $\angle A$. _____

    b. Compute $\angle T$. _____

11. Given: $\angle A = 41.20°$
    $\angle T = 52.00°$

    a. Compute $\angle R$. _____

    b. Compute $\angle B$. _____

Use this figure for #9, #10, and #11.

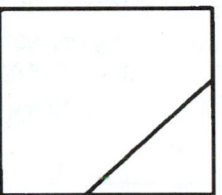

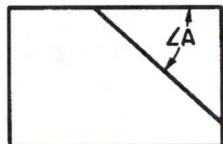 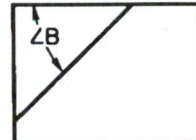

12. Given: $\angle A = 19°00'$
    $\angle B = 23°10'$

    a. Compute $\angle R$. _____

    b. Compute $\angle T$. _____

13. Given: $\angle B = 18°00'$
    $\angle R = 22°00'$

    a. Compute $\angle A$. _____

    b. Compute $\angle T$. _____

14. Given: $\angle A = 15.60°$
    $\angle T = 26.50°$

    a. Compute $\angle R$. _____

    b. Compute $\angle B$. _____

Use this figure for #12, #13, and #14.

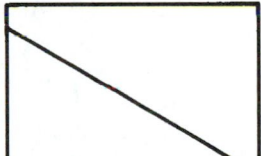

 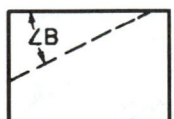

UNIT 65 Computing Angles Made by the Intersection of Two Angular Surfaces

Objectives After studying this unit you should be able to

- Compute the true angles of compound-angular edges made by the intersection of two angular surfaces as given in rectangular solids.
- Sketch and label compound-angular surface edge components within rectangular solids and compute true angles.
- Compute true angles, front view angles and side view angles by the use of formulas.

Computing Angles Made by the Intersection of Two Angular Surfaces

For design or inspection purposes it may be required to compute angles which are made by the intersection of two cut surfaces in reference to the horizontal plane.

Example Three views of a part are shown. A pictorial view of the angular portion of the part with auxiliary lines required for computations is also shown. The surfaces are to be machined in reference to the horizontal plane at angles of 32°00′ and 40°00′ as shown in the front and right side views.

a. Compute ∠R.

b. Compute ∠C.

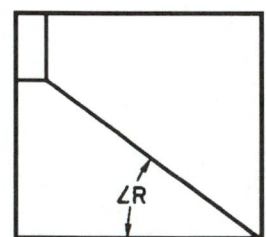
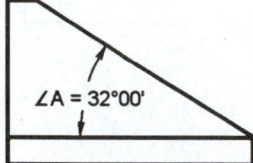
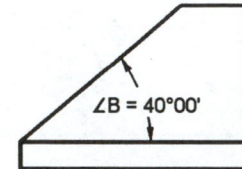
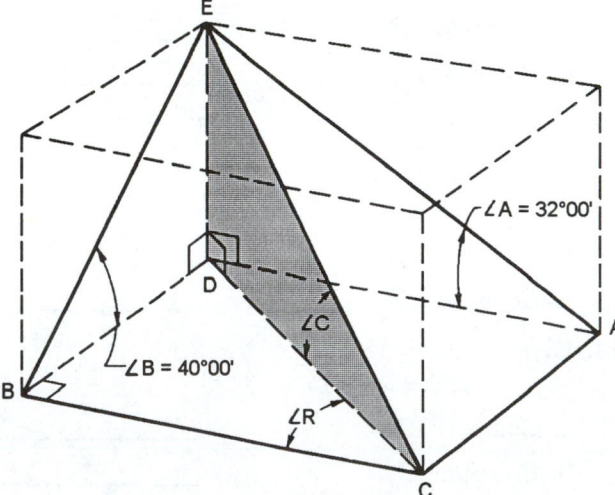

Since DE is a side of right △BDE, right △ADE, and right △CDE, make DE = 1.

a. In right △BDE, DE = 1, ∠B = 40°00′.

Compute BD:

$$\cot 40°00′ = \frac{BD}{DE}$$

$$\cot 40°00′ = \frac{BD}{1}$$

BD = 40 $\boxed{\text{tan}}$ $\boxed{1/x}$ $\rightarrow$ 1.191753593

or   BD = $\boxed{\text{tan}}$ 40 $\boxed{\text{EXE}}$ $\boxed{1/x}$ or $\boxed{x^{-1}}$ $\rightarrow$ 1.191753593

BD = 1.19175

In right △ADE, DE = 1, ∠A = 32°.

Compute DA:

$$\cot 32°00' = \frac{DA}{DE}$$

$$\cot 32°00' = \frac{DA}{1}$$

DA = 32 $\boxed{\text{tan}}$ $\boxed{1/x}$ $\rightarrow$ 1.600334529

or   DA = $\boxed{\text{tan}}$ 32 $\boxed{\text{EXE}}$ $\boxed{1/x}$ or $\boxed{x^{-1}}$ $\rightarrow$ 1.600334529

DA = 1.60033

In right △CBD, BD = 1.1918, BC = DA = 1.6003.

Compute ∠R:

$$\tan \angle R = \frac{BD}{BC}$$

$$\tan \angle R = \frac{1.19175}{1.60033}$$

$$\tan \angle R = 0.744690158$$

∠R = .74469 $\boxed{\text{2nd}}$ $\boxed{\text{tan}^{-1}}$ $\boxed{\text{3rd}}$ $\boxed{\blacktriangleright\text{DMS}}$ $\boxed{\longrightarrow}$ 36°40′29″

or   ∠R = $\boxed{\text{SHIFT}}$ $\boxed{\text{tan}^{-1}}$ .74469 $\boxed{\text{EXE}}$ $\boxed{\text{SHIFT}}$ $\boxed{\longleftarrow}$ $\rightarrow$ 36°40′29″

∠R = 36°40′   Ans

b. In right △CBD, ∠R = 36°40′, BD = 1.1918.

Compute DC:

$$\sin 36°40' = \frac{BD}{DC}$$

$$\sin 36°40' = \frac{1.1918}{DC}$$

$$DC = \frac{1.1918}{\sin 36°40'}$$

DC = 1.1918 $\boxed{+}$ 36.40 $\boxed{\text{2nd}}$ $\boxed{\blacktriangleright\overset{DD}{\underset{=}{}}}$ $\boxed{\text{sin}}$ $\boxed{=}$ 1.995784732

or   DC = 1.1918 $\boxed{+}$ $\boxed{\text{sin}}$ 36 $\boxed{°′″}$ 40 $\boxed{°′″}$ $\boxed{\text{SHIFT}}$ $\boxed{\longleftarrow}$ $\boxed{\text{EXE}}$ 1.995784732

DC = 1.9958

In right △CDE, DE = 1, DC = 1.9958.

Compute ∠C:

$$\cot \angle C = \frac{DC}{DE}$$

$$\cot \angle C = \frac{1.9958}{1}$$

∠C = 1.9958 $\boxed{1/x}$ $\boxed{\text{2nd}}$ $\boxed{\text{tan}^{-1}}$ $\boxed{\text{3rd}}$ $\boxed{\text{DMS}}$ 26°36′48″

or   ∠C = $\boxed{\text{SHIFT}}$ $\boxed{\text{tan}^{-1}}$ 1.9950 $\boxed{1/x}$ or $\boxed{x^{-1}}$ $\boxed{\text{EXE}}$ $\boxed{\text{SHIFT}}$ $\boxed{\longleftarrow}$ $\rightarrow$ 26°36′48″

∠C = 26°37′   Ans

Formulas for Computing Angles of Intersecting Angular Surfaces

Apply the formulas for intersecting angular surfaces only after a problem has been completely visualized and the previous method of solution is fully understood.

To use formulas for intersecting angular surfaces, $\angle A$ and $\angle B$ must be identified.

$\angle A$ is the given angle in the front view (frontal plane) in relation to the horizontal plane.

$\angle B$ is the given angle in the side view (profile plane) in relation to the horizontal plane.

Formula for $\angle R$ in Relation to the Frontal Plane (Plane of $\angle A$) Used for Intersecting Angular Surfaces

$$\tan \angle R = \frac{\cot \angle B}{\cot \angle A}$$

Formula for $\angle C$ Used for Intersecting Angular Surfaces

$$\cot \angle C = \sqrt{\cot^2 \angle A + \cot^2 \angle B}$$

In using the formula given for the angle of rotation, $\angle R$ must be determined in relation to the frontal plane (plane of $\angle A$). If $\angle R$ is to be determined in relation to the profile plane (plane of $\angle B$), the complement of the computed formula $\angle R$ must be used.

Example 1   Three conventional views and a pictorial view of the intersection of two angular surfaces are shown. (These are the same intersecting angular surfaces used in the previous example.)

   a. Compute $\angle R$.
   b. Compute $\angle C$.

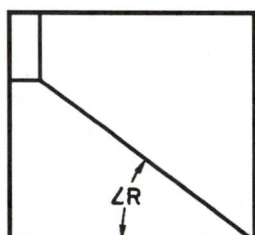
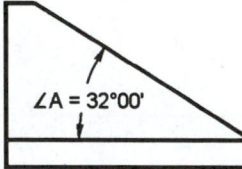
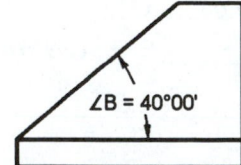
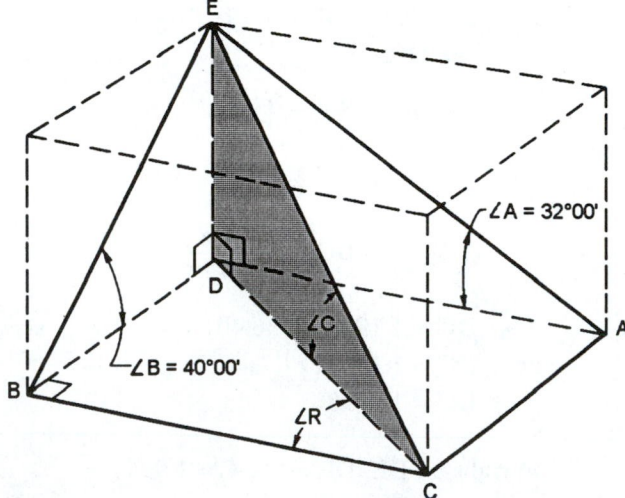

   a. $\tan \angle R = \dfrac{\cot \angle B}{\cot \angle A}$

   $\tan \angle R = \dfrac{\cot 40°00'}{\cot 32°00'}$

   $\tan \angle R =$ 40 $\boxed{\tan}$ $\boxed{1/x}$ $\boxed{+}$ 32 $\boxed{\tan}$ $\boxed{1/x}$ $\boxed{=}$ 0.744690295

   or $\tan \angle R =$ $\boxed{(}$ $\boxed{\tan}$ 40 $\boxed{)}$ $\boxed{1/x}$ or $\boxed{x^{-1}}$ $\boxed{+}$ $\boxed{(}$ $\boxed{\tan}$ 32 $\boxed{)}$
   $\boxed{1/x}$ or $\boxed{x^{-1}}$ $\boxed{EXE}$ 0.744690295

$\angle R = .744690295$ [2nd] [tan⁻¹] [3rd] [▶DMS] → 36°40′29″

or $\angle R =$ [SHIFT] [tan⁻¹] .744690295 [EXE] [SHIFT] [←] → 36°40′29″

$\angle R = 36°40'$ Ans

b. $\cot \angle C = \sqrt{\cot^2 \angle A + \cot^2 \angle B}$

$\cot \angle C = \sqrt{\cot^2 32°00' + \cot^2 40°00'}$

$\cot \angle C =$ [(] [32] [tan] [¹/ₓ] [X²] [+] [40] [tan] [¹/ₓ] [X²] [)] [√x] → 1.995331359

or $\cot \angle C =$ [√‾] [(] [(] [tan] [32] [)] [¹/ₓ] or [x⁻¹] [X²] [+] [(] [tan] [40] [)] [¹/ₓ]

or [x⁻¹] [X²] [)] [EXE] 1.995331359

$\angle C = 1.995331359$ [¹/ₓ] [2nd] [tan⁻¹] [3rd] [▶DMS] 26°37′07″

or $\angle C = 1.995331359$ [¹/ₓ] or [x⁻¹] [SHIFT] [tan⁻¹] [EXE] [SHIFT] [←] → 26°37′07″

$\angle C = 26°37'$ Ans

Observe that the values of $\angle R = 36°40'$ and $\angle C = 26°37'$ are the same as those computed in the previous example.

The same formulas for $\angle R$ and $\angle C$ may be used to compute an unknown front view angle, $\angle A$, when a side view angle, $\angle B$, and $\angle R$ or $\angle C$ are known. An unknown side view angle, $\angle B$, may be computed if the front view angle, $\angle A$, and $\angle R$ or $\angle C$ are known.

Example 2   Given: $\angle B = 35.50°$ and $\angle R = 28.30°$.

Compute: $\angle A$.

$$\tan \angle R = \frac{\cot \angle B}{\cot \angle A}$$

$$\tan 28.30° = \frac{\cot 35.50°}{\cot \angle A}$$

$$0.53844 = \frac{1.4020}{\cot \angle A}$$

$$\cot \angle A = 2.6038$$

$$\angle A = 21.01°   \text{Ans}$$

Example 3   Given: $\angle A = 23°10'$ and $\angle C = 17°40'$.

Compute: $\angle B$.

$$\cot \angle C = \sqrt{\cot^2 \angle A + \cot^2 \angle B}$$

$$\cot 17°40' = \sqrt{\cot^2 23°10' + \cot^2 \angle B}$$

$$3.1397 = \sqrt{2.3369^2 + \cot^2 \angle B}$$

$$3.1397^2 = 2.3369^2 + \cot^2 \angle B$$

$$\cot \angle B = \sqrt{4.3966}$$

$$\cot \angle B = 2.0968$$

$$\angle B = 25°30'   \text{Ans}$$

APPLICATION

For problems 1–4 compute angles to nearer minute or hundredth degree.

Computing Angles Without Using Formulas for Intersecting Angular Surfaces

Three views of a part are shown in which surfaces are to be machined in reference to the horizontal plane at ∠A and ∠B as shown in the front and right side views. A pictorial view of the angular portion of the part with auxiliary lines required for computations is also shown. Do **not** use intersecting angular surface formulas in solving these problems. For each of the following problems, 1–4:

a. Compute ∠R.

b. Compute ∠C.

1. Given: ∠A = 42°00′
            ∠B = 55°00′

    a. _____   b. _____

2. Given: ∠A = 40°00′
            ∠B = 48°00′

    a. _____   b. _____

3. Given: ∠A = 50°10′
            ∠B = 61°40′

    a. _____   b. _____

4. Given: ∠A = 43.35°
            ∠B = 52.70°

    a. _____   b. _____

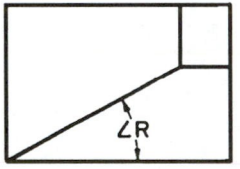

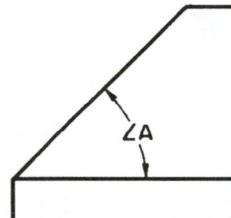
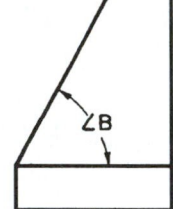

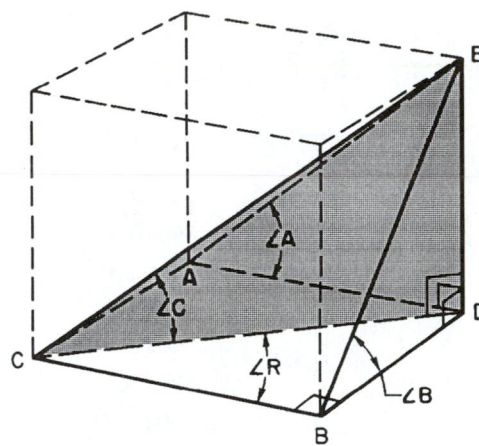

In each of the following problems, 5–8, three views of a part are shown. Two surfaces are to be machined in reference to the horizontal plane at the angles shown in the front and right side views. Do **not** use intersecting angular surface formulas in solving these problems. For each problem:

a. Sketch and label a rectangular solid and the pyramid formed by the angular surface edges. Show the right triangle which contains ∠C and the right triangles which contain the given angles and ∠R. Identify ∠C, ∠R, and the given angles.

b. Compute ∠R.

c. Compute ∠C.

5. a. *(sketch)*

   b. _____ c. _____

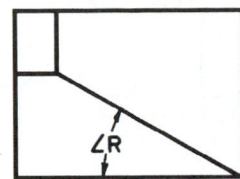

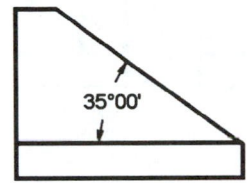 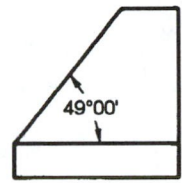

7. a. *(sketch)*

   b. _____ c. _____

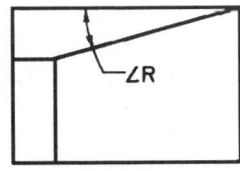

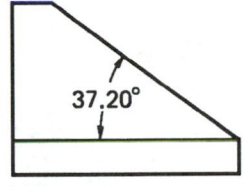 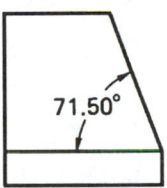

6. a. *(sketch)*

   b. _____ c. _____

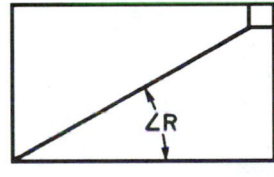

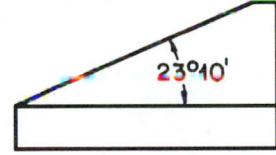 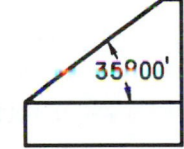

8. a. *(sketch)*

   b. _____ c. _____

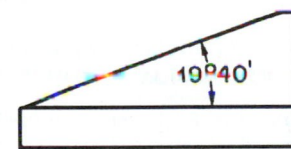 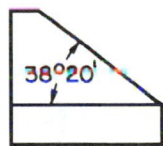

Computing Angles Using Formulas for Intersecting Angular Surfaces

In each of the following problems, three views of a part are shown. The angular surfaces are to be machined in reference to the horizontal plane at $\angle A$ and $\angle B$ as shown in the front and right side views. Compute the required angles using these formulas.

$$\tan \angle R = \frac{\cot \angle B}{\cot \angle A}$$

$$\cot \angle C = \sqrt{\cot^2 \angle A + \cot^2 \angle B}$$

Use this figure for #9, #10, and #11.

9. Given: $\angle A = 36°00'$

          $\angle B = 43°50'$

  a. Compute $\angle R$. _____

  b. Compute $\angle C$. _____

10. Given: $\angle B = 48°10'$

          $\angle R = 40°00'$

  a. Compute $\angle A$. _____

  b. Compute $\angle C$. _____

11. Given: $\angle A = 31.60°$

          $\angle C = 28.00°$

  a. Compute $\angle B$. _____

  b. Compute $\angle R$. _____

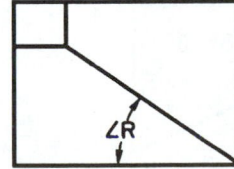

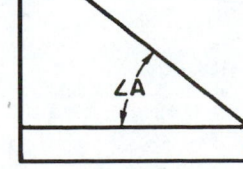 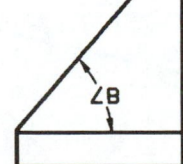

12. Given: $\angle A = 17°40'$
    $\angle B = 25°00'$

    a. Compute $\angle R$. _____

    b. Compute $\angle C$. _____

13. Given: $\angle B = 31°00'$
    $\angle R = 27°50'$

    a. Compute $\angle A$. _____

    b. Compute $\angle C$. _____

14. Given: $\angle A = 16.40°$
    $\angle C = 14.00°$

    a. Compute $\angle B$. _____

    b. Compute $\angle R$. _____

Use this figure for #12, #13, and #14.

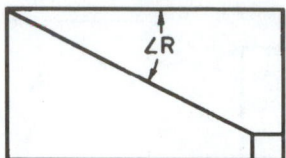

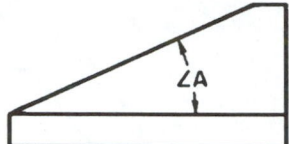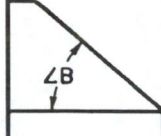

UNIT 66 Computing Compound Angles on Cutting and Forming Tools

Objectives After studying this unit you should be able to

- Compute compound angles required for cutting and forming tools as given in rectangular solids.

- Sketch and label tool angular-surface-edge components within rectangular solids and compute true angles.

- Compute true angles by use of formulas.

Computing True Angles for Cutting and Forming Tools

The following examples show methods of computing compound angles that are often required in making die sections, cutting tools and forming tools.

Example Three conventional views and a pictorial view of the angular portion of a tool are shown. Compute $\angle C$.

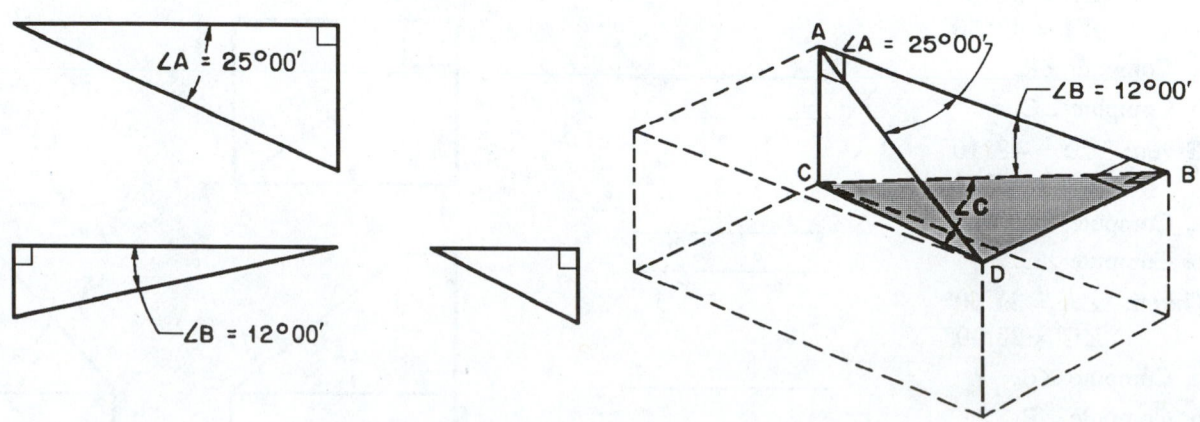

Since AB is a side of both right △ABD and right △CAB which contain given angles of 25°00′ and 12°00′, make AB = 1.

In right △ABD, AB = 1, ∠A = 25°00′. Compute DB:

$$\tan 25°00' = \frac{DB}{AB}$$

$$\tan 25°00' = \frac{DB}{1}$$

DB = 25 [tan] → 0.466307658
or DB = [tan] 25 [EXE] 0.466307658
DB = 0.46631

In right △CAB, AB = 1, ∠B = 12°00′. Compute CB:

$$\cos 12°00' = \frac{AB}{CB}$$

$$\cos 12°00' = \frac{1}{CB}$$

$$CB = \frac{1}{\cos 12°00'}$$

CB = 1 [÷] 12 [cos] [=] 1.022340595
or CB = 1 [÷] [cos] 12 [EXE] 1.022340595
CB = 1.0223

In right △CBD, DB = 0.46631, CB = 1.0223. Compute ∠C:

$$\tan \angle C = \frac{DB}{CB} = \frac{0.46631}{1.0223} = 0.45614$$

∠C = .45614 [2nd] [tan⁻¹] [3rd] [►DMS] → 24°31′11″
or ∠C = [SHIFT] [tan⁻¹] .45614 [EXE] [SHIFT] [◄—] → 24°31′11″
∠C = 24°31′   Ans

Formula for Computing ∠C Used for Cutting and Forming Tools

Apply the formula for finding ∠C used for cutting and forming tools only after a problem has been completely visualized and the previous solution is fully understood.

To use the formula for cutting and forming tools, ∠A and ∠B must be identified.

∠A is the given angle in the top view (horizontal plane) in relation to the frontal plane.

∠B is the given angle in the front view (frontal plane) in relation to the horizontal plane.

In the front view, a right angle is made with either the left or right edge and the horizontal plane.

Formula for ∠C Used for Cutting and Forming Tools

$$\tan \angle C = (\tan \angle A)(\cos \angle B)$$

Example 1   Three conventional views and a pictorial view of an angular portion of a tool are shown. Compute ∠C. (This is the same tool used in the previous example.)

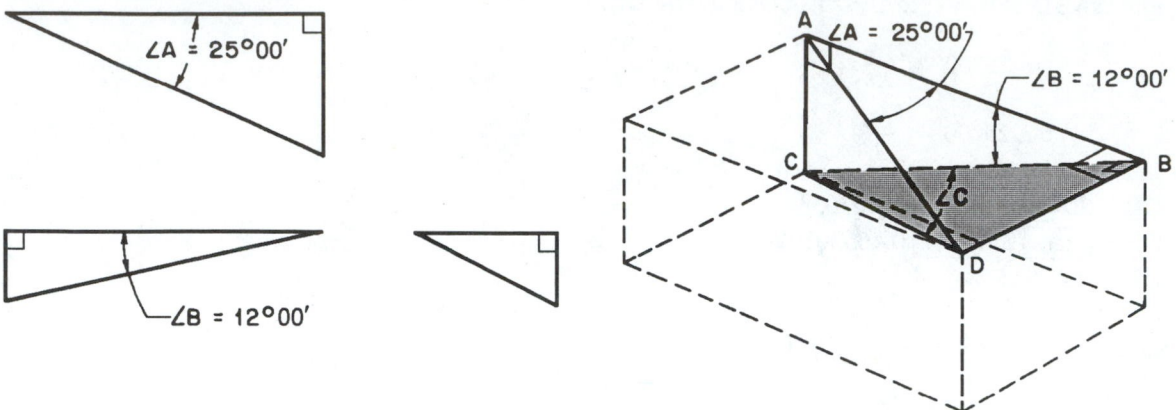

$$\tan \angle C = (\tan \angle A)(\cos \angle B)$$

$$\tan \angle C = (\tan 25°00')(\cos 12°00')$$

tan ∠C = 25 ⟦tan⟧ ⟦x⟧ 12 ⟦cos⟧ ⟦=⟧ 0.456117717

or tan ∠C = ⟦tan⟧ 25 ⟦x⟧ ⟦cos⟧ 12 ⟦EXE⟧ 0.456117717

∠C = .456117717 ⟦2nd⟧ ⟦tan⁻¹⟧ ⟦3rd⟧ ⟦►DMS⟧ → 24°31'7″

or ∠C = ⟦SHIFT⟧ ⟦tan⁻¹⟧ .456117717 ⟦EXE⟧ ⟦SHIFT⟧ ⟦◄—⟧ → 24°31'7″

∠C = 24°31'   Ans

Observe that the value of ∠C = 24°31' is the same value as computed in the previous example.

The same formula may be used to compute an unknown top view angle, ∠A, when the front view angle, ∠B, and ∠C are known. The front view angle, ∠B, may be computed when the top view angle, ∠A, and ∠C are known.

Example 2   Given: ∠B = 14.00°, ∠C = 28.75°.

Compute: ∠A.

$$\tan \angle C = (\tan \angle A)(\cos \angle B)$$

$$\tan 28.75° = (\tan \angle A)(\cos 14.00°)$$

$$0.54862 = (\tan \angle A)(0.97030)$$

$$\tan \angle A = 0.56541$$

$$\angle A = 29.48°   \text{Ans}$$

Computing True Angles in Front-Clearance-Angle Applications

The following type of compound angle problem is often found in cutting tool situations where a front clearance angle is required, such as in thread cutting.

Example   Three conventional views and a pictorial view of the angular portion of a cutting tool with a front clearance angle of 10°00' are shown. Compute ∠C.

Since AB is a side of both right △ACB and right △ABD which contain given angles of 10°00′ and 30°00′, make AB = 1.

In right △ABD, AB = 1, ∠A = 30°00′. Compute DB:

$$\tan 30°00′ = \frac{DB}{AB}$$

$$\tan 30°00′ = \frac{DB}{1}$$

DB = 30 ⟦tan⟧ → 0.577350269
or DB = ⟦tan⟧ 30 ⟦EXE⟧ 0.577350269
DB = 0.57735

In right △ACB, AB = 1, ∠B = 10°00′. Compute CB:

$$\cos 10°00′ = \frac{CB}{AB}$$

$$\cos 10°00′ = \frac{CB}{1}$$

CB = 10 ⟦cos⟧ → 0.984807753
or CB = ⟦cos⟧ 10 ⟦EXE⟧ 0.984807753
CB = 0.98481

In right △CBD, DB = 0.57735, CB = 0.98481. Compute ∠C:

$$\tan \angle C = \frac{DB}{CB} = \frac{0.57735}{0.98481}$$

tan ∠C = .57735 ⟦÷⟧ .98481 ⟦=⟧ 0.586255217
∠C = .586255217 ⟦2nd⟧ ⟦tan⁻¹⟧ ⟦3rd⟧ ⟦▸DMS⟧ → 30°22′53″
or ∠C = ⟦SHIFT⟧ ⟦tan⁻¹⟧ .586255217 ⟦EXE⟧ ⟦SHIFT⟧ ⟦◂—⟧ → 30°22′53″
∠C = 30°23′   Ans

Formula for Computing ∠C in Front-Clearance-Angle Applications

Appy the formula for finding ∠C in front-clearance-angle applications only after a problem has been completely visualized and the previous method of solution is fully understood.

To use the formula, ∠A and ∠B must be identified.

∠A is the given angle in the top view (horizontal plane) in relation to the frontal plane.

∠B is the given angle in the front view (frontal plane) in relation to the horizontal plane. ∠B is also the front clearance angle made with the vertical.

Formula for ∠C Used for Front-Clearance-Angle Applications

$$\tan \angle C = \frac{\tan \angle A}{\cos \angle B}$$

Example   Three conventional views and a pictorial view of the angular portion of a cutting tool with a front clearance angle of 10°00′ are shown. Compute ∠C. (This is the same front-clearance-angle application used on the previous example.)

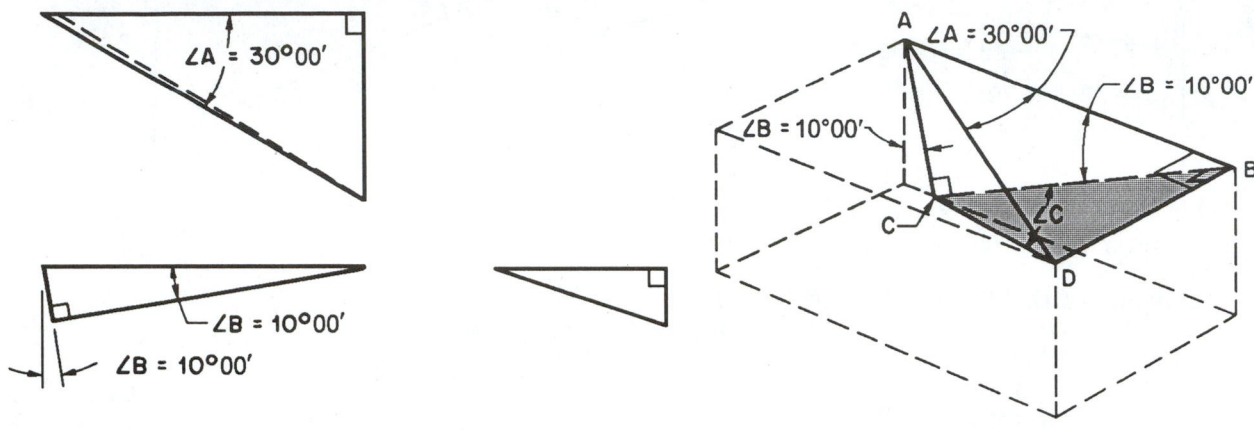

$$\tan \angle C = \frac{\tan \angle A}{\cos \angle B} = \frac{\tan 30°00'}{\cos 10°00'}$$

tan ∠C = 30 | tan | + | 10 | cos | = | 0.58625683

or tan ∠C = | tan | 30 | + | | cos | 10 | EXE | 0.58625683

∠C = .58625683 | 2nd | tan⁻¹ | 3rd | ►DMS | → 30°22′53″

or   ∠C = | SHIFT | tan⁻¹ | .58625683 | EXE | SHIFT | ← | → 30°22′53″

∠C = 30°23′   Ans

Observe that the value of ∠C = 30°23′ is the same value as computed in the previous example.

APPLICATION

For problems 1–24, compute angles to the nearer minute or hundredth decimal degree.

Computing Angles Without Using Formulas for Cutting and Forming Tools

Three views of the angular portion of a tool are shown. A pictorial view with auxiliary lines forming the right triangles which are required for computations is also shown. Do **not** use cutting and forming tool formulas in solving these problems. For each of the following problems, 1–4, compute ∠C.

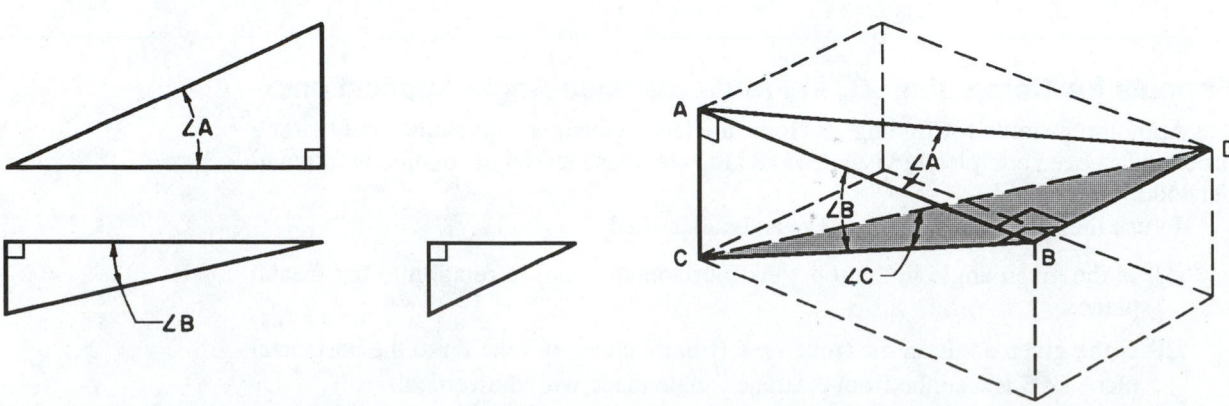

1. Given: ∠A = 30°00′
            ∠B = 15°00′ _____

2. Given: ∠A = 26°00′
            ∠B = 12°00′ _____

3. Given: ∠A = 27°00′
            ∠B = 11°00′ _____

4. Given: ∠A = 30°00′
            ∠B = 24°00′ _____

In each of the following problems, 5–8, three views of the angular portion of a tool are shown. Do **not** use cutting and forming tool formulas in solving these problems. For each problem:

a. Sketch and label a rectangular solid and the pyramid formed by the angular surface edges. Show the right triangles which contain ∠A, ∠B, and ∠C. Identify the angles.

b. Compute ∠C.

Use this figure for #5 and #6.

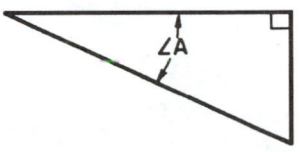

5. Given: ∠A = 24°00′
            ∠B = 15°00′
   a. *(sketch)*   b. _____

6. Given: ∠A = 20°00′
            ∠B = 8°00′
   a. *(sketch)*   b. _____

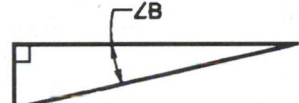

Use this figure for #7 and #8.

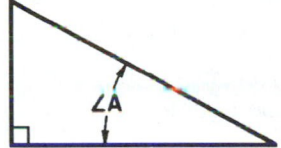

7. Given: ∠A = 30°00′
            ∠B = 12°00′
   a. *(sketch)*   b. _____

8. Given: ∠A = 23°00′
            ∠B = 10°00′
   a. *(sketch)*   b. _____

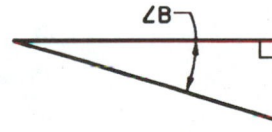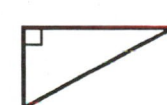

Computing Angles Using Formulas for Cutting and Forming Tools

For each of the following problems, 9–12, compute the required angle using this formula.

$$\tan \angle C = (\tan \angle A)(\cos \angle B)$$

9. Given: ∠A = 33°00′
            ∠B = 14°00′
   Compute: ∠C.   _____

Use this figure for #9 – #12.

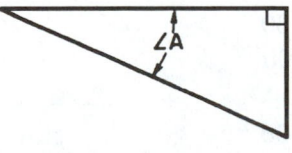

10. Given: ∠B = 10°00′
             ∠C = 28°30′
    Compute: ∠A.   _____

11. Given: ∠A = 26°00′
             ∠B = 14°00′
    Compute: ∠C.   _____

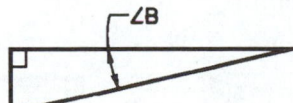

12. Given: ∠A = 28.00°
             ∠C = 27.20°
    Compute: ∠B.   _____

Computing Angles Without Using Front-Clearance-Application Formulas

Three views of the angular portion of a tool with front clearance are shown. A pictorial view with auxiliary lines forming the right triangles which are required for computations is also shown. Do **not** use front-clearance-application formulas for solving these problems. Compute ∠C for each of the following problems, 13–16.

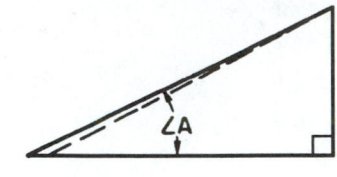
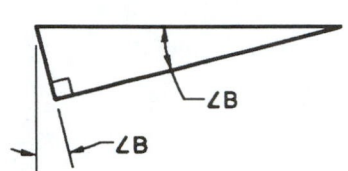
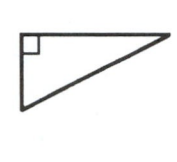
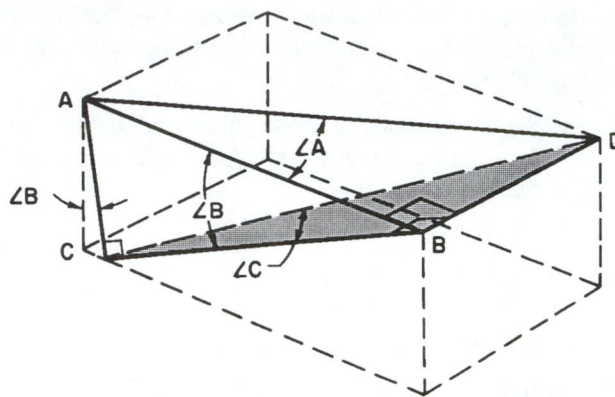

13. Given: ∠A = 30°00′
    ∠B = 10°00′   _____

14. Given: ∠A = 28°00′
    ∠B = 15°00′   _____

15. Given: ∠A = 33°00′
    ∠B = 11°00′   _____

16. Given: ∠A = 25°00′
    ∠B = 14°00′   _____

In each of the following problems, 17–20, three views of the angular portion of a tool with front clearance are shown. Do **not** use front-clearance-application formulas for solving these problems. For each problem:

a. Sketch and label a rectangular solid and the pyramid formed by the angular surface edges. Show the right triangles which contain ∠A, ∠B, and ∠C. Identify the angles.

b. Compute ∠C.

Use this figure for #17 and #18.

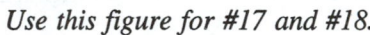
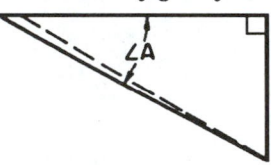

17. Given: ∠A = 30°00′
    ∠B = 15°00′
    a. *(sketch)*   b. _____

18. Given: ∠A = 38°00′
    ∠B = 12°00′
    a. *(sketch)*   b. _____

19. Given: ∠A = 32°00′
    ∠B = 15°00′
    a. *(sketch)*   b. _____

20. Given: ∠A = 25°00′
    ∠B = 12°00′
    a. *(sketch)*   b. _____

Use this figure for #19 and #20.

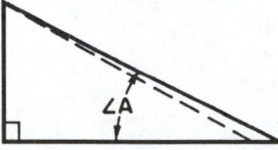
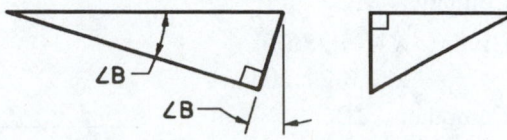

Computing Angles Using Front-Clearance-Application Formulas

For each of the following problems, 21–24, compute the required angle using this formula.

$$\tan \angle C = \frac{\tan \angle A}{\cos \angle B}$$

21. Given: $\angle A = 25°00'$
             $\angle B = 9°00'$

    Compute: $\angle C$. _____

22. Given: $\angle B = 15°00'$
             $\angle C = 30°40'$

    Compute: $\angle A$. _____

23. Given: $\angle A = 34°00'$
             $\angle B = 8°00'$

    Compute: $\angle C$. _____

24. Given: $\angle A = 28.00°$
             $\angle C = 28.60°$

    Compute: $\angle B$. _____

Use this figure for #21 – #24.

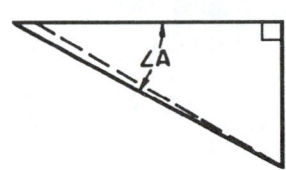

UNIT 67 Achievement Review— Section Six

Objective

You should be able to solve the problems in this Achievement Review by applying the principles and methods covered in units 61–66.

For problems 1–6, compute angles to the nearer minute or hundredth degree.

1. Three views of a compound-angular hole are shown. All dimensions are in inches.

   a. Compute the angle of rotation,
      $\angle R$. _____

   b. Compute the angle of tilt,
      $\angle T$. _____

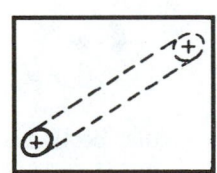

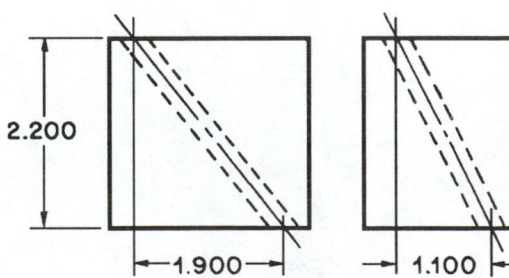

2. Three views of a compound-angular hole are shown.

   a. Compute the angle of rotation, ∠R. _____

   b. Compute the angle of tilt, ∠T. _____

37.30°   22.00°

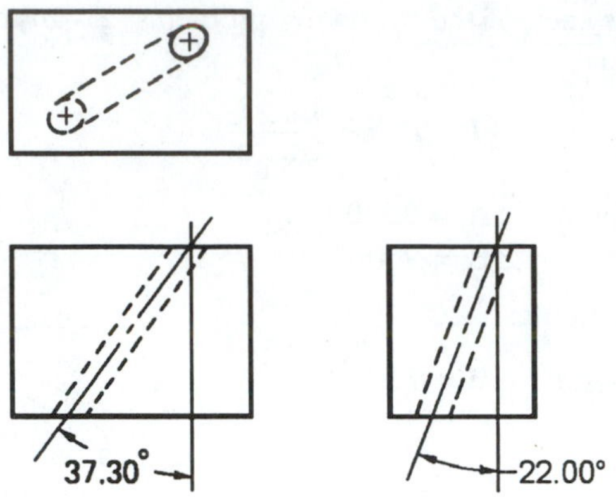

3. Three views of a rectangular solid block are shown in which a compound-angular surface is to be machined.

   a. Compute the angle of rotation, ∠R. _____

   b. Compute the angle of tilt, ∠T. _____

34.00°   44.50°

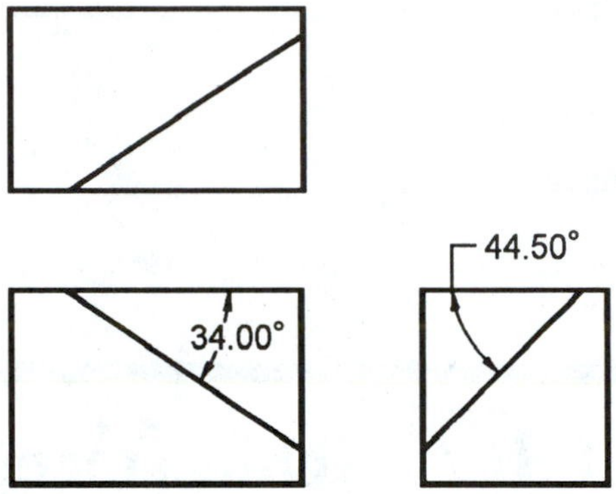

4. Three views of a part are shown. Two surfaces are to be machined in reference to the horizontal plane at the angles shown in the front and right side views.

   a. Compute ∠R. _____

   b. Compute ∠C. _____

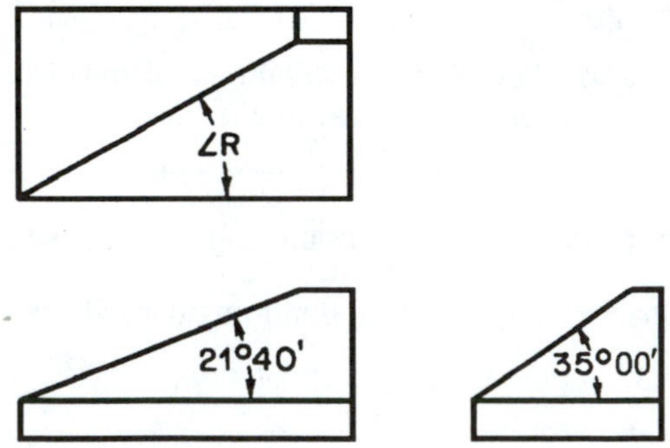

∠R   21°40′   35°00′

5. Three views of the angular portion of a tool are shown.

   Compute ∠C. _____

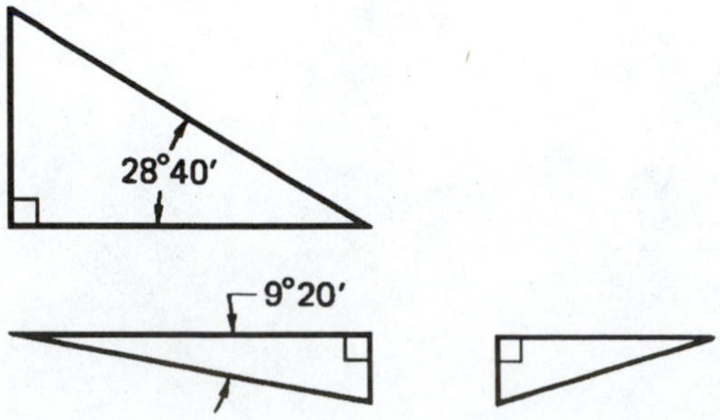

28°40′   9°20′

6. Three views of the angular portion of a tool with front clearance are shown.

   Compute ∠C.   _____

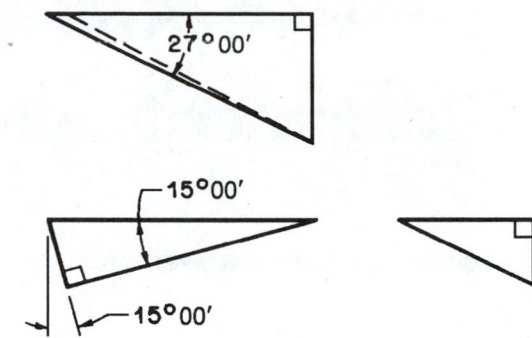

Section Seven
Computer Numerical Control (CNC)

UNIT 68 Introduction to Computer Numerical Control (CNC)

Objectives   After studying this unit you should be able to

- Locate points in a two-axis Cartesian coordinate system.
- Plot points in a two-axis Cartesian coordinate system.
- Sketch point locations in a three-axis Cartesian coordinate system.

Numerical control is the operating of a machine using numerical commands. Computer numerical control machines (CNC) are widely used in the manufacture of machined parts. CNC machines have largely replaced manually operated machines and earlier numerical control (NC) machines. Although a machinist does not usually write a program of operations, some basics of numerical control program tool locating should be understood.

CNC machines are designed for a wide range of applications. Before CNC machines, NC machine programs were usually coded on punched paper tape. The program had to be loaded into the machine control each time the program was run. Now, paper tape and magnetic tape are seldom used. Usually with CNC, the program is edited and stored in CNC control memory. Regardless of the method of instructing the CNC control, the programs can be read from the control memory.

The most common type of CNC machines are machining centers and turning centers. A machining center is a large CNC milling machine with either a vertical or a horizontal spindle. It is capable of performing multiple operations with automatic tool changers. The machines usually have from three to five axes. A turning center is a large CNC lathe capable of performing multiple operations with automatic tool changers. They have from two to four axes. Some other types of CNC machines are grinding, flame cutting, inspection, and electrical discharge machines.

Programming

A program is a complete set of instructions for tool motion and for preparatory functions such as feed rate, type of operation, and mode of operation. Auxiliary operations such as tool changes, spindle control, and coolant control are also programmed. Before the program is written, the programmer selects the machine or machines to be used and determines the operations that will be done on a machine. The programmer determines how the part is going to be held, the tooling required, and the order of operations including calculating feeds and speeds. The programmer then writes the programs with or without the assistance of the computer. CNC machines are either manually programmed or computer-assisted programmed. In manual programming, the programmer makes the mathematical calculations required in writing the program. The CNC machine computer does not perform calculations or coding with manual programming.

With computer-assisted programming, a computer performs program calculations. Computer-assisted programming uses either language-based systems or graphic-based systems. With language-based systems, geometry and tool paths are described using a specific descriptive language. Graphic-based systems are menu-driven with the part and tool path drawn on the computer screen. Selecting from a menu, part geometry and tool paths are described. Cutting locations and offsets are calculated by the computer.

Graphic-based systems are easier to operate and are more economical than language-based systems. Programs are usually executed on personal computer-based systems. Because of relative ease of operations and cost, graphic-based systems are widely in use. Language-based systems are seldom used.

Location of Points: Two-Axis Cartesian Coordinate System

Programming is based on locating points within the Cartesian coordinate system, which is discussed in unit 58. In a plane, a point can be located from a fixed point by two dimensions. For example, a point can be located by stating that it is three units up and five units to the right of a fixed point. In machine technology applications, generally the units are either inches or millimeters. The Cartesian coordinate system gives point locations by using positive and negative values rather than locations stated as being up or down and left or right from a fixed point.

The figure shows a two-axis Cartesian coordinate system with an x-axis and a y-axis. On a milling machine the x and y axes are perpendicular to the spindle. A point is located in reference to the origin by giving the point an x and y value. The x value is always given first. The x and y values are called the *coordinates* of the point. The following examples locate points in the Cartesian coordinate system shown.

Example 1 Locate point A which has coordinates of (3, 5).

The x value is +3 units and the y value is +5 units. Therefore, point A is located in Quadrant I.

Example 2 Locate point B which has coordinates of (−6, 4).

The x value is −6 units and the y value is +4 units. Therefore, point B is located in Quadrant II.

Example 3 Locate point C which has coordinates of (−7, −3).

The x value is −7 units and the y value is −3 units. Therefore, point C is located in Quadrant III.

Example 4 Locate point D which has coordinates of (2, −5).

The x value is +2 units and the y value is −5 units. Therefore, point D is located in Quadrant IV.

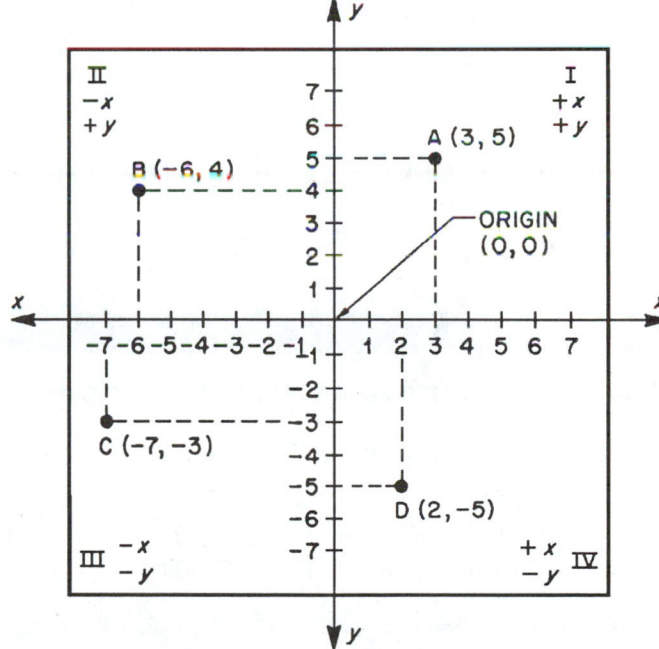

Two-Axis Cartesian Coordinate System

Location of Points: Three-Axis Cartesian Coordinate System

Figure (A) shows a three-axis Cartesian coordinate system in which a point is considered to be in space. A point is located from a fixed point by three dimensions, x, y, and z. The x-axis and y-axis are identical to that of the two-axis coordinate system. The z-axis is perpendicular to the x and y axes. Most systems consider the z value as a positive value if it is in an upward direction from the origin. On a milling machine, the z-axis is parallel to the spindle; a z-location determines the depth of cut. On a lathe, two axes are used, the x-axis and z-axis. The x-axis is perpendicular to the spindle and determines part diameters. As with milling machines, the z-axis is parallel to the spindle and determines part lengths.

In the three-axis coordinate system, the x value is given first, the y value second, and the z value third. The x, y, and z values are the coordinates of the point. Figure (B) shows two points, point A (7, 5, 3) and point B (– 6.8, –3.5, –2.2) in a three-axis coordinate system. This figure illustrates the x, y, and z locations on a vertical spindle milling machine. In machine technology applications, the x, y, and z units are usually either inches or millimeters.

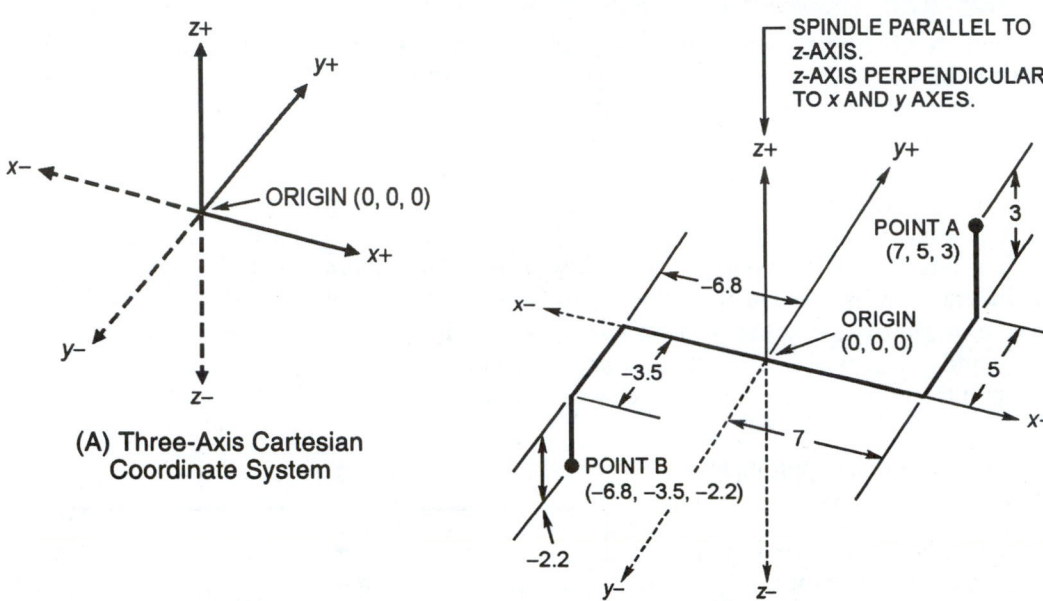

(A) Three-Axis Cartesian Coordinate System

(B) Point Locations on a Three-Axis Cartesian Coordinate System

APPLICATION

Plotting Points: Two-Axis Cartesian Coordinate System

1. Using graph paper, plot the following coordinates.

A = (–2, 5)	D = (0, 3)	G = (6, –8)	J = (0, 0)
B = (2, 8)	E = (–4, 0)	H = (–7, 5)	K = (–3, –4)
C = (–7, –2)	F = (–2, –2)	I = (–1, 0)	L = (9, –3)

2. Graph the following points: (–5, –5), (–3, –3), (0, 0), (2, 2), (4, 4), (7, 7). Connect these points.

   a. What kind of geometric figure is formed?   _____

   b. What is the value of the angle formed in reference to the x-axis?   _____

3. Graph the following points. Connect these points in the order that they are given. What kind of a geometric figure is formed?   _____

Point 1: (–9, –7)	Point 6: (7, 2.5)	Point 11: (–2, 7)
Point 2: (–6, –5.3)	Point 7: (6, 3)	Point 12: (–3, 5)
Point 3: (–3, –3.5)	Point 8: (4, 4)	Point 13: (–5, 1)
Point 4: (1, –1)	Point 9: (2, 5)	Point 14: (–6.5, –2)
Point 5: (4.5, 1)	Point 10: (0, 6)	Point 15: (–8, –5)

Coordinates of Points: Two-Axis Cartesian Coordinate System

4. Refer to the points plotted on the illustrated Cartesian Coordinate plane. Give coordinates of the following points.

A _____   S _____
B _____   T _____
C _____   U _____
D _____   V _____
E _____   5 _____
F _____   6 _____
G _____   7 _____
H _____   8 _____
J _____   9 _____
K _____   10 _____
L _____   12 _____
M _____   13 _____
N _____   14 _____
P _____   15 _____
Q _____   16 _____
R _____   17 _____

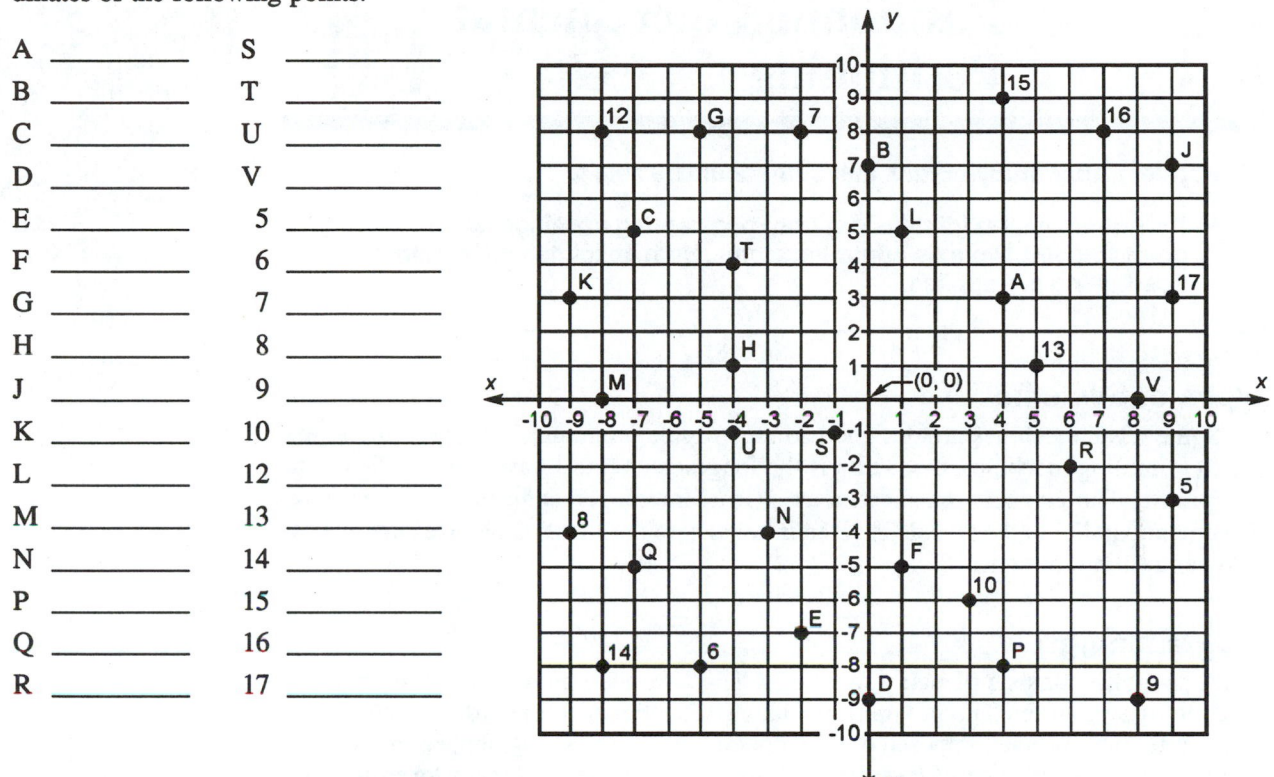

Sketching Point Locations: Three-Axis Cartesian Coordinate System

5. Make sketches similar to the three-axis Cartesian coordinate system shown in Figure (B). For each of the following points, A–F, show the approximate distances of x, y, and z and the approximate locations of the points (x, y, and z coordinates) as shown in Figure (B). The coordinates can be considered as either inch or millimeter units with approximate distances sketched to any scale.

Point A: (6, 3, 2)   Point C: (−7, 5, 3)   Point E: (8.3, −7.6, 2.6)
Point B: (−5, −2, 2)   Point D: (−5.5, 3.6, −1.8)   Point F: (6.4, 4.4, −3.8)

UNIT 69 Control Systems, Absolute Positioning, Incremental Positioning

Objective After studying this unit you should be able to

- **Program position (dimension) from engineering drawings using point-to-point two-axis control systems. Both absolute and incremental positioning are applied.**

Types of Systems

Some CNC machines are designed so they can be programmed for as many as six axes. On milling machines, the x-axis is the longest axis (most travel) perpendicular to the spindle. The y-axis is the shortest axis (least travel) perpendicular to the spindle. Movement parallel to the spindle is given in relation to the z-axis. Other axes involve rotation and tilting motions.

Programming

A program consists of many functions. The x, y, and z movements are only one function of a program. The program includes all functions required to machine a part. Using a definite format composed of all numbers and letters, a programmer writes commands such as sequencing of tools and cutting speeds and feeds. Preparatory functions are coded, such as modal commands specifying inch and metric units and absolute or incremental positioning. Miscellaneous functions such as tool changes, turning the spindle on and off, and calling for coolant are also coded.

An extensive study of CNC programming is required to write a complete manual or computer-assisted program. In this text, the purpose of presenting programming is to provide a very basic understanding of tool locations. The topic is limited to basic principles of location programming for vertical spindle milling machines controlling only x-axis and y-axis motions. The z-axis motions and all other functions are not considered.

Control systems are either continuous path or point-to-point. Machining centers are continuous path machines. They are capable of linear and circular interpolation. With linear interpolation the motions of two or more axes are coordinated with each other for angular milling cuts. Circular interpolation is the coordination of axes to give path in cutting an arc.

Some milling machines are point-to-point machines. They are usually restricted to drilling, boring, and non-angular milling. The continuous path system is more complex than the point-to-point system. Continuous path programming is also more complex than point-to-point programming. The part to be machined is positioned on the machine table. The movement of a CNC milling machine as it machines holes in a part is similar to conventional machinery. As previously stated, the purpose of the simple program applications in this book is to provide only a very basic understanding of programming tool locations. An in-depth presentation of programming is a study within itself and requires a textbook dealing exclusively with programming.

Tool Positioning (Coordinate) Systems

Most machine controls can operate with both incremental and absolute positioning commands. By means of a code, the program tells the control the type of positioning to be used. Most programs are written with absolute positioning. Absolute positioning, also called absolute coordinates, always directs the control where tool locations are relative to the origin (program zero point). Incremental positioning, also called incremental coordinates, always directs the control where the tool is located from the tool's immediate previous location.

Absolute Positioning (Absolute Coordinates)

Tool locations (coordinates) are given from a reference point called the origin or zero point. The origin is the point where the x, y, and z coordinates all are zero (0, 0, 0). The location of the origin is determined by the programmer. The origin is often located at the corner of a part or the center of a hole. To repeat, absolute positioning always directs the control where the tool locations are relative to the origin (program zero point). The following examples show point-to-point systems using absolute positioning with two axes on a vertical spindle milling machine.

Example 1   The figure shows a part as it is location dimensioned on an engineering drawing before programming for CNC. All dimensions are in inches. The hole locations (x and y coordinates) are to be programmed using the absolute positioning system. The bottom left edge of the workpiece is established as the origin (0, 0).

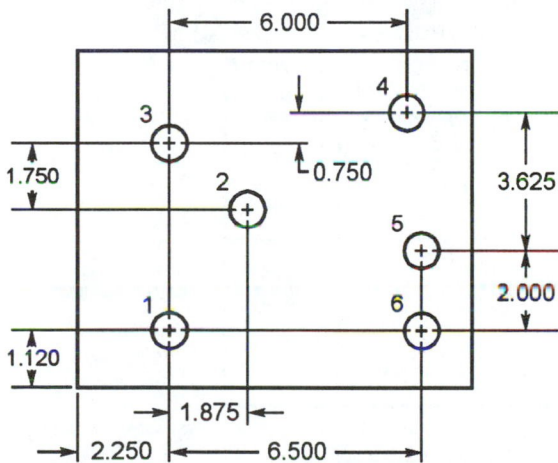

The position of the workpiece origin and the machine table are shown in the following figure.

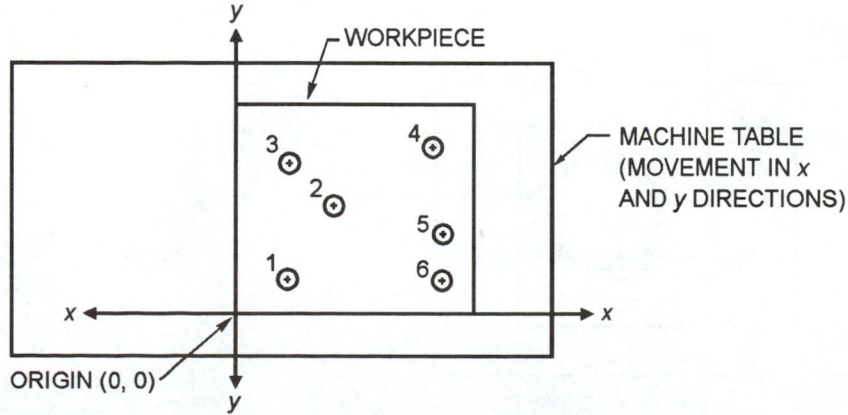

All the hole locations are programmed from the origin. Generally, the order in which the holes are machined is that which requires the least amount of machine movement. The coordinates of the hole locations from (0, 0) are listed in the following table.

Hole 1: $x = 2.250''$
$y = 1.120''$

Hole 2: $x = 2.250'' + 1.875'' = 4.125''$
$y = 1.120'' + 2.000'' + 3.625'' - 0.750'' - 1.750'' = 4.245''$

Hole 3: $x = 2.250''$
$y = 4.245'' + 1.750'' = 5.995''$

Hole 4: $x = 2.250'' + 6.000'' = 8.250''$
$y = 5.995'' + 0.750'' = 6.745''$

Hole 5: $x = 2.250'' + 6.500'' = 8.750''$
$y = 1.120'' + 2.000'' = 3.120''$

Hole 6: $x = 8.750''$
$y = 1.120''$

Hole	x	y
1	2.250″	1.120″
2	4.125″	4.245″
3	2.250″	5.995″
4	8.250″	6.745″
5	8.750″	3.120″
6	8.750″	1.120″

Example 2 The figure shows a part as it is location dimensioned before programming for CNC. All dimensions are in millimeters. The hole locations (x and y coordinates) are to be programmed using the absolute positioning system. The center of the bored hole is established as the origin (0, 0).

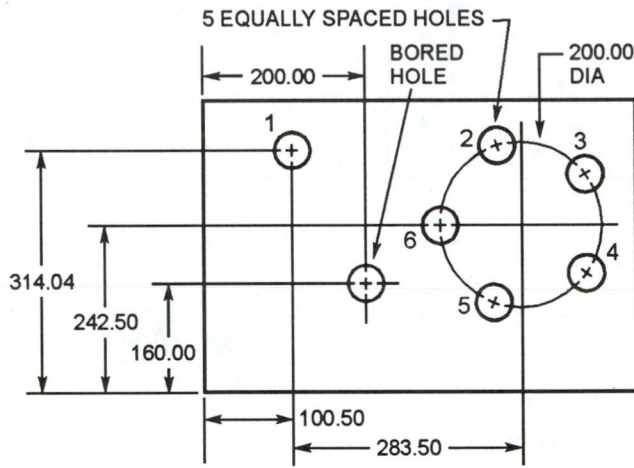

The position of the workpiece origin and the machine table are shown in the following figure.

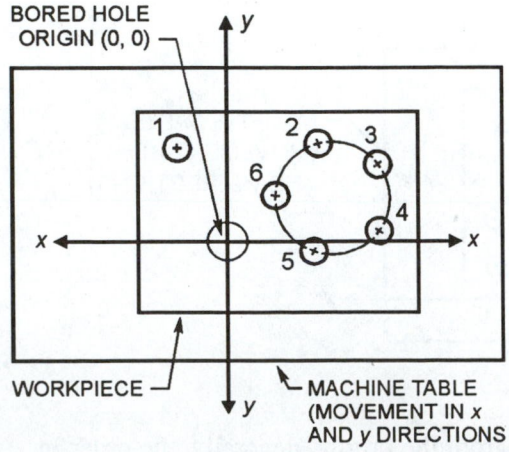

All hole locations are programmed from the origin. The coordinates of the hole locations from (0, 0) are listed in the table following **Hole 6** on page 436.

Hole 1: $x = -200.00$ mm $+ 100.50$ mm $= -99.50$ mm
$y = 314.04$ mm $- 160.00$ mm $= 154.04$ mm

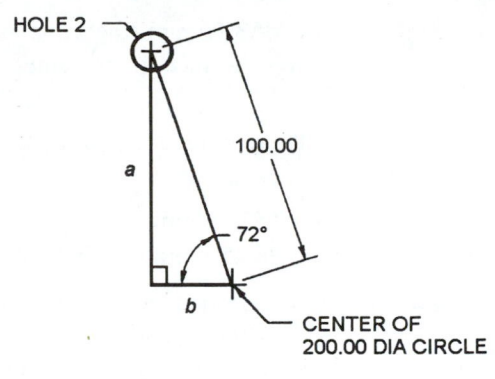

Hole 2: $x = x$ distance to the center of the 200.00-mm
diameter circle $- b$
$y = y$ distance to the center of the 200.00-mm
diameter circle $+ a$

Calculate the number of degrees between two consecutive
holes on the 200.00-mm diameter circle.

$$\frac{360°}{5} = 72°$$

From the center of the 200.00-mm diameter circle calculate a and b dimensions.

$$\sin 72° = \frac{a}{100.00 \text{ mm}} \qquad \cos 72° = \frac{b}{100.00 \text{ mm}}$$
$$0.95106 = \frac{a}{100.00 \text{ mm}} \qquad 0.30902 = \frac{b}{100.00 \text{ mm}}$$
$$a = 95.11 \text{ mm} \qquad b = 30.90 \text{ mm}$$

$x = -200.00$ mm $+ 100.50$ mm $+ 283.50$ mm $- 30.90$ mm $= 153.10$ mm
$y = 242.50$ mm $- 160.00$ mm $+ 95.11$ mm $= 177.61$ mm

Hole 3: $x = x$ distance to the center of the 200.00-mm diameter
circle $+ b$
$y = y$ distance to the center of the 200.00-mm diameter
circle $+ a$

From the center of the 200.00-mm diameter circle calculate the
angle formed by the horizontal centerline and Hole 3.

$$180° - 2(72°) = 36°$$

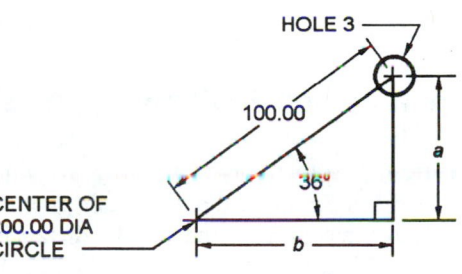

Calculate a and b dimensions.

$$\sin 36° = \frac{a}{100.00 \text{ mm}} \qquad \cos 36° = \frac{b}{100.00 \text{ mm}}$$
$$0.58779 = \frac{a}{100.00 \text{ mm}} \qquad 0.80902 = \frac{b}{100.00 \text{ mm}}$$
$$a = 58.78 \text{ mm} \qquad b = 80.90 \text{ mm}$$

$x = -200.00$ mm $+ 100.50$ mm $+ 283.50$ mm $+ 80.90$ mm $= 264.90$ mm
$y = 242.50$ mm $- 160.00$ mm $+ 58.78$ mm $= 141.28$ mm

Hole 4: $x = 264.90$ mm (the same as x of Hole 3)
$y = y$ distance to the center of the 200.00-mm
diameter circle $- a$

From the center of the 200.00-mm diameter circle calculate the
angle formed by the horizontal centerline and Hole 4.

$$3(72°) - 180° = 36°$$

Since both Hole 4 and Hole 3 are projected 36° from the horizontal, the a and b
dimensions of Hole 4 are the same as Hole 3.

$x = 264.90$ mm
$y = 242.50$ mm $- 160.00$ mm $- 58.78$ mm $= 23.72$ mm

Hole 5: $x = 153.10$ mm (the same as x of Hole 2)

$y = y$ distance to the center of the 200.00-mm diameter circle $- a$

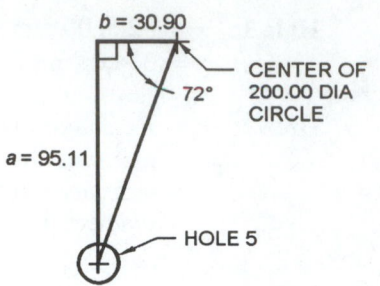

$b = 30.90$

$a = 95.11$

72°

CENTER OF 200.00 DIA CIRCLE

HOLE 5

Since both Hole 5 and Hole 2 are projected 72° from the horizontal, the a and b dimensions of Hole 5 are the same as Hole 2.

$x = 153.10$ mm

$y = 242.50$ mm $- 160.00$ mm $- 95.11$ mm $= -12.61$ mm

Hole 6: $x = x$ distance to the center of the 200.00-mm diameter circle $- 100.00$ mm

$x = -200.00$ mm $+ 100.50$ mm $+ 283.50$ mm $- 100.00$ mm $= 84.00$ mm

$y = 242.50$ mm $- 160.00$ mm $= 82.50$ mm

This table lists the coordinates of the hole locations from (0, 0).

Hole	x	y
1	−99.50 mm	154.04 mm
2	153.10 mm	177.61 mm
3	264.90 mm	141.28 mm
4	264.90 mm	23.72 mm
5	153.10 mm	−12.61 mm
6	84.00 mm	82.50 mm

Incremental Positioning (Incremental Coordinates)

In incremental positioning, each location is given from the immediate previous location. The location of a hole is considered the origin (0, 0) of the x and y axes. From this origin, x and y distances are given to the next hole. Each new location in turn becomes the origin for the x and y distances to the next hole. The direction of travel, positive and negative, must be noted and is based upon the Cartesian coordinate system just as it was with absolute positioning. The first hole is located from the established first origin, while each subsequent hole is located from the hole directly preceding it. Each hole becomes the origin for the next hole to be machined.

Example The adjacent figure shows a part as it is location dimensioned before programming for CNC. All dimensions are in inches. The hole locations (x and y coordinates) are to be programmed using the incremental positioning system. The bottom left edge is the established first origin (0, 0). (This same part was used to illustrate absolute positioning).

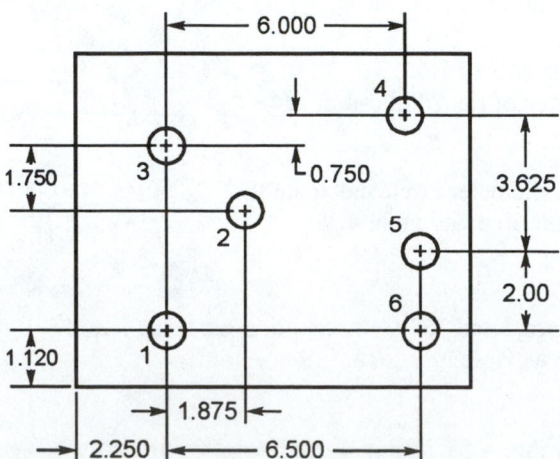

The position of the workpiece origin and the machine table are shown in the following figure.

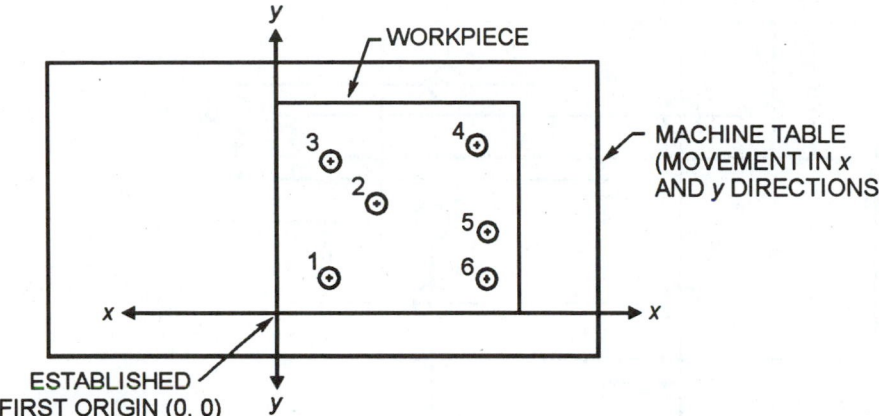

The first hole is located from the established first origin. Then each hole is the origin for the next hole to be machined. The x and y locations for Hole 1 are identical to those using absolute positioning. This is true for the first hole only. The following table lists the coordinates using incremental positioning.

Hole 1: $x = 2.250''$
$ y = 1.120''$

Hole 2: $x = 1.875''$
$ y = 2.000'' + 3.625'' - 0.750'' - 1.750'' = 3.125''$

Hole 3: $x = -1.875''$
$ y = 1.750''$

Hole 4: $x = 6.000''$
$ y = 0.750''$

Hole 5: $x = 6.500'' - 6.000'' = 0.500''$
$ y = -3.625''$

Hole 6: $x = 0''$
$ y = -2.000''$

Hole	x	y
1	2.250″	1.120″
2	1.875″	3.125″
3	−1.875″	1.750″
4	6.000″	0.750″
5	0.500″	−3.625″
6	0″	−2.000″

APPLICATION

Program Absolute and Incremental Positioning

Program the hole locations of the following part drawings. The location dimensions given in the tables are taken from drawings before programming for CNC. The origins (0, 0) used for programming are shown on the drawings. Use the hole location dimensions in the tables to write program hole locations. Write the program hole locations (coordinates) in table form listing the holes in sequence similar to the tables in this unit using:

a. Absolute Positioning

b. Incremental Positioning

Use this figure for #1, #2, and #3.

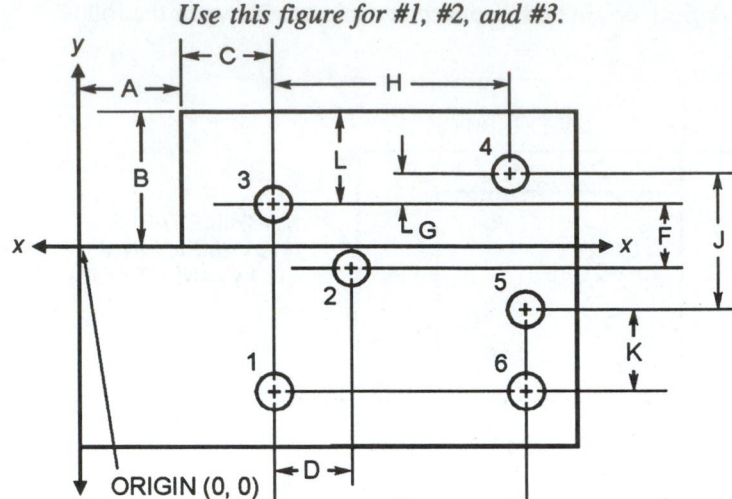

LOCATION DIMENSIONS

	A	B	C	D	E	F	G	H	J	K	L
1*	4.000	5.000	2.400	2.000	6.600	1.725	0.800	6.050	3.750	2.100	2.320
2*	6.000	5.000	2.710	2.615	7.010	2.070	0.920	6.475	4.307	2.416	2.300
3**	100.00	100.00	60.00	56.24	148.06	45.06	20.16	132.32	92.24	52.30	51.00

* All dimensions are in inches.
** All dimensions are in millimeters.

Use this figure for #4, #5, and #6.

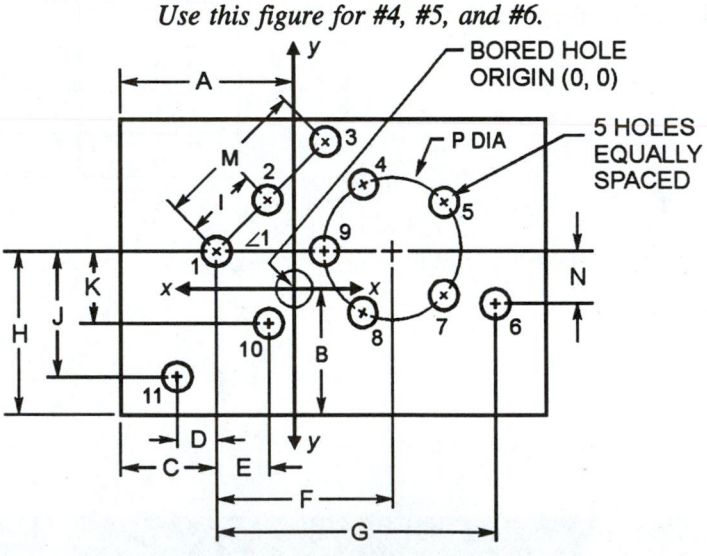

LOCATION DIMENSIONS

	A	B	C	D	E	F	G	H	J	K	L	M
4*	10.000	8.000	5.175	1.300	3.250	14.250	22.100	12.500	9.150	5.150	5.500	11.250
5*	10.000	8.000	5.250	1.412	3.562	14.400	22.250	12.750	9.375	5.270	5.600	11.300
6**	170.00	130.00	122.40	30.00	70.00	300.40	450.00	249.30	187.30	108.70	104.00	228.00

	N	P DIA	∠1
4*	4.625	10.000	42°0'
5*	4.850	10.200	43°0'
6**	95.10	196.00	41.75°

* All dimensions are in inches.
** All dimensions are in millimeters.

Use this figure for #7, #8, and #9.

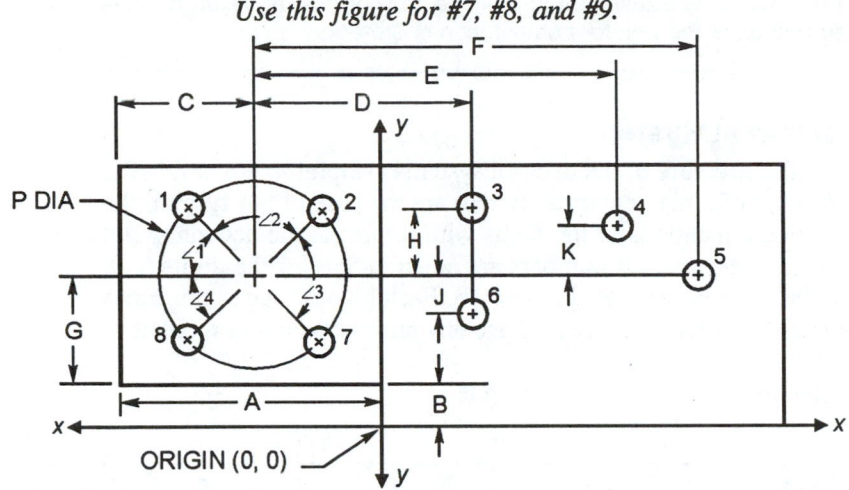

					LOCATION DIMENSIONS									
	A	**B**	**C**	**D**	**E**	**F**	**G**	**H**	**J**	**K**	**P DIA**	**∠1**	**∠2**	**∠3**	**∠4**
7*	18.000	5.000	10.185	13.700	19.215	26.750	7.500	5.750	3.170	4.250	12.200	75°45′	55°30′	95°15′	20°10′
8*	19.000	5.000	10.520	14.020	19.570	27.380	7.615	5.912	2.602	4.508	12.400	77°10′	57°15′	93°25′	15°0′
9**	400.00	80.00	196.30	255.00	378.34	521.40	142.50	106.80	59.50	79.68	224.00	72.67°	61.50°	98.83°	18.33°

* All dimensions are in inches.
** All dimensions are in millimeters.

UNIT 70 Binary Numeration System

Objectives After studying this unit you should be able to

- **Express binary numbers as decimal numbers.**
- **Express decimal numbers as binary numbers.**

The mathematical system which uses only the digits 0 and 1 is called the *binary numeration system*. The two symbols, zeros and ones, are the base of any digital computer from personal computers to mainframe computers. The binary numeration system is fundamental to all electronic computers regardless of their size or purpose.

On early NC machines, program data was transferred to either punched paper tape or magnetic tape. Holes were punched in the paper tape in rows and columns. The tape was fed through a tape reader which converted the tape codes to electrical signals. A hole punched in the tape (binary digit one) signaled an open circuit. The absence of a hole in the tape (binary digit zero) signaled a closed circuit. Part of a simplified binary-decimal system tape is shown in the figure immediately preceding the Application portion of this unit. Early NC control systems which used tape commands have been replaced by CNC controls. A few tape command NC control systems are still in use.

The computer of all CNC machines is based on the binary system as are computer aided drafting/design (CAD), computer aided manufacturing (CAM) and computer integrated manufacturing (CIM). The microscopic electronic switches in a computer's central processor assume only two states, ON (binary system one) or OFF (binary system zero). The switches are called transistors. If no charge is applied, current cannot flow and the transistor is OFF. If a positive charge is applied, the transistor is turned ON. An integrated circuit, usually a silicon chip, is made up of thousands of transistors. The smallest unit of information in a computer, which is equivalent to a single zero or one, is called a bit. A sequence of bits, called a byte, is the unit for computation or storage.

Structure of the Decimal System

An understanding of the structure of the decimal system is helpful in discussing the binary system. The elements of a mathematical system are the base of the system, the particular digits used, and the locations of the digits with respect to the decimal point (place value). In the decimal system, all numbers are combinations of the digits 0–9. The decimal system is built on powers of the base 10. Each place value is ten times greater than the place value directly to its right. Since any number with an exponent of 0 equals 1, 10^0 equals 1.

An analysis of the number 64,216 shows this structure.

6	4	2	1	6	Number
$10^4 = 10,000$	$10^3 = 1000$	$10^2 = 100$	$10^1 = 10$	$10^0 = 1$	Place Value
$6 \times 10^4 =$ $6 \times 10,000 =$ $60,000$	$4 \times 10^3 =$ $4 \times 1000 =$ 4000	$2 \times 10^2 =$ $2 \times 100 =$ 200	$1 \times 10^1 =$ $1 \times 10 =$ 10	$6 \times 10^0 =$ $6 \times 1 =$ 6	Value
60,000 +	4000 +	200 +	10 +	6 =	64,216

Examples   Analyze the following numbers.

1. $16 = 1(10^1) + 6(10^0) = 10 + 6$ Ans

2. $216 = 2(10^2) + 1(10^1) + 6(10^0) = 200 + 10 + 6$ Ans

3. $4216 = 4(10^3) + 2(10^2) + 1(10^1) + 6(10^0) = 4000 + 200 + 10 + 6$ Ans

4. $64,216 = 6(10^4) + 4(10^3) + 2(10^2) + 1(10^1) + 6(10^0) = 60,000 + 4000 + 200 + 10 + 6$ Ans

The same principles of structure hold true for numbers that are less than one. A number less than one can be expressed by using negative exponents. A number with a negative exponent is equal to its positive reciprocal. When the number is inverted and the negative exponent changed to a positive exponent, the result is as follows.

$$10^{-1} = \frac{1}{10^1} = 0.1$$

$$10^{-2} = \frac{1}{10^2} = \frac{1}{100} = 0.01$$

$$10^{-3} = \frac{1}{10^3} = \frac{1}{1000} = 0.001$$

$$10^{-4} = \frac{1}{10^4} = \frac{1}{10,000} = 0.0001$$

An analysis of the number 0.8502 shows this structure.

• 8	5	0	2	Number
$10^{-1} = 0.1$	$10^{-2} = 0.01$	$10^{-3} = 0.001$	$10^{-4} = 0.0001$	Place Value
$8 \times 10^{-1} =$ $8 \times 0.1 =$ 0.8	$5 \times 10^{-2} =$ $5 \times 0.01 =$ 0.05	$0 \times 10^{-3} =$ $0 \times 0.001 =$ 0	$2 \times 10^{-4} =$ $2 \times 0.0001 =$ 0.0002	Value
0.8 +	0.05 +	0 +	0.0002 =	0.8502

Structure of the Binary System

The same principles of structure apply to the binary system as to the decimal system. The binary system is built upon the base 2 and uses only the digits 0 and 1. Numbers are shown as binary numbers by putting a 2 to the right and below the number (subscript) as shown; 11_2, 100_2, 1_2, 10001_2 are binary numbers. As with the decimal system, the elements which must be considered are the base, the particular digits used, and the place value of the digits. The binary system is built on the powers of the base 2, each place value is twice as large as the place value directly to its right.

Place Values of Binary Numbers											
2^6	2^5	2^4	2^3	2^2	2^1	2^0		2^{-1}	2^{-2}	2^{-3}	2^{-4}
64	32	16	8	4	2	1	$\bullet$	0.5	0.25	0.125	0.0625

Expressing Binary Numbers as Decimal Numbers

Numbers in the decimal system are usually shown without a subscript. It is understood the number is in the decimal system. In certain instances, for clarity, decimal numbers are shown with the subscript 10. The following examples show the method of expressing binary numbers as equivalent decimal numbers. Remember that 0 and 1 are the only digits in the binary system.

Examples   Express each binary number as an equivalent decimal number.

1. $11_2 = 1(2^1) + 1(2^0) = 2 + 1 = 3_{10}$ Ans

2. $111_2 = 1(2^2) + 1(2^1) + 1(2^0) = 4 + 2 + 1 = 7_{10}$ Ans

3. $11101_2 = 1(2^4) + 1(2^3) + 1(2^2) + 0(2^1) + 1(2^0)$
$= 16 + 8 + 4 + 0 + 1 = 29_{10}$ Ans

4. $101.11_2 = 1(2^2) + 0(2^1) + 1(2^0) + 1(2^{-1}) + 1(2^{-2})$
$= 4 + 0 + 1 + 0.5 + 0.25 = 5.75_{10}$ Ans

Expressing Decimal Numbers as Binary Numbers

The following examples show the method of expressing decimal numbers as equivalent binary numbers.

Example 1   Express 25_{10} as an equivalent binary number.

Determine the largest power of 2 in 25; $2^4 = 16$. There is one 2^4.
Subtract 16 from 25;

$$25 - 16 = 9.$$

Determine the largest power of 2 in 9; $2^3 = 8$. There is one 2^3.
Subtract 8 from 9;

$$9 - 8 = 1.$$

Determine the largest power of 2 in 1; $2^0 = 1$. There is one 2^0.
Subtract 1 from 1;

$$1 - 1 = 0.$$

There are no 2^2 and 2^1. The place positions for these values must be shown as zeros.

$$25_{10} = 1(2^4) + 1(2^3) + 0(2^2) + 0(2^1) + 1(2^0)$$
$$25_{10} = \quad 1 \qquad 1 \qquad 0 \qquad 0 \qquad 1$$
$$25_{10} = 11001_2 \quad \text{Ans}$$

Example 2   Express 11.625_{10} as an equivalent binary number.

$$2^3 = 8; \ 11.625 - 8 = 3.625$$
$$2^1 = 2; \ 3.625 - 2 = 1.625$$
$$2^0 = 1; \ 1.625 - 1 = 0.625$$
$$2^{-1} = 0.5; \ 0.625 - 0.5 = 0.125$$
$$2^{-3} = 0.125; \ 0.125 - 0.125 = 0$$

There are no 2^2 and 2^{-2}.

$$11.625_{10} = 1(2^3) + 0(2^2) + 1(2^1) + 1(2^0)\bullet + 1(2^{-1}) + 0(2^{-2}) + 1(2^{-3})$$
$$11.625_{10} = \quad 1 \qquad 0 \qquad 1 \qquad 1 \ \bullet \ 1 \qquad 0 \qquad 1$$
$$11.625_{10} = 1011.101_2 \quad \text{Ans}$$

Part of a simplified binary-decimal system tape in the vertical form of an early NC control system for the decimal number 243 is shown in the following figure. The decimal system is used for place location, but each digit of the vertically positioned decimal number is converted to a binary number.

	Decimal Number	Vertical Binary-Decimal Number
10^2	2	10
10^1	4	100
10^0	3	11

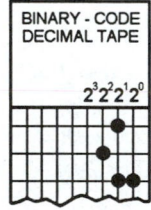

BINARY - CODE DECIMAL TAPE

$2^3 2^2 2^1 2^0$

APPLICATION

Stucture of the Decimal System

Analyze the following numbers.

1. 265
2. 2855
3. 90,500
4. 0.802
5. 23.023
6. 105.009
7. 4751.107
8. 3006.0204
9. 163.0643

Expressing Binary Numbers as Decimal Numbers

Express the following binary numbers as decimal numbers.

10. 10_2 _____
11. 1_2 _____
12. 100_2 _____
13. 101_2 _____
14. 1101_2 _____

15. 1111_2 _____
16. 10100_2 _____
17. 1011_2 _____
18. 11000_2 _____
19. 10101_2 _____

20. 101010_2 _____

21. 110101_2 _____

22. 111010_2 _____

23. 0.1_2 _____

24. 0.1011_2 _____

25. 11.11_2 _____

26. 11.01_2 _____

27. 10.000_2 _____

28. 1111.11_2 _____

29. 1001.0101_2 _____

30. 10011.0101_2 _____

Expressing Decimal Numbers as Binary Numbers

Express the following decimal numbers as binary numbers.

31. 14 _____

32. 100 _____

33. 87 _____

34. 23 _____

35. 43 _____

36. 4 _____

37. 105 _____

38. 98 _____

39. 1 _____

40. 6 _____

41. 51 _____

42. 270 _____

43. 0.5 _____

44. 0.125 _____

45. 0.375 _____

46. 10.5 _____

47. 81.75 _____

48. 19.0625 _____

49. 101.25 _____

50. 1.125 _____

51. 163.875 _____

UNIT 71 Achievement Review— Section Seven

Objective

You should be able to solve the exercises and problems in this Achievement Review by applying the principles and methods covered in units 68–70.

---

1. Using graph paper, draw an x- and a y-axis and plot the following coordinates.

A = (6, −8)	C = (−2, 0)	E = (−7, −7)
B = (−3, 9)	D = (0, −8)	F = (3, 3)

2. Refer to the points plotted on the illustrated Cartesian Coordinate plane. Write the *x* and *y* coordinates of the following points, A–M.

A = _____
B = _____
C = _____
D = _____
E = _____
F = _____
G = _____
H = _____
I = _____
J = _____
K = _____
L = _____
M = _____

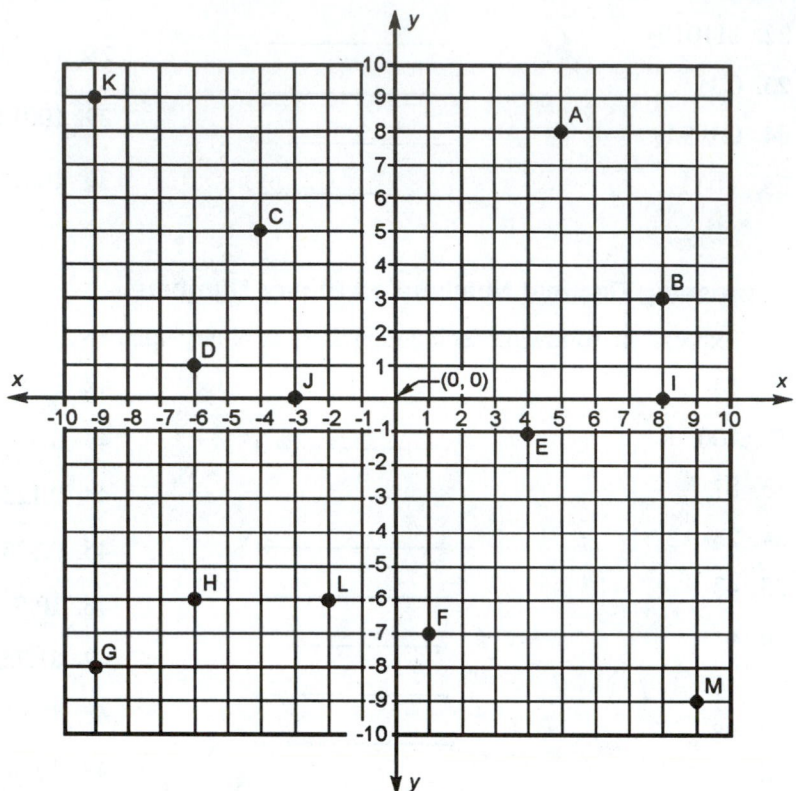

3. Write the program hole locations (coordinates) in table form. List the holes in sequence similar to the tables in Unit 69. All dimensions are in inches. Use

   a. absolute dimensioning

   b. incremental dimensioning

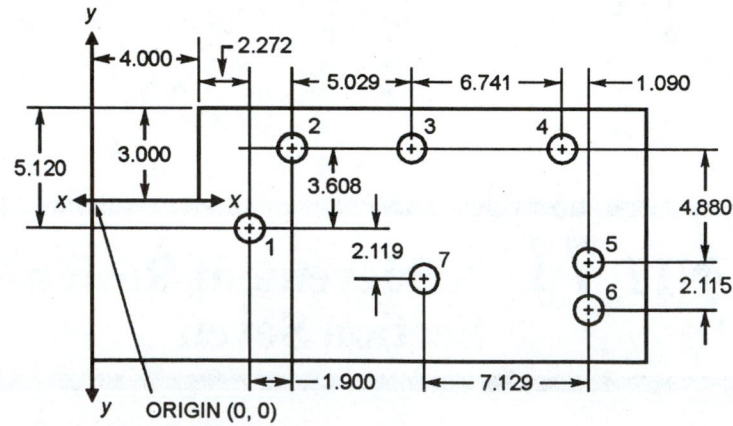

4. Write the program hole locations (coordinates) in table form. List the holes in sequence similar to the tables in Unit 69. All dimensions are in millimeters. Use

   a. absolute dimensioning

   b. incremental dimensioning

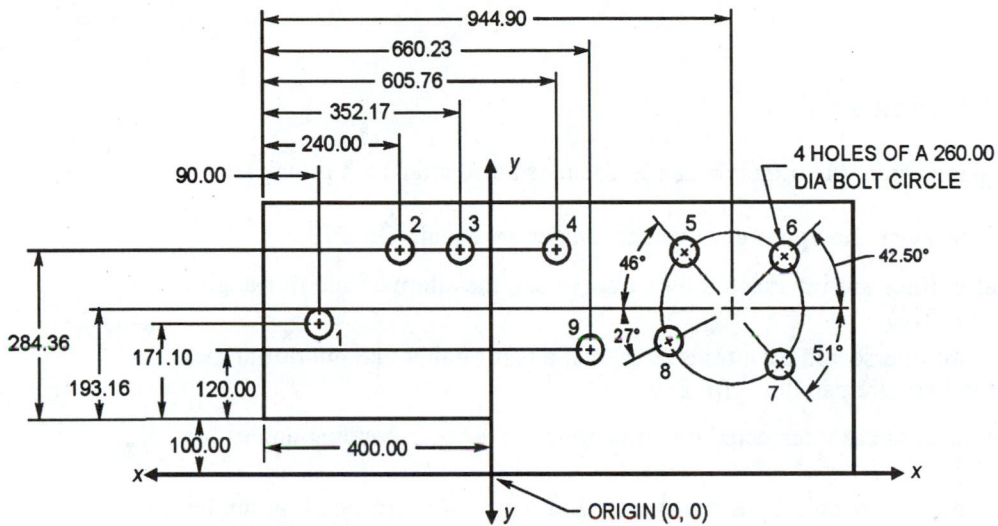

5. Express the following binary numbers as decimal numbers.

   a. 1_2 _____

   b. 111_2 _____

   c. 10101_2 _____

   d. 11.01_2 _____

   e. 1001.1001_2 _____

6. Express the following decimal numbers as binary numbers.

   a. 7 _____

   b. 32 _____

   c. 157 _____

   d. 0.125 _____

   e. 74.25 _____

APPENDIX

Geometric Principles

Note: The page where the principle can be found is noted after each principle.

1. If two lines intersect, the opposite or vertical angles are equal.   (p. 278)

2. If two parallel lines are intersected by a transversal, the alternate interior angles are equal.   (p. 278)

   If two lines are intersected by a transversal and a pair of alternate interior angles are equal, the lines are parallel.   (p. 279)

3. If two parallel lines are intersected by a transversal, the corresponding angles are equal.   (p. 279)

   If two lines are intersected by a transversal and a pair of corresponding angles are equal, the lines are parallel.

4. Two angles are either equal or supplementary if their corresponding sides are parallel.   (p. 279)

5. Two angles are either equal or supplementary if their corresponding sides are perpendicular.   (p. 279)

6. The sum of the angles of any triangle is equal to 180°.   (p. 285)

7. Two triangles are similar if their sides are respectively parallel.   (p. 292)

   - Two triangles are similar if their sides are respectively perpendicular.

   - Within a triangle, if a line is drawn parallel to one side, the triangle formed is similar to the original triangle.

   - In a right triangle, if a line is drawn from the vertex of the right angle perpendicular to the opposite side, the two triangles formed and the original triangle are similar.

8. In an isosceles triangle, an altitude to the base bisects the base and the vertex angle. (p. 292)

   In an equilateral triangle, an altitude to any side bisects the side and the vertex angle.

9. In a right triangle, the square of the hypotenuse is equal to the sum of the squares of the other two sides or legs.   (p. 293)

10. The sum of the interior angles of a polygon of N sides is equal to $(N - 2)$ times 180°.   (p. 294)

11. In the same circle or in equal circles, equal chords cut off equal arcs.   (p. 303)

12. In the same circle or in equal circles, equal central angles cut off equal arcs. (p. 303)

13. In the same circle or in equal circles, two central angles have the same ratio as the arcs which are cut off by the angles.   (p. 303)

14. A line drawn from the center of a circle perpendicular to a chord bisects the chord and the arc cut off by the chord. The perpendicular bisector of a chord passes through the center of a circle.   (p. 303)

15. A line perpendicular to a radius at its extremity is tangent to the circle. A tangent is perpendicular to a radius at its tangent point.   (p. 304)

16. Two tangents drawn to a circle from a point outside the circle are equal. The angle at the outside point is bisected by a line drawn from the point to the center of the circle.   (p. 304)

17. If two chords intersect inside a circle, the product of the two segments of one chord is equal to the product of the two segments of the other chord.   (p. 305)

18. A central angle is equal to its intercepted arc.   (p. 309)

19. An angle formed by a tangent and a chord at the tangent point is equal to one-half of its intercepted arc.   (p. 311)

20. An angle formed at a point outside a circle by two secants, two tangents, or a secant and a tangent is equal to one-half the difference of the intercepted arcs.   (p. 312)

21. If two circles are either internally or externally tangent, a line connecting the centers of the circles passes through the point of tangency and is perpendicular to the tangent line.   (p. 313)

DECIMAL EQUIVALENT TABLE

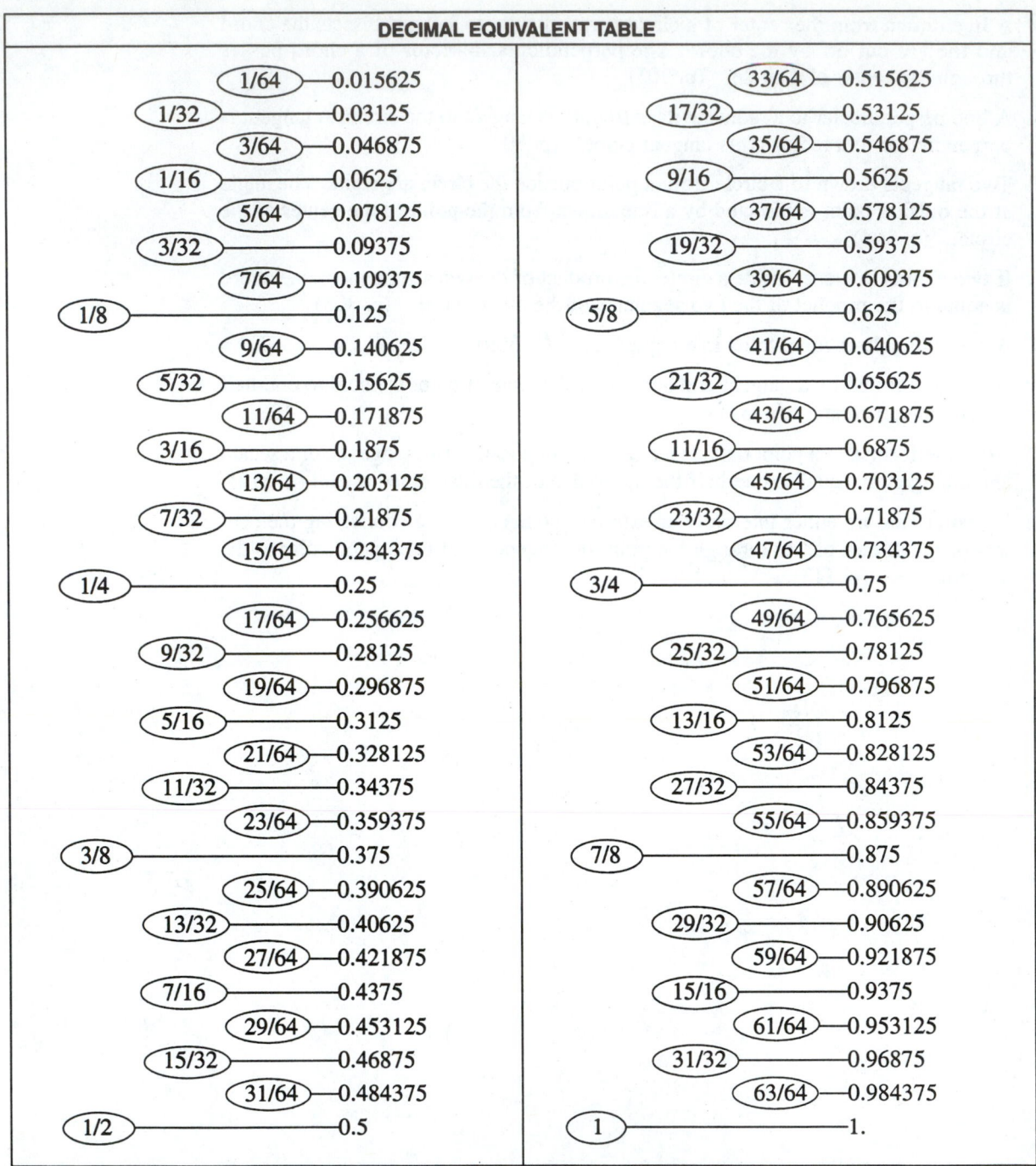

1/64 — 0.015625	33/64 — 0.515625
1/32 — 0.03125	17/32 — 0.53125
3/64 — 0.046875	35/64 — 0.546875
1/16 — 0.0625	9/16 — 0.5625
5/64 — 0.078125	37/64 — 0.578125
3/32 — 0.09375	19/32 — 0.59375
7/64 — 0.109375	39/64 — 0.609375
1/8 — 0.125	5/8 — 0.625
9/64 — 0.140625	41/64 — 0.640625
5/32 — 0.15625	21/32 — 0.65625
11/64 — 0.171875	43/64 — 0.671875
3/16 — 0.1875	11/16 — 0.6875
13/64 — 0.203125	45/64 — 0.703125
7/32 — 0.21875	23/32 — 0.71875
15/64 — 0.234375	47/64 — 0.734375
1/4 — 0.25	3/4 — 0.75
17/64 — 0.256625	49/64 — 0.765625
9/32 — 0.28125	25/32 — 0.78125
19/64 — 0.296875	51/64 — 0.796875
5/16 — 0.3125	13/16 — 0.8125
21/64 — 0.328125	53/64 — 0.828125
11/32 — 0.34375	27/32 — 0.84375
23/64 — 0.359375	55/64 — 0.859375
3/8 — 0.375	7/8 — 0.875
25/64 — 0.390625	57/64 — 0.890625
13/32 — 0.40625	29/32 — 0.90625
27/64 — 0.421875	59/64 — 0.921875
7/16 — 0.4375	15/16 — 0.9375
29/64 — 0.453125	61/64 — 0.953125
15/32 — 0.46875	31/32 — 0.96875
31/64 — 0.484375	63/64 — 0.984375
1/2 — 0.5	1 — 1.

ENGLISH UNITS OF LINEAR MEASURE

1 yard (yd) = 3 feet (ft)

1 yard (yd) = 36 inches (in)

1 foot (ft) = 12 inches (in)

1 mile (mi) = 1760 yards (yd)

1 mile (mi) = 5280 feet (ft)

METRIC UNITS OF LINEAR MEASURE

1 millimeter (mm)	= 0.001 meter (m)	1000 millimeters (mm)	= 1 meter (m)
1 centimeter (cm)	= 0.01 meter (m)	100 centimeters (cm)	= 1 meter (m)
1 decimeter (dm)	= 0.1 meter (m)	10 decimeters (dm)	= 1 meter (m)
1 meter (m)	= 1 meter (m)	1 meter (m)	= 1 meter (m)
1 dekameter (dam)	= 10 meters (m)	0.1 dekameter (dam)	= 1 meter (m)
1 hectometer (hm)	= 100 meters (m)	0.01 hectometer (hm)	= 1 meter (m)
1 kilometer (km)	= 1000 meters (m)	0.001 kilometer (km)	= 1 meter (m)

METRIC-ENGLISH LINEAR EQUIVALENTS (CONVERSION FACTORS)

Metric to English Units		English to Metric Units	
1 millimeter (mm)	= 0.03937 inch (in)	1 inch (in)	= 25.4 millimeters (mm)
1 centimeter (cm)	= 0.3937 inch (in)	1 inch (in)	= 2.54 centimeters (cm)
1 meter (m)	= 39.37 inches (in)	1 foot (ft)	= 0.3048 meter (m)
1 meter (m)	= 3.2808 feet (ft)	1 yard (yd)	= 0.9144 meter (m)
1 kilometer (km)	= 0.6214 mile (mi)	1 mile (mi)	= 1.609 kilometers (km)

TRIGONOMETRIC FUNCTIONS

Function	Symbol	Definition of Function
sine of Angle A	sin A	$\sin A = \dfrac{\text{opp side}}{\text{hyp}} = \dfrac{a}{c}$
cosine of Angle A	cos A	$\cos A = \dfrac{\text{adj side}}{\text{hyp}} = \dfrac{b}{c}$
tangent of Angle A	tan A	$\tan A = \dfrac{\text{opp side}}{\text{adj side}} = \dfrac{a}{b}$
cotangent of Angle A	cot A	$\cot A = \dfrac{\text{adj side}}{\text{opp side}} = \dfrac{b}{a}$
secant of Angle A	sec A	$\sec A = \dfrac{\text{hyp}}{\text{adj side}} = \dfrac{c}{b}$
cosecant of Angle A	csc A	$\csc A = \dfrac{\text{hyp}}{\text{opp side}} = \dfrac{c}{a}$

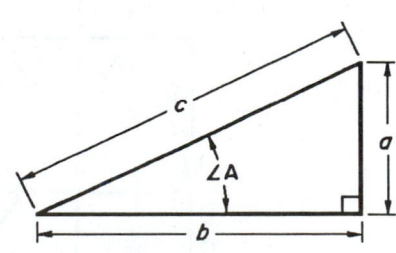

ANSWERS TO ODD-NUMBERED APPLICATIONS

SECTION 1 Common Fractions, Decimal Fractions, and Percentage

UNIT 1 Introduction to Common Fractions and Mixed Numbers

1. A $= \frac{3}{32}$
   B $= \frac{7}{32}$
   C $= \frac{3}{8}$
   D $= \frac{19}{32}$
   E $= \frac{27}{32}$
   F $= 1$

3. a. $\frac{1}{16}$
   b. $\frac{3}{16}$
   c. $\frac{7}{16}$
   d. $\frac{5}{16}$

3. e. $\frac{16}{16} = 1$
   f. $\frac{1}{32}$
   g. $\frac{1}{48}$
   h. $\frac{3}{64}$
   i. $\frac{1}{60}$
   j. $\frac{1}{256}$

5. a. $\frac{3}{4}$
   b. 3
   c. $\frac{3}{5}$
   d. 6

5. e. $\frac{1}{4}$
   f. $\frac{7}{3}$
   g. 3
   h. $\frac{13}{3}$
   i. $\frac{1}{6}$
   j. $\frac{2}{15}$

7. a. $\frac{6}{8}$
   b. $\frac{21}{36}$
   c. $\frac{24}{60}$
   d. $\frac{51}{42}$

7. e. $\frac{100}{45}$
   f. $\frac{84}{18}$
   g. $\frac{56}{128}$
   h. $\frac{78}{48}$
   i. $\frac{210}{16}$

9. a. $1\frac{2}{3}$
   b. $10\frac{1}{2}$
   c. $1\frac{1}{8}$
   d. $21\frac{3}{4}$

9. e. 8
   f. $1\frac{3}{124}$
   g. $3\frac{31}{32}$
   h. $3\frac{12}{15} = 3\frac{4}{5}$
   i. $16\frac{2}{3}$
   j. $14\frac{11}{16}$
   k. $128\frac{1}{2}$
   l. $6\frac{17}{64}$

11.

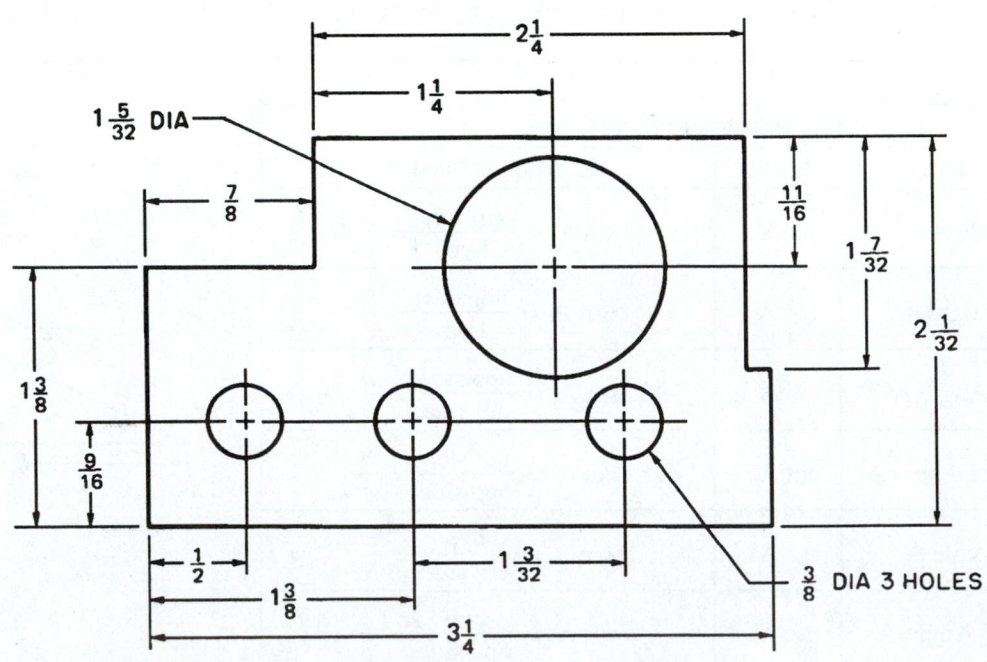

450

UNIT 2   Addition of Common Fractions and Mixed Numbers

1. 12

3. 48

5. $\frac{6}{12}, \frac{9}{12}, \frac{5}{12}$

7. $\frac{18}{20}, \frac{5}{20}, \frac{12}{20}, \frac{4}{20}$

9. $A = 1\frac{1}{16}''$

9. $B = \frac{59}{64}''$

$C = 1\frac{9}{16}''$

$D = \frac{11}{16}''$

$E = 3\frac{11}{64}''$

$F = \frac{13}{16}''$

11. $A = 2\frac{9}{32}''$

$B = 3\frac{7}{8}''$

$C = 4\frac{1}{4}''$

$D = 2\frac{19}{32}''$

11. $E = 3\frac{3}{8}''$

$F = 1\frac{21}{64}''$

$G = 4\frac{45}{64}''$

13. $5\frac{17}{60}$ h

UNIT 3   Subtraction of Common Fractions and Mixed Numbers

1. a. $\frac{11}{32}$

b. $\frac{1}{4}$

c. $\frac{13}{25}$

d. $\frac{31}{64}$

e. $\frac{23}{64}$

1. f. $\frac{29}{48}$

3. $A = \frac{7}{32}''$

$B = \frac{19}{32}''$

$C = \frac{25}{64}''$

$D = \frac{3}{16}''$

3. $E = \frac{7}{16}''$

$F = \frac{13}{32}''$

5. $A = \frac{15}{32}''$

$B = \frac{21}{32}''$

5. $C = \frac{7}{16}''$

$D = \frac{15}{32}''$

$E = \frac{9}{32}''$

$F = \frac{5}{16}''$

5. $G = 1\frac{7}{32}''$

$H = \frac{33}{64}''$

$I = \frac{19}{32}''$

UNIT 4   Multiplication of Common Fractions and Mixed Numbers

1. a. $\frac{1}{9}$

b. $\frac{1}{8}$

c. $\frac{65}{512}$

1. d. $\frac{3}{10}$

e. $13\frac{1}{2}$

f. $\frac{1}{8}$

3. a. $\frac{1595}{2048}''$

b. $\frac{7}{256}''$

5. a. $10\frac{1}{2}$

5. b. $25\frac{43}{64}$

c. $11\frac{9}{16}$

d. $6\frac{13}{32}$

5. e. $\frac{201}{256}$

f. $37\frac{1}{3}$

UNIT 5   Division of Common Fractions and Mixed Numbers

1. $\frac{8}{7}$

3. $\frac{8}{25}$

5. $A = 6$ threads

$B = 8\frac{7}{16}$ threads

5. $C = 3\frac{1}{2}$ threads

$D = 7$ threads

$E = 6\frac{7}{8}$ threads

$F = 6\frac{1}{2}$ threads

5. $G = 2\frac{13}{16}$ threads

7. 5 cuts

9. 240 revolutions

11. $\frac{4}{5}$ foot

13. $4\frac{70}{93}$ lb

UNIT 6   Combined Operations of Common Fractions and Mixed Numbers

1. a. $\frac{7}{16}$

b. $2\frac{1}{16}$

c. $5\frac{33}{50}$

d. $28\frac{1}{2}$

e. $33\frac{49}{64}$

f. $3\frac{1}{2}$

1. g. $5\frac{3}{4}$

h. $28\frac{3}{8}$

i. $4\frac{8}{21}$

j. $20\frac{77}{87}$

3. a. $B = 4\frac{7}{8}''$

$E = 5\frac{1}{16}''$

3. b. $A = \frac{9}{64}''$

$G = 6\frac{3}{4}''$

c. $C = 1\frac{1}{16}''$

$D = 5\frac{19}{32}''$

d. $B = 3\frac{29}{32}''$

$E = 5\frac{5}{8}''$

3. e. $A = \frac{19}{32}''$

$G = 7\frac{27}{64}''$

f. $C = \frac{63}{64}''$

$D = 5\frac{55}{64}''$

5. $7\frac{29}{32}''$

7. $3\frac{1}{5}$ min

9. $\frac{57}{64}''$

11. $3\frac{3}{4}$ lb

UNIT 7   Computing with a Calculator: Fractions and Mixed Numbers

(All answers are given in the unit)

UNIT 8   Introduction to Decimal Fractions

1. A = 0.3
   B = 0.5
   C = 0.8
   D = 0.92
   E = 0.04
3. A = 0.0025
   B = 0.006
   C = 0.007
   D = 0.0073
   E = 0.0004
5. 0.01

7. 10
9. 0.1
11. 100
13. 0.001
15. seven thousandths
17. thirty-five ten-thousandths
19. one and five tenths
21. sixteen and seven ten-thousandths
23. thirteen and one hundred three thousandths

25. 0.3
27. 4.00005
29. 10.2
31. 20.71
33. 0.0007
35. 0.43
37. 0.0999
39. 0.01973

UNIT 9   Rounding Decimal Fractions and Equivalent Decimal and Common Fractions

1. 0.632
3. 0.240
5. 0.04
7. 0.7201
9. 0.000
11. 0.6875
13. 0.6250

15. 0.6667
17. 0.0800
19. 0.2188
21. 0.5714
23. 0.3333
25. a. 0.125
    b. 0.6154

27. $\frac{1}{8}$
29. $\frac{3}{4}$
31. $\frac{11}{16}$
33. $\frac{3}{1000}$
35. $\frac{251}{500}$
37. $\frac{7}{16}$

39. $\frac{8717}{10,000}$
41. $\frac{3}{100}$
43. $\frac{237}{1000}$
45. $\frac{9}{200}$
47. $\frac{7}{8}$

49. a. $\frac{1}{4}$
    b. $\frac{9}{16}$
    c. $\frac{1}{10}$
    d. $\frac{3}{80}$
    e. $\frac{3}{8}$

UNIT 10   Addition and Subtraction of Decimal Fractions

1. a. 15.775
   b. 0.14095
   c. 1.295
   d. 5.129
   e. 381.357
   f. 4.444
   g. 94.2539
   h. 0.1101
   i. 5.7787
   j. 328.963
3. 4.1758"

5. (Other combinations may total certain thicknesses)
   a. 0.010" + 0.004"
   b. 0.015" + 0.010" + 0.008"
   c. 0.015" + 0.006"
   d. 0.015" + 0.012" + 0.008" + 0.003"
   e. 0.008" + 0.003"
   f. 0.015" + 0.012" + 0.010" + 0.003" + 0.002"
   g. 0.015" + 0.012" + 0.002"
   h. 0.015" + 0.012" + 0.010 " + 0.008" + 0.004"

7. A = 12.82 mm
   B = 27.02 mm
   C = 6.58 mm
   D = 20.00 mm
   E = 10.58 mm
   F = 7.39 mm

UNIT 11 Multiplication of Decimal Fractions

1. a. 0.0563
   b. 3.3
   c. 6
   d. 1.6718

3. Dia A = 31.763 mm
   Dia B = 19.199 mm
   Dia C = 12.847 mm

3. Dia D = 22.571 mm
   Dia E = 6.741 mm

UNIT 12 Division of Decimal Fractions

1. a. 1.597
   b. 2.56
   c. 0.0100
   d. 10,000.000
   e. 11.367
   f. 4.29

1. g. 135.53
   h. 0.0062

3. A = 11.75 mm
   B = 5.91 mm
   C = 12.46 mm
   D = 10.95 mm

5. 26 complete bushings

7. 0.063 mm

9. 0.125″

11. 38.50 mm

UNIT 13 Powers

1. 39.304

3. 100,000,000

5. $2\frac{2}{3}$

7. 4.41

9. 64

11. 532.23 mm²

13. 114.49 mm²

15. $\frac{9}{16}$ sq in

17. $14\frac{1}{16}$ sq in

19. $189\frac{1}{16}$ sq in

21. 8741.82 mm³

23. 2744 mm³

25. $\frac{1}{27}$ cu in

27. $3\frac{3}{8}$ cu in

29. $\frac{27}{64}$ cu in

31. 764 mm²

33. 270 sq in

35. 0.1 cu in

37. 0.2 cu in

39. 184.2 mm³

41. 16 cu in

43. 10 cu in

45. 11 cu in

47. 385 mm³

49. 0 cu in

51. 329 mm²

53. 1 cu in

55. 3600 cu in

UNIT 14 Roots

1. 6

3. $\frac{2}{9}$

5. $\frac{3}{4}$

7. 12

9. 4

11. a. 6 mm
    b. 4 in
    c. 8 in
    d. 10 mm
    e. 1 in

13. a. D = 3 in
    b. D = 6 mm

13. c. D = 2 in
    d. D = 1 in
    e. D = 10 mm

15. 19.77

17. 1.871

19. 4.42

21. 0.0857

23. a. D = 7.45 mm
    b. D = 6.08 in
    c. D = 21.91 mm
    d. D = 1.07 in

25. D = 1.5 in

UNIT 15 Table of Decimal Equivalents and Combined Operations of Decimal Fractions

1. 0.78125

3. 0.34375

5. 0.078125

7. $\frac{5}{16}$

9. $\frac{13}{64}$

11. $\frac{49}{64}$

13. $\frac{1}{2}$

15. $\frac{13}{16}$

17. 14.1

19. 25.12

21. 7.24

23. 10.57

25. 16.21

27. 0.084 mm

29. a. C = 9.02 mm
    b. C = 8.74 mm
    c. C = 5.48 mm

31. H = 0.077″

UNIT 16 Computing with a Calculator: Decimals

(All answers are given in the unit)

UNIT 17 Introduction to Percents

1. 44%	13. 0.02%	25. 0.82	37. 0.02375	47. $1\frac{9}{10}$
3. 25%	15. 25%	27. 0.03	39. 0.3725	49. $\frac{9}{500}$
5. 35%	17. 15%	29. 0.2776	41. $\frac{1}{2}$	51. $\frac{9}{1000}$
7. 4%	19. 53.125%	31. 2.249		
9. 0.8%	21. 159%	33. 0.0473	43. $\frac{5}{8}$	
11. 207.6%	23. 1462.5%	35. 0.0075	45. $\frac{4}{25}$	

UNIT 18 Basic Calculations of Percentages, Percents, and Rates

1. 16	17. 0.99	31. 30.77%	45. 170	59. 28.87
3. 120	19. 5.38	33. 42.86%	47. 184.55	61. 75%
5. 78.15	21. 50%	35. 154.55%	49. 4.17	63. 15.60
7. 101.4	23. 37%	37. 24.49%	51. 270.57	65. 19.05%
9. 37.47	25. 118.95%	39. 50%	53. 42.93	67. 153.99
11. 392	27. 155.46%	41. 150	55. 0.5	69. 57.99
13. 7.14	29. 40%	43. 320	57. 3.90	71. 3.38
15. 0.13				

UNIT 19 Percent Practical Applications

1. 56%

3. 6.8 hr

5. 1,500 units

7. 12%

9. 262.5 ft

11. Copper: 725 lb
    Tin: 500 lb
    Manganese: 19 lb
    Other: 6 lb

13. 1%

15. 9/16   Rework: 2.7%
    9/16   Scrap: 3.6%

15. 9/17   Rework: 1.8%
    9/17   Scrap: 3.0%
    9/18   Rework: 3.1%
    9/18   Scrap: 2.0%

17. $3,264

19. a. 13%
    b. 69%
    c. 110%

21. 936 castings

23. 18 lb

25. 5.6%

27. 0.63%

29. 8.7 hp

31. 3,189 pieces

33. 1,750 pieces

35. **Job 1** Labor Cost: 47%
    Material Cost: 22%
    Overhead Cost: 31%
    Job 2 Labor Cost: 32%
    Material Cost: 37%
    Overhead Cost: 31%
    Job 3 Labor Cost: 42%
    Material Cost: 31%
    Overhead Cost: 27%

UNIT 20 Achievement Review—Section One

1. a. $\frac{12}{32}$

   b. $\frac{70}{100}$

   c. $\frac{16}{64}$

   d. $\frac{72}{128}$

3. a. $2\frac{1}{2}$

   b. $4\frac{1}{5}$

   c. $18\frac{3}{4}$

   d. $3\frac{19}{32}$

   e. $5\frac{9}{64}$

5. a. $\frac{8}{32}, \frac{6}{32}, \frac{9}{32}$

   b. $\frac{28}{64}, \frac{10}{64}, \frac{9}{64}$

   c. $\frac{70}{100}, \frac{75}{100}, \frac{36}{100}, \frac{65}{100}$

7. a. $\frac{5}{16}$

   b. $\frac{2}{5}$

   c. $1\frac{245}{256}$

   d. $25\frac{23}{40}$

   e. $20\frac{5}{8}$

   f. $\frac{3}{4}$

   g. $3\frac{1}{3}$

7. h. 48

   i. $6\frac{3}{4}$

   j. $\frac{93}{280}$

9. 51 complete pieces

11. 8 min

13. A $= 3\frac{5}{16}''$

    B $= 2\frac{15}{16}''$

    C $= 3\frac{15}{64}''$

    D $= 3\frac{15}{16}''$

    E $= 3\frac{15}{32}''$

15. a. 0.3

    b. 0.026

    c. 9.026

    d. 5.0081

17. a. 0.75

    b. 0.875

    c. 0.667

    d. 0.08

    e. 0.65

19. a. 1.587

    b. 6.4274

    c. 12.3069

    d. 9.1053

    e. 23.4077

19. f. 0.356

    g. 0.1444

    h. 0.001

    i. 0.0022

    j. 0.002

21. a. 6.76

    b. 0.125

    c. 0.000036

    d. $\frac{9}{25}$

    e. 32.768

23. a. 19.47

    b. 0.935

    c. 0.632

    d. 6.780

25. a. $\frac{15}{32}$

    b. $\frac{49}{64}$

    c. $\frac{1}{32}$

    d. $\frac{31}{32}$

27. A $= 1.299$ mm

    B $= 0.812$ mm

    C $= 0.325$ mm

    D $= 0.162$ mm

    E $= 0.188$ mm

    F $= 0.375$ mm

29. 0.12 mm

31. 12.6 in

33. a. 0.19%

    b. 0.007%

    c. 0.0075%

    d. 3.103%

35. a. 9

    b. 1.27

    c. 87.36

    d. 5.68

    e. 22.90

    f. 275.6

    g. 4

    h. 1.57

37. a. 33.33

    b. 16.47

    c. 223.68

    d. 59.97

    e. 3.90

    f. 41.18

    g. 0.61

39. a. 0.38 kg

    b. 0.63 kg

41. a. 23 kg

    b. 8.6 kg

    c. 11.5 kg

43. 17 lb

SECTION 2   Linear Measurement: English and Metric

UNIT 21   English and Metric Units of Measure

1. a. 8 ft
   b. 10.25 ft
   c. 42 in
   d. 14.4 in
   e. 45 in
   f. 4 yd
   g. 6.25 ft
   h. 24 ft
   i. 12.6 ft
   j. 9 yd
   k. 17 yd
   l. 12 in
   m. 21.5 ft
   n. 92 in
   o. 7.2 in
   p. 46.75 yd
   q. 9.25 yd

1. r. 15.5 ft
   s. 62 ft
   t. 111 in

3. 6 complete lengths

5. a. 29 mm
   b. 157.8 mm
   c. 21.975 cm
   d. 9.783 cm
   e. 97 cm
   f. 170 mm
   g. 0.153 m
   h. 6.73 m
   i. 0.093 cm
   j. 0.8 mm
   k. 8.6 mm
   l. 104.6 cm
   m. 300.3 mm

5. n. 87.684 cm
   o. 2.039 m
   p. 0.0347 m
   q. 49 mm
   r. 732.1 cm
   s. 63.77 mm
   t. 934 mm

7. 52 mm

9. a. 1.457 in
   b. 4.992 in
   c. 6.811 in
   d. 0.331 in
   e. 94.488 in
   f. 3.543 in
   g. 26.246 ft
   h. 33.464 ft
   i. 28.976 in

9. j. 1.341 in
   k. 22.165 in
   l. 2.187 yd
   m. 1.476 ft
   n. 2.559 ft

11. 53.7 in

13. A = 15.75 mm
    B = 28.58 mm
    C = 327.03 mm
    D = 25.10 mm
    E = 3.30 mm
    F = 12.70 mm
    G = 25.00 mm
    H = 2.38 mm
    I = 17.46 mm
    J = 9.53 mm

UNIT 22   Degree of Precision and Greatest Possible Error

1. a. 0.1"
   b. 4.25"
   c. 4.35"

3. a. 0.1"
   b. 4.25"
   c. 4.35"

5. a. 0.001"
   b. 15.8845"
   c. 15.8855"

7. a. 0.001"
   b. 11.0025"
   c. 11.0035"

9. a. 0.01"
   b. 7.005"
   c. 7.015"

11. a. 0.1"
    b. 6.05"
    c. 6.15"

13. a. 0.01 mm
    b. 26.865 mm
    c. 26.875 mm

15. a. 0.01 mm
    b. 117.055 mm
    c. 117.065 mm

17. a. 0.01 mm
    b. 48.005 mm
    c. 48.015 mm

19. a. 0.01 mm
    b. 6.995 mm
    c. 7.005 mm

21. a. 0.001 mm
    b. 8.0005 mm
    c. 8.0015 mm

	Greatest Possible Error (inches)	ACTUAL LENGTH	
		Smallest Possible (inches)	Largest Possible (inches)
23.	0.025	5.275	5.325
25.	0.0005	0.7525	0.7535
27.	0.00005	0.93685	0.93695

	Greatest Possible Error (millimeters)	ACTUAL LENGTH	
		Smallest Possible (millimeters)	Largest Possible (millimeters)
29.	0.5	63.5	64.5
31.	0.25	98.25	98.75
33.	0.005	13.365	13.375

35. Absolute Error: 0.020 in
    Relative Error: 0.052%

37. Absolute Error: 0.200°
    Relative Error: 1.575%

39. Absolute Error: 0.140 mm
    Relative Error: 0.587%

41. Absolute Error: 0.030 mm
    Relative Error: 2.857%

43. Absolute Error: 0.010°
    Relative Error: 0.995%

45. Absolute Error: 0.026 in
    Relative Error: 0.142%

UNIT 23   Tolerance, Clearance, and Interference

1. a. $\frac{1''}{32}$

   b. $\frac{1''}{8}$

   c. 16.73″

   d. 0.911″

   e. 0.0003″

   f. 11.003″

3. a. Max. Limit = 4.643″
      Min. Limit = 4.640″

   b. Max. Limit = 5.932″
      Min. Limit = 5.927″

   c. Max. Limit = 2.004″
      Min. Limit = 2.000″

   d. Max. Limit = 4.6729″
      Min. Limit = 4.6717″

3. e. Max. Limit = 1.0884″
      Min. Limit = 1.0875″

   f. Max. Limit = 28.16 mm
      Min. Limit = 28.10 mm

   g. Max. Limit = 43.98 mm
      Min. Limit = 43.94 mm

   h. Max. Limit = 118.73 mm
      Min. Limit = 118.66 mm

   i. Max. Limit = 73.398 mm
      Min. Limit = 73.386 mm

   j. Max. Limit = 45.115 mm
      Min. Limit = 45.106 mm

5. a. 0.943″ ± 0.005″

   b. 1.687″ ± 0.001″

   c. 2.998″ ± 0.002″

5. d. 0.069″ ± 0.004″

   e. 4.1880″ ± 0.0007″

   f. 0.9984″ ± 0.0037″

   g. 0.0006″ ± 0.0004″

   h. 8.4660″ ± 0.0011″

   i. 44.31 mm ± 0.01 mm

   j. 10.02 mm ± 0.04 mm

   k. 64.92 mm ± 0.03 mm

   l. 38.016 mm ± 0.028 mm

   m. 124.9915 ± 0.0085 mm

   n. 43.078 mm ± 0.013 mm

   o. 98.8835 mm ± 0.0045 mm

7. All dimensions are in millimeters.

		Basic Dimension	Maximum Diameter (Max. Limit)	Minimum Diameter (Min. Limit)	Maximum Interference (Allowance)	Minimum Interference
a.	DIA A	20.73	20.75	20.71	0.09	0.01
	DIA B	20.68	20.70	20.66		
b.	DIA A	32.07	32.09	32.05	0.10	0.02
	DIA B	32.01	32.03	31.99		
c.	DIA A	12.72	12.74	12.70	0.11	0.03
	DIA B	12.65	12.67	12.63		

9. All dimensions are in millimeters.

		Basic Dimension	Maximum Diameter (Max. Limit)	Minimum Diameter (Min. Limit)	Maximum Interference (Allowance)	Minimum Interference
a.	DIA A	87.58	87.61	87.55	0.14	0.02
	DIA B	87.50	87.53	87.47		
b.	DIA A	9.94	9.97	9.91	0.15	0.03
	DIA B	9.85	9.88	9.82		
c.	DIA A	130.03	130.06	130.00	0.13	0.01
	DIA B	129.96	129.99	129.93		

11. 18.20 mm

13. Max. thickness = 2.88 mm
    Min. thickness = 2.82 mm

15. Holes 5 and 6 are out of tolerance.

UNIT 24　English and Metric Steel Rules

1. a. $\frac{3''}{32}$

   b. $\frac{5''}{16}$

   c. $\frac{1''}{2}$

   d. $\frac{5''}{8}$

   e. $\frac{3''}{4}$

   f. $\frac{29''}{32}$

   g. $1\frac{3''}{32}$

   h. $1\frac{5''}{16}$

   i. $\frac{5''}{64}$

   j. $\frac{7''}{32}$

   k. $\frac{3''}{8}$

1. l. $\frac{35''}{64}$

   m. $\frac{51''}{64}$

   n. $\frac{31''}{32}$

   o. $1\frac{15''}{64}$

   p. $1\frac{29''}{64}$

3. a. $\frac{1''}{4}$

   b. $\frac{9''}{16}$

   c. $\frac{1''}{2}$

   d. $\frac{1''}{2}$

   e. $\frac{11''}{32}$

   f. $2\frac{7''}{32}$

3. g. $\frac{7''}{32}$

   h. $\frac{3''}{4}$

   i. $\frac{23''}{32}$

   j. $\frac{1''}{2}$

   k. $\frac{15''}{32}$

   l. $\frac{3''}{8}$

   m. $\frac{5''}{32}$

   n. $4\frac{29''}{32}$

5. a. 0.12″

   b. 0.22″

   c. 0.40″

   d. 0.62″

   e. 0.80″

5. f. 1.04″

   g. 1.32″

   h. 1.42″

   i. 0.11″

   j. 0.23″

   k. 0.38″

   l. 0.57″

   m. 0.84″

   n. 1.07″

   o. 1.29″

   p. 1.45″

7. A = 0.54″

   B = 0.42″

   C = 1.38″

   D = 1.18″

   E = 0.34″

   F = 0.28″

7. G = 1.00″

   H = 0.22″

   I = 0.10″

9. a. 46 mm

   b. 70 mm

   c. 20 mm

   d. 82 mm

   e. 10 mm

   f. 23 mm

   g. 25 mm

   h. 121 mm

   i. 17 mm

   j. 22 mm

   k. 36 mm

   l. 10 mm

   m. 52 mm

   n. 6 mm

UNIT 25　English Vernier Calipers and Height Gages

1. a. 2.641″

   b. 3.376″

   c. 2.021″

   d. 0.508″

   e. 4.788″

   f. 2.991″

   g. 1.581″

   h. 1.098″

3.

	A (inches)	B (inches)	C
a.	3.225	3.250	17
b.	2.875	2.900	2
c.	4.825	4.850	14
d.	0.600	0.625	11
e.	4.350	4.375	19
f.	0.075	0.100	9
g.	7.850	7.875	7
h.	1.625	1.650	21
i.	4.025	4.050	9
j.	0.000	0.025	22
k.	3.325	3.350	8
l.	5.975	6.000	24
m.	0.275	0.300	3
n.	0.950	0.975	15

5. a. 1.909″

   b. 4.620″

   c. 7.969″

   d. 0.439″

   e. 2.779″

   f. 6.459″

   g. 3.612″

   h. 8.391″

UNIT 26   Metric Vernier Calipers and Height Gages

1. a. 30.82 mm
    b. 60.52 mm
    c. 11.76 mm
    d. 78.82 mm
    e. 52.42 mm
    f. 18.16 mm

3.

	A (millimeters)	B (millimeters)	C
a.	37.5	38.0	9
b.	18.5	19.0	3
c.	42.0	42.5	2
d.	88.5	89.0	16
e.	56.5	57.0	9
f.	10.0	10.5	19
g.	43.0	43.5	3
h.	77.0	77.5	20
i.	81.0	81.5	11
j.	93.5	94.0	24

5. a. 30.22 mm
    b. 48.62 mm
    c. 78.60 mm
    d. 65.82 mm
    e. 52.18 mm
    f. 14.76 mm
    g. 8.82 mm
    h. 34.42 mm

UNIT 27   English Micrometers

1. 0.589″
3. 0.736″
5. 0.808″
7. 0.738″
9. 0.157″
11. 0.949″
13. 0.441″
15. 0.153″
17. 0.424″
19. 0.038″
21. 0.983″

	Barrel Scale Setting (inches)	Thimble Scale Setting (inches)
23.	0.375–0.400	0.012
25.	0.950–0.975	0.023
27.	0.075–0.100	0.004
29.	0.025–0.050	0.013
31.	0.425–0.450	0.002

33. 0.3637″
35. 0.0982″
37. 0.3105″
39. 0.1448″
41. 0.5157″
43. 0.2749″

45. 0.3928″
47. 0.9717″
49. 0.3004″
51. 0.0009″
53. 0.8594″

	Barrel Scale Setting (inches)	Thimble Scale Setting (inches)	Vernier Scale Setting (inches)
55.	0.775–0.800	0.009–0.010	0.0006
57.	0.000–0.025	0.008–0.009	0.0003
59.	0.300–0.325	0.000–0.001	0.0001
61.	0.800–0.825	0.000–0.001	0.0008
63.	0.975–1.000	0.014–0.015	0.0004

UNIT 28   Metric Micrometers

1. 7.09 mm
3. 5.69 mm
5. 9.78 mm
7. 0.34 mm
9. 3.12 mm
11. 24.93 mm

	Barrel Scale Setting Is Between: (millimeters)	Thimble Scale Setting (millimeters)
13.	12.5–13.0	0.36
15.	15.0–15.5	0.08
17.	0.5–1.0	0.28
19.	18.0–18.5	0.12
21.	8.0–8.5	0.44
23.	23.0–23.5	0.08
25.	21.5–22.0	0.32

27. 4.268 mm
29. 7.218 mm
31. 2.132 mm

33. 8.308 mm
35. 9.484 mm
37. 11.114 mm

	Barrel Scale Setting Is Between: (millimeters)	Thimble Scale Setting Is Between: (millimeters)	Vernier Scale Setting (millimeters)
39.	14.5–15.0	0.37–0.38	0.004
41.	9.0–9.5	0.23–0.24	0.008
43.	4.0–4.5	0.05–0.06	0.006
45.	7.0–7.5	0.00–0.01	0.004
47.	5.5–6.0	0.20–0.21	0.008
49.	9.0–9.5	0.23–0.24	0.004
51.	14.5–15.0	0.08–0.09	0.002

UNIT 29 English and Metric Gage Blocks

One combination for each dimension is given. A number of different combinations will produce the given dimensions.

1. 0.1008″, 0.113″, 0.650″, 3.000″
3. 0.1002″, 0.122″, 0.900″, 2.000″
5. 0.1009″, 0.125″, 0.050″
7. 0.123″, 0.850″, 3.000″, 4.000″
9. 0.250″, 1.000″, 2.000″, 3.000″, 4.000″
11. 0.1007″, 0.125″, 0.650″, 4.000″
13. 0.1001″, 0.112″, 0.050″
15. 0.140″, 0.950″, 4.000″
17. 0.1009″, 0.128″, 0.750″, 2.000″
19. 0.1007″, 0.127″, 0.550″, 3.000″, 4.000″
21. 0.1006″, 0.134″, 0.200″, 2.000″, 3.000″, 4.000″
23. 0.103″, 0.900″, 1.000″, 4.000″
25. 0.1008″, 0.149″, 0.450″

27. 1.003 mm, 1.07 mm, 2 mm, 10 mm
29. 1.09 mm, 5 mm, 60 mm, 90 mm
31. 1.007 mm, 1.7 mm, 1 mm, 40 mm
33. 1.06 mm, 1.4 mm, 4 mm, 70 mm
35. 1.06 mm, 1.8 mm, 1 mm, 10 mm
37. 1.001 mm, 1.07 mm, 4 mm
39. 1.009 mm, 1.09 mm, 7 mm, 30 mm
41. 1.005 mm, 6 mm, 60 mm
43. 1.007 mm, 1 mm
45. 1.03 mm, 2 mm, 20 mm, 80 mm, 90 mm
47. 1.004 mm, 1.8 mm, 8 mm
49. 1.005 mm, 1.05 mm, 1.5 mm, 2 mm, 50 mm

UNIT 30 Achievement Review—Section Two

1. a. 6.75 ft
   b. 75 in
   c. 28.8 ft
   d. 27 mm
   e. 800 mm
   f. 21.8 cm

3. 5 complete lengths

5.

	Greatest Possible Error	ACTUAL LENGTH	
		Smallest Possible	Largest Possible
a.	0.01″	4.27″	4.29″
b.	0.00005″	0.83665″	0.83675″
c.	0.01 mm	46.15 mm	46.17 mm
d.	0.005 mm	16.445 mm	16.455 mm

7. a. Max. Limit: 1.719″
      Min. Limit: 1.709″
   b. Max. Limit: 4.0688″
      Min. Limit: 4.0676″
   c. Max. Limit: 5.9055″
      Min. Limit: 5.9047″
   d. Max. Limit: 64.99 mm
      Min. Limit: 64.83 mm
   e. Max. Limit: 173.003 mm
      Min. Limit: 172.990 mm

9. a. 0.0040″
   b. 0.0012″
   c. 0.0020″
   d. 0.0006″
   e. 0.0028″
   f. 0.0018″
   g. 0.0021″
   h. 0.0009″

11. a. $\frac{3}{32}''$
    b. $\frac{11}{32}''$
    c. $\frac{9}{16}''$
    d. $\frac{25}{32}''$
    e. $\frac{29}{32}''$
    f. $1\frac{3}{32}''$
    g. $1\frac{7}{32}''$
    h. $1\frac{11}{32}''$
    i. $\frac{3}{64}''$
    j. $\frac{13}{64}''$
    k. $\frac{23}{64}''$
    l. $\frac{17}{32}''$
    m. $\frac{49}{64}''$
    n. $\frac{61}{64}''$
    o. $1\frac{7}{32}''$
    p. $1\frac{25}{64}''$

13. a. 6 mm
    b. 19 mm
    c. 29 mm
    d. 43 mm
    e. 49 mm
    f. 57 mm
    g. 66 mm
    h. 74 mm
    i. 4.5 mm
    j. 11 mm
    k. 21.5 mm
    l. 28.5 mm
    m. 45.5 mm

UNIT 30 (*continued*)

13. n. 54 mm
    o. 65.5 mm
    p. 72.5 mm

15. a. (1) 0.558″
       (2) 0.089″
       (3) 0.679″
       (4) 0.638″
    b. (1) 0.3023″
       (2) 0.2855″

15. b. (3) 0.0732″
       (4) 0.4180″
    c. (1) 8.62 mm
       (2) 0.02 mm
       (3) 22.23 mm
       (4) 11.88 mm
    d. (1) 7.838 mm
       (2) 12.474 mm
       (3) 3.772 mm

17. a. 1.03 mm, 1.5 mm, 5 mm, 60 mm
    b. 1.02 mm, 1.2 mm, 3 mm, 30 mm, 90 mm
    c. 1.002 mm, 1.09 mm, 3 mm, 80 mm
    d. 1.004 mm, 1.07 mm, 1.2 mm, 10 mm
    e. 1.006 mm, 1.06 mm, 4 mm, 60 mm
    f. 1.004 mm, 1.3 mm, 1 mm, 40 mm
    g. 1.008 mm, 1.09 mm, 1.9 mm, 6 mm, 90 mm
    h. 1.001 mm, 1.07 mm, 5 mm, 10 mm, 90 mm

SECTION 3   Fundamentals of Algebra

UNIT 31 Symbolism

1. $6x + y$

3. $21 - b$

5. r/s

7. xy/m^2

9. a. $2\frac{1}{2}R$
   b. $2\frac{3}{4}R$
   c. $2\frac{1}{4}R$

9. d. $6\frac{1}{4}R$

11. $n - p - t$

13. a. 52
    b. 20
    c. 6
    d. 4
    e. 1

15. a. 151
    b. 96
    c. 14.5
    d. 8
    e. 126

17. a. 112.5 sq in
    b. 10.6065 in

19. a. 14.7 in
    b. 35.8 sq in

21. 32.4 sq in

23. 27.2 mm^2

25. 62.7 sq in

27. a. 5.1 in
    b. 215.5 cu in

UNIT 32 Signed Numbers

1. a. (+)9
   b. (+)5
   c. (+)6
   d. (−)6
   e. (−)10
   f. (+)7
   g. (−)20
   h. (−)10
   i. (+)3
   j. (−)8
   k. (−)11
   l. (−)6
   m. (+)17.5
   n. (−)17.5
   o. (+)6.5

1. p. (+)1.5
   q. (−)5$\frac{1}{4}$
   r. (−)5$\frac{3}{4}$

3. a. −25, −18, −1, 0, +2, +4, +17
   b. −21, −19, −5, −2, 0, +5, +13, +27
   c. −25, −10, −7, 0, +7, +10, +14, +25
   d. −14.9, −3.6, −2.5, 0, +0.3, +15, +17
   e. −16, −13$\frac{7}{8}$, −3$\frac{5}{8}$, +6, +14$\frac{1}{8}$

5. a. 23
   b. 30
   c. 25
   d. −23
   e. −33
   f. 7

5. g. −8
   h. −1
   i. −6
   j. −22
   k. −13
   l. −3$\frac{1}{8}$
   m. −13$\frac{5}{16}$
   n. −14.47
   o. 0.43
   p. 1
   q. −39.62
   r. 31.25
   s. −28.9
   t. −14.06

7. a. −24
   b. 24
   c. −30
   d. 30
   e. −35
   f. 28
   g. 0
   h. −32.5
   i. 0.32
   j. 0.036
   k. −1$\frac{1}{8}$
   l. 0
   m. −8
   n. −8
   o. 0

UNIT 32 *(continued)*

7. p. 7350.488
   q. 10.6
   r. −0.221
   s. 0.384
   t. −0.3

9. a. 4
   b. 8
   c. −8
   d. −64
   e. 16
   f. −32
   g. 36
   h. −125

9. i. 64
   j. 2.56
   k. −0.064
   l. 0.647
   m. 2.496
   n. −0.614
   o. 0.389
   p. $-\frac{8}{27}$
   q. −0.830
   r. 0.003
   s. −1.749
   t. 0.001

11. a. 3
    b. 9
    c. 2
    d. 4
    e. −2
    f. 2
    g. −5
    h. 5
    i. 42.103
    j. 0.155
    k. 4.002
    l. 0.060

13. 14

15. 4
17. 21
19. 142
21. 9
23. 9.672
25. 0.009
27. 2
29. −0.5
31. 14
33. 4.569
35. 0.135

UNIT 33   Algebraic Operations of Addition, Subtraction, and Multiplication

1. $19y$
3. $-22xy$
5. 0
7. $-10pt$
9. $15.2a^2b$
11. $1\frac{1}{4}xy$
13. $2.91gh^3$
15. $11P$
17. $-1\frac{7}{8}xy$
19. $6.666M$
21. A: $2.3x$
    B: $3.8x$
    C: $6.1x$
    D: $4.0x$
    E: $7.2x$
    F: $3.1x$
    G: $1.1x$
23. $2a - 11m$

25. $3xy^2 + 3x^2y$
27. $-2x^3 - 7x^2 + 4x + 12$
29. 0
31. $-0.4c + 3.6cd + 3.7d$
33. $2xy$
35. $-2xy$
37. $-10a^2$
39. $12mn^3$
41. $1\frac{1}{4}x^2$
43. $-13a + 7a^2$
45. $-2ax^2$
47. d^2t^2
49. $3x - 21$
51. 0
53. $x^2 + 3xy$
55. 0
57. $3a^3 - 1.3a^2 + a$

59. $-d^2 - 2dt + dt^2 + 4$
61. $8.08e + 15.76f + 10.03$
63. x^3
65. $56a^4b^3c^3$
67. 0
69. $3d^8r^4$
71. $0.21x^7y^4$
73. 0
75. $-3.36bc$
77. $-2x^8y^6$
79. $-49a^4b^4$
81. $-x^4y^2$
83. $-10x^2y^3 + 15x^5y$
85. $-8a^4b^5 + 2a^3b^4 + 4a^3b^2$
87. $-4dt - 4t^2 + 4$
89. $3x^3 + 27x + 7x^2 + 63$
91. $10a^3x^6 + 5ab^2x^4 + 2a^2bx^4 + b^3x^2$

UNIT 34 Algebraic Operations of Division, Powers, and Roots

1. $2x$

3. -1

5. 0

7. $-6H$

9. 3.7

11. $5cd$

13. $8g^2h$

15. xz^2

17. $4P^2V$

19. $\frac{1}{4}FS^2$

21. $8x^2 + 12x$

23. $-3x^5y + 2xy^3$

25. $-15a - 25a^4$

27. $-2cd + 5c^2d + 1$

29. $3a^2x + ax^2 - 2$

31. $4a - 6a^2c - 8c^2$

33. $9a^2b^2$

35. $8x^6y^3$

37. $-27c^9d^6e^{12}$

39. $49x^8y^{10}$

41. $a^9b^3c^6$

43. $-x^{12}y^{15}z^3$

45. $0.064x^9y^3$

47. $18.49M^4N^4P^2$

49. $-512a^{12}b^{18}c^3$

51. $0.36d^6e^6f^{12}$

53. $9x^4 - 30x^2y^3 + 25y^6$

55. $25t^4 - 60t^2x + 36x^2$

57. $0.16d^4t^6 - 0.16d^2t^4 + 0.04t^2$

59. $\frac{4}{9}c^4d^2 + c^3d^3 + \frac{9}{16}c^2d^4$

61. $a^{16}b^4 + 2a^8b^2x^6y^3 + x^{12}y^6$

63. m^3n^2s

65. $9x^4y^3$

67. $-3x^2y^4$

69. $0.4a^4cf^3$

71. $\frac{1}{4}xy$

73. $-4d^2t^3$

75. $2h^2$

77. $4a^3\sqrt{c}$

79. $\frac{3}{4}ac\sqrt{b}$

81. $-2a\sqrt[5]{b^3}$

83. $9b - 15b^2 + c - d$

85. $-ab - a^2b + a$

87. $-16 - xy$

89. $1 - r$

91. $-2x + 24$

93. $6 + c^2d$

95. $6a^2 - 6b$

97. $3b$

99. $7y^6 + 15$

101. $2\frac{2}{3}d$

103. $100a - 5a^4b^6$

105. $5f^4 + 6f^2h$

107. 8×10^4

109. 9.76×10^5

111. 1.5×10^{-2}

113. 2×10^{-1}

115. 3.9×10^{-4}

117. 1.75×10^{-3}

119. $160,000$

121. $5,090,000$

123. 0.0000632

125. 0.000003123

127. 0.0007321

129. 0.0209

131. 1.61×10^{-6}

133. 3.20×10^{-10}

135. 1.01×10^6

137. -4.77×10^{13}

139. 4.61×10^7

141. -4.38×10^{10}

143. -2.61×10^{-7}

145. 4.30×10^{12}

147. 1.02×10^3

UNIT 35 Introduction to Equations

1. 12

3. 11

5. 4

7. 12

9. 5

11. $0.5", 1", 3"$

13. 1.115 mm

15. 50 mm

17. $12°$

19. $36°$

21. $\frac{1}{2}"$

23. a. $\frac{3}{4}"$

 b. $\frac{1}{2}"$

 c. $1\frac{3}{4}"$

25. 3

27. 7

29. 6

31. 16

33. 84

35. 3

37. 48

39. 20

UNIT 36 Solution of Equations by the Subtraction, Addition, and Division Principles of Equality

1. 7	29. −17.101	59. −3.69	89. −19
3. 19	31. 18″	61. −0.005	91. 0
5. 4	33. $\frac{13''}{16}$	63. 0.09	93. 20
7. 43	35. 37.61 mm	65. −4.89	95. −1.8
9. −22	37. 0.1008″	67. $\frac{1}{2}$	97. 19.75
11. −53	39. $7\frac{11''}{32}$	69. $-16\frac{5}{32}$	99. 11
13. 43	41. 4.4286″	71. 18.052	101. 32
15. −50	43. 0.1653″	73. 48.1995	103. −72
17. 18.8	45. −10	75. $4\frac{1}{2}''$	105. $-4\frac{1}{2}$
19. 16.14	47. 135	77. 53.3 mm	107. 0.2
21. 0	49. 28	79. 830 mm	109. $\frac{3}{17}$
23. $-1\frac{5}{8}$	51. 83	81. −3	111. 21.75°
25. $-1\frac{1}{4}$	53. 14	83. 6	113. 124.94 mm
27. $-23\frac{1}{8}$	55. 78	85. 9	115. 63.33 r/min
	57. 9.3	87. 2.3	

UNIT 37 Solution of Equations by the Multiplication, Root, and Power Principles of Equality

1. 30	31. 435.12 mm	61. 1.659	83. −32
3. 63	33. 7.0711″	63. 0.497	85. −0.216
5. 27	35. 0.032″	65. 1.673	87. 0.001
7. 0	37. 163.8 mm	67. 0.340	89. $\frac{9}{64}$
9. 36	39. 4	69. a. 6 in	91. $\frac{1}{256}$
11. 21.5	41. 9	b. $\frac{5}{8}$ ft	93. $\frac{25}{64}$
13. 23.4	43. 4	c. 1.2 m	95. 23.591
15. −6	45. 12	d. 8.044 m	97. 0.480
17. 0	47. −5	e. 0.221 ft	99. −26.016
19. 0.001	49. 100	71. 36	101. a. 11.56 sq in
21. 0.0624	51. $\frac{3}{5}$	73. 1.44	b. 0.563 sq ft
23. $3\frac{3}{4}$	53. $\frac{3}{5}$	75. 0.6724	c. 0.425 m²
25. 2	55. $-\frac{1}{2}$	77. 4.913	d. 4.674 mm²
27. $\frac{3}{4}$	57. $\frac{4}{5}$	79. −0.001	e. 1.664 in
29. 0.9	59. 0.2	81. 0	

UNIT 38   Solution of Equations Consisting of Combined Operations and Rearrangement of Formulas

1. 9
3. 7
5. 2
7. 1
9. 9
11. −0.67
13. 4.8
15. 7
17. 30.5
19. 3
21. 4

23. −1
25. 9
27. 3
29. 6
31. 0.788
33. 27,066.929
35. 0.5
37. 1939.655
39. 0.093
41. 133.690
43. 12.341

45. a. $a = \frac{A}{b}$

    b. $b = \frac{A}{a}$

    c. $a = \sqrt{d^2 - b^2}$

    d. $b = \sqrt{d^2 - a^2}$

47. a. $D_O = \sqrt{FW^2 + D^2}$

    b. $D = \sqrt{D_O^2 - FW^2}$

    c. $d = 2a + 2C - D_O$

    d. $a = (D_O - 2C + d) \div 2$

49. a. $D = M + 1.5155P - 3W$

    b. $P = \frac{D + 3W - M}{1.5155}$

49. c. $W = \frac{M - D + 1.5155P}{3}$

51. a. $D = \frac{L - 1.57d - 2x}{1.57}$

    b. $d = \frac{L - 1.57D - 2x}{1.57}$

    c. $x = \frac{L - 1.57D - 1.57d}{2}$

53. a. $S = \frac{Ca}{C - F}$

    b. $C = \frac{Ca + SF}{S}$

55. a. h = 5.87 cm

    b. h = 0.847 cm

UNIT 39   Ratio and Proportion

1. $\frac{2}{7}$
3. $\frac{2}{11}$
5. $\frac{6}{23}$
7. $\frac{13}{9}$
9. $\frac{a}{3}$
11. $\frac{4}{3}$
13. $\frac{1}{10}$
15. a. $\frac{2}{1}$
    b. $\frac{2}{3}$

15. c. $\frac{3}{2}$
    d. $\frac{3}{5}$
    e. $\frac{2}{7}$
    f. $\frac{7}{1}$
    g. $\frac{7}{3}$
    h. $\frac{5}{2}$
    i. $\frac{2}{1}$
    j. $\frac{3}{7}$
17. 0.5

19. 35
21. 12
23. 17.5
25. 2.25
27. 8.2
29. 4
31. $\frac{5}{12}$
33. $-31\frac{1}{2}$
35. 13.5
37. 2.855
39. 0.244

41. a. 12 in
    b. $1\frac{1}{8}$ in
    c. 72.9 mm
    d. 32.4 mm
43. a. 8.031 in
    b. 1.124 in
    c. 4.016 in
    d. 2.720 in
    e. 2.808 in
    f. 7.950 in
    g. 1.125 in
    h. 1.575 in

43. i. 5.300 in
    j. 1.686 in
    k. 9.450 in
    l. 2.040 in
    m. 6.300 in
    n. 0.843 in
    o. 4.016 in
    p. 3.744 in
45. A = 0.85 in
    B = 1.02 in
    C = 1.36 in
    D = 1.53 in

UNIT 40   Direct and Inverse Proportions

1. a. 1.50 mm
    b. 2.59 mm
    c. 1.16 mm
    d. 2.33 mm
    e. 2.33 mm

3. a. 0.990 in
    b. 0.763 in
    c. 79.403 mm
    d. 12.966 mm
    e. 0.429 in

5. 1650 parts
7. 0.48 kg
9. a. 240 rpm
    b. 157.5 rpm

9. c. 28 teeth
    d. 25 teeth
    e. 166.2 rpm

UNIT 41 Applications of Formulas as to Cutting Speed, Revolutions Per Minute, and Cutting Time

1. 57 fpm
3. 90 fpm
5. 100 fpm
7. 130 m/min
9. 106 m/min
11. 111 rpm
13. 43 rpm

15. 2037 rpm
17. 477 r/min
19. 3626 r/min
21. 4.8 min
23. 11.9 min
25. 45 m/min
27. 153 fpm
29. 183 rpm

31. 3820 r/min
33. 87 rpm
35. 1.6 min
37. 264 min
39. 0.29 inch per revolution
41. 45 h
43. 14.61 h

45. 154 rpm
47. 1493 rpm
49. 230 rpm
51. 800 rpm
53. 414 rpm
55. 436 rpm
57. 343 rpm

UNIT 42 Applications of Formulas to Spur Gears

1. 2
3. 0.6283 inch
5. 3.7143 inches
7. 0.1745 inch
9. 14 teeth
11. 7.25 inches
13. 1.2047 inches
15. 0.1818 inch
17. 0.0785 inch
19. 0.0351 inch
21. 0.3082 inch
23. 0.2112 inch

25. 0.2222 inch
27. 15
29. 7
31. 26
33. 0.1429 inch
35. 0.0964 inch
37. 0.2857 inch
39. 2.7239 inches
41. 0.0143 inch
43. 23 teeth
45. 0.0015 inch
47. 0.0038 inch

49. 0.0060 inch
51. 6.7821 inches
53. 1.9375 inches
55. a. 117 mm
    b. 20.421 mm
    c. 130 mm
    d. 6.5 mm
    e. 13 mm
    f. 10.210 mm
57. a. 25 mm
    b. 7.854 mm
    c. 30 mm

57. d. 2.5 mm
    e. 5 mm
    f. 3.927 mm
59. a. 260 mm
    b. 31.417 mm
    c. 280 mm
    d. 10 mm
    e. 20 mm
    f. 15.708 mm
61. 30 teeth
63. 8.169 mm

UNIT 43 Achievement Review—Section Three

1. a. $x + y - c$
   b. $ab + d$
   c. $2M - P^2$

3. a. -37
   b. 16
   c. -14.4
   d. -72
   e. 0.78
   f. -6
   g. 32
   h. 36
   i. -125
   j. -3
   k. $\frac{1}{16}$

3. l. 1.448
   m. 40.085
   n. -0.510

5. a. 21
   b. 38
   c. 14
   d. 32.2
   e. -39.3
   f. -1.4
   g. -12
   h. 5.8
   i. 74.052
   j. -6.784
   k. -0.333
   l. 3

5. m. 9
   n. $\frac{-2}{3}$
   o. 166.204
   p. 202.572
   q. -0.305
   r. 0.887

7. a. 4.243
   b. 1.56
   c. 290.948
   d. 6.089
   e. 2.709

9. a. 19.2
   b. 10.8
   c. $\frac{1}{3}$

9. d. 7
   e. 32.5
   f. 0.778
   g. 8.282
   h. 4.050

11. a. 150 fpm
   b. 1273 r/min
   c. 3.57 min
   d. 3.5
   e. 0.5393 in

13. a. 8.19×8^{-8}
   b. -5.51×10^1
   c. 6.32×10^5
   d. 1.25×10^{-9}

SECTION 4 Fundamentals of Plane Geometry

UNIT 44 Introduction to Geometric Figures

1. a. parallel
   b. perpendicular
   c. oblique
3. a. ‖
   b. ⊥
   c. °
   d. ′
   e. ″
5. 67°51′
7. 117°42′
9. 93°09′
11. 6°28′

13. 77°40′
15. 212°04′16″
17. 44°26′38″
19. 103°0′32″
21. 89°54′20″
23. 19°53′50″
25. 107.75°
27. 87.27°
29. 56.80°
31. 2.32°
33. 79.98°

35. 57.1458°
37. 98.3403°
39. 2.1203°
41. 61.2017°
43. 76°52′
45. 244°08′
47. 46°42′12″
49. 540°
51. 16°09′
53. 109°21′09″
55. 44°

57. 21°59′35″
59. 97°03′59″
61. 87°57′
63. 110°51′05″
65. 84°
67. 270°15′
69. 43°30′
71. 68°30′
73. 51°25′43″
75. 161°10′25″

UNIT 45 Protractors—Simple and Vernier

1. ∠A = 25°
   ∠B = 42°
   ∠C = 57°
   ∠D = 77°
   ∠E = 93°
   ∠F = 11°
   ∠G = 27°
   ∠H = 46°
   ∠I = 76°
   ∠J = 87°

3. The third angle measures 28°.

5. ∠1 = 29°
   ∠2 = 133°
   ∠3 = 29°
   ∠4 = 58°
   ∠5 = 39°
   ∠6 = 27°
   ∠7 = 122°
   ∠8 = 31°

5. ∠9 = 72°
   ∠10 = 103°
   ∠11 = 64°
   ∠12 = 48°
   ∠13 = 150°
   ∠14 = 19°
7. 19°45′
9. 50°15′
11. 20°15′

13. 20°30′
15. a. 47°
    b. 14°
    c. 73°
    d. 85°
    e. 22°11′
    f. 44°41′
    g. 68°17′
    h. 11°40′33″
    i. 30°59′1″

UNIT 46 Angles

1. a. ∠A, ∠BAF, ∠FAB
   b. ∠B, ∠ABC, ∠CBA
   c. ∠3, ∠BCD, ∠DCB
   d. ∠4, ∠CDE, ∠EDC
   e. ∠5, ∠DEF, ∠FED
   f. ∠6, ∠AFE, ∠EFA
3. a. acute
   b. right
   c. right
   d. acute

3. e. acute
   f. obtuse
   g. straight
   h. acute
   i. right
   j. reflex
   k. straight

5. a. ∠3 and ∠6, ∠4 and ∠5
   b. ∠1 and ∠6, ∠2 and ∠5, ∠3 and ∠8, ∠4 and ∠7
7. a. ∠2 = 148°, ∠3 = 32°, ∠4 = 148°
   b. ∠2 = 144°41′, ∠3 = 35°19′, ∠4 = 144°41′
9. a. ∠1, ∠2, ∠5, ∠7, ∠9, ∠11, ∠13, ∠15 = 109°
      ∠3, ∠4, ∠6, ∠8, ∠10, ∠12, ∠14 = 71°
   b. ∠1, ∠2, ∠5, ∠7, ∠9, ∠11, ∠13, ∠15 = 93°08′
      ∠3, ∠4, ∠6, ∠8, ∠10, ∠12, ∠14 = 86°52′
11. a. ∠2 = 67°, ∠3 = 113°
    b. ∠2 = 74°12′, ∠3 = 105°48′

UNIT 47 Introduction to Triangles

1. isosceles

3. scalene

5. right

7. equilateral

9. 180°

11. a. 28°
    b. 29°42′47″

13. a. 17.3″
    b. 17.3″

15. a. 81°30′
    b. 77°20′30″

17. a. 11°
    b. 46°

19. a. 48°
    b. 79°

21. a. ∠A
    b. ∠C
    c. ∠B

23. a. ∠D
    b. ∠E
    c. ∠2

UNIT 48 Geometric Principles for Triangles and Other Common Polygons

1. Pairs A, B, D, and F

3. a. 55.8 mm
   b. 93.85 mm

5. a. 52°42′
   b. 37°18′

7. a. 79°
   b. 11°
   c. 11°

9. a. 4.909 in
   b. 2.640 in

11. a. 72.5 mm
    b. 113.6 mm

13. a. 118.30 mm
    b. 118.30 mm

15. a. 15 in
    b. 5 in

17. a. 960 mm
    b. 576 mm

19. a. 3.779 in
    b. 4.281 in

21. $x = 187.75$ mm
    $y = 191.96$ mm

23. a. 65°
    b. 96°

UNIT 49 Introduction to Circles

1. a. Chord
   b. Diameter or Chord
   c. Radius
   d. Center

3. a. Sector
   b. Segment
   c. Radius
   d. Radius
   e. Chord
   f. Arc

5. a. 20.420 in
   b. 94.248 mm
   c. 116.868 mm
   d. 18.410 in
   e. 11.141 in
   f. 69.391 mm
   g. 52.044 mm
   h. 1.222 in

7. 211.12 mm

9. 167.98 in

11. a. 4.090″
    b. 3.980″

13. a. 100°
    b. 60°

15. 21.23 mm

17. a. 118°56′
    b. 114°29′

19. a. 80°
    b. 1.44″

21. a. (1) 48°39′
       (2) 24°03′
    b. (1) 41°40′
       (2) 31°02′

23. a. 0.114 in
    b. 0.472 in

25. a. 168.75 mm
    b. 210.94 mm

UNIT 50 Arcs and Angles of Circles

1. 5.498 in

3. 243°30′ or 243.506°

5. 14.921 in

7. a. (1) 76°
      (2) 31°
      (3) 134°
   b. (1) 63.76°
      (2) 43.24°
      (3) 99°

9. a. (1) 41°
      (2) 139°
   b. (1) 37°30′
      (2) 142°30′

11. a. (1) 94°
       (2) 16°
    b. (1) 51°30′
       (2) 35°46′

13. a. (1) 58°
       (2) 90°

13. b. (1) 56°28′
       (2) 90°

15. a. (1) 46°25′
       (2) 28°38′
    b. (1) 43°56′
       (2) 30°0′

17. a. (1) 11°
       (2) 24°
    b. (1) 1°
       (2) 16°

19. a. 41.82 mm
    b. 84.48 mm

21. a. 1.911 in
    b. 1.844 in

23. a. 22°
    b. 30°54′

25. a. 75°
    b. 82°30′ or 82.50°

UNIT 51   Fundamental Geometric Constructions

All problems are constructed.

UNIT 52   Achievement Review—Section Four

1. a. 123°41′
    b. 62°29′
    c. 42°19′13″
    d. 13°16′
    e. 109°32′
    f. 22°11′9″
    g. 43°30′
    h. 25°50′

3. 145°38′51″

5. 64°8′31″

7. 103.6453°

9. a. 10°30′
    b. 19°45′
    c. 29°30′

11. a. 139°
    b. 80°28′

11. c. 76°56′33″

13. a. (1) 39°43′
        (2) 1 ft
    b. (1) 60°
        (2) 9.6 in
        (3) 4.8 in
    c. (1) 17°30′
        (2) 72°30′

15. 198°

17. 114.59 mm

19. a. 3.262 in
    b. 8.200 in

21. a. 9.023 in
    b. 261.17°

23. Problem to be layed out using construction methods.

SECTION 5   Trigonometry

UNIT 53   Introduction to Trigonometric Functions

1. r is hyp
    x is adj
    y is opp

3. a is adj
    b is opp
    c is hyp

5. a is hyp
    b is opp
    c is adj

7. d is hyp
    m is opp
    p is adj

9. h is adj
    k is hyp
    l is opp

11. m is opp
    p is hyp
    s is adj

13. m is hyp
    r is adj
    t is opp

15. f is opp
    g is hyp
    h is adj

17. $\sin \angle 1 = \frac{y}{r}$
    $\cos \angle 1 = \frac{x}{r}$
    $\tan \angle 1 = \frac{y}{x}$
    $\cot \angle 1 = \frac{x}{y}$
    $\sec \angle 1 = \frac{r}{x}$
    $\csc \angle 1 = \frac{r}{y}$

19. $\sin \angle 1 = \frac{k}{g}$
    $\cos \angle 1 = \frac{h}{g}$
    $\tan \angle 1 = \frac{k}{h}$
    $\cot \angle 1 = \frac{h}{k}$
    $\sec \angle 1 = \frac{g}{h}$
    $\csc \angle 1 = \frac{g}{k}$

21. $\sin \angle 1 = \frac{r}{s}$
    $\cos \angle 1 = \frac{p}{s}$
    $\tan \angle 1 = \frac{r}{p}$

21. $\cot \angle 1 = \frac{p}{r}$
    $\sec \angle 1 = \frac{s}{p}$
    $\csc \angle 1 = \frac{s}{r}$

23. Group 1: a, b, d
    Group 2: a, c
    Group 3: a, c, d

25. 0.60182
27. 0.95106
29. 0.99756
31. 0.45492
33. 0.99674
35. 0.76661
37. 0.01501
39. 0.49083
41. 0.18173
43. 0.11985
45. 0.95097
47. 0.34966
49. 0.76557
51. 0.06086

53. 2.88168
55. 1.78829
57. 1.49448
59. 2.92910
61. 4.47246
63. 1.43756
65. 1.65896
67. 1.37974
69. 6.28853
71. 1.03059
73. 1.75552
75. 56.82°
77. 30.68°
79. 75.84°
81. 1.45°
83. 5.50°
85. 28.67°
87. 67.30°
89. 39.52°
91. 50.95°

93. 48.16°
95. 9.70°
97. 11.70°
99. 76°39′
101. 67°7′
103. 87°43′
105. 47°3′
107. 89°43′
109. 17°47′
111. 46°8′
113. 45°0′
115. 46°10′
117. 74°22′
119. 84°12′
121. 19°14′

UNIT 54 Analysis of Trigonometric Functions

1. a. side y and side r are almost the same length
   b. side x is very small compared to side r
   c. side x is very small compared to side y

3. a. side y is very small compared to side r
   b. side x and side r are almost the same length
   c. side x is very large compared to side y

5. a. 45°
   b. 1.000 . . .
   c. 1.000 . . .

7. a. 1.000 . . .
   b. 0
   c. 0
   d. 1.000 . . .

9. tan 18°
11. cot 36°
13. csc 22°
15. cos 81°19′
17. csc 40.45°
19. sec 55°
21. cos 41°
23. csc 8°
25. sec 39°
27. cos 90°
29. sin 77.8°

31. csc 53°54′
33. cos 84.11°
35. tan 90°
37. csc 46°41′
39. cos 0°1′
41. sin 40°
43. cot 45°
45. sec 43°
47. cos 75°
49. tan 2°40′
51. csc 89.0°

UNIT 55 Basic Calculations of Angles and Sides of Right Triangles

1. 36°28′
3. 42°2′
5. 59.24°
7. 52.88°
9. a. 22°21′
   b. 67°39′
11. a. 31°30′
    b. 58°30′

13. 5.706 in
15. 50.465 in
17. 136.16 mm
19. 75.51 mm
21. a. 2.229 in
    b. 2.049 in
23. a. 7.285 in
    b. 2.480 in

25. a. 17°30′
    b. 55.237 in
    c. 52.680 in
27. a. 82.79 mm
    b. 28.02 mm
    c. 71.30°

29. a. 16°30′
    b. 1.579 in
    c. 5.559 in
31. a. 9°10′
    b. 1.099 in
    c. 6.812 in

UNIT 56 Simple Practical Machine Applications

1. a. 5.7358 in
   b. 2.2778 in
   c. 5.9949 in
   d. 1.6906 in
   e. 4.7767 in
   f. 6.9883 in

1. g. 0.0582 in
   h. 0.4246 in
   i. 3.3956 in
3. 8°48′
5. 13.12 mm

7. 51°25′
9. 255.48 mm
11. 28.19 mm
13. 0.897 in
15. 5.42 mm

17. 2.763 in
19. 2.577 in
21. 54.28 mm
23. 259.05 mm

UNIT 57 Complex Practical Machine Applications

1. 3.748 in
3. 14°13′
5. 0.564 in

7. 42°43′
9. 298.85 mm
11. 4.499 in

13. 5.408 in
15. 37°26′
17. 4.227 in

19. 0.667 in
21. 29°40′
23. 1.433 in

UNIT 58 The Cartesian Coordinate System

1. $\sin 120° = 0.8660$
   $\cos 120° = -0.5000$
   $\tan 120° = -1.7321$
   $\cot 120° = -0.5774$
   $\sec 120° = -2.0000$
   $\csc 120° = 1.1547$

3. $\sin 260° = -0.9848$
   $\cos 260° = -0.1736$
   $\tan 260° = 5.6713$
   $\cot 260° = 0.1763$
   $\sec 260° = -5.7588$
   $\csc 260° = -1.0154$

5. $\sin 300° = -0.8660$
   $\cos 300° = 0.5000$
   $\tan 300° = -1.7321$
   $\cot 300° = -0.5774$
   $\sec 300° = 2.0000$
   $\csc 300° = -1.1547$

7. $\sin 208°50' = -0.4823$
   $\cos 208°50' = -8.760$
   $\tan 208°50' = 0.5505$
   $\cot 208°50' = 1.8165$
   $\sec 208°50' = -1.1415$
   $\csc 208°50' = -2.0735$

9. $\sin 146°10' = 0.5568$
   $\cos 146°10' = -0.8307$
   $\tan 146°10' = -0.67028$
   $\cot 146°10' = -1.4919$
   $\sec 146°10' = -1.2039$
   $\csc 146°10' = 1.7960$

11. $\sin 313.17° = -0.7293$
    $\cos 313.17° = 0.6842$
    $\tan 313.17° = -1.0660$
    $\cot 313.17° = -0.9381$
    $\sec 313.17° = 1.4616$
    $\csc 313.17° = -1.3711$

UNIT 59 Oblique Triangles: Law of Sines and Law of Cosines

1. 5.078 in
3. 3.533 in
5. 33°54'
7. 18°16'
9. 128.73 mm
11. two solutions
13. two solutions
15. one solution
17. two solutions
19. 97.70 mm
21. 71.48°
23. 27°2'
25. 36.96 mm
27. a. 1.163 in
    b. 42°19'
29. a. 364.34 mm
    b. 14.52°
31. a. 94°44'
    b. 49°20'
33. 71.98°
35. 9.046 in
37. 44°38'
39. 2.202 in
41. 70°36'

UNIT 60 Achievement Review—Section Five

1. a. a is opp
      b is adj
      c is hyp
   b. a is adj
      b is opp
      c is hyp
   c. m is hyp
      s is adj
      p is opp
   d. r is hyp
      x is adj
      y is opp

3. a. 45°50'
   b. 26°50'

3. c. 80°10'
   d. 41°24'
   e. 50°36'
   f. 80°51'

5. a. $\cos 54°$
   b. $\cot 41°41'$
   c. $\sin 73°7'$
   d. $\tan 9.53°$

7. a. (1) 5.537 in
      (2) 3.075 in
   b. 6.60°

7. c. 118.936 mm
   d. 0.594 in
   e. (1) 64.16°
      (2) 195.553 mm
   f. 0.278 in
   g. 39°35'
   h. 58.50°

9. a. 185.999 mm
   b. 49°57'
   c. (1) 110°30'
      (2) 14.784 in
      (3) 8.558 in

9. d. 127.153 mm
   e. 125.10°
   f. (1) 8.484 in
      (2) 69°22'
      (3) 75°48'
   g. 7.859 in
   h. 59.74°
   i. 107°49'
   j. 6.288 in

SECTION 6 Compound Angles

UNIT 61 Introduction to Compound Angles

1. a. 3.680 in
   b. 65°57′

3. a. 7.777 in
   b. 56°5′

5. a. 1.897 in
   b. 65°4′

UNIT 62 Drilling and Boring Compound-Angular Holes: Computing Angles of Rotation and Tilt Using Given Lengths

1. a. 38°22′
   b. 49°39′

3. a. 43°9′
   b. 49°6′

5. a. 41°5′
   b. 49°38′

7. a.

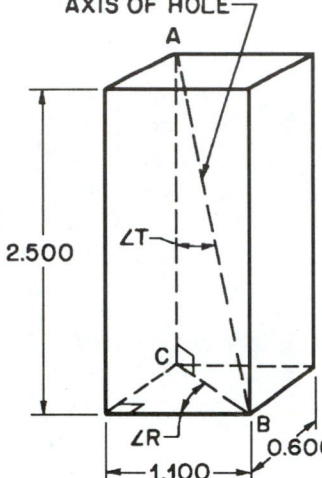

9. a.

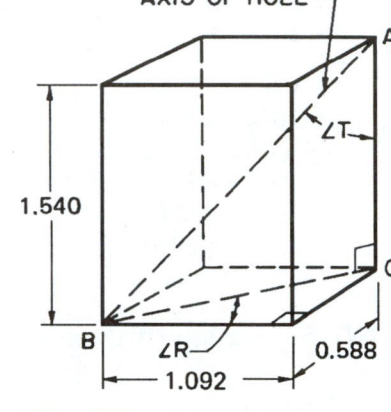

9. b. 28°18′
   c. 38°51′

7. b. 28°37′
   c. 26°37′

UNIT 63 Drilling and Boring Compound-Angular Holes: Computing Angles of Rotation and Tilt Using Given Angles

1. a. 37°52′
   b. 48°46′

3. a. 28.76°
   b. 34.43°

5. a. 32°31′
   b. 47°3′

7. a.

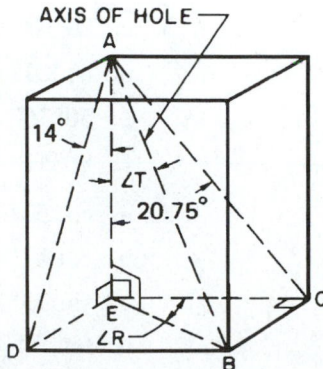

9. a.

11. a. 20°30′
    b. 42°52′

13. a. 21.89°
    b. 27.21°

15. a. 28.76°
    b. 32.94°

7. b. 30°49′
   c. 33°54′

9. b. 33.35°
   c. 24.40°

UNIT 64 **Machining Compound-Angular Surfaces: Computing Angles of Rotation and Tilt**

1. a. 57°6′
   b. 49°0′
3. a. 58°8′
   b. 31°52′

7. a.

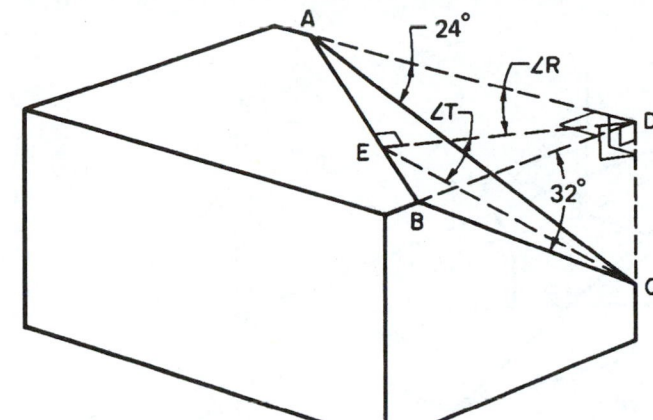

9. a. 50°01′
   b. 51°34′
11. a. 46.84°
    b. 43.03°
13. a. 38°48′
    b. 40°56′

7. b. 54.53°
   c. 37.50°

5. a.

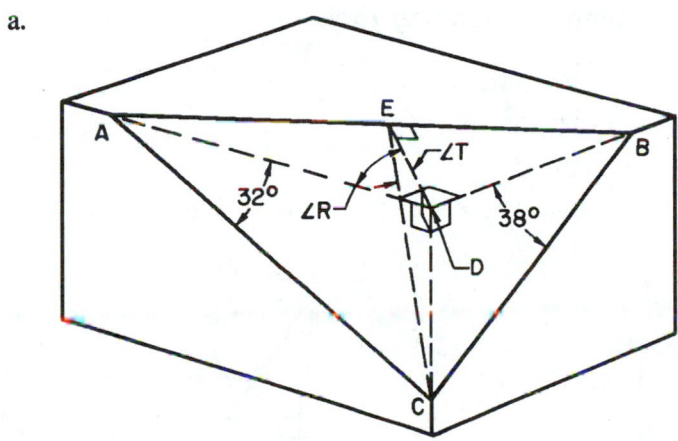

5. b. 51°21′
   c. 45°1′

UNIT 65 **Computing Angles Made by the Intersection of Two Angular Surfaces**

1. a. 32°14′
   b. 37°18′
3. a. 32°53′
   b. 45°12′

5. a.

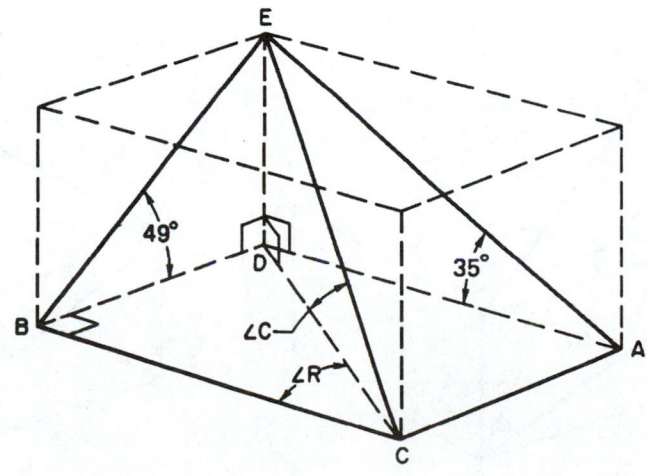

5. b. 31°20′
   c. 30°53′

UNIT 65 (continued)

7. a.

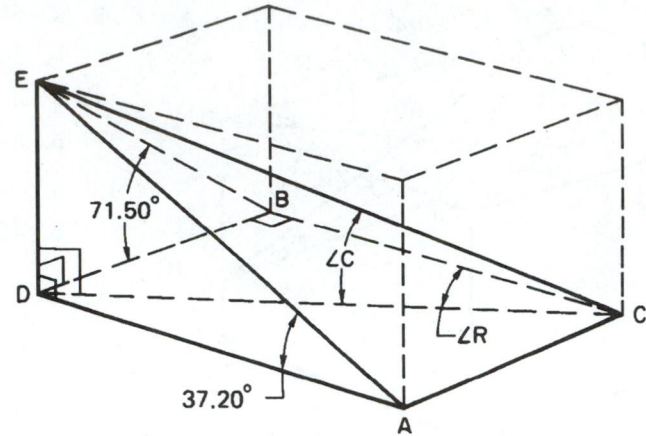

7. b. 14.25°
   c. 36.34°

9. a. 37°7′
   b. 30°5′
11. a. 46.59°
    b. 30.20°
13. a. 17°36′
    b. 15°40′

UNIT 66 Computing Compound Angles on Cutting and Forming Tools

1. 29°9′
3. 26°34′
5. a.

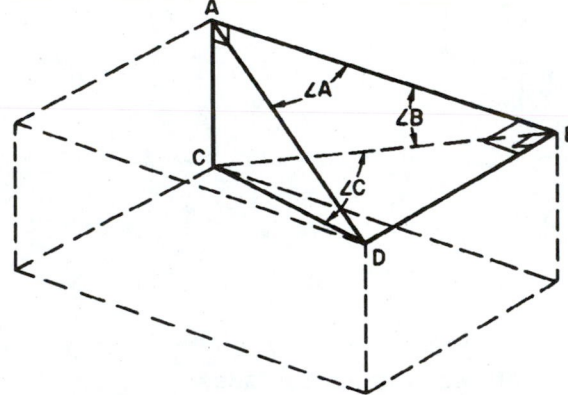

9. 32°13′
11. 25°20′
13. 30°23′
15. 33°29′
17. a.

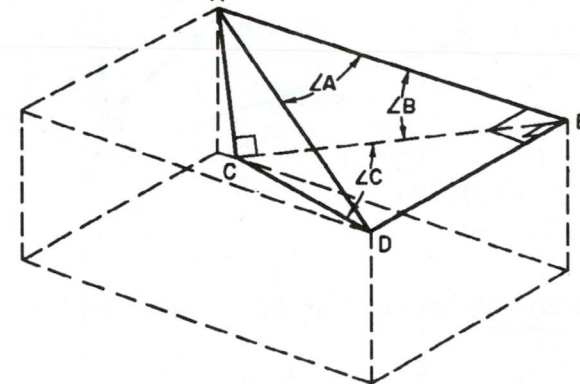

21. 25°16′
23. 34°16′

5. b. 23°16′

7. a.

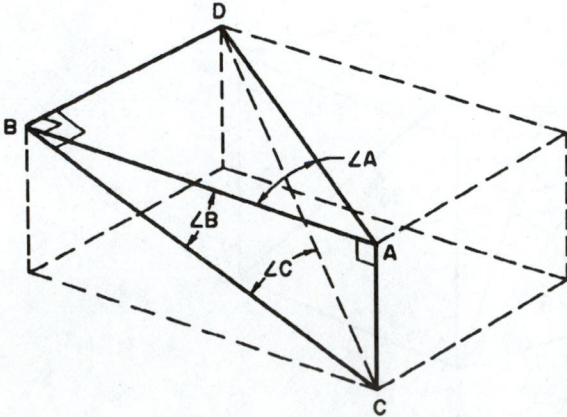

17. b. 30°52′
19. a.

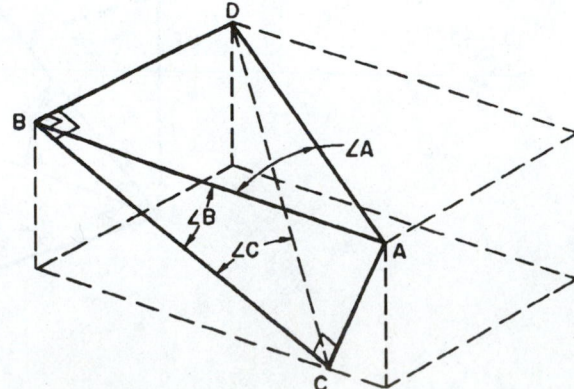

7. b. 29°27′

19. b. 32°54′

UNIT 67 Achievement Review—Section Six

1. a. 30°4′

   b. 44°57′

3. a. 55.54°

   b. 50.00°

5. 28°21′

SECTION 7 Computer Numerical Control (CNC)

UNIT 68 Introduction to Computer Numerical Control (CNC)

1. See Instructor's Guide

3. See Instructor's Guide. A triangle is formed

5. Point locations to be sketched

UNIT 69 Control Systems, Absolute Positioning, Incremental Positioning

1. a.

Hole	x	y
1	6.400″	−2.370″
2	8.400″	0.955″
3	6.400″	2.680″
4	12.450″	3.480″
5	13.000″	−0.270″
6	13.000″	−2.370″

1. b.

Hole	x	y
1	6.400″	−2.370″
2	2.000″	3.325″
3	−2.000″	1.725″
4	6.050″	0.800″
5	0.550″	−3.750″
6	0	−2.100″

3. a.

Hole	x	y
1	160.00 mm	−75.38 mm
2	216.24 mm	3.94 mm
3	160 mm	49.00 mm
4	292.32 mm	69.16 mm
5	308.06 mm	−23.08 mm
6	308.06 mm	−75.38 mm

3. b.

Hole	x	y
1	160.00 mm	−75.38 mm
2	56.24 mm	79.32 mm
3	−56.24 mm	45.06 mm
4	132.32 mm	20.16 mm
5	15.74 mm	−92.24 mm
6	0	−52.30 mm

5. a.

Hole	x	y
1	−4.750″	4.750″
2	−0.654″	8.569″
3	3.514″	12.457″
4	8.074″	9.600″
5	13.776″	7.748″
6	17.500″	−0.100″
7	13.776″	1.752″
8	8.074″	−0.100″
9	4.550″	4.750″
10	−1.188″	−0.520″
11	−6.162″	−4.625″

5. b.

Hole	x	y
1	−4.750″	4.750″
2	4.096″	3.819″
3	4.168″	3.888″
4	4.560″	−2.857″
5	5.702″	−1.852″
6	3.724″	−7.848″
7	−3.724″	1.852″
8	−5.702″	−1.852″
9	−3.524″	4.850″
10	−5.738″	−5.270″
11	−4.974	−4.105″

UNIT 69 (continued)

7. a.

Hole	x	y
1	−9.317″	18.412″
2	−3.793″	17.086″
3	5.885″	18.250″
4	11.400″	16.750″
5	18.935″	12.500″
6	5.885″	9.330″
7	−3.616″	8.075″
8	−13.541″	10.397″

9. a.

Hole	x	y
1	−237.06 mm	329.42 mm
2	−125.66 mm	302.84 mm
3	51.30 mm	329.30 mm
4	174.64 mm	302.18 mm
5	317.70 mm	222.50 mm
6	51.30 mm	163.00 mm
7	−136.30 mm	133.05 mm
8	−310.02 mm	187.28 mm

7. b.

Hole	x	y
1	−9.317″	18.412″
2	5.524″	−1.326″
3	9.678″	1.164″
4	5.515″	−1.500″
5	7.535″	−4.250″
6	−13.050″	−3.170″
7	−9.501″	−1.255″
8	−9.925″	2.322″

9. b.

Hole	x	y
1	−237.06 mm	329.42 mm
2	111.40 mm	−26.58 mm
3	176.96 mm	26.46 mm
4	123.34 mm	−27.12 mm
5	143.06 mm	−79.68 mm
6	−266.40 mm	−59.50 mm
7	−187.60 mm	−29.95 mm
8	−173.72 mm	54.23 mm

UNIT 70   Binary Numeration System

1. $2(10^2) + 6(10^1) + 5(10^0)$
$200 + 60 + 5 = 265$

3. $9(10^4) + 0(10^3) + 5(10^2) + 0(10^1) + 0(10^0)$
$90,000 + 0 + 500 + 0 + 0 + = 90,500$

5. $2(10^1) + 3(10^0) + 0(10^{-1}) + 2(10^{-2}) + 3(10^{-3})$
$20 + 3 + 0 + 0.02 + 0.003 = 23.023$

7. $4(10^3) + 7(10^2) + 5(10^1) + 1(10^0) + 1(10^{-1}) + 0(10^{-2}) + 7(10^{-3})$
$4000 + 700 + 50 + 1 + 0.1 + 0 + 0.007 = 4751.107$

9. $1(10^2) + 6(10^1) + 3(10^0) + 0(10^{-1}) + 6(10^{-2}) + 4(10^{-3}) + 3(10^{-4})$
$100 + 60 + 3 + 0 + 0.06 + 0.004 + 0.0003 = 163.0643$

11. 1_{10}

13. 5_{10}

15. 15_{10}

17. 11_{10}

19. 21_{10}

21. 53_{10}

23. 0.5_{10}

25. 3.75_{10}

27. 2.000_{10}

29. 9.3125_{10}

31. 1110_2

33. 1010111_2

35. 101011_2

37. 1101001_2

39. 1_2

41. 110011_2

43. 0.1_2

45. 0.011_2

47. 1010001.11_2

49. 1100101.01_2

51. 10100011.111_2

UNIT 71   Achievement Review—Section Seven

1. Coordinates to be plotted

3. a.

Hole	x	y
1	6.272″	−2.120″
2	8.172″	1.488″
3	13.201″	1.488″
4	19.942″	1.488″
5	21.032″	−3.392″
6	21.032″	−5.507″
7	13.903″	−4.239″

3. b.

Hole	x	y
1	6.272″	−2.120″
2	1.900″	3.608″
3	5.029″	0
4	6.741″	0
5	1.090″	−4.880″
6	0	−2.115″
7	−7.129″	1.268″

5. a. 1
   b. 7
   c. 21
   d. 3.25
   e. 9.5625

INDEX

A

Abscissa, 372
Absolute coordinates, 433–436
Absolute error, 104–105
Absolute positioning, 432–436
Absolute value, 163–171
Acute angle, 277, 347, 349–351, 372
Acute triangle, 375–378
Addendum, 246
Addition
  in algebra, 157, 176–177, 185
  in angular measurement, 263–264
  on calculator, 28–30, 68
  in combined operations, 23–24
  of decimal fractions, 41, 68
  of fractions, 8, 28–30
  of fractions and mixed and whole
    numbers, 9
  of mixed numbers, 9, 28–30
  in order of operations, 63
  powers and, 52
  of signed numbers, 164–165
Addition principles of equality,
  202–203
Adjacent angles, 278
Adjacent side (in triangle), 334,
  347–351, 372
Algebra, 155–251
  addition in, 176–177, 202–203
  combined operations in, 185–186,
    216–218
  combined with trigonometry, 333
  common letter designations in, 193
  for cutting speed, 237–242
  for cutting time, 238–242
  definitions in, 176
  direct proportions in, 232–233
  division in, 182–183, 203–204
  ratios in, 224–225
  equations in, 192–251
  formulas in, 218–221. *See also*
    Formulas
  for gears, 246–251
  inverse proportions in, 233–234
  multiplication in, 178–179, 209
  parentheses in, 185
  powers in, 183–184, 210–211
  principles of equality in, 199–204
  proportions in, 226–227, 231–234
  rearranging formulas in, 219–221
  for revolutions per minute, 238–242
  roots in, 184–185, 209–210
  scientific notation in, 186–188

  signed numbers in, 162–171
  signs of operation in, 155, 163–171
  for spur gears, 246–251
  substituting values in, 218–219
  subtraction in, 177–178, 200–201
  symbolism in, 155–158
  terms vs. factors in, 176
  transposition in, 201–202
  unknown quantity in, 193
  word statements and, 193–196
Algebraic expressions, 155–158, 179
Allowance, 109
  steel rules and, 115
  vernier height gage and, 122
Alternate interior angles, 278, 279, 280
Altitude, 292–293, 356
Ambiguous cases, 376–377
American National Standards Institute
  (ANSI), 95
Angle cuts, 357–358
Angles, 259–280
  acute, 277, 347, 349–351, 372
  adjacent, 278
  alternate interior, 278, 279, 280
  base, 284–285, 292, 356
  bisecting, 322–323
  calculator and, 262–266, 335–338,
    347–351
  coordinates and, 371–373. *See also*
    Cartesian coordinate system
  central, 301, 303, 309–311
  of circles, 309–310
  complementary, 274, 279, 344
  compound, 390–422
  corresponding. *See* Corresponding
    angles
  English measurement of, 260, 261,
    335, 347
  formed on circles, 311
  formed outside circles, 312
  functions and, 335–338, 342–385,
    See also Cartesian coordinate
    system
  geometric principles and, 278–280
  increasing, 342–344
  inscribed, 301, 302, 310
  mean, 337–338
  metric measurement of, 260,
    261–263, 335, 347
  naming, 277
  obtuse, 277
  of polygons, 294
  protractors and, 271–274
  reference, 334, 372

  reflex, 277
  right, 277
  of right triangles, 347–351. *See*
    also Right triangle
  of rotation, 393–415
  sine bar/plate and, 355
  straight, 277
  supplementary, 274, 279
  taper, 355–356
  of tilt, 393–415
  of triangles, 285–286
  trigonometry and, 333–385. *See*
    also Right triangle;
    Trigonometry
  true, 391–392, 418–419
  types of, 277–278
  units of measure of, 260–261
  vertex of, 271, 292, 372
ANSI. *See* American National
  Standards Institute (ANSI)
Anvil micrometers, 135
Arc, 301, 303, 309–313
  intercepted, 309–312
  trigonometry and, 337–338,
    363–367
Arccos, 337–338
Arcsin, 337–338
Arctan, 337–338
Area measurements, 95, 157–158. *See*
  also Powers; Roots
Arithmetic
  algebra and, 155, 179, 182–184
  in angular measurement, 263–266
Arithmetic numbers, 155
Auxiliary lines, 254–257, 361–367,
  398–401, 405–422
Axes, 371–373. *See also* Cartesian
  coordinate system
Axioms, 258–259

B

Base (in percentage problem), 77–84
Base angles, 284–285, 356
  isosceles triangle and, 292
Bench micrometers, 134
Bevel protractor, 273–274
Bevels, 355–356
Bilateral tolerance, 107–109
Binary numeration system, 439–442
Bisecting
  of angles, 322–323
  of circles, 303, 304, 321
  of triangles, 292–293

Bisector, perpendicular, 320–321
Bit, 440
Boring. *See* Drilling/boring holes
Bow micrometers, 135
Brackets, 63, 157
Bureau of Standards, 95
Byte, 440

C

CAD. *See* Computer aided
 drafting/design (CAD)
Calculator
 angular measurement on, 262–266,
  335–338, 347–351
 clearing, 28
 combined operations and, 28,
  30–31, 71–74
 decimals and, 68–74
 fractions and, 28–30
 mixed numbers and, 27–30
 positive vs. negative numbers on,
  164–171, 186–187
 powers and, 69, 167–168
 roots and, 59, 70, 169–170
 scientific notation and, 187–188
 trigonometry on, 335–338, 347–351
Calipers
 dial, 103
 gage blocks and, 146
 vernier, 103, 121–124, 129–131
CAM. *See* Computer aided
 manufacturing (CAM)
Cancellation, 15–16
Cartesian coordinate system, 371–373
 absolute positioning and, 432–436
 and computer numerical control
  (CNC), 428, 429–437
 incremental positioning and, 432,
  436–437
 Law of Cosines and, 378, 380
Centimeters, 97–98, 99
Central angle, 301, 303, 309–311
Chord, 301, 303, 305, 310–312
 perpendicular bisector of, 321
CIM. *See* Computer integrated
 manufacturing (CIM)
Circle, 301–315
 angle formed on, 311
 angle formed outside, 312
 angles of, 260, 309–310
 area of, 157–158
 definitions involving, 301
 degrees and, 260. *See also* Degrees
 polygon inscribed in, 302
 tangent circle to another, 301, 304,
  313–315, 366–367
 tangent lines to, 323–324
Circular pitch, 247–251
Circular sections, computing areas of,
  51, 57. *See also* Arcs; Powers;
  Roots

Circular tooth thickness, 247–251
Circumference, 301, 301–305
 in arc length formula, 310–311
Clearance, 108–109, 247–251
 cutting/forming tools and, 420–422
CNC. *See* Computer numerical
 control (CNC)
Cofunctions, 335–338, 342–385
Combined operations
 in algebra, 185–186, 216–218
 on calculator, 28, 30–31, 71–74
 of common fractions and mixed
  numbers, 22–25
 of decimal fractions, 63–65
 powers and, 52
 of signed numbers, 170–171
Common factors, dividing by, 15–16
Comparison measurement, 103
Complementary angles, 274, 344
Complex fractions, 2, 24–25
Compound angles, 390–422
 angles of rotation/tilt and, 393–415
 clearance and, 420–422
 cutting/forming tools and, 418–422
 diagonal of rectangular solid and,
  390–391
 drilling/boring holes and, 393–401
 holes and, 393–401
 machining surfaces and, 404–409
 made by intersecting surfaces,
  412–415
 surfaces and, 404–422
 formulas for, 390. *See also*
  Formulas
 true lengths/angles and, 391–392,
  412–422
Computer aided drafting/design
 (CAD), 439
Computer aided manufacturing
 (CAM), 439
Computer integrated manufacturing
 (CIM), 439
Computer numerical control (CNC),
  162, 428–442
 absolute positioning in, 432–436
 binary numeration system in,
  439–442
 incremental positioning in, 432,
  436–437
 language-based vs. graphic-based
  programs for, 428–429
 programming for, 428–429, 432
 three-axis system in, 429–430
 two-axis system in, 429
Computers, bits and bytes in, 440
Congruent triangles, 290
Continuous path, 432
Coordinates
 absolute, 433–436
 incremental, 436–437
 See also Cartesian coordinate
  system

Corresponding angles, 278, 279, 286,
  291
 congruent triangles and, 290
Corresponding sides, 286, 291–292
Cos. *See* Cosine (cos)
Cosecant, 334, 336–338, 342–385
 Cartesian coordinate system and,
  373
Cosine (cos), 334, 335–338, 342–385
 Cartesian coordinate system and,
  372–373
 Law of Cosines and, 378–380
Cotangent (tan), 334, 336–338,
  342–385
 Cartesian coordinate system and,
  373
Cross multiplying, 226
Cutting speed, 237–242
Cutting time, 238–242
Cutting/forming tools, 418–422

D

Decimal degrees, 260, 261–263
Decimal-inches
 on steel rules, 117
 on vernier calipers, 121–124
 on vernier height gages, 121–122,
  124–125
Decimals, 32–49
 addition of, 41
 binary numbers and, 439–442
 on calculator, 68–74
 combined operations of, 63–65
 division of, 47–49, 69
 equivalent, 62–63
 expressing as common fractions, 38
 expressing as percents, 74–75
 expressing common fractions as,
  33, 37–38
 expressing percents as, 75
 multiplication of, 45–46, 69
 reading and writing, 32–33
 rounding, 36–38, 68
 in scientific notation, 186–188
 on steel rules, 117
 structure of system of, 440
 subtraction of, 42
 terminating vs. nonterminating, 37
Decimeters, 97–98
Dedendum, 246
Degrees, 260–266
 in arc length formula, 310–311
 calculator and, 335–338
 protractors and, 271–274
 trigonometry and, 335–338,
  342–385
 See also Angles
Degrees of precision, 103–104, 320
 angles and, 260
 gage blocks and, 146
 pi and, 302

steel rules and, 115
vernier calipers and, 121, 129
Denominate numbers, 78, 95
Denominator, 2
cancellation and, 15–16
least common, 6–8
in parentheses, 22, 52, 63
Depth gage attachment, 124
Depth micrometers, 134, 135
Diagonals, 390–392
Dial calipers, 103
Dial indicators, 103, 124, 129
gage blocks and, 146
Diameter, 246, 247–251
circumference and, 302
definition of, 301
Diametral pitch, 247–251
Disc and blade micrometers, 134
Dividers, 320
Division
in algebra, 157, 182–183, 185, 187
in angular measurement, 265–266
on calculator, 28–30, 69
in combined operations, 23–24
by common factors, 15–16
of decimal fractions, 47–49, 69
of fractions, 19–20, 28–30
of mixed numbers, 20, 28–30
in order of operations, 63
powers and, 52
scientific notation and, 187
of signed numbers, 167
Division principles of equality,
203–204
Dovetails, 357–358
Drill press, cutting speed of, 237–239
Drilling/boring holes, 393–401
Dual dimensioning, 98

E

Electrical discharge machines, 428
Electronic comparators, 103
Ellipse, perimeter of, 158
English measurement
angular, 260, 261, 335, 347
conversion between metric and,
98–99
cutting speed using, 237–239
expressing larger units as smaller
in, 96
expressing smaller units as larger
in, 96–97
gage blocks and, 146, 147
metric vs., 95
micrometers and, 134–138
revolutions per minute using, 238
on steel rules, 115–117
units of, 96
vernier calipers and, 121–124
vernier height gages and, 121–122,
124–125

Equal sign, 192–193
Equality
expression of, 192–193
principles of, 199–204, 209–211
Equations, 175, 192–251
checking, 196
combined operations in, 216–218
definition of, 192
and principles of equality,
199–204, 209–211
solving, 199–204
transposition and, 201–202
writing from word statements,
193–196
See also Formulas
Equidistant lines, 259
Equilateral triangle, 285, 292–293
Equivalent decimals, 62–63
Equivalent fractions, 3, 7–8
Errors (in measurement), 104–105
gage blocks and, 147
Exponents, 51
algebra and, 176–179, 182–184
fractional, 170–171
positive vs. negative, 167–171,
186–187, 440
scientific notation and, 186–188
Extremes, 226

F

Factors
in algebra, 176–179, 182–185
conversion, 99
dividing by common, 15–16
literal, 176–179, 182–184
powers and, 51
Feet, 96–97, 99
Flame cutting, 428
Forming tools. *See* Cutting/forming
tools
Formulas, 51
algebraic, 155–158
for angle of rotation in relation to
frontal plane used in drilling,
399–401
for angle of rotation in relation to
frontal plane used in
machining, 408–409
for angle of rotation/tilt, 399–415
for angle of tilt used in drilling,
399–401
for angle of tilt used in machining,
408–409
for angles of intersecting angular
surfaces, 414–415
arc length, 310–311
circumference, 302–305
compound angles and, 390,
399–422
for cutting/forming tools, 419–422
for cutting speed, 237–242

for cutting time, 238–242
definition of, 192
for front-clearance-angle
applications, 421–422
for gears, 246–251
Law of Cosines and, 378–380
Law of Sines and, 375–378
for percentages, 78–84
rearranging, 175, 219–221, 248, 380
for revolutions per minute, 238–242
for spur gears, 246–251
square roots and, 57, 58
substituting values in, 218–219, 248
See also Equations
Fraction bar, 63, 157
Fractional degrees, 260. *See also*
Degrees
Fractional exponents, 170–171. *See
also* Exponents
Fractions, 1–49
addition of, 6–11, 28–30
calculator and, 28–30
combined operations involving,
22–25
common, 1–38
complex, 2, 24–25
decimal. *See* Decimals
definitions of, 2
division of, 19–20, 28–30
equivalent, 3, 7–8
expressing as percents, 75
expressing common fractions as
decimal, 33, 37–38
expressing decimal fractions as
common, 38
expressing percents as common, 76
improper, 2, 3–4
inverting, 19–20
multiplication of, 15–17, 28–30
reduced to lowest terms, 3
roots of, 58
subtraction of, 11–12, 28–30
Functions
angles and, 335–385. *See also*
Cartesian coordinate system
co-, 335–338, 342–385
of complementary angles, 344
inverse, 337–338
in trigonometry, 335–338, 342–385
variations of, 342–344

G

Gage blocks, 146–148
trigonometry and, 355
Gages, vernier. *See* Vernier height
gages
Gearing-diametral pitch system,
247–249
Gearing-metric module system,
249–251

Gears, 246–251
Geometric constructions, 319–324
Geometric principles, 278–280
   applications of, 354–358
   circles and, 303–305
Geometry, 51, 258–328
   algebra and, 155, 192
   angles in, 259–266
   axioms and postulates in, 258–259
   circles in, 301–315
   combined with trigonometry, 333
   plane, 258–328
   points and lines in, 259
   polygons in, 284–295
   triangles in, 284–295
Greatest possible error, 104
Grinder
   computer numerical control (CNC)
      for, 428
   cutting speed of, 237–239
Grouping symbols, 63, 64. *See also*
      Fraction bar; Parentheses;
      Radical symbol

H

Hectometers, 97–98
Hexagon, 294
High amplification comparators, 103
Holes
   absolute positioning system for,
      433–436
   compound-angular, 393–401
   distance between, 356
   drilling/boring, 393–401
Horizontal coordinates, 371–373. *See*
     also Cartesian coordinate system
Hypotenuse, 285, 293, 334, 347–351
   Cartesian coordinate system and,
      372–373
   sine bar/plate and, 355

I

Improper fractions, 2, 3–4
Inches, 96–97, 99, 347
   computer numerical control (CNC)
      and, 429, 430
   decimal-. *See* Decimal-inches
   degrees of precision and, 104
   on steel rules, 115–117
Incremental coordinates, 436–437
Incremental positioning, 432, 436–437
Indicators, dial. *See* Dial indicators
Infinity, 344
Inscribed angle, 301, 302, 310
Inside micrometers, 134, 135
Inspection, 428
Intercepted arc, 309–312
Interference, 109
International System of Units (SI), 95
Intersecting surfaces, 412–415

Inverse functions, 337–338
Inversion, 19–20, 440
Involute curve, 246, 248
Irrational number, 302
Isosceles triangle, 284, 285, 292–293
   applications of, 356–357

K

Kilometers, 97–98, 99

L

Lathe
   computer numerical control (CNC)
      for, 428, 429–430
   cutting speed of, 237–239
Law of Cosines, 378–380
Law of Sines, 375–378
Least common denominator, 6–8
Legs (of triangle), 284, 285
Lengths
   computing angles using given,
      393–395
   true, and compound angles, 390.
      See also Compound angles
   See also Linear measurement;
      Powers; Roots; Unknown
      sides/angles
Like terms, 176–179, 182–185,
   225–227
Limits (in tolerance), 107–109
Linear measurement
   absolute error in, 104–105
   allowance in, 109, 115
   basic dimension in, 107–108
   clearance in, 108–109. *See also*
      Clearance
   conversion between English and
      metric, 98–99
   decimal-inch rules for, 117–118
   definition of, 95
   degrees of precision in, 103–104.
      See also Degrees of precision
   English, 96–97. *See also* English
      measurement
   errors in, 104–105
   fit of mating parts in, 108–109
   fractional-inch rules for, 116–117
   gage blocks and, 146–148
   greatest possible error in, 104
   interference in, 109
   limits in, 107–109
   mating parts and, 108–109
   mean dimension in, 107–108
   metric, 97–98. *See also* Metric
      measurement
   relative error in, 104–105
   tolerance in, 107–109. *See also*
      Tolerance
Lines, 259

auxiliary, 254–257, 361–367,
   398–401, 405–422
   dividing into equal parts, 324
   geometric constructions of, 320–324
   oblique, 259
   parallel. *See* Parallel lines
   perpendicular. *See* Perpendicular
      lines
   projected, 354–358, 361–367,
      398–401, 405–422
Literal factors, 176–179, 182–184
Literal numbers, 155
Location of points, 259, 429–430

M

Machining, of compound-angular
      surfaces, 404–409
Machining centers, 428
Marking tools, 320–324
Mating parts, 108–109
Mean, 226
   in angles, 337–338
   in dimensions, 107–108
Measurement
   absolute error in, 104–105
   angular, 260–284, 335, 347. *See*
      also Angles
   calipers for, 103, 121–124, 129–131
   comparison, 103
   decimal-inch rules for, 117–118
   definitions of, 95–96
   degrees of precision in, 103–104.
      See also Degrees of precision
   English vs. metric, 95. *See also*
      English measurement; Metric
      measurement
   fractional-inch rules for, 116–117
   greatest possible error in, 104
   height gages for, 121–122,
      124–125, 130–131
   high amplification comparators
      for, 103
   instruments of, 103
   of length. *See* Linear measurement
   limitations in, 103
   linear. *See* Linear measurement
   micrometers for, 103, 134–138,
      141–143, 146
   protractors for, 271–274
   relative error in, 104–105
   of squares. *See* Squares, computing
      areas of
   of volume. *See* Volumes, computing
   steel rules for, 103, 115–118
   vernier bevel protractor for,
      273–274
   vernier calipers for, 103
   vernier height gages for, 121–122,
      124–125, 130–131
   vernier micrometers for, 137–138,
      142–143

Mechanical comparators, 103
Meters, 97–98, 99
Metric measurement
  angular, 260, 261–263, 335, 347
  conversion between English and,
    98–99
  cutting speed using, 240
  English vs., 95
  gage blocks and, 147, 148
  micrometers and, 141–143
  revolutions per minute using, 240
  on steel rules, 115–118
  units of linear, 97
  vernier calipers and, 129–131
Micrometers, 103
  English, 134–138
  gage blocks and, 146
  metric, 141–143
Miles, 96–97, 99
Millimeters, 97–98, 99, 347
  computer numerical control (CNC)
    and, 429, 430, 432
Milling machine
  computer numerical control (CNC)
    for, 428, 429
  continuous-path, 432
  cutting speed of, 237–239
  point-to-point, 432, 433–436
Minus sign, 163–171
Minutes (angles), 260–266
Mixed numbers
  addition of, 9, 28–30
  calculator and, 27–30
  combined operations involving,
    22–25
  definition of, 2
  division of, 20, 28–30
  expressed as improper fractions, 3–4
  expressed as percents, 75
  improper fractions expressed as, 4
  multiplication of, 16–17
  subtraction of, 12–13
Module system, 249–251
Multiplication
  in algebra, 155, 157, 178–179, 185,
    187
  in angular measurement, 265
  on calculator, 28–30, 69
  in combined operations, 23–24
  cross, 226
  of decimal fractions, 45–46, 69
  decimal places and, 65
  division as inverse of, 19–20
  of fractions, 15–17, 28–30
  of mixed numbers, 16–17
  in order of operations, 63
  powers and, 52, 65
  scientific notation and, 187
  of signed numbers, 166
Multiplication principles of equality,
  209

N

NC. *See* Numerical control (NC)
Negative exponents, 167–171,
  186–187, 440
Negative functions/values, 372–373.
  See also Cartesian coordinate
  system
Negative numbers, 163–171, 186–187
Number scale, 163–171
Numbers
  arithmetic vs. literal, 155
  binary, 439–442
  degree of precision of, 103–104
  directed, 163
  irrational, 302
  positive vs. negative, 163–171,
    186–187
  signed, 162–171
  whole. *See* Whole numbers
Numerator, 2
  cancellation and, 15–16
  in parentheses, 22, 52, 64
Numerical coefficients, 176–179,
  182–184
Numerical control (NC), 162, 439.
  See also Computer numerical
  control (CNC)

O

Oblique lines, 259
Oblique triangles, 371, 375–380
Obtuse angle, 277, 375–378
Obtuse triangles, 375–378
ON/OFF, 439–442
Opposite side (in triangle), 334,
  347–351, 372
  gage blocks and, 355
  Law of Sines and, 375–378
Optical comparators, 103
Order of operations, 22–23, 63
  in algebra, 157–158, 185–186
  calculator and, 30
  decimal fractions and, 63–65
  powers and, 52, 63
  radical symbol and, 58
  roots and, 58, 63
  signed numbers and, 170
Order of terms, 225
Ordinate, 372
Origin, 371–373, 433
Outside diameter, 246

P

Parallel lines, 259, 292
  geometric constructions and, 322
  transversals and, 279–280
Parallelogram, 294
Parentheses, 22–23
  in algebra, 155, 157, 185
  in combined operations, 25

in order of operations, 25, 63–65
  parentheses within, 63
  powers and, 51
Percentages
  calculating, 77–84
  practical applications involving,
    82–84
  types of complex problems
    involving, 84
  types of simple problems
    involving, 77–78
Percents, 74–84
  definition of, 74
  expressing common fractions as, 75
  expressing decimal fractions as,
    74–75, 75
  expressing mixed numbers as, 75
Perimeter
  calculating, 157, 158
  circumference and, 302
Perpendicular bisector, 320–321
Perpendicular lines, 259, 279, 292
  circles and, 303, 304
  geometric constructions and,
    320–321
Pi, 157–158, 302
Pinions, 246–251
Pitch, 246, 247–251
Pitch circles, 246
Pitch diameter, 246
Plane geometry, 258–328
  circles in, 301–315
  polygons in, 284–295
  triangles in, 284–295
Planes
  in angles of rotation/tilt, 404–415
  compound angles and, 390–409
  cutting/forming tools and, 421–422
  See also Surfaces
Plotting points, 259, 429–430
Plus sign, 163–171
Pneumatic comparators, 103
Point-to-point system, 432, 433–436
Points (geometric), 259
  computer numerical control (CNC)
    and, 429–430
Polygons, 284, 290, 294–295
  inscribed in circles, 302
  regular, 294, 302
  similar, 290
  types of, 294
Positioning
  absolute, 432–436
  incremental, 432, 436–437
  for machining, 407
  system for hole, 433–436
  system for tool, 432–437
Positive exponents, 167–171,
  186–187, 440
Positive functions/values, 372–373.
  See also Cartesian coordinate
  system

Positive numbers, 163–171, 186–187
Postulates, 258–259, 278. *See also* Geometric principles
Power principles of equality, 210–211
Powers, 51–53
  algebra and, 157, 183–184
  on calculator, 69, 167–168
  decimal places and, 65
  in order of operations, 63
  of signed numbers, 167–168
Precision, degrees of. *See* Degrees of precision
Principal root, 169
Principles, geometric. *See* Geometric principles
Principles of equality, 199–204, 209–211
  addition, 202–203
  division, 203–204
  multiplication, 209
  power, 210–211
  root, 209–210
  subtraction, 200–201
Projected lines, 354–358, 361–367, 398–401, 405–422
Proportions, 226–227
  circles and, 303
  direct, 232–233
  inverse, 233–234
Protractors, 271–274
  bevel, with vernier scale, 273–274
  simple semicircular, 271–273
Pyramids, 405
Pythagorean Theorem, 293, 304, 351

Q

Quadrants, 371–373. *See also* Cartesian coordinate system

R

Radical symbol, 57, 58, 157
  in order of operations, 63
Radius, 301, 303
  in arc length formula, 311
Rate (percent), 77–84
Ratio
  in arc length formula, 310–311
  description of, 224–225
  expressed in lowest terms, 225
  proportion in, 226, 231–234
  right triangle and, 333
  trigonometry and, 333–335
Rectangle, 294
  perimeter of, 157
  solid, 390–392, 398–422
Rectangular coordinate system, 371–373. *See also* Cartesian coordinate system

Rectangular solid, 390–392, 398–422
Reference angle, 334, 372
Reflex angle, 277
Regular hexagon, 294
Regular polygons, 294, 302
Relative error, 104–105
Revolutions per minute, 238–242
Right angle, 277
  greater than, 371, 375. *See also* Obtuse angle
  less than, 375. *See also* Acute angle
Right triangle, 285, 292–293
  complex applications of, 361–367
  compound angles and, 397, 405–415
  ratio of sides of, 333, 347–351
  simple applications of, 354–358
  terms for sides of, 334
  trigonometry and, 333–334, 347–367
Root circle, 246
Root diameter, 246
Root principles of equality, 209–210
Roots, 57–59
  algebra and, 157
  on calculator, 59, 70, 169–170
  in order of operations, 63
  principal, 169
  of signed numbers, 169–170
  that are not whole numbers, 59
Rotation, double, 404
Rotation, angle of, 292–415
Rounding, 36–38, 68
Rules, steel, 103, 115–118

S

Scalene triangle, 284
Scientific notation, 186–188
Screw thread micrometers, 134, 135
Scriber, 124
Secant, 301, 312
  Cartesian coordinate system and, 373
  in trigonometry, 334, 336–338, 342–385
Seconds (angles), 260–266
Sector, 301
Segment, 301, 305
SI. *See* International System of Units (SI)
Signed numbers, 162–171
  calculator and, 164–165
Signs of operation, 155, 163–171
Similar figures, 290
Similar polygons, 290
Similar triangles, 291, 292
Sine (sin), 335–338, 348–385
  Cartesian coordinate system and, 372–373
  Law of Sines and, 375–378
Sine bar, 355
Sine plate, 355

Solid, rectangular, 390–392, 398–422
Solid trigonometry, 390. *See also* Compound angles
Spur gears, 246–251
Square, 294
  computing area of, 51–53, 57. *See also* Powers; Roots
Square roots, 57–59, 69, 70. *See also* Roots
Squared numbers, 51–53, 69. *See also* Powers
Steel rules, 103, 115–118
Straight angle, 277
Subtraction
  in algebra, 157, 177–178, 185, 200–201
  in angular measurement, 264–265
  on calculator, 28–30, 68
  in combined operations, 23–24
  of decimals, 42, 68
  of fractions, 11–12, 28–30
  of mixed numbers, 12–13, 28–30
  in order of operations, 63
  powers and, 52
  of signed numbers, 165–166
Subtraction principles of equality, 200–201
Supplementary angles, 274, 279
Surfaces, compound-angular, 404–422
  angles made by intersecting, 412–415
Symbol(s)
  in algebra, 155–158
  fraction bar, 63, 157
  grouping, 63, 64
  infinity, 344
  parentheses as. *See* Parentheses
  radical, 57, 58, 157

T

Tangent (tan), 301, 304, 311–315
  Cartesian coordinate system and, 373
  to circle, 323–324, 357–358
  circles that are, 301, 304, 313–315, 366–367
  constructing, 323–324
  in trigonometry, 334–338, 342–385
Tapers, 355–356
Terms (in algebra), 176–179, 182–185
  order of, 225
  ratios and, 225–227
Thread cutting, 420–422
Thread wire checking dimensions, 357–358
Tilt, angle of, 393–415
Tilt, double, 404
Tolerance, 107–109, 320
Tool-positioning systems, 432–437
Tools/instruments
  calipers, 103, 121–124, 129–131

cutting/forming, 418–422
depth gage attachment, 124
dial indicators, 103, 124, 129
gage blocks. *See* Gage blocks
height gages, 121–122, 124–125, 130–131
high amplification comparators, 103
marking, 320–324
measuring, 103
micrometers, 103, 134–138, 141–143, 146
positioning systems for, 432–437
protractors for, 271–274
steel rules, 103, 115–118
vernier bevel protractor, 273–274
vernier calipers, 103
vernier height gages, 121–122, 124–125, 130–131
vernier micrometers, 137–138, 142–143
Tooth thickness, 247–251
Trammels, 320
Transposition, 201–202
Transversal, 278–280
Triangles, 284–295
    acute, 375–378
    adjacent side in, 334, 347–351, 372
    coordinates and, 371–373. *See also* Cartesian coordinate system
    congruent, 290
    equilateral, 285, 292–293
    hypotenuse of. *See* Hypotenuse
    isosceles. *See* Isosceles triangle
    legs of, 284, 285
    oblique, 371, 375–380
    obtuse, 375–378
    opposite side in, 334, 347–351, 372
    right. *See* Right triangle
    scalene, 284
    trigonometry and, 333–385
    types of, 284–285
Trigonometry, 333–385
    algebra and, 155, 192
    analysis of functions in, 342–344

Cartesian coordinate system in, 371–373. *See also* Cartesian coordinate system
complex applications of, 361–367
powers and, 51
ratio method in, 334–335
rectangular coordinate system in, 371–373. *See also* Cartesian coordinate system
right-angle, 354–358. *See also* Right triangle
signed numbers and, 162
simple applications of, 354–358
solid, 390. *See also* Compound angles
three-dimensional, 390. *See also* Compound angles
unity or unit circle method in, 334
True length/angle, 391–392
    cutting/forming tools and, 418–422
True value, 104–105
Turning center, 428

U

Unilateral tolerance, 107–109
Unity/unit circle method, 334
Unknown quantity, 193, 226
    gears and, 248–251
Unknown sides/angles, 291–292, 333, 347–367
    compound angles and, 390–422
    Law of Cosines and, 378–380
    Law of Sines and, 375–378
Unlike terms, 176–179, 182–185, 225–227

V

V-blocks, 357–358
V-slots, distance between, 356–357
Vernier bevel protractor, 273–274
Vernier calipers, 103
    English, 121–124
    gage blocks and, 146
    metric, 129–131

Vernier height gages
    English, 121–122, 124–125
    metric, 130–131
Vernier micrometers
    English, 137–138
    metric, 142–143
Vertex, 271, 292, 372
Vertical coordinates, 371–373. *See also* Cartesian coordinate system
Volumes, computing, 51–53
    English vs. metric measurements for, 95. *See also* English measurement; Metric measurement

W

Whole depth, 246
Whole numbers
    addition of fractions and mixed and, 9
    dividing decimals by, 48–49
    roots of, 57–59
    roots that are not, 59
Word statements (writing equations from), 193–196
Working depth, 247–251

X

x-axis, 371–373, 429–430, 432. *See also* Cartesian coordinate system

Y

y-axis, 371–373, 429–430, 432. *See also* Cartesian coordinate system
Yards, 96–97, 99

Z

z-axis, 429–430, 432
Zero/one (binary system), 439–442
Zero point, 433. *See also* Origin